7/e

Fundamentals of Physics

Part 2

7/e

Fundamentals of Physics

Part 2

David Halliday
University of Pittsburgh

Robert Resnick
Rensselaer Polytechnic Institute

Jearl Walker
Cleveland State University

WILEY

John Wiley & Sons, Inc.

ACQUISITIONS EDITOR   Stuart Johnson
SENIOR DEVELOPMENT EDITOR   Ellen Ford
MARKETING MANAGER   Bob Smith
SENIOR PRODUCTION EDITOR   Elizabeth Swain
TEXT/COVER DESIGNER   Madelyn Lesure
DUMMY DESIGNER   Lee Goldstein
PHOTO EDITOR   Hilary Newman
PHOTO RESEARCHER   Elyse Rieder
ILLUSTRATION EDITOR   Anna Melhorn
ILLUSTRATION STUDIO   Radiant
COVER PHOTO   © Jeff Hunter/The Image Bank/Getty Images.

This book was set in 10/12 Times Ten by Progressive Information Technologies and printed and bound by Von Hoffmann Press. The cover was printed by Von Hoffmann Press.

This book is printed on acid free paper. ∞

Library of Congress Cataloging-in-Publication Data

Halliday, David
    Fundamentals of physics / David Halliday, Robert Resnick, Jearl Walker—7th ed.
       p. cm.
    Includes index.
    ISBN 0-471-42962-7 (pt. 2 : pbk. : acid-free paper)
    1. Physics. I. Resnick, Robert II. Walker, Jearl III. Title.

QC21.3.H35 2005
530—dc21   2003049663

Printed in the United States of America

10 9 8 7 6 5 4 3 2 1

Brief Contents

v

Contents

Chapter 17

Waves — II   445

How can an emperor penguin find its mate among thousands of huddled penguins?

Chapter 18

Temperature, Heat, and the First Law of Thermodynamics   476

How can a dead rattlesnake detect and strike a reaching hand?

Chapter 19

The Kinetic Theory of Gases   507

How can cooling steam inside a railroad tank car cause the car to be crushed?

Chapter 20

Entropy and the Second Law of Thermodynamics   536

Why is the popping of popcorn irreversible?

Appendices

Answers to Checkpoints and Odd-Numbered Questions and Problems

Index

Preface

The Goal

The principle goal of *Fundamentals of Physics*, seventh edition, is the one established by David Halliday and Robert Resnick in their first edition over 40 years ago: to provide instructors with a tool by which they can teach students how to effectively read scientific material, identify fundamental concepts, reason through scientific questions, and solve quantitative problems.

This process is not easy for either students or instructors. Indeed, the course associated with this book may be one of the most demanding of all the courses taken by a student. However, it can also be one of the most rewarding, because it reveals the world's fundamental clockwork from which all scientific and engineering applications spring.

Why a New Edition?

Some people question: "Because *Fundamentals of Physics* is so widely used and has been so well accepted, why write a new edition?" There are two primary reasons why this seventh edition was written:

➡ Many users of the sixth edition (both instructors and students) sent in comments and suggestions to improve the book. These improvements are now incorporated into the narrative and problems throughout the book. We (the current author Jearl Walker and the publisher John Wiley & Sons) regard the book as an ongoing project and encourage more input from users. You can send suggestions, corrections, and positive or negative comments to John Wiley & Sons (http:www.wiley.com/college/halliday) or Jearl Walker (mail address: Physics Department, Cleveland State University, Cleveland OH 44115 USA; fax number: (USA) 216 687 2424; or email address: physics@wiley.com). We may not be able to respond to all suggestions, but we keep and study each of them.

➡ In his full-time teaching, Jearl Walker gains first-hand experience with students as they learn from *Fundamentals of Physics*. After every lecture and exam, he carefully notes what concepts and procedures worked well with the students and (even more important for writing a book) which items confused the students, and improvements based on his notes are now incorporated throughout the book. His goal in his HRW courses is the same as the goal in the HRW book: To show students that they *can* reason from basic physics concepts all the way to valid conclusions.

Major Content Changes

➡ Several thousand of the end-of-chapter problems have been rewritten to streamline both the presentation and the answer.

➡ All of the chapter-opening puzzlers (the examples of applied physics designed to entice a reader into each chapter) are new and come straight from research journals in many different fields.

➡ The first section in every chapter introduces the subject of the chapter by asking and answering the question "What is physics?" as the answer pertains to the physics of the chapter.

➡ Chapters 9 and 10 in the sixth edition have been combined so that the current Chapter 9 contains the full discussion of momentum and the conservation of momentum for both explosions and collisions.

➡ The chapter on angular momentum (Chapter 11) now contains the physics of a gyroscope.

➡ The chapter on waves (Chapter 16) now contains a derivation of the wave equation.

➡ The chapter on AC circuits (Chapter 31) is now presented before the chapter on Maxwell's equations and magnetic materials (Chapter 32).

➡ Chapter 39 (in the extended version of the book) now contains a derivation of the Bohr model of the hydrogen atom.

Chapter Features

Opening puzzlers. A curious puzzling situation opens each chapter and is explained somewhere within the chapter, to entice a student to read the chapter. These features, which are a hallmark of *Fundamentals of Physics*, are based on current research as reported in scientific, engineering, medical, and legal journals.

What is physics? The narrative of every chapter now begins with this question, and with an answer that pertains to the subject of the chapter. (A plumber once asked Jearl Walker, "What do you do for a living?" Walker replied, "I teach physics." The plumber thought for several minutes and then asked, "What is physics?" The plumber's career was entirely based on physics, yet he did not even know what physics is. Many students in introductory physics do not know what physics is but assume that it is irrelevant to their chosen career.)

Checkpoints are stopping points that effectively ask the student, "Can you answer this question with some reasoning based on the narrative or sample problem that you just read?" If not, then the student should go back over that previous material before traveling deeper into the chapter. For example, see Checkpoint 2 on page 62 and Checkpoint 2 on page 280. *Answers to all checkpoints are in the back of the book.*

Sample problems have been chosen to help a student organize the basic concepts of the narrative and to develop problem-solving skills. They are chosen to demonstrate how problems can be solved with reasoned solutions rather than quick and simplistic plugging of numbers into an equation with no regard for what the equation means.

Key Ideas in the sample problems focus a student on the basic concepts at the root of the solution to a problem. In effect, these key ideas say, "We start our solution by using this basic concept, a procedure that prepares us for solving many other problems. We don't start by grabbing an equation for a quick plug-and-chug, a procedure that prepares us for nothing."

Problem-solving tactics contain helpful instructions to guide the beginning physics student as to how to solve problems and avoid common errors.

Review & Summary is a brief outline of the chapter contents that contains the essential concepts but which is not a substitute for reading the chapter.

Questions are like the checkpoints and require reasoning and understanding rather than calculations. *Answers to the odd-numbered questions are in the back of the book.*

Problems are grouped under section titles and are labeled according to difficulty. *Answers to the odd-numbered problems are in the back of the book.*

Icons for additional help. When worked-out solutions are provided either in print or electronically for certain of the odd-numbered problems, the statements for those problems include a trailing icon to alert both student and instructor as to where the solutions are located. An icon guide is provided here and at the beginning of each set of problems:

SSM	Solution is in the Student Solutions Manual.
WWW	Solution is at http://www.wiley.com/college/halliday
ILW	Interactive LearningWare solution is at http://www.wiley.com/college/halliday

Additional problems. These problems are not ordered or sorted in any way so that a student must determine which parts of the chapter apply to any given problem.

On-line simulation problems. Nearly all the chapters have on-line problems involving simulations of the physics within the chapter. For those chapters, a notice with an icon appears at the very end of the homework problems.

Additional Features

Reasoning versus plug-and-chug. A primary goal of this book is to teach students to reason through challenging situations, from basic principles to a solution. Although some plug-and-chug homework problems remain in the book (on purpose), most homework problems emphasize reasoning.

Chapters of reasonable length. To avoid producing a book thick enough to stop a bullet (and thus also a student), we have made the chapters of reasonable length. We explain enough to get a student going but not so much that a student no longer must analyze and fuse ideas. After all, a student will need the skill of analyzing and fusing ideas long after this book is read and the course is completed.

Use of vector-capable calculators. When vector calculations in a sample problem can be performed directly on-screen with a vector-capable calculator, the solution of the sample problem indicates that fact but still carries through the traditional component analysis. When vector calculations cannot be performed directly on-screen, the solution explains why.

Graphs as puzzles. These are problems that give a graph and ask for a result that requires much more than just reading off a data point from the graph. Rather, the solution requires an understanding of the physical arrangement in a problem and the principles behind the associated equations. These problems are more like Sherlock Holmes puzzles because a student must decide what data are important. For examples, see problem 38 on page 79, problem 8 on page 108, and problem 21 on page 230.

Problems with applied physics, based on published research, appear in many places, either as the opening puzzler of a chapter, a sample problem, or a homework problem. For example, see the opening puzzler for Chapter 9 on page 201, Sample Problem 2-8 on page 28, and homework problem 30 on page 231. For an example of homework problems that build on a continuing story, see problems 4, 35, and 53 on pages 131, 134, and 135.

Problems with novel situations. Here is one of several hundred such problems: Problem 57 on page 112 relates a true story of how Air Canada flight 143 ran out of fuel at an altitude of 7.9 km because the crew and airport personnel did not consider the units for the fuel (an important lesson for students who tend to "blow off" units).

eGrade Plus *(powered by EduGen)*

eGrade Plus is a powerful, web-based course-management tool that allows instructors to provide their students with a dynamic, interactive learning environment.

eGrade Plus automates the process of assigning and grading homework, quizzes, and exams. It provides professors and students with numerous benefits including the ability to keep students on task and provide them with immediate feedback and scoring on their work without having to invest hours every week grading homework and quizzes. Every homework problem has a link to an on-line version of the text, giving the students immediate, structured information to help with problem solving. This context-sensitive help gives students quick access to a set of focused problem-solving aids.

For Instructors:

An Administration tool allows you to manage your class roster on-line.

A "Prepare and Present" tool contains a variety of the Wiley-provided resources to help make your preparation time more efficient. The specific resources include all of the *Instructor's Solutions Manual* in LaTex, MSWord, and pdf formats and all of the text illustrations in electronic format suitable both for classroom projection and for printing. You may easily adapt, customize, and add to this content to meet the needs of your course.

An "Assignment" tool allows you to create student homework and quizzes. You can use dynamic versions of end-of-chapter problems from *Fundamentals of Physics* or you can write your own dynamic questions. You may also assign readings, activities, and other work you want your students to complete. One of the most powerful features of eGrade Plus is that student assignments will be automatically graded and recorded in your gradebook. This will not only save you time but will provide your students with immediate feedback on their work, allowing them to determine right away how well they understand the course material.

A Gradebook tool will keep track of your students' progress and allow you to analyze individual and overall class results to determine their progress and level of understanding.

For Students

A "Study and Practice" area links directly to the multimedia version of *Fundamentals of Physics*, allowing students to review the text while they study and complete homework assignments. In addition to the complete text, students can also access the Student Solutions Manual, the Student Study Guide, interactive simulations, and the Interactive LearningWare Program.

An "Assignment" area keeps all the work you want your students to complete in one location, making it easy for them to stay "on task." Students will have access to a variety of interactive problem-solving tools, as well as other resources for building their confidence and understanding. In addition, the homework problems contain a link to the relevant section of the multimedia book, providing students with context-sensitive help that allows them to conquer problem-solving obstacles as they arise.

A Personal Gradebook for each student will allow students to view their results from past assignments at any time.

Versions of the Text

To accommodate the individual needs of instructors and students, the seventh edition of *Fundamentals of Physics* is available in a number of different versions.

The Regular Edition consists of Chapters 1 through 37 (ISBN 0-471-21643-7).

The Extended Edition contains six additional chapters on quantum physics and cosmology, Chapters 1–44 (ISBN 0-471-23231-9).

Both editions are available as single, hard-cover books, or in the following alternative versions:

➡ Volume 1 - Chapters 1–20 (Mechanics and Thermodynamics), hardcover, ISBN 0-471-42959-7

➡ Volume 2 - Chapters 21–44 (E&M, Optics, and Quantum Physics), hardcover, ISBN 0-471-42960-0

➡ Part 1 - Chapters 1–11, paperback, ISBN 0-471-42961-9

➡ Part 2 - Chapters 12–20, paperback, ISBN 0-471-42962-7

➡ Part 3 - Chapters 21–32, paperback, ISBN 0-471-42963-5

➡ Part 4 - Chapters 33–37, paperback, ISBN 0-471-42964-3

➡ Part 5 - Chapters 38–44, paperback, 0-471-42965-1

Instructor's Supplements

Instructor's Manual by J. Richard Christman, U.S. Coast Guard Academy. This manual contains lecture notes outlining the most important topics of each chapter; demonstration experiments; laboratory and computer projects; film and video sources; answers to all Questions, Exercises, Problems, and Checkpoints; and a correlation guide to the Questions, Exercises, and Problems in the previous edition. It also contains a complete list of all problems for which solutions are available to students (SSM, WWW, and ILW).

Instructor's Solutions Manual by Sen-Ben Liao, Massachusetts Institute of Technology. This manual provides worked-out solutions for all problems found at the end of each chapter.

Test Bank by J. Richard Christman, U.S. Coast Guard Academy. This manual includes more than 2200 multiple-choice questions. These items are also available in the Computerized Test Bank (see below).

Instructor's Resource CD. This CD contains:
• All of the *Instructor's Solutions Manual* in MSWord, and pdf files

• Computerized Test Bank, in both IBM and Macintosh versions, with full editing features to help you customize tests.
• All text illustrations, suitable for both classroom projection and printing.

Transparencies. More than 200 four-color illustrations from the text are provided in a form suitable for projection in the classroom.

On-line homework and quizzing. In addition to eGrade, *Fundamentals of Physics*, seventh edition also supports WebAssign and CAPA, which are other programs that give instructors the ability to deliver and grade homework and quizzes over the Internet.

WebCT and Blackboard. A variety of materials have been prepared for easy incorporation in either WebCT or Blackboard. WebCT and Blackboard are powerful and easy-to-use web-based course-management systems that allow instructors to set up complete on-line courses with chat rooms, bulletin boards, quizzing, student tracking, etc.

Student's Supplements

Student Web Site. This web site
    http://www.wiley.com/college/halliday
was developed specifically for *Fundamentals of Physics*, seventh edition, and is designed to further assist students in the study of physics. The site includes solutions to selected end-of-chapter problems (which are identified with a **www** icon in the text); self-quizzes; simulation exercises; links to other helpful websites; and the Interactive LearningWare tutorials that are described below.

A Student's Companion by J. Richard Christman, U.S. Coast Guard Academy. This student study guide consists of an overview of the chapter's important concepts, hints for solving end-of-chapter questions and problems, and a practice quiz.

Student's Solutions Manual by J. Richard Christman, U.S. Coast Guard Academy and Edward Derringh,

Wentworth Institute. This manual provides student with complete worked-out solutions to 30 percent of the problems found at the end of each chapter within the text. These problems are indicated with an **ssm** icon.

Interactive LearningWare. This software guides students through solutions to 200 of the end-of-chapter problems. These problems are indicated with an **ilw** icon. The solutions process is developed interactively, with appropriate feedback and access to error-specific help for the most common mistakes.

Multimedia Book. This web-based version of *Fundamentals of Physics*, seventh edition contains the complete, extended version of the text, *A Student's Companion*, the *Student's Solutions Manual*, the *Interactive LearningWare*, and numerous simulations, all connected with extensive hyperlinking.

Acknowledgments

A great many people have contributed to this book. J. Richard Christman, of the U.S. Coast Guard Academy, has once again created many fine supplements for us; his recommendations to this book have been invaluable. Sen-Ben Liao of MIT, James Whitenton of Southern Polytechnic State University, and Jerry Shi, of Pasadena City College, performed the Herculean task of working out solutions for every one of the homework problems in the book. We thank Renee M. Goertzen, now at the University of Maryland, for her checking many of the problems from a student perspective.

At John Wiley publishers, we have been fortunate to receive strong coordination and support from our editor Stuart Johnson, who constantly encourages our efforts and champions our work, Ellen Ford, who coordinated the developmental editing and multilayered preproduction process, Martin Batey, who coordinated the state-of-the-art media package, and Geraldine Osnato, who managed a super team to create an impressive supplements package.

We thank Lucille Buonocore and Pam Kennedy, our production managers, and Elizabeth Swain, our production editor, for pulling all the pieces together and guiding us through the complex production process. We also thank Maddy Lesure, for her design of both the text and the book cover; Helen Walden for her copyediting; Anna Melhorn for managing the illustration program; and Lilian Brady for her proofreading. Hilary Newman and Elyse Rieder were inspired in the search for unusual and interesting photographs.

We are very grateful to Irene Nunes who read every word in this book and made countless improvements to the clarity of the message in those words. Jearl Walker is also extremely grateful to Edward Millman who guided Walker through his first three editions of the book. (Walker dedicates this edition of the book to Ed Millman as a token of Walker's gratitude.)

Finally, our external reviewers have been outstanding and we acknowledge here our debt to each member of that team.

Maris A. Abolins
Michigan State University

Edward Adelson
Ohio State University

Barbara Andereck
Ohio Wesleyan University

Mark Arnett
Kirkwood Community College

Arun Bansil
Northeastern University

Albert Bartlett
University of Colorado

Michael E. Browne
University of Idaho

Timothy J. Burns
Leeward Community College

Joseph Buschi
Manhattan College

Philip A. Casabella
Rensselaer Polytechnic Institute

Randall Caton
Christopher Newport College

J. Richard Christman
U.S. Coast Guard Academy

Roger Clapp
University of South Florida

W. R. Conkie
Queen's University

Mike Crivello
San Diego State University

Peter Crooker
University of Hawaii at Manoa

William P. Crummett
Montana College of Mineral Science and Technology

Robert N. Davie, Jr.
St. Petersburg Junior College

Cheryl K. Dellai
Glendale Community College

Eric R. Dietz
California State University at Chico

N. John DiNardo
Drexel University

Eugene Dunnam
University of Florida

Robert Endorf
University of Cincinnati

F. Paul Esposito
University of Cincinnati

Jerry Finkelstein
San Jose State University

Alexander Firestone
Iowa State University

Alexander Gardner
Howard University

Andrew L. Gardner
Brigham Young University

John Gieniec
Central Missouri State University

Robert H. Good
California State University-Hayward

John B. Gruber
San Jose State University

Ann Hanks
American River College

Randy Harris
University of California-Davis

Samuel Harris
Purdue University

Harold B. Hart
Western Illinois University

Rebecca Hartzler
Edmonds Community College

Emily Haught
Georgia Institute of Technology

Laurent Hodges
Iowa State University

John Hubisz
North Carolina State University

Joey Huston
Michigan State University

Darrell Huwe
Ohio University

Shawn Jackson
University of Tulsa

Hector Jimenez
University of Puerto Rico

Sudhakar B. Joshi
York University

Claude Kacser
University of Maryland

Leonard M. Kahn
University of Rhode Island

Leonard Kleinman
University of Texas at Austin

Earl Koller
Stevens Institute of Technology

Arthur Z. Kovacs
Rochester Institute of Technology

Kenneth Krane
Oregon State University

Sol Krasner
University of Illinois at Chicago

Yuichi Kubota
Cornell University

Priscilla Laws
Dickinson College

Edbertho Leal
Polytechnic University of Puerto Rico

Peter Loly
University of Manitoba

Dale Long
Virginia Tech

Andreas Mandelis
University of Toronto

Robert R. Marchini
Memphis State University

David Markowitz
University of Connecticut

Paul Marquard
Caspar College

Howard C. McAllister
University of Hawaii at Manoa

W. Scott McCullough
Oklahoma State University

James H. McGuire
Tulane University

David M. McKinstry
Eastern Washington University

Joe P. Meyer
Georgia Institute of Technology

Roy Middleton
University of Pennsylvania

Irvin A. Miller
Drexel University

Eugene Mosca
United States Naval Academy

James Napolitano
Rensselaer Polytechnic Institute

Michael O'Shea
Kansas State University

Patrick Papin
San Diego State University

George Parker
North Carolina State University

Robert Pelcovits
Brown University

Des Penny
Southern Utah University

Oren P. Quist
South Dakota State University

Joe Redish
University of Maryland

Jonathan Reichart
SUNY—Buffalo

Timothy M. Ritter
University of North Carolina at Pembroke

Gerardo A. Rodriguez
Skidmore College

John Rosendahl
University of California at Irvine

Michael Schatz
Georgia Institute of Technology

Manuel Schwartz
University of Louisville

Darrell Seeley
Milwaukee School of Engineering

Bruce Arne Sherwood
Carnegie Mellon University

John Spangler
St. Norbert College

Ross L. Spencer
Brigham Young University

Harold Stokes
Brigham Young University

Michael G. Strauss
University of Oklahoma

Jay D. Strieb
Villanova University

Dan Styer
Oberlin College

Marshall Thomsen
Eastern Michigan University

Fred F. Tomblin
New Jersey Institute of Technology

David Toot
Alfred University

J. S. Turner
University of Texas at Austin

T. S. Venkataraman
Drexel University

Gianfranco Vidali
Syracuse University

Fred Wang
Prairie View A & M

Robert C. Webb
Texas A & M University

B. R. Weinberger
Trinity College

William M. Whelan
Ryerson Polytechnic University

George Williams
University of Utah

David Wolfe
University of New Mexico

William Zimmerman, Jr.
University of Minnesota

Equilibrium and Elasticity

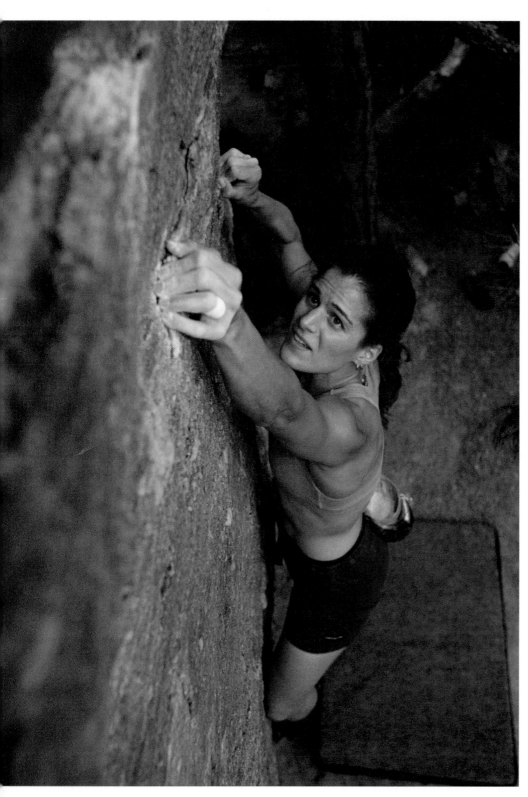

Many skilled rock climbers suffer from chronic pain that runs along their fingers, and some climbers also display a noticeable bulge along the palm side of an injured finger when they draw the finger in toward the palm.

What is the connection between the bulge, the pain, and the physics of climbing?

The answer is in this chapter.

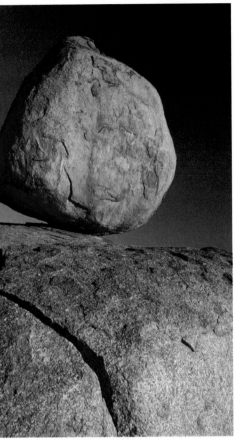

Fig. 12-1 A balancing rock. Although its perch seems precarious, the rock is in static equilibrium.

12-1 What Is Physics?

Human constructions are supposed to be stable in spite of the forces that act on them. A building, for example, should be stable in spite of the gravitational force and wind forces on it, and a bridge should be stable in spite of the gravitational force pulling it downward and the repeated jolting it receives from cars and trucks.

One focus of physics is on what allows an object to be stable in spite of any forces acting on it. In this chapter we examine the two main aspects of stability: the *equilibrium* of the forces and torques acting on rigid objects and the *elasticity* of nonrigid objects, a property that governs how such objects can deform. When this physics is done correctly, it is the subject of countless articles in physics and engineering journals; when it is done incorrectly, it is the subject of countless articles in newspapers and legal journals.

12-2 Equilibrium

Consider these objects: (1) a book resting on a table, (2) a hockey puck sliding with constant velocity across a frictionless surface, (3) the rotating blades of a ceiling fan, and (4) the wheel of a bicycle that is traveling along a straight path at constant speed. For each of these four objects,

1. The linear momentum $\vec{P}$ of its center of mass is constant.
2. Its angular momentum $\vec{L}$ about its center of mass, or about any other point, is also constant.

We say that such objects are in **equilibrium.** The two requirements for equilibrium are then

$$\vec{P} = \text{a constant} \quad \text{and} \quad \vec{L} = \text{a constant.} \tag{12-1}$$

Our concern in this chapter is with situations in which the constants in Eq. 12-1 are zero; that is, we are concerned largely with objects that are not moving in any way—either in translation or in rotation—in the reference frame from which we observe them. Such objects are in **static equilibrium.** Of the four objects mentioned at the beginning of this section, only one—the book resting on the table—is in static equilibrium.

The balancing rock of Fig. 12-1 is another example of an object that, for the present at least, is in static equilibrium. It shares this property with countless other structures, such as cathedrals, houses, filing cabinets, and taco stands, that remain stationary over time.

As we discussed in Section 8-6, if a body returns to a state of static equilibrium after having been displaced from that state by a force, the body is said to be in *stable* static equilibrium. A marble placed at the bottom of a hemispherical bowl is an example. However, if a small force can displace the body and end the equilibrium, the body is in *unstable* static equilibrium.

For example, suppose we balance a domino with the domino's center of mass vertically above the supporting edge, as in Fig. 12-2a. The torque about the supporting edge due to the gravitational force $\vec{F}_g$ on the domino is zero because the line of action of $\vec{F}_g$ is through that edge. Thus, the domino is in equilibrium. Of course, even a slight force on it due to some chance disturbance ends the equilibrium. As the line of action of $\vec{F}_g$ moves to one side of the supporting edge (as in Fig. 12-2b), the torque due to $\vec{F}_g$ increases the rotation of the domino. Therefore, the domino in Fig. 12-2a is in unstable static equilibrium.

The domino in Fig. 12-2c is not quite as unstable. To topple this domino, a force would have to rotate it through and then beyond the balance position of Fig. 12-2a, in which the center of mass is above a supporting edge. A slight force will not topple this domino, but a vigorous flick of the finger against the domino

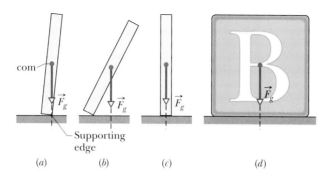

Fig. 12-2 (*a*) A domino balanced on one edge, with its center of mass vertically above that edge. The gravitational force $\vec{F}_g$ on the domino is directed through the supporting edge. (*b*) If the domino is rotated even slightly from the balanced orientation, then $\vec{F}_g$ causes a torque that increases the rotation. (*c*) A domino upright on a narrow side is somewhat more stable than the domino in (*a*). (*d*) A square block is even more stable.

certainly will. (If we arrange a chain of such upright dominos, a finger flick against the first can cause the whole chain to fall.)

The child's square block in Fig. 12-2*d* is even more stable because its center of mass would have to be moved even farther to get it to pass above a supporting edge. A flick of the finger may not topple the block. (This is why you never see a chain of toppling square blocks.) The worker in Fig. 12-3 is like both the domino and the square block: Parallel to the beam, his stance is wide and he is stable; perpendicular to the beam, his stance is narrow and he is unstable (and at the mercy of a chance gust of wind).

The analysis of static equilibrium is very important in engineering practice. The design engineer must isolate and identify all the external forces and torques that may act on a structure and, by good design and wise choice of materials, ensure that the structure will remain stable under these loads. Such analysis is necessary to ensure, for example, that bridges do not collapse under their traffic and wind loads and that the landing gear of aircraft will function after the shock of rough landings.

12-3 *The Requirements of Equilibrium*

The translational motion of a body is governed by Newton's second law in its linear momentum form, given by Eq. 9-27 as

$$\vec{F}_{\text{net}} = \frac{d\vec{P}}{dt}. \tag{12-2}$$

If the body is in translational equilibrium—that is, if $\vec{P}$ is a constant—then $d\vec{P}/dt = 0$ and we must have

$$\vec{F}_{\text{net}} = 0 \quad \text{(balance of forces)}. \tag{12-3}$$

The rotational motion of a body is governed by Newton's second law in its angular momentum form, given by Eq. 11-29 as

$$\vec{\tau}_{\text{net}} = \frac{d\vec{L}}{dt}. \tag{12-4}$$

If the body is in rotational equilibrium—that is, if $\vec{L}$ is a constant—then $d\vec{L}/dt = 0$ and we must have

$$\vec{\tau}_{\text{net}} = 0 \quad \text{(balance of torques)}. \tag{12-5}$$

Thus, the two requirements for a body to be in equilibrium are as follows:

1. The vector sum of all the external forces that act on the body must be zero.

2. The vector sum of all the external torques that act on the body, measured about *any* possible point, must also be zero.

Fig. 12-3 A construction worker balanced on a steel beam is in static equilibrium but is more stable parallel to the beam than perpendicular to it.

These requirements obviously hold for *static* equilibrium. They also hold for the more general equilibrium in which $\vec{P}$ and $\vec{L}$ are constant but not zero.

Equations 12-3 and 12-5, as vector equations, are each equivalent to three independent component equations, one for each direction of the coordinate axes:

Balance of forces	Balance of torques
$F_{\text{net},x} = 0$	$\tau_{\text{net},x} = 0$
$F_{\text{net},y} = 0$	$\tau_{\text{net},y} = 0$
$F_{\text{net},z} = 0$	$\tau_{\text{net},z} = 0$

(12-6)

We shall simplify matters by considering only situations in which the forces that act on the body lie in the *xy* plane. This means that the only torques that can act on the body must tend to cause rotation around an axis parallel to the *z* axis. With this assumption, we eliminate one force equation and two torque equations from Eqs. 12-6, leaving

$$F_{\text{net},x} = 0 \quad \text{(balance of forces),} \qquad (12\text{-}7)$$

$$F_{\text{net},y} = 0 \quad \text{(balance of forces),} \qquad (12\text{-}8)$$

$$\tau_{\text{net},z} = 0 \quad \text{(balance of torques).} \qquad (12\text{-}9)$$

Here, $\tau_{\text{net},z}$ is the net torque that the external forces produce either about the *z* axis or about *any* axis parallel to it.

A hockey puck sliding at constant velocity over ice satisfies Eqs. 12-7, 12-8, and 12-9 and is thus in equilibrium *but not in static equilibrium*. For static equilibrium, the linear momentum $\vec{P}$ of the puck must be not only constant but also zero; the puck must be at rest on the ice. Thus, there is another requirement for static equilibrium:

➤ **3.** The linear momentum $\vec{P}$ of the body must be zero.

✔ **CHECKPOINT 1** The figure gives six overhead views of a uniform rod on which two or more forces act perpendicularly to the rod. If the magnitudes of the forces are adjusted properly (but kept nonzero), in which situations can the rod be in static equilibrium?

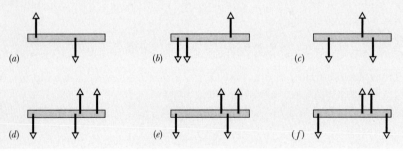

12-4 The Center of Gravity

The gravitational force on an extended body is the vector sum of the gravitational forces acting on the individual elements (the atoms) of the body. Instead of considering all those individual elements, we can say that

➤ The gravitational force $\vec{F}_g$ on a body effectively acts at a single point, called the **center of gravity** (cog) of the body.

Here the word "effectively" means that if the forces on the individual elements were somehow turned off and force $\vec{F_g}$ at the center of gravity were turned on, the net force and the net torque (about any point) acting on the body would not change.

Until now, we have assumed that the gravitational force $\vec{F_g}$ acts at the center of mass (com) of the body. This is equivalent to assuming that the center of gravity is at the center of mass. Recall that, for a body of mass M, the force $\vec{F_g}$ is equal to $M\vec{g}$, where $\vec{g}$ is the acceleration that the force would produce if the body were to fall freely. In the proof that follows, we show that

> If $\vec{g}$ is the same for all elements of a body, then the body's center of gravity (cog) is coincident with the body's center of mass (com).

This is approximately true for everyday objects because $\vec{g}$ varies only a little along Earth's surface and decreases in magnitude only slightly with altitude. Thus, for objects like a mouse or a moose, we have been justified in assuming that the gravitational force acts at the center of mass. After the following proof, we shall resume that assumption.

Proof

First, we consider the individual elements of the body. Figure 12-4a shows an extended body, of mass M, and one of its elements, of mass m_i. A gravitational force $\vec{F_{gi}}$ acts on each such element and is equal to $m_i \vec{g_i}$. The subscript on $\vec{g_i}$ means $\vec{g_i}$ is the gravitational acceleration *at the location of the element i* (it can be different for other elements).

In Fig. 12-4a, each force $\vec{F_{gi}}$ produces a torque τ_i on the element about the origin O, with moment arm x_i. Using Eq. 10-41 ($\tau = r_\perp F$), we can write torque τ_i as

$$\tau_i = x_i F_{gi}. \tag{12-10}$$

The net torque on all the elements of the body is then

$$\tau_{\text{net}} = \sum \tau_i = \sum x_i F_{gi}. \tag{12-11}$$

Next, we consider the body as a whole. Figure 12-4b shows the gravitational force $\vec{F_g}$ acting at the body's center of gravity. This force produces a torque τ on the body about O, with moment arm x_{cog}. Again using Eq. 10-41, we can write this torque as

$$\tau = x_{\text{cog}} F_g. \tag{12-12}$$

The gravitational force $\vec{F_g}$ on the body is equal to the sum of the gravitational forces $\vec{F_{gi}}$ on all its elements, so we can substitute ΣF_{gi} for F_g in Eq. 12-12 to write

$$\tau = x_{\text{cog}} \sum F_{gi}. \tag{12-13}$$

Now recall that the torque due to force $\vec{F_g}$ acting at the center of gravity is equal to the net torque due to all the forces $\vec{F_{gi}}$ acting on all the elements of the body. (That is how we defined the center of gravity.) Thus, τ in Eq. 12-13 is equal to τ_{net} in Eq. 12-11. Putting those two equations together, we can write

$$x_{\text{cog}} \sum F_{gi} = \sum x_i F_{gi}.$$

Substituting $m_i g_i$ for F_{gi} gives us

$$x_{\text{cog}} \sum m_i g_i = \sum x_i m_i g_i. \tag{12-14}$$

Now here is a key idea: If the accelerations g_i at all the locations of the elements are the same, we can cancel g_i from this equation to write

$$x_{\text{cog}} \sum m_i = \sum x_i m_i. \tag{12-15}$$

Fig. 12-4 (a) An element of mass m_i in an extended body. The gravitational force $\vec{F_{gi}}$ on the element has moment arm x_i about the origin O of the coordinate system. (b) The gravitational force $\vec{F_g}$ on a body is said to act at the center of gravity (cog) of the body. Here $\vec{F_g}$ has moment arm x_{cog} about origin O.

The sum Σm_i of the masses of all the elements is the mass M of the body. Therefore, we can rewrite Eq. 12-15 as

$$x_{\text{cog}} = \frac{1}{M} \Sigma x_i m_i. \qquad (12\text{-}16)$$

The right side of this equation gives the coordinate x_{com} of the body's center of mass (Eq. 9-4). We now have what we sought to prove:

$$x_{\text{cog}} = x_{\text{com}}. \qquad (12\text{-}17)$$

✔ **CHECKPOINT 2** Suppose that you skewer an apple with a thin rod, missing the apple's center of gravity. When you hold the rod horizontally and allow the apple to rotate freely, where does the center of gravity end up and why?

12-5 Some Examples of Static Equilibrium

In this section we examine four sample problems involving static equilibrium. In each, we select a system of one or more objects to which we apply the equations of equilibrium (Eqs. 12-7, 12-8, and 12-9). The forces involved in the equilibrium are all in the xy plane, which means that the torques involved are parallel to the z axis. Thus, in applying Eq. 12-9, the balance of torques, we select an axis parallel to the z axis about which to calculate the torques. Although Eq. 12-9 is satisfied for *any* such choice of axis, you will see that certain choices simplify the application of Eq. 12-9 by eliminating one or more unknown force terms.

Sample Problem 12-1

In Fig. 12-5a, a uniform beam, of length L and mass $m = 1.8$ kg, is at rest on two scales. A uniform block, with mass $M = 2.7$ kg, is at rest on the beam, with its center a distance $L/4$ from the beam's left end. What do the scales read?

Solution: The first steps in the solution of *any* problem about static equilibrium are these: Clearly define the system to be analyzed and then draw a free-body diagram of it, indicating all the forces on the system. Here, let us choose the system as the beam and block taken together. Then the forces on the system are shown in the free-body diagram of Fig. 12-5b. (Choosing the system takes experience, and often there can be more than one good choice; see item 2 of Problem-Solving Tactic 1 below.)

The normal forces on the beam from the scales are $\vec{F}_l$ on the left and $\vec{F}_r$ on the right. The scale readings that we want are equal to the magnitudes of those forces. The gravitational force $\vec{F}_{g,\text{beam}}$ on the beam acts at the beam's center of mass and is equal to $m\vec{g}$. Similarly, the gravitational force $\vec{F}_{g,\text{block}}$ on the block acts at the block's center of mass and is equal to $M\vec{g}$. However, to simplify Fig. 12-5b, the block is represented by a dot within the boundary of the beam and $\vec{F}_{g,\text{block}}$ is drawn with its tail on that dot. (This shift of vector $\vec{F}_{g,\text{block}}$ along its line of action does not alter the torque due to $\vec{F}_{g,\text{block}}$ about any axis perpendicular to the figure.)

The **Key Idea** here is that, because the system is in static equilibrium, we can apply the balance of forces equations (Eqs. 12-7 and 12-8) and the balance of torques equation (Eq. 12-9) to it. The forces have no x components, so Eq. 12-7 ($F_{\text{net},x} = 0$) provides no information. For the y components, Eq. 12-8 ($F_{\text{net},y} = 0$) gives us

$$F_l + F_r - Mg - mg = 0. \qquad (12\text{-}18)$$

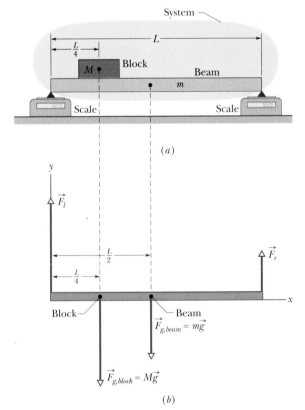

Fig. 12-5 (a) A beam of mass m supports a block of mass M. (b) A free-body diagram, showing the forces that act on the system *beam* + *block*.

This equation contains two unknowns, the forces F_l and F_r, so we also need to use Eq. 12-9, the balance of torques equation. We can apply it to *any* rotation axis perpendicular to the plane of Fig. 12-5. Let us choose a rotation axis through the left end of the beam. We shall also use our general rule for assigning signs to torques: If a torque would cause an initially stationary body to rotate clockwise about the rotation axis, the torque is negative. If the rotation would be counterclockwise, the torque is positive. Finally, we shall write the torques in the form $r_\perp F$, where the moment arm $r_\perp$ is 0 for $\vec{F}_l$, $L/4$ for $M\vec{g}$, $L/2$ for $m\vec{g}$, and L for $\vec{F}_r$.

We now can write the balancing equation ($\tau_{\text{net},z} = 0$) as

$$(0)(F_l) - (L/4)(Mg) - (L/2)(mg) + (L)(F_r) = 0,$$

which gives us

$$F_r = \tfrac{1}{4}Mg + \tfrac{1}{2}mg$$
$$= \tfrac{1}{4}(2.7 \text{ kg})(9.8 \text{ m/s}^2) + \tfrac{1}{2}(1.8 \text{ kg})(9.8 \text{ m/s}^2)$$
$$= 15.44 \text{ N} \approx 15 \text{ N}. \qquad \text{(Answer)}$$

Now, solving Eq. 12-18 for F_l and substituting this result, we find

$$F_l = (M + m)g - F_r$$
$$= (2.7 \text{ kg} + 1.8 \text{ kg})(9.8 \text{ m/s}^2) - 15.44 \text{ N}$$
$$= 28.66 \text{ N} \approx 29 \text{ N}. \qquad \text{(Answer)}$$

Notice the strategy in the solution: When we wrote an equation for the balance of force components, we got stuck with two unknowns. If we had written an equation for the balance of torques around some *arbitrary* axis, we would have again gotten stuck with those two unknowns. However, because we chose the axis to pass through the point of application of one of the unknown forces, here $\vec{F}_l$, we did not get stuck. Our choice neatly eliminated that force from the torque equation, allowing us to solve for the other unknown force magnitude F_r. Then we returned to the equation for the balance of force components to find the remaining unknown force magnitude.

CHECKPOINT 3 The figure gives an overhead view of a uniform rod in static equilibrium. (a) Can you find the magnitudes of unknown forces $\vec{F}_1$ and $\vec{F}_2$ by balancing the forces? (b) If you wish to find the magnitude of force $\vec{F}_2$ by using a single equation, where should you place a rotational axis? (c) The magnitude of $\vec{F}_2$ turns out to be 65 N. What then is the magnitude of $\vec{F}_1$?

Sample Problem 12-2

In Fig. 12-6a, a ladder of length $L = 12$ m and mass $m = 45$ kg leans against a slick (frictionless) wall. The ladder's upper end is at height $h = 9.3$ m above the pavement on which the lower end rests (the pavement is not frictionless). The ladder's center of mass is $L/3$ from the lower end. A firefighter of mass $M = 72$ kg climbs the ladder until her center of mass is $L/2$ from the lower end. What then are the magnitudes of the forces on the ladder from the wall and the pavement?

Solution: First, we choose our system as being the firefighter and ladder, together, and then we draw the free-body diagram of Fig. 12-6b. The firefighter is represented with a dot within the boundary of the ladder. The gravitational force on her is represented with its equivalent $M\vec{g}$, and that vector has been shifted along its line of action, so that its tail is on the dot. (The shift does not alter a torque due to $M\vec{g}$ about any axis perpendicular to the figure.)

The only force on the ladder from the wall is the horizontal force $\vec{F}_w$ (there cannot be a frictional force along a frictionless wall). The force $\vec{F}_p$ on the ladder from the pavement has a horizontal component $\vec{F}_{px}$ that is a static frictional force and a vertical component $\vec{F}_{py}$ that is a normal force.

A **Key Idea** here is that the system is in static equilibrium, so the balancing equations (Eqs. 12-7 through 12-9) apply to it. Let us start with Eq. 12-9 ($\tau_{\text{net},z} = 0$). To choose an axis about which to calculate the torques, note that we have unknown forces ($\vec{F}_w$ and $\vec{F}_p$) at the two ends of the ladder. To eliminate, say, $\vec{F}_p$ from the calculation, we place the axis at point O, perpendicular to the figure. We also place the origin of an xy coordinate system at O. We can find torques about O with any of Eqs. 10-39 through 10-41, but Eq. 10-41 ($\tau = r_\perp F$) is easiest to use here.

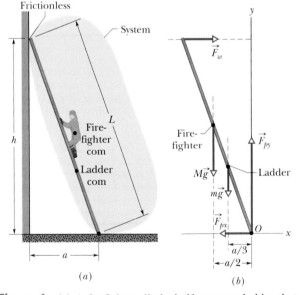

Fig. 12-6 (a) A firefighter climbs halfway up a ladder that is leaning against a frictionless wall. The pavement beneath the ladder is not frictionless. (b) A free-body diagram, showing the forces that act on the *firefighter + ladder* system. The origin O of a coordinate system is placed at the point of application of the unknown force $\vec{F}_p$ (whose vector components $\vec{F}_{px}$ and $\vec{F}_{py}$ are shown).

To find the moment arm $r_\perp$ of $\vec{F}_w$, we draw a line of action through that vector (horizontal dashed line in Fig. 12-6b). Then $r_\perp$ is the perpendicular distance between O and the line of action. In Fig. 12-6b, $r_\perp$ extends along the y axis and is equal to the height h. We similarly draw lines of action for $M\vec{g}$ and $m\vec{g}$ and see that their moment arms extend along the x axis. For the distance a shown in Fig. 12-6a, the moment arms are $a/2$ (the firefighter is halfway up the ladder) and $a/3$ (the ladder's center of mass is one-third of the way up the ladder), respectively. The moment arms for $\vec{F}_{px}$ and $\vec{F}_{py}$ are zero.

Now, with torques written in the form $r_\perp F$, the balancing equation $\tau_{net,z} = 0$ becomes

$$-(h)(F_w) + (a/2)(Mg) + (a/3)(mg)$$
$$+ (0)(F_{px}) + (0)(F_{py}) = 0. \quad (12\text{-}19)$$

(Recall our rule: A positive torque corresponds to counter-clockwise rotation and a negative torque corresponds to clockwise rotation.)

Using the Pythagorean theorem, we find that

$$a = \sqrt{L^2 - h^2} = 7.58 \text{ m}.$$

Then Eq. 12-19 gives us

$$F_w = \frac{ga(M/2 + m/3)}{h}$$
$$= \frac{(9.8 \text{ m/s}^2)(7.58 \text{ m})(72/2 \text{ kg} + 45/3 \text{ kg})}{9.3 \text{ m}}$$
$$= 407 \text{ N} \approx 410 \text{ N}. \quad \text{(Answer)}$$

Now we need to use the force balancing equations. The equation $F_{net,x} = 0$ gives us

$$F_w - F_{px} = 0, \quad (12\text{-}20)$$

so

$$F_{px} = F_w = 410 \text{ N}. \quad \text{(Answer)}$$

The equation $F_{net,y} = 0$ gives us

$$F_{py} - Mg - mg = 0, \quad (12\text{-}21)$$

so

$$F_{py} = (M + m)g = (72 \text{ kg} + 45 \text{ kg})(9.8 \text{ m/s}^2)$$
$$= 1146.6 \text{ N} \approx 1100 \text{ N}. \quad \text{(Answer)}$$

Sample Problem 12-3

Figure 12-7a shows a safe, of mass $M = 430$ kg, hanging by a rope from a boom with dimensions $a = 1.9$ m and $b = 2.5$ m. The boom consists of a hinged beam and a horizontal cable. The uniform beam has a mass m of 85 kg; the masses of the cable and rope are negligible.

(a) What is the tension T_c in the cable? In other words, what is the magnitude of the force $\vec{T}_c$ on the beam from the cable?

Solution: The system here is the beam alone, and the forces on it are shown in the free-body diagram of Fig. 12-7b. The force from the cable is $\vec{T}_c$. The gravitational force on the beam acts at the beam's center of mass (at the beam's center) and is represented by its equivalent $m\vec{g}$. The vertical component of the force on the beam from the hinge is $\vec{F}_v$, and the horizontal component of the force from the hinge is $\vec{F}_h$. The force from the rope supporting the safe is $\vec{T}_r$. Because beam, rope, and safe are stationary, the magnitude of $\vec{T}_r$ is equal to the weight of the safe: $T_r = Mg$. We place the origin O of an xy coordinate system at the hinge.

One **Key Idea** here is that our system is in static equilibrium, so the balancing equations apply to it. Let us start with Eq. 12-9 ($\tau_{net,z} = 0$). Note that we are asked for the magnitude of force $\vec{T}_c$ and not of forces $\vec{F}_h$ and $\vec{F}_v$ acting at the hinge, at point O. Thus, a second **Key Idea** is that, to eliminate $\vec{F}_h$ and $\vec{F}_v$ from the torque calculation, we should calculate torques about an axis that is perpendicular to the figure at point O. Then $\vec{F}_h$ and $\vec{F}_v$ will have moment arms of zero. The lines of action for $\vec{T}_c$, $\vec{T}_r$, and $m\vec{g}$ are dashed in Fig. 12-7b. The corresponding moment arms are a, b, and $b/2$.

Writing torques in the form of $r_\perp F$ and using our rule about signs for torques, the balancing equation $\tau_{net,z} = 0$ becomes

$$(a)(T_c) - (b)(T_r) - (\tfrac{1}{2}b)(mg) = 0.$$

Substituting Mg for T_r and solving for T_c, we find that

$$T_c = \frac{gb(M + \tfrac{1}{2}m)}{a}$$
$$= \frac{(9.8 \text{ m/s}^2)(2.5 \text{ m})(430 \text{ kg} + 85/2 \text{ kg})}{1.9 \text{ m}}$$
$$= 6093 \text{ N} \approx 6100 \text{ N}. \quad \text{(Answer)}$$

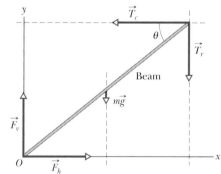

Fig. 12-7 (a) A heavy safe is hung from a boom consisting of a horizontal steel cable and a uniform beam. (b) A free-body diagram for the beam.

(b) Find the magnitude F of the net force on the beam from the hinge.

Solution: Now we want F_h and F_v so we can combine them to get F. Because we know T_c, our **Key Idea** here is to apply the force balancing equations to the beam. For the horizontal balance, we write $F_{net,x} = 0$ as

$$F_h - T_c = 0,$$

and so

$$F_h = T_c = 6093 \text{ N}.$$

For the vertical balance, we write $F_{net,y} = 0$ as

$$F_v - mg - T_r = 0.$$

Substituting Mg for T_r and solving for F_v, we find that

$$F_v = (m + M)g = (85 \text{ kg} + 430 \text{ kg})(9.8 \text{ m/s}^2)$$
$$= 5047 \text{ N}.$$

From the Pythagorean theorem, we now have

$$F = \sqrt{F_h^2 + F_v^2}$$
$$= \sqrt{(6093 \text{ N})^2 + (5047 \text{ N})^2} \approx 7900 \text{ N}. \quad \text{(Answer)}$$

Note that F is substantially greater than either the combined weights of the safe and the beam, 5000 N, or the tension in the horizontal wire, 6100 N.

CHECKPOINT 4   In the figure, a stationary 5 kg rod AC is held against a wall by a rope and friction between rod and wall. The uniform rod is 1 m long, and angle $\theta = 30°$. (a) If you are to find the magnitude of the force $\vec{T}$ on the rod from the rope with a single equation, at what labeled point should a rotational axis be placed? With that choice of axis and counterclockwise torques positive, what is the sign of (b) the torque τ_w due to the rod's weight and (c) the torque τ_r due to the pull on the rod by the rope? (d) Is the magnitude of τ_r greater than, less than, or equal to the magnitude of τ_w?

Sample Problem 12-4

Rock climbers who complain of acute pain along their fingers often remember exactly when the pain first appeared: They were using a *crimp hold* on an overhead support that may have been only a few millimeters deep. In this type of hold, the climber presses down with four fingers to gain purchase.

(a) Figure 12-8a shows a 70 kg climber hanging by only the crimp hold of one hand. Her feet touch the rock wall directly below her crimped fingers. Assume that the force from the horizontal ledge supporting her fingers is equally shared by the four fingers. What are the values of the horizontal component F_h and vertical component F_v of the force on each fingertip?

Solution: Our system is the climber, and Fig. 12-8a shows the forces that act on her. Force $\vec{F}$ acts on each fingertip, with horizontal component F_h and vertical component F_v. Thus, the net force on the four fingertips has horizontal component $4F_h$ and vertical component $4F_v$. The gravitational force $\vec{F}_g = m\vec{g}$ acts at her center of mass, and a normal force $\vec{F}_N$ from the wall acts on her feet at point O. (There is no vertical force from the wall on her feet.)

A **Key Idea** here is that, because the system is in static equilibrium, we can apply Eqs. 12-7, 12-8, and 12-9 for the balance of forces and the balance of torques. The balance of forces equation $F_{net,x} = 0$ leads to

$$-F_N + 4F_h = 0.$$

However, we cannot solve this equation for F_h because we do not know the value of F_N. The balance of forces equation $F_{net,y} = 0$ leads to

$$4F_v - mg = 0,$$

which we can solve for F_v:

$$F_v = mg/4 = (70 \text{ kg})(9.8 \text{ m/s}^2)/4$$
$$= 171.5 \text{ N} \approx 1.7 \times 10^2 \text{ N}. \quad \text{(Answer)}$$

The third equation we need is the balance of torques equation, $\tau_{net,z} = 0$. Because we do not know or want the magnitude of $\vec{F}_N$, let's calculate torques about an axis that is perpendicular to the figure at point O. To do so, we need the moment arm $r_\perp$ of each force acting on the climber. The moment arm for torques about O is 0 for $\vec{F}_N$, 0.20 m for $m\vec{g}$, 2.0 m for F_h, and 0 for F_v. Applying Eq. 11-17 ($\tau = r_\perp F$), we can

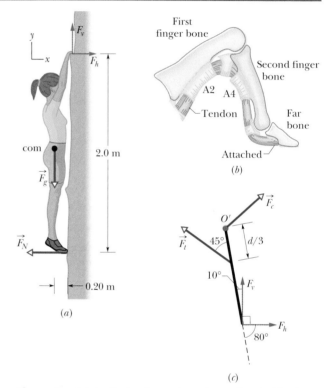

Fig. 12-8   (a) A climber hangs by only a crimp hold of one hand. The forces on her are indicated. (b) A tendon runs through the A2 and A4 pulleys on the finger bones. (c) A simplifed drawing of the forces acting on the second finger bone in a crimp hold.

now rewrite $\tau_{net,z} = 0$ as

$$(0)F_N + (0.20 \text{ m})(mg) - (2.0 \text{ m})(4F_h) + (0)(4F_v) = 0.$$

Solving for F_h and substituting known data give us

$$F_h = \frac{(0.20 \text{ m})(70 \text{ kg})(9.8 \text{ m/s}^2)}{4(2.0 \text{ m})}$$
$$= 17.15 \text{ N} \approx 17 \text{ N}. \quad \text{(Answer)}$$

(b) Figure 12-8b shows details of a finger. A tendon that runs from muscles in the forearm is attached to the far bone in the finger. Along the way, the tendon runs through several guiding sheaths called *pulleys*. The A2 pulley is attached to the first finger bone; the A4 pulley is attached to the second finger bone. To pull the finger toward the palm, the forearm muscles pull the tendon through the pulleys, much like strings on a marionette can be pulled to move parts of the marionette.

Figure 12-8c is a simplified diagram of the second finger bone, which has length d. The tendon's pull $\vec{F}_t$ on the bone acts at the point where the tendon enters the A4 pulley, at distance $d/3$ along the bone. For the crimp-hold results of part (a) and for the angles shown in Fig. 12-8c, what is the magnitude of $\vec{F}_t$?

Solution: Our system is now the second finger bone. The **Key Idea** is that because the system is in equilibrium, we can again apply the balance equations—in particular, the equation for the balance of torques ($\tau_{net,z} = 0$). Because we do not know or want force $\vec{F}_c$ at point O', where the first and second bones meet, we calculate torques about that point. We are letting the unknown length of the second bone be d; let us use Eq. 11-15 ($\tau = rF \sin \phi$) to calculate torques. The values of r

are 0 for $\vec{F}_c$, $d/3$ for $\vec{F}_t$, and d for F_h and F_v. We can now write $\tau_{net,z} = 0$ as

$$(0)F_c - \tfrac{d}{3}F_t \sin 45° + dF_v \sin 10° + dF_h \sin 80° = 0.$$

Solving for F_t and substituting values for F_v and F_h, we find

$$F_t = \frac{3[(171.5 \text{ N})(\sin 10°) + (17.15 \text{ N})(\sin 80°)]}{\sin 45°}$$

$$= 198 \text{ N} \approx 2.0 \times 10^2 \text{ N}. \qquad \text{(Answer)}$$

Medical researchers believe that this magnitude of F_t is tolerable. However, if a climber does not distribute the forces equally on all four fingers, the magnitude of F_t for one or more of the fingers may exceed what the A2 and A4 pulleys can withstand. When this happens, one or both pulleys rupture. When a climber suffering from this type of rupture injury pulls the finger toward the palm, the tendon produces a noticeable bulge under the skin because it is no longer constrained by the pulleys.

PROBLEM-SOLVING TACTICS

TACTIC 1: *Static Equilibrium Problems*
Here is a list of steps for solving static equilibrium problems:

1. Draw a *sketch* of the problem.
2. Select the *system* to which you will apply the laws of equilibrium, drawing a closed curve around the system on your sketch to fix it clearly in your mind. In some situations you can select a single object as the system; it is the object you wish to be in equilibrium (such as the rock climber in Sample Problem 12-4). In other situations, you might include additional objects in the system *if* their inclusion simplifies the calculations for equilibrium. For example, suppose in Sample Problem 12-2 you select only the ladder as the system. Then in Fig. 12-6b you will have to account for additional unknown forces exerted on the ladder by the hands and feet of the firefighter. These additional unknowns complicate the equilibrium calculations. The system of Fig. 12-6 was chosen to include the firefighter so that those unknown forces are *internal* to the system and thus need not be found in order to solve Sample Problem 12-2.
3. Draw a *free-body diagram* of the system. Show all the forces that act on the system, labeling them and making sure that their points of application and lines of action are correctly shown.

4. Draw in the *x and y axes* of a coordinate system with at least one axis parallel to one or more unknown force. Resolve into components the forces that do not lie along one of the axes. In all our sample problems it made sense to choose the x axis horizontal and the y axis vertical.
5. Write the two *balance of forces equations,* using symbols throughout.
6. Choose one or more rotation axes perpendicular to the plane of the figure and write the *balance of torques equation* for each axis. If you choose an axis that passes through the line of action of an unknown force, the equation will be simplified because that force will not appear in it.
7. *Solve* your equations *algebraically* for the unknowns. Some students feel more confident in substituting numbers with units in the independent equations at this stage, especially if the algebra is particularly involved. However, experienced problem solvers prefer the algebra, which reveals the dependence of solutions on the various variables.
8. Finally, *substitute numbers* with units in your algebraic solutions, obtaining numerical values for the unknowns.
9. Look at your answer—does it make sense? Is it obviously too large or too small? Is the sign correct? Are the units appropriate?

12-6 Indeterminate Structures

For the problems of this chapter, we have only three independent equations at our disposal, usually two balance of forces equations and one balance of torques equation about a given rotation axis. Thus, if a problem has more than three unknowns, we cannot solve it.

It is easy to find such problems. In Sample Problem 12-2, for example, we could have assumed that there is friction between the wall and the top of the ladder. Then there would have been a vertical frictional force acting where the ladder touches the wall, making a total of four unknown forces. With only three equations, we could not have solved this problem.

Consider also an unsymmetrically loaded car. What are the forces—all different—on the four tires? Again, we cannot find them because we have only three independent equations with which to work. Similarly, we can solve an equilibrium problem for a table with three legs but not for one with four legs. Problems like these, in which there are more unknowns than equations, are called **indeterminate.**

Yet solutions to indeterminate problems exist in the real world. If you rest the tires of the car on four platform scales, each scale will register a definite reading, the sum of the readings being the weight of the car. What is eluding us in our efforts to find the individual forces by solving equations?

The problem is that we have assumed—without making a great point of it—that the bodies to which we apply the equations of static equilibrium are perfectly rigid. By this we mean that they do not deform when forces are applied to them. Strictly, there are no such bodies. The tires of the car, for example, deform easily under load until the car settles into a position of static equilibrium.

We have all had experience with a wobbly restaurant table, which we usually level by putting folded paper under one of the legs. If a big enough elephant sat on such a table, however, you may be sure that if the table did not collapse, it would deform just like the tires of a car. Its legs would all touch the floor, the forces acting upward on the table legs would all assume definite (and different) values as in Fig. 12-9, and the table would no longer wobble. How do we find the values of those forces acting on the legs?

To solve such indeterminate equilibrium problems, we must supplement equilibrium equations with some knowledge of *elasticity,* the branch of physics and engineering that describes how real bodies deform when forces are applied to them. The next section provides an introduction to this subject.

Fig. 12-9 The table is an indeterminate structure. The four forces on the table legs differ from one another in magnitude and cannot be found from the laws of static equilibrium alone.

✓CHECKPOINT 5 A horizontal uniform bar of weight 10 N is to hang from a ceiling by two wires that exert upward forces $\vec{F}_1$ and $\vec{F}_2$ on the bar. The figure shows four arrangements for the wires. Which arrangements, if any, are indeterminate (so that we cannot solve for numerical values of $\vec{F}_1$ and $\vec{F}_2$)?

12-7 Elasticity

When a large number of atoms come together to form a metallic solid, such as an iron nail, they settle into equilibrium positions in a three-dimensional *lattice,* a repetitive arrangement in which each atom is a well-defined equilibrium distance from its nearest neighbors. The atoms are held together by interatomic forces that are modeled as tiny springs in Fig. 12-10. The lattice is remarkably rigid, which is another way of saying that the "interatomic springs" are extremely stiff. It is for this reason that we perceive many ordinary objects, such as metal ladders, tables, and spoons, as perfectly rigid. Of course, some ordinary objects, such as garden hoses or rubber gloves, do not strike us as rigid at all. The atoms that make up these objects *do not* form a rigid lattice like that of Fig. 12-10 but are aligned in long, flexible molecular chains, each chain being only loosely bound to its neighbors.

Fig. 12-10 The atoms of a metallic solid are distributed on a repetitive three-dimensional lattice. The springs represent interatomic forces.

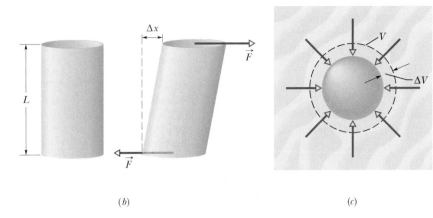

(a) (b) (c)

Fig. 12-11 (a) A cylinder subject to *tensile stress* stretches by an amount ΔL. (b) A cylinder subject to *shearing stress* deforms by an amount Δx, somewhat like a pack of playing cards would. (c) A solid sphere subject to uniform *hydraulic stress* from a fluid shrinks in volume by an amount ΔV. All the deformations shown are greatly exaggerated.

Fig. 12-12 A test specimen used to determine a stress–strain curve such as that of Fig. 12-13. The change ΔL that occurs in a certain length L is measured in a tensile stress–strain test.

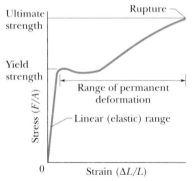

Fig. 12-13 A stress–strain curve for a steel test specimen such as that of Fig. 12-12. The specimen deforms permanently when the stress is equal to the *yield strength* of the specimen's material. It ruptures when the stress is equal to the *ultimate strength* of the material.

All real "rigid" bodies are to some extent **elastic,** which means that we can change their dimensions slightly by pulling, pushing, twisting, or compressing them. To get a feeling for the orders of magnitude involved, consider a vertical steel rod 1 m long and 1 cm in diameter attached to a factory ceiling. If you hang a subcompact car from the free end of such a rod, the rod will stretch but only by about 0.5 mm, or 0.05%. Furthermore, the rod will return to its original length when the car is removed.

If you hang two cars from the rod, the rod will be permanently stretched and will not recover its original length when you remove the load. If you hang three cars from the rod, the rod will break. Just before rupture, the elongation of the rod will be less than 0.2%. Although deformations of this size seem small, they are important in engineering practice. (Whether a wing under load will stay on an airplane is obviously important.)

Figure 12-11 shows three ways in which a solid might change its dimensions when forces act on it. In Fig. 12-11a, a cylinder is stretched. In Fig. 12-11b, a cylinder is deformed by a force perpendicular to its long axis, much as we might deform a pack of cards or a book. In Fig. 12-11c, a solid object placed in a fluid under high pressure is compressed uniformly on all sides. What the three deformation types have in common is that a **stress,** or deforming force per unit area, produces a **strain,** or unit deformation. In Fig. 12-11, *tensile stress* (associated with stretching) is illustrated in (a), *shearing stress* in (b), and *hydraulic stress* in (c).

The stresses and the strains take different forms in the three situations of Fig. 12-11, but—over the range of engineering usefulness—stress and strain are proportional to each other. The constant of proportionality is called a **modulus of elasticity,** so that

$$\text{stress} = \text{modulus} \times \text{strain}. \qquad (12\text{-}22)$$

In a standard test of tensile properties, the tensile stress on a test cylinder (like that in Fig. 12-12) is slowly increased from zero to the point at which the cylinder fractures, and the strain is carefully measured and plotted. The result is a graph of stress versus strain like that in Fig. 12-13. For a substantial range of applied stresses, the stress–strain relation is linear, and the specimen recovers its original dimensions when the stress is removed; it is here that Eq. 12-22 applies. If the stress is increased beyond the **yield strength** S_y of the specimen, the specimen becomes permanently deformed. If the stress continues to increase, the specimen eventually ruptures, at a stress called the **ultimate strength** S_u.

Tension and Compression

For simple tension or compression, the stress on the object is defined as F/A, where F is the magnitude of the force applied perpendicularly to an area A on the object. The strain, or unit deformation, is then the dimensionless quantity $\Delta L/L$, the fractional (or sometimes percentage) change in a length of the specimen. If the specimen is a long rod and the stress does not exceed the yield strength, then not only the entire rod but also every section of it experiences the same strain when a given stress is applied. Because the strain is dimensionless, the modulus in Eq. 12-22 has the same dimensions as the stress—namely, force per unit area.

The modulus for tensile and compressive stresses is called the **Young's modulus** and is represented in engineering practice by the symbol E. Equation 12-22 becomes

Fig. 12-14 A strain gage of overall dimensions 9.8 mm by 4.6 mm. The gage is fastened with adhesive to the object whose strain is to be measured; it experiences the same strain as the object. The electrical resistance of the gage varies with the strain, permitting strains up to 3% to be measured. Courtesy Vishay Micro-Measurements Group, Inc., Raleigh, NC, USA.

$$\frac{F}{A} = E \frac{\Delta L}{L}. \tag{12-23}$$

The strain $\Delta L/L$ in a specimen can often be measured conveniently with a *strain gage* (Fig. 12-14). This simple and useful device, which can be attached directly to operating machinery with an adhesive, is based on the principle that its electrical properties are dependent on the strain it undergoes.

Although the Young's modulus for an object may be almost the same for tension and compression, the object's ultimate strength may well be different for the two types of stress. Concrete, for example, is very strong in compression but is so weak in tension that it is almost never used in that manner. Table 12-1 shows the Young's modulus and other elastic properties for some materials of engineering interest.

Shearing

In the case of shearing, the stress is also a force per unit area, but the force vector lies in the plane of the area rather than perpendicular to it. The strain is the dimensionless ratio $\Delta x/L$, with the quantities defined as shown in Fig. 12-11*b*. The corresponding modulus, which is given the symbol G in engineering practice, is called the **shear modulus.** For shearing, Eq. 12-22 is written as

$$\frac{F}{A} = G \frac{\Delta x}{L}. \tag{12-24}$$

Shearing stresses play a critical role in the buckling of shafts that rotate under load and in bone fractures caused by bending.

Hydraulic Stress

In Fig. 12-11*c*, the stress is the fluid pressure p on the object, and, as you will see in Chapter 14, pressure is a force per unit area. The strain is $\Delta V/V$, where V is the original volume of the specimen and ΔV is the absolute value of the change in volume. The corresponding modulus, with symbol B, is called the **bulk modulus**

TABLE 12-1
Some Elastic Properties of Selected Materials of Engineering Interest

Material	Density ρ (kg/m³)	Young's Modulus E (10^9 N/m²)	Ultimate Strength S_u (10^6 N/m²)	Yield Strength S_y (10^6 N/m²)
Steel[a]	7860	200	400	250
Aluminum	2710	70	110	95
Glass	2190	65	50[b]	—
Concrete[c]	2320	30	40[b]	—
Wood[d]	525	13	50[b]	—
Bone	1900	9[b]	170[b]	—
Polystyrene	1050	3	48	—

[a]Structural steel (ASTM-A36).   [c]High strength.
[b]In compression.   [d]Douglas fir.

of the material. The object is said to be under *hydraulic compression*, and the pressure can be called the *hydraulic stress*. For this situation, we write Eq. 12-22 as

$$p = B \frac{\Delta V}{V}. \tag{12-25}$$

The bulk modulus is 2.2×10^9 N/m² for water and 1.6×10^{11} N/m² for steel. The pressure at the bottom of the Pacific Ocean, at its average depth of about 4000 m, is 4.0×10^7 N/m². The fractional compression $\Delta V/V$ of a volume of water due to this pressure is 1.8%; that for a steel object is only about 0.025%. In general, solids—with their rigid atomic lattices—are less compressible than liquids, in which the atoms or molecules are less tightly coupled to their neighbors.

Sample Problem 12-5

A steel rod has a radius R of 9.5 mm and a length L of 81 cm. A 62 kN force $\vec{F}$ stretches it along its length. What are the stress on the rod and the elongation and strain of the rod?

Solution: The first **Key Idea** here has to do with what is meant by the second sentence in the problem statement. We assume the rod is held stationary by, say, a clamp or vise at one end. Then force $\vec{F}$ is applied at the other end, parallel to the length of the rod and thus perpendicular to the end face there. Therefore, the situation is like that in Fig. 12-11a.

The next **Key Idea** is that we assume the force is applied uniformly across the end face and thus over an area $A = \pi R^2$. Then the stress on the rod is given by the left side of Eq. 12-23:

$$\text{stress} = \frac{F}{A} = \frac{F}{\pi R^2} = \frac{6.2 \times 10^4 \text{ N}}{(\pi)(9.5 \times 10^{-3} \text{ m})^2}$$
$$= 2.2 \times 10^8 \text{ N/m}^2. \qquad \text{(Answer)}$$

The yield strength for structural steel is 2.5×10^8 N/m², so this rod is dangerously close to its yield strength.

Another **Key Idea** is that the elongation of the rod depends on the stress, the original length L, and the type of material in the rod. The last determines which value we use for Young's modulus E (from Table 12-1). Using the value for steel, Eq. 12-23 gives us

$$\Delta L = \frac{(F/A)L}{E} = \frac{(2.2 \times 10^8 \text{ N/m}^2)(0.81 \text{ m})}{2.0 \times 10^{11} \text{ N/m}^2}$$
$$= 8.9 \times 10^{-4} \text{ m} = 0.89 \text{ mm}. \qquad \text{(Answer)}$$

The last **Key Idea** we need here is that strain is the ratio of the change in length to the original length, so we have

$$\frac{\Delta L}{L} = \frac{8.9 \times 10^{-4} \text{ m}}{0.81 \text{ m}}$$
$$= 1.1 \times 10^{-3} = 0.11\%. \qquad \text{(Answer)}$$

Sample Problem 12-6

A table has three legs that are 1.00 m in length and a fourth leg that is longer by $d = 0.50$ mm, so that the table wobbles slightly. A steel cylinder with mass $M = 290$ kg is placed on the table (which has a mass much less than M) so that all four legs are compressed but unbuckled and the table is level but no longer wobbles. The legs are wooden cylinders with cross-sectional area $A = 1.0$ cm²; Young's modulus is $E = 1.3 \times 10^{10}$ N/m². What are the magnitudes of the forces on the legs from the floor?

Solution: We take the table plus steel cylinder as our system. The situation is like that in Fig. 12-9, except now we have a steel cylinder on the table. One **Key Idea** is that if the tabletop remains level, the legs must be compressed in the following ways: Each of the short legs must be compressed by the same amount (call it ΔL_3) and thus by the same force of magnitude F_3. The single long leg must be compressed by a larger amount ΔL_4 and thus by a force with a larger magnitude F_4. In other words, for a level tabletop, we must have

$$\Delta L_4 = \Delta L_3 + d. \tag{12-26}$$

A second **Key Idea** is that, from Eq. 12-23, we can relate a change in length to the force causing the change with $\Delta L = FL/AE$, where L is the original length of a leg. We can use this relation to replace ΔL_4 and ΔL_3 in Eq. 12-26. However,

note that we can approximate the original length L as being the same for all four legs. The replacements then give us

$$\frac{F_4 L}{AE} = \frac{F_3 L}{AE} + d. \tag{12-27}$$

We cannot solve this equation because it has two unknowns, F_4 and F_3.

To get a second equation containing F_4 and F_3, we can use a vertical y axis and then write the balance of vertical forces ($F_{\text{net},y} = 0$) as

$$3F_3 + F_4 - Mg = 0, \tag{12-28}$$

where Mg is equal to the magnitude of the gravitational force on the system. (*Three* legs have force $\vec{F}_3$ on them.) To solve the simultaneous equations 12-27 and 12-28 for, say, F_3, we first use Eq. 12-28 to find that $F_4 = Mg - 3F_3$. Substituting that into Eq. 12-27 then yields, after some algebra,

$$F_3 = \frac{Mg}{4} - \frac{dAE}{4L}$$

$$= \frac{(290 \text{ kg})(9.8 \text{ m/s}^2)}{4}$$

$$- \frac{(5.0 \times 10^{-4} \text{ m})(10^{-4} \text{ m}^2)(1.3 \times 10^{10} \text{ N/m}^2)}{(4)(1.00 \text{ m})}$$

$$= 548 \text{ N} \approx 5.5 \times 10^2 \text{ N}. \qquad \text{(Answer)}$$

From Eq. 12-28 we then find

$$F_4 = Mg - 3F_3 = (290 \text{ kg})(9.8 \text{ m/s}^2) - 3(548 \text{ N})$$
$$\approx 1.2 \text{ kN}. \qquad \text{(Answer)}$$

You can show that to reach their equilibrium configuration, the three short legs are each compressed by 0.42 mm and the single long leg by 0.92 mm.

✔CHECKPOINT 6   The figure shows a horizontal block that is suspended by two wires, *A* and *B*, which are identical except for their original lengths. The center of mass of the block is closer to wire *B* than to wire *A*. (a) Measuring torques about the block's center of mass, state whether the magnitude of the torque due to wire *A* is greater than, less than, or equal to the magnitude of the torque due to wire *B*. (b) Which wire exerts more force on the block? (c) If the wires are now equal in length, which one was originally shorter?

Review & Summary

Static Equilibrium   A rigid body at rest is said to be in **static equilibrium.** For such a body, the vector sum of the external forces acting on it is zero:

$$\vec{F}_{net} = 0 \qquad \text{(balance of forces)}. \qquad (12\text{-}3)$$

If all the forces lie in the *xy* plane, this vector equation is equivalent to two component equations:

$$F_{net,x} = 0 \text{ and } F_{net,y} = 0 \qquad \text{(balance of forces)}. \qquad (12\text{-}7, 12\text{-}8)$$

Static equilibrium also implies that the vector sum of the external torques acting on the body about *any* point is zero, or

$$\vec{\tau}_{net} = 0 \qquad \text{(balance of torques)}. \qquad (12\text{-}5)$$

If the forces lie in the *xy* plane, all torque vectors are parallel to the *z* axis, and Eq. 12-5 is equivalent to the single component equation

$$\tau_{net,z} = 0 \qquad \text{(balance of torques)}. \qquad (12\text{-}9)$$

Center of Gravity   The gravitational force acts individually on each element of a body. The net effect of all individual actions may be found by imagining an equivalent total gravitational force $\vec{F}_g$ acting at the **center of gravity.** If the gravitational acceleration $\vec{g}$ is the same for all the elements of the body, the center of gravity is at the center of mass.

Elastic Moduli   Three **elastic moduli** are used to describe the elastic behavior (deformations) of objects as they respond to forces that act on them. The **strain** (fractional change in length) is linearly related to the applied **stress** (force per unit area) by the proper modulus, according to the general relation

$$\text{stress} = \text{modulus} \times \text{strain}. \qquad (12\text{-}22)$$

Tension and Compression   When an object is under tension or compression, Eq. 12-22 is written as

$$\frac{F}{A} = E\frac{\Delta L}{L}, \qquad (12\text{-}23)$$

where $\Delta L/L$ is the tensile or compressive strain of the object, *F* is the magnitude of the applied force $\vec{F}$ causing the strain, *A* is the cross-sectional area over which $\vec{F}$ is applied (perpendicular to *A*, as in Fig. 12-11a), and *E* is the **Young's modulus** for the object. The stress is *F/A*.

Shearing   When an object is under a shearing stress, Eq. 12-22 is written as

$$\frac{F}{A} = G\frac{\Delta x}{L}, \qquad (12\text{-}24)$$

where $\Delta x/L$ is the shearing strain of the object, Δx is the displacement of one end of the object in the direction of the applied force $\vec{F}$ (as in Fig. 12-11b), and *G* is the **shear modulus** of the object. The stress is *F/A*.

Hydraulic Stress   When an object undergoes *hydraulic compression* due to a stress exerted by a surrounding fluid, Eq. 12-22 is written as

$$p = B\frac{\Delta V}{V}, \qquad (12\text{-}25)$$

where *p* is the pressure (*hydraulic stress*) on the object due to the fluid, $\Delta V/V$ (the strain) is the absolute value of the fractional change in the object's volume due to that pressure, and *B* is the **bulk modulus** of the object.

Questions

1   Figure 12-15 shows an overhead view of a uniform stick on which four forces act. Suppose we choose a rotational axis through point *O*, calculate the torques about that axis due to the forces, and find that these torques balance. Will the torques balance if, instead, the rotational axis is chosen to be

Fig. 12-15   Question 1.

at (a) point *A*, (b) point *B*, or (c) point *C*? (d) Suppose, instead, that we find that the torques about point *O* do not balance. Is there another point about which the torques will balance?

2   Figure 12-16 shows four overhead views of rotating uniform disks that are sliding across a frictionless floor. Three forces, of magnitude *F*, 2*F*, or 3*F*, act on each disk, either at the rim, at the center, or halfway between rim and center. The

force vectors rotate along with the disks, and, in the "snapshots" of Fig. 12-16, point left or right. Which disks are in equilibrium?

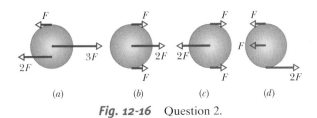

Fig. 12-16 Question 2.

3 Figure 12-17 shows three situations in which the same horizontal rod is supported by a hinge on a wall at one end and a cord at its other end. Without written calculation, rank the situations according to the magnitudes of (a) the force on the rod from the cord, (b) the vertical force on the rod from the hinge, and (c) the horizontal force on the rod from the hinge, greatest first.

Fig. 12-17 Question 3.

4 In Fig. 12-18, a rigid beam is attached to two posts that are fastened to a floor. A small but heavy safe is placed at the six positions indicated, in turn. Assume that the mass of the beam is negligible compared to that of the safe. (a) Rank the positions according to the force on post A due to the safe, greatest compression first, greatest tension last, and indicate where, if anywhere, the force is zero. (b) Rank them according to the force on post B.

Fig. 12-18 Question 4.

5 Figure 12-19 shows a mobile of toy penguins hanging from a ceiling. Each crossbar is horizontal, has negligible mass, and extends three times as far to the right of the wire supporting it as to the left. Penguin 1 has mass $m_1 = 48$ kg. What are the masses of (a) penguin 2, (b) penguin 3, and (c) penguin 4?

Fig. 12-19 Question 5.

6 Three piñatas hang from the (stationary) assembly of massless pulleys and cords seen in Fig. 12-20. One long cord runs from the ceiling at the right to the lower pulley at the left. Several shorter cords suspend pulleys from the ceiling or piñatas from the pulleys. The weights (in newtons) of two piñatas are given. (a) What is the weight of the third piñata? (*Hint:* A cord that loops halfway around a pulley pulls on the pulley with a net force that is twice the tension in the cord.) (b) What is the tension in the short cord labeled with T?

Fig. 12-20 Question 6.

7 In Fig. 12-21, a vertical rod is hinged at its lower end and attached to a cable at its upper end. A horizontal force $\vec{F}_a$ is to be applied to the rod as shown. If the point at which the force is applied is moved up the rod, does the tension in the cable increase, decrease, or remain the same?

Fig. 12-21 Question 7.

8 A ladder leans against a frictionless wall but is prevented from falling because of friction between it and the ground. Suppose you shift the base of the ladder toward the wall. Determine whether the following become larger, smaller, or stay the same (in magnitude): (a) the normal force on the ladder from the ground, (b) the force on the ladder from the wall, (c) the static frictional force on the ladder from the ground, and (d) the maximum value $f_{s,\text{max}}$ of the static frictional force.

9 Four cylindrical rods are stretched as in Fig. 12-11a. The force magnitudes, the areas of the end faces, the changes in length, and the initial lengths are given here. Rank the rods according to their Young's moduli, greatest first.

Rod	Force	Area	Length Change	Initial Length
1	F	A	ΔL	L
2	$2F$	$2A$	$2\Delta L$	L
3	F	$2A$	$2\Delta L$	$2L$
4	$2F$	A	ΔL	$2L$

10 The table gives the areas of three surfaces and the magnitude of a force that is applied perpendicular to the surface and uniformly across it. Rank the surfaces according to the stress on them, greatest first.

	Area	Force
Surface A	$0.5A_0$	$2F_0$
Surface B	$2A_0$	$4F_0$
Surface C	$3A_0$	$6F_0$

Problems

SSM   Solution is in the Student Solutions Manual.
WWW   Solution is at http://www.wiley.com/college/halliday
ILW   Interactive LearningWare solution is at
                 http://www.wiley.com/college/halliday
• – ••• Number of dots indicates level of problem difficulty.

sec. 12-4   The Center of Gravity

•1   Because g varies so little over the extent of most structures, any structure's center of gravity effectively coincides with its center of mass. Here is a fictitious example where g varies more significantly. Figure 12-22 shows an array of six particles, each with mass m, fixed to the edge of a rigid structure of negligible mass. The distance between adjacent particles along the edge is 2.00 m. The following table gives the value of g (m/s^2) at each particle's location. Using the coordinate system shown, find (a) the x coordinate x_{com} and (b) the y coordinate y_{com} of the center of mass of the six-particle system. Then find (c) the x coordinate x_{cog} and (d) the y coordinate y_{cog} of the center of gravity of the six-particle system.

Particle	g	Particle	g
1	8.00	4	7.40
2	7.80	5	7.60
3	7.60	6	7.80

Fig. 12-22   Problem 1.

sec. 12-5   Some Examples of Static Equilibrium

•2   A physics Brady Bunch, whose weights in newtons are indicated in Fig. 12-23, is balanced on a seesaw. What is the number of the person who causes the largest torque about the rotation axis at *fulcrum f* directed (a) out of the page and (b) into the page?

Fig. 12-23   Problem 2.

•3   A rope of negligible mass is stretched horizontally between two supports that are 3.44 m apart. When an object of weight 3160 N is hung at the center of the rope, the rope is observed to sag by 35.0 cm. What is the tension in the rope? **ILW**

•4   An archer's bow is drawn at its midpoint until the tension in the string is equal to the force exerted by the archer. What is the angle between the two halves of the string?

•5   In Fig. 12-24, a uniform sphere of mass $m = 0.85$ kg and radius $r = 4.2$ cm is held in place by a massless rope attached to a frictionless wall a distance $L = 8.0$ cm above the center of the sphere. Find (a) the tension in the rope and (b) the force on the sphere from the wall. **SSM WWW**

Fig. 12-24   Problem 5.

•6   A scaffold of mass 60 kg and length 5.0 m is supported in a horizontal position by a vertical cable at each end. A window washer of mass 80 kg stands at a point 1.5 m from one end. What is the tension in (a) the nearer cable and (b) the farther cable?

•7   A diver of weight 580 N stands at the end of a diving board of length $L = 4.5$ m and negligible mass (Fig. 12-25). The board is fixed to two pedestals separated by distance $d = 1.5$ m. Of the forces acting on the board, what are the (a) magnitude and (b) direction (up or down) of the force from the left pedestal and the (c) magnitude and (d) direction (up or down) of the force from the right pedestal? (e) Which pedestal (left or right) is being stretched, and (f) which is being compressed? **SSM**

Fig. 12-25   Problem 7.

•8   An automobile with a mass of 1360 kg has 3.05 m between the front and rear axles. Its center of gravity is located 1.78 m behind the front axle. With the automobile on level ground, determine the magnitude of the force from the ground on (a) each front wheel (assuming equal forces on the front wheels) and (b) each rear wheel (assuming equal forces on the rear wheels).

•9   A meter stick balances horizontally on a knife-edge at the 50.0 cm mark. With two 5.00 g coins stacked over the 12.0 cm mark, the stick is found to balance at the 45.5 cm mark. What is the mass of the meter stick? **SSM**

•10   In Fig. 12-26, a man is trying to get his car out of mud on the shoulder of a road. He ties one end of a rope tightly around the front bumper and the other end tightly around a utility pole 18 m away. He then pushes sideways on the rope at its midpoint with a force of 550 N, displacing the center of the rope 0.30 m from its previous position, and the car barely moves. What is the magnitude of the force on the car from the rope? (The rope stretches somewhat.)

Fig. 12-26   Problem 10.

•11 A 75 kg window cleaner uses a 10 kg ladder that is 5.0 m long. He places one end on the ground 2.5 m from a wall, rests the upper end against a cracked window, and climbs the ladder. He is 3.0 m up along the ladder when the window breaks. Neglect friction between the ladder and window and assume that the base of the ladder does not slip. When the window is on the verge of breaking, what are (a) the magnitude of the force on the window from the ladder, (b) the magnitude of the force on the ladder from the ground, and (c) the angle (relative to the horizontal) of that force on the ladder? SSM

•12 A uniform cubical crate is 0.750 m on each side and weighs 500 N. It rests on a floor with one edge against a very small, fixed obstruction. At what least height above the floor must a horizontal force of magnitude 350 N be applied to the crate to tip it?

•13 Forces $\vec{F}_1$, $\vec{F}_2$, and $\vec{F}_3$ act on the structure of Fig. 12-27, shown in an overhead view. We wish to put the structure in equilibrium by applying a fourth force, at a point such as P. The fourth force has vector components $\vec{F}_h$ and $\vec{F}_v$. We are given that $a = 2.0$ m, $b = 3.0$ m, $c = 1.0$ m, $F_1 = 20$ N, $F_2 = 10$ N, and $F_3 = 5.0$ N. Find (a) F_h, (b) F_v, and (c) d. ILW

Fig. 12-27 Problem 13.

•14 The system in Fig. 12-28 is in equilibrium, with the string in the center exactly horizontal. Block A weighs 40 N, block B weighs 50 N, and angle ϕ is 35°. Find (a) tension T_1, (b) tension T_2, (c) tension T_3, and (d) angle θ.

Fig. 12-28 Problem 14.

•15 Figure 12-29 shows the anatomical structures in the lower leg and foot that are involved in standing on tiptoe, with the heel raised slightly off the floor so that the foot effectively contacts the floor only at point P. Assume distance $a = 5.0$ cm, distance $b = 15$ cm, and the person's weight $W = 900$ N. Of the forces acting on the foot, what are the (a) magnitude and (b) direction (up or down) of the force at point A from the calf muscle and the (c) magnitude and (d) direction (up or down) of the force at point B from the lower leg bones?

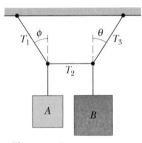

Fig. 12-29 Problem 15.

•16 In Fig. 12-30, a horizontal scaffold, of length 2.00 m and uniform mass 50.0 kg, is suspended from a building by two cables. The scaffold has dozens of paint cans stacked on it at various points. The total mass of the paint cans is 75.0 kg. The tension in the cable at the right is 722 N. How far horizontally from *that* cable is the center of mass of the system of paint cans?

Fig. 12-30 Problem 16.

•17 In Fig. 12-31, a uniform beam of weight 500 N and length 3.0 m is suspended horizontally. On the left it is hinged to a wall; on the right it is supported by a cable bolted to the wall at distance D above the beam. The least tension that will snap the cable is 1200 N. (a) What value of D corresponds to that tension? (b) To prevent the cable from snapping, should D be increased or decreased from that value?

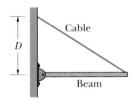

Fig. 12-31 Problem 17.

•18 In Fig. 12-32, horizontal scaffold 2, with uniform mass $m_2 = 30.0$ kg and length $L_2 = 2.00$ m, hangs from horizontal scaffold 1, with uniform mass $m_1 = 50.0$ kg. A 20.0 kg box of nails lies on scaffold 2, centered at distance $d = 0.500$ m from the left end. What is the tension T in the cable indicated?

Fig. 12-32 Problem 18.

••19 In Fig. 12-33, what magnitude of force $\vec{F}$ applied horizontally at the axle of the wheel is necessary to raise the wheel over an obstacle of height $h = 3.00$ cm? The wheel's radius is $r = 6.00$ cm, and its mass is $m = 0.800$ kg. SSM WWW

••20 In Fig. 12-34, a 15 kg block is held in place via a pulley system. The person's upper arm is vertical; the forearm

Fig. 12-33 Problem 19.

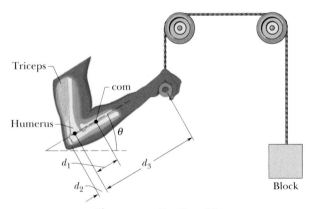

Fig. 12-34 Problem 20.

makes angle $\theta = 30°$ with the horizontal. Forearm and hand together have a mass of 2.0 kg, with a center of mass at distance $d_1 = 15$ cm from the contact point of the forearm bone and the upper-arm bone (humerus). The triceps muscle pulls vertically upward on the forearm at distance $d_2 = 2.5$ cm behind that contact point. Distance d_3 is 35 cm. What are the (a) magnitude and (b) direction (up or down) of the force on the forearm from the triceps muscle and the (c) magnitude and (d) direction (up or down) of the force on the forearm from the humerus?

••21 In Fig. 12-35, one end of a uniform beam of weight 222 N is hinged to a wall; the other end is supported by a wire that makes angles $\theta = 30.0°$ with both wall and beam. Find (a) the tension in the wire and the (b) horizontal and (c) vertical components of the force of the hinge on the beam. **SSM**

Fig. 12-35 Problem 21.

••22 In Fig. 12-36, a 50.0 kg uniform square sign, of edge length $L = 2.00$ m, is hung from a horizontal rod of length $d_h = 3.00$ m and negligible mass. A cable is attached to the end of the rod and to a point on the wall at distance $d_v = 4.00$ m above the point where the rod is hinged to the wall. (a) What is the tension in the cable? What are the (b) magnitude and (c) direction (left or right) of the horizontal component of the force on the rod from the wall, and the (d) magnitude and (e) direction (up or down) of the vertical component of this force?

Fig. 12-36 Problem 22.

••23 The system in Fig. 12-37 is in equilibrium. A concrete block of mass 225 kg hangs from the end of the uniform strut of mass 45.0 kg. For angles $\phi = 30.0°$ and $\theta = 45.0°$, find (a) the tension T in the cable and the (b) horizontal and (c) vertical components of the force on the strut from the hinge. **ILW**

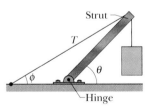

Fig. 12-37 Problem 23.

••24 In Fig. 12-38, a 55 kg rock climber is in a lie-back climb along a fissure, with hands pulling on one side of the fissure and feet pressed against the opposite side. The fissure has width $w = 0.20$ m, and the center of mass of the climber is a horizontal distance $d = 0.40$ m from the fissure. The coefficient of static friction

Fig. 12-38 Problem 24.

between hands and rock is $\mu_1 = 0.40$, and between boots and rock it is $\mu_2 = 1.2$. (a) What is the least horizontal pull by the hands and push by the feet that will keep the climber stable? (b) For the horizontal pull of (a), what must be the vertical distance h between hands and feet? If the climber encounters wet rock, so that μ_1 and μ_2 are reduced, what happens to (c) the answer to (a) and (d) the answer to (b)?

••25 A nonuniform bar is suspended at rest in a horizontal position by two massless cords as shown in Fig. 12-39. One cord makes the angle $\theta = 36.9°$ with the vertical; the other makes the angle $\phi = 53.1°$ with the vertical. If the length L of the bar is 6.10 m, compute the distance x from the left end of the bar to its center of mass. **SSM**

Fig. 12-39 Problem 25.

••26 A door has a height of 2.1 m along a y axis that extends vertically upward and a width of 0.91 m along an x axis that extends outward from the hinged edge of the door. A hinge 0.30 m from the top and a hinge 0.30 m from the bottom each support half the door's mass, which is 27 kg. In unit-vector notation, what are the forces on the door at (a) the top hinge and (b) the bottom hinge?

••27 In Fig. 12-40, suppose the length L of the uniform bar is 3.00 m and its weight is 200 N. Also, let the block's weight $W = 300$ N and the angle $\theta = 30.0°$. The wire can withstand a maximum tension of 500 N. (a) What is the maximum possible distance x before the wire breaks? With the block placed at this maximum x, what are the (b) horizontal and (c) vertical components of the force on the bar from the hinge at A?

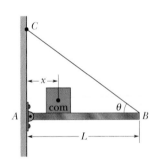

Fig. 12-40 Problems 27 and 28.

••28 In Fig. 12-40, a thin horizontal bar AB of negligible weight and length L is hinged to a vertical wall at A and supported at B by a thin wire BC that makes an angle θ with the horizontal. A block of weight W can be moved anywhere along the bar; its position is defined by the distance x from the wall to its center of mass. As a function of x, find (a) the tension in the wire, and the (b) horizontal and (c) vertical components of the force on the bar from the hinge at A.

••29 A cubical box is filled with sand and weighs 890 N. We wish to "roll" the box by pushing horizontally on one of the upper edges. (a) What minimum force is required? (b) What minimum coefficient of static friction between box and floor is required? (c) If there is a more efficient way to roll the box, find the smallest possible force that would have to be applied directly to the box to roll it. (*Hint*: At the onset of tipping, where is the normal force located?) **SSM WWW**

••30 Figure 12-41*a* shows a vertical uniform beam of length L that is hinged at its lower end. A horizontal force $\vec{F}_a$ is applied to the beam at distance y from the lower end. The beam remains vertical because of a cable attached at the upper end, at angle θ with the horizontal. Figure 12-41*b* gives the tension T in the cable as a function of the position of the applied force given as a fraction y/L of the beam length. Figure 12-41c gives the magnitude F_h of the horizontal force on the beam from the hinge, also as a function of y/L. Evaluate (a) angle θ and (b) the magnitude of $\vec{F}_a$.

(a)

(b)   (c)

Fig. 12-41 Problem 30.

••31 In Fig. 12-42, a uniform plank, with a length L of 6.10 m and a weight of 445 N, rests on the ground and against a frictionless roller at the top of a wall of height $h = 3.05$ m. The plank remains in equilibrium for any value of $\theta \geq 70°$ but slips if $\theta < 70°$. Find the coefficient of static friction between the plank and the ground. **SSM**

Fig. 12-42 Problem 31.

••32 In Fig. 12-43, uniform beams A and B are attached to a wall with hinges and loosely bolted together. Beam A has length $L_A = 2.40$ m and mass 54.0 kg; beam B has mass 68.0 kg. The two hinge points are separated by distance $d = 1.80$ m. In unit-vector notation, what is the force on (a) beam A due to its hinge, (b) beam A due to the bolt, (c) beam B due to its hinge, and (d) beam B due to the bolt?

Fig. 12-43 Problem 32.

•••33 A crate, in the form of a cube with edge lengths of 1.2 m, contains a piece of machinery; the center of mass of the crate and its contents is located 0.30 m above the crate's geometrical center. The crate rests on a ramp that makes an angle θ with the horizontal. As θ is increased from zero, an angle will be reached at which the crate will either tip over or start to slide down the ramp. If the coefficient of static friction μ_s between ramp and crate is 0.60, (a) does the crate tip or slide

and (b) at what angle θ does this occur? If $\mu_s = 0.70$, (c) does the crate tip or slide and (d) at what angle θ does this occur? (*Hint:* At the onset of tipping, where is the normal force located?) **SSM**

•••34 Figure 12-44a shows a horizontal uniform beam of mass m_b and length L that is supported on the left by a hinge attached to a wall and on the right by a cable at angle θ with the horizontal. A package of mass m_p is positioned on the beam at a distance x from the left end. The total mass is $m_b + m_p = 61.22$ kg. Figure 12-44b gives the tension T in the cable as a function of the package's position given as a fraction x/L of the beam length. Evaluate (a) angle θ, (b) mass m_b, and (c) mass m_p.

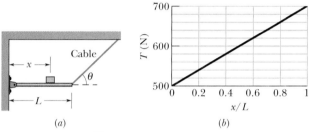

(a)   (b)

Fig. 12-44 Problem 34.

•••35 For the stepladder shown in Fig. 12-45, sides AC and CE are each 2.44 m long and hinged at C. Bar BD is a tie-rod 0.762 m long, halfway up. A man weighing 854 N climbs 1.80 m along the ladder. Assuming that the floor is frictionless and neglecting the mass of the ladder, find (a) the tension in the tie-rod and the magnitudes of the forces on the ladder from the floor at (b) A and (c) E. (*Hint:* Isolate parts of the ladder in applying the equilibrium conditions.) **SSM**

Fig. 12-45 Problem 35.

sec. 12-7 **Elasticity**

•36 Figure 12-46 shows the stress–strain curve for quartzite. What are (a) the Young's modulus and (b) the approximate yield strength for this material?

•37 A horizontal aluminum rod 4.8 cm in diameter projects 5.3 cm from a wall. A 1200 kg object is suspended from the end of the rod. The shear modulus of aluminum is 3.0×10^{10} N/m². Neglecting the rod's mass, find (a) the shear stress on the rod and (b) the vertical deflection of the end of the rod. **SSM ILW**

•38 Figure 12-47 shows the stress versus strain plot for an

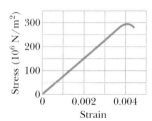

Fig. 12-46 Problem 36.

aluminum wire that is stretched by a machine pulling in opposite directions at the two ends of the wire. The wire has an initial length of 0.800 m and an initial cross-sectional area of 2.00×10^{-6} m². How much work does the force from the machine do on the wire to produce a strain of 1.00×10^{-3}?

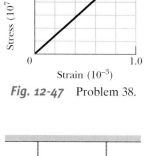

Fig. 12-47 Problem 38.

••39 In Fig. 12-48, a 103 kg uniform log hangs by two steel wires, A and B, both of radius 1.20 mm. Initially, wire A was 2.50 m long and 2.00 mm shorter than wire B. The log is now horizontal. What are the magnitudes of the forces on it from (a) wire A and (b) wire B? (c) What is the ratio d_A/d_B? **SSM**

Fig. 12-48 Problem 39.

••40 In Fig. 12-49, a lead brick rests horizontally on cylinders A and B. The areas of the top faces of the cylinders are related by $A_A = 2A_B$; the Young's moduli of the cylinders are related by $E_A = 2E_B$. The cylinders had identical lengths before the brick was placed on them. What fraction of the brick's mass is supported (a) by cylinder A and (b) by cylinder B? The horizontal distances between the center of mass of the brick and the centerlines of the cylinders are d_A for cylinder A and d_B for cylinder B. (c) What is the ratio d_A/d_B?

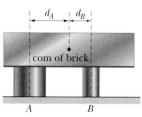

Fig. 12-49 Problem 40.

••41 A tunnel of length $L = 150$ m, height $H = 7.2$ m, and width 5.8 m (with a flat roof) is to be constructed at distance $d = 60$ m beneath the ground. (See Fig. 12-50.) The tunnel roof is to be supported entirely by square steel columns, each with a cross-sectional area of 960 cm². The mass of 1.0 cm³ of the ground material is 2.8 g. (a) What is the total weight of the ground material the columns must support? (b) How many columns are needed to keep the compressive stress on each column at one-half its ultimate strength? **SSM**

•••42 Figure 12-51 is an overhead view of a rigid rod that turns about a vertical axle until the identical rubber stoppers A and B are forced against rigid walls at distances $r_A = 7.0$ cm and $r_B = 4.0$ cm from the axle. Initially the stoppers touch the walls without being compressed. Then force $\vec{F}$ of magnitude 220 N is applied perpendicular to the rod at a distance $R = 5.0$ cm from the axle. Find the magnitude of the force compressing (a) stopper A and (b) stopper B.

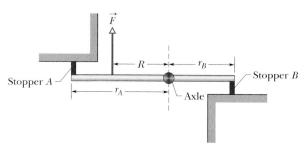

Fig. 12-51 Problem 42.

Additional Problems

43 In Fig. 12-52, a 10 kg sphere is supported on a frictionless plane inclined at angle $\theta = 45°$ from the horizontal. Angle ϕ is 25°. Calculate the tension in the cable.

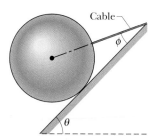

Fig. 12-52 Problem 43.

44 Figure 12-53a shows a uniform ramp between two buildings that allows for motion between the buildings due to strong winds. At its left end, it is hinged to the building wall; at its right end, it has a roller that can roll along the building wall. There is no vertical force on the roller from the building, only a horizontal force with magnitude F_h. The horizontal distance between the buildings is $D = 4.00$ m. The rise of the ramp is $h = 0.490$ m. A man walks across the ramp from the left. Figure 12-53b gives F_h as a function of the horizontal distance x of the man from the building at the left. What are the masses of (a) the ramp and (b) the man?

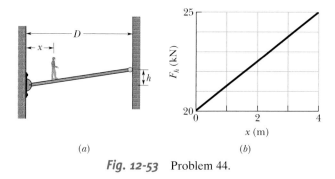

Fig. 12-53 Problem 44.

45 In Fig. 12-54, an 817 kg construction bucket is suspended by a cable A that is attached at O to two other cables B and

Fig. 12-50 Problem 41.

C, making angles $\theta_1 = 51.0°$ and $\theta_2 = 66.0°$ with the horizontal. Find the tensions in (a) cable *A*, (b) cable *B*, and (c) cable *C*. (*Hint:* To avoid solving two equations in two unknowns, position the axes as shown in the figure.)

46 In Fig. 12-55*a*, a uniform 40.0 kg beam is centered over two rollers. Vertical lines across the beam mark off equal lengths. Two of the lines are centered over the rollers; a 10.0 kg package of tamale is centered over roller *B*. What are the magnitudes of the forces on the beam from (a) roller *A* and (b) roller *B*? The beam is then rolled to the left until the right-hand end is centered over roller *B* (Fig. 12-55*b*). What now are the magnitudes of the forces on the beam from (c) roller *A* and (d) roller *B*? Next, the beam is rolled to the right. Assume that it has a length of 0.800 m. (e) What horizontal distance between the package and roller *B* puts the beam on the verge of losing contact with roller *A*?

47 The force $\vec{F}$ in Fig. 12-56 keeps the 6.40 kg block and the pulleys in equilibrium. The pulleys have negligible mass and friction. Calculate the tension *T* in the upper cable. (*Hint:* When a cable wraps halfway around a pulley as here, the magnitude of its net force on the pulley is twice the tension in the cable.)

48 In Figure 12-57, a package of mass *m* hangs from a short cord that is tied to the wall via cord 1 and to the ceiling via cord 2. Cord 1 is at angle $\phi = 40°$ with the horizontal; cord 2 is at angle θ. (a) For what value of θ is the tension in cord 2 minimized? (b) In terms of *mg*, what is the minimum tension in cord 2?

49 Figure 12-58 shows a 300 kg horizontal cylinder. Three steel wires support the cylinder from a ceiling. Wires 1 and 3 are at the ends of the cylinder, and wire 2 is at the center. The wires each have a cross-sec-

Fig. 12-54 Problem 45.

(a)

(b)

Fig. 12-55 Problem 46.

Fig. 12-56 Problem 47.

Fig. 12-57 Problem 48.

Fig. 12-58 Problem 49.

tional area of 2.00×10^{-6} m². Initially (before the cylinder was put in place) wires 1 and 3 were 2.0000 m long and wire 2 was 6.00 mm longer than that. Now all three wires have been stretched. What is the tension in (a) wire 1 and (b) wire 2?

50 In Fig. 12-59, two identical, uniform, and frictionless spheres, each of mass *m*, rest in a rigid rectangular container. A line connecting their centers is at 45° to the horizontal. Find the magnitudes of the forces on the spheres from (a) the bottom of the container, (b) the left side of the container, (c) the right side of the container, and (d) each another. (*Hint:* The force of one sphere on the other is directed along the center-center line.)

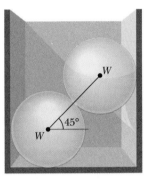

Fig. 12-59 Problem 50.

51 A uniform cube of side length 8.0 cm rests on a horizontal floor. The coefficient of static friction between cube and floor is μ. A horizontal pull $\vec{P}$ is applied perpendicular to one of the vertical faces of the cube, at a distance 7.0 cm above the floor on the vertical midline of the cube face. The magnitude of $\vec{P}$ is gradually increased. For what values of μ will the cube eventually (a) begin to slide and (b) begin to tip? (*Hint:* At the onset of tipping, where is the normal force located?)

52 The rigid square frame in Fig. 12-60 consists of the four side bars *AB*, *BC*, *CD*, and *DA* plus two diagonal bars *AC* and *BD*, which pass each other freely at *E*. By means of the turnbuckle *G*, bar *AB* is put under tension, as if its ends were subject to horizontal, outward forces $\vec{T}$ of magnitude 535 N. (a) Which of the other bars are in tension? What are the magnitudes of (b) the forces causing the tension in those bars and (c) the forces causing compression in the other bars? (*Hint:* Symmetry considerations can lead to considerable simplification in this problem.)

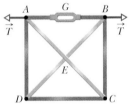

Fig. 12-60 Problem 52.

53 Four bricks of length *L*, identical and uniform, are stacked on top of one another (Fig. 12-61) in such a way that part of each extends beyond the one beneath. Find, in terms of *L*, the maximum values of (a) a_1, (b) a_2, (c) a_3, (d) a_4, and (e) *h*, such that the stack is in equilibrium.

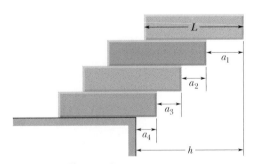

Fig. 12-61 Problem 53.

54 In Fig. 12-62, the driver of a car on a horizontal road makes an emergency stop by applying the brakes so that all four wheels lock and skid along the road. The coefficient of kinetic friction between tires and road is 0.40. The separation between the front and rear axles is $L = 4.2$ m, and the center of mass of the car is located at distance $d = 1.8$ m behind the front axle and distance $h = 0.75$ m above the road. The car weighs 11 kN. Find the magnitude of (a) the braking acceleration of the car, (b) the normal force on each rear wheel, (c) the normal force on each front wheel, (d) the braking force on each rear wheel, and (e) the braking force on each front wheel. (*Hint:* Although the car is not in translational equilibrium, it *is* in rotational equilibrium.)

Fig. 12-62 Problem 54.

55 In Fig. 12-63, block A (mass 10 kg) is in equilibrium, but it would slip if block B (mass 5.0 kg) were any heavier. For angle $\theta = 30°$, what is the coefficient of static friction between block A and the surface below it?

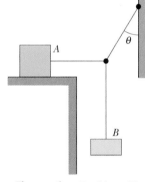

Fig. 12-63 Problem 55.

56 A pan balance is made up of a rigid, massless rod with a hanging pan attached at each end. The rod is supported at and free to rotate about a point not at its center. It is balanced by unequal masses placed in the two pans. When an unknown mass m is placed in the left pan, it is balanced by a mass m_1 placed in the right pan; when the mass m is placed in the right pan, it is balanced by a mass m_2 in the left pan. Show that $m = \sqrt{m_1 m_2}$.

57 To crack a certain nut in a nutcracker, forces with magnitudes of at least 40 N must act on its shell from both sides. For the nutcracker of Fig. 12-64, with distances $L = 12$ cm and $d = 2.6$ cm, what are the force components $F_\perp$ (perpendicular to the handles) corresponding to that 40 N?

Fig. 12-64 Problem 57.

58 A cylindrical aluminum rod, with an initial length of 0.8000 m and radius 1000.0 μm, is clamped in place at one end and then stretched by a machine pulling parallel to its length at its other end. Assuming that the rod's density (mass per

unit volume) does not change, find the force magnitude that is required of the machine to decrease the radius to 999.9 μm. (The yield strength is not exceeded.)

59 The system in Fig. 12-65 is in equilibrium. The angles are $\theta_1 = 60°$ and $\theta_2 = 20°$, and the ball has mass $M = 2.0$ kg. What is the tension in (a) string ab and (b) string bc?

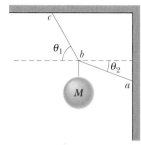

Fig. 12-65 Problem 59.

60 Figure 12-66 shows a stationary arrangement of two crayon boxes and three cords. Box A has a mass of 11.0 kg and is on a ramp at angle $\theta = 30.0°$; box B has a mass of 7.00 kg and hangs on a cord. The cord connected to box A is parallel to the ramp, which is frictionless. (a) What is the tension in the upper cord, and (b) what angle does that cord make with the horizontal?

Fig. 12-66 Problem 60.

61 In Fig. 12-67, a uniform rod of mass m is hinged to a building at its lower end, while its upper end is held in place by a rope attached to the wall. If angle $\theta_1 = 60°$, what value must angle θ_2 have so that the tension in the rope is equal to $mg/2$?

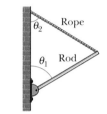

Fig. 12-67 Problem 61.

62 A construction worker attempts to lift a uniform beam off the floor and raise it to a vertical position. The beam is 2.50 m long and weighs 500 N. At a certain instant the worker holds the beam momentarily at rest with one end at distance $d = 1.50$ m above the floor, as shown in Fig. 12-68, by exerting a force $\vec{P}$ on the beam, perpendicular to the beam. (a)

Fig. 12-68 Problem 62.

What is the magnitude P? (b) What is the magnitude of the (net) force of the floor on the beam? (c) What is the minimum value the coefficient of static friction between beam and floor can have in order for the beam not to slip at this instant?

63 A solid copper cube has an edge length of 85.5 cm. How much stress must be applied to the cube to reduce the edge length to 85.0 cm? The bulk modulus of copper is 1.4×10^{11} N/m².

64 A mine elevator is supported by a single steel cable 2.5 cm in diameter. The total mass of the elevator cage and occupants is 670 kg. By how much does the cable stretch when the elevator hangs by (a) 12 m of cable and (b) 362 m of cable? (Neglect the mass of the cable.)

65 A beam of length L is carried by three men, one man at one end and the other two supporting the beam between them on a crosspiece placed so that the load of the beam is equally divided among the three men. How far from the beam's free end is the crosspiece placed? (Neglect the mass of the crosspiece.)

Fig. 12-69 Problem 66.

66 A uniform beam is 5.0 m long and has a mass of 53 kg. In Fig. 12-69, the beam is supported in a horizontal position by a hinge and a cable, with angle $\theta = 60°$. In unit-vector notation, what is the force on the beam from the hinge?

67 In Fig. 12-70, a uniform beam with a weight of 60 N and a length of 3.2 m is hinged at its lower end, and a horizontal force $\vec{F}$ of magnitude 50 N acts at its upper end. The beam is held vertical by a cable that makes angle $\theta = 25°$ with the ground and is attached to the beam at height $h = 2.0$ m. What are (a) the tension in the cable and (b) the force on the beam from the hinge in unit-vector notation?

Fig. 12-70 Problem 67.

68 After a fall, a 95 kg rock climber finds himself dangling from the end of a rope that had been 15 m long and 9.6 mm in diameter but has stretched by 2.8 cm. For the rope, calculate (a) the strain, (b) the stress, and (c) the Young's modulus.

69 In Fig. 12-71, a rectangular slab of slate rests on a bedrock surface inclined at angle $\theta = 26°$. The slab has length $L = 43$ m, thickness $T = 2.5$ m, and width $W = 12$ m, and 1.0 cm^3 of it has a mass of 3.2 g. The coefficient of static friction between slab and bedrock is

Fig. 12-71 Problem 69.

0.39. (a) Calculate the component of the gravitational force on the slab parallel to the bedrock surface. (b) Calculate the magnitude of the static frictional force on the slab. By comparing (a) and (b), you can see that the slab is in danger of sliding. This is prevented only by chance protrusions of bedrock. (c) To stabilize the slab, bolts are to be driven perpendicular to the bedrock surface (two bolts are shown). If each bolt has a cross-sectional area of 6.4 cm^2 and will snap under a shearing stress of 3.6×10^8 N/m^2, what is the minimum number of bolts needed? Assume that the bolts do not affect the normal force.

70 A uniform ladder whose length is 5.0 m and whose weight is 400 N leans against a frictionless vertical wall. The coefficient of static friction between the level ground and the foot of the ladder is 0.46. What is the greatest distance the foot of the ladder can be placed from the base of the wall without the ladder immediately slipping?

71 A 73 kg man stands on a level bridge of length L. He is at distance $L/4$ from one end. The bridge is uniform and weighs 2.7 kN. What are the magnitudes of the vertical forces on the bridge from its supports at (a) the end farther from him and (b) the nearer end?

72 A trap door in a ceiling is 0.91 m square, has a mass of 11 kg, and is hinged along one side, with a catch at the opposite side. If the center of gravity of the door is 10 cm toward the hinged side from the door's center, what are the magnitudes of the forces exerted by the door on (a) the catch and (b) the hinge?

73 A gymnast with mass 46.0 kg stands on the end of a uniform balance beam as shown in Fig. 12-72. The beam is 5.00 m long and has a mass of 250 kg (excluding the mass of the two supports). Each support is 0.540 m from its end of the beam. In unit-vector notation, what are the forces on the beam due to (a) support 1 and (b) support 2?

Fig. 12-72 Problem 73.

74 In Fig. 12-73, a uniform beam of length 12.0 m is supported by a horizontal cable and a hinge at angle $\theta = 50.0°$. The tension in the cable is 400 N. In unit-vector notation, what are (a) the gravitational force on the beam and (b) the force on the beam from the hinge?

Fig. 12-73 Problem 74.

75 A uniform ladder is 10 m long and weighs 200 N. In Fig. 12-74, the ladder leans against a vertical, frictionless wall at height $h = 8.0$ m above the ground. A horizontal force $\vec{F}$ is applied to the ladder at distance $d = 2.0$ m from its base (measured along the ladder). (a) If $F = 50$ N, what is the force of the ground on the ladder, in unit-vector notation? (b) If $F = 150$ N, what is the force of the ground on the ladder? (c) Suppose the coefficient of static friction between the ladder and the ground is 0.38; for what minimum value of F will the base of the ladder just start to move toward the wall?

76 If the (square) beam in Fig. 12-7a is of Douglas fir, what must be its thickness to keep the compressive stress on it to $\frac{1}{6}$ of its ultimate strength? (See Sample Problem 12-3.)

Fig. 12-74 Problem 75.

77 In Fig. 12-75, a uniform diving board of mass 40 kg is attached to two supports separated by distance $d_1 = 1.0$ m. When a diver stands at the far end of the board, at distance

$d_2 = 2.5$ m from support 2, support 1 pulls down on the board with a force of magnitude 1200 N. How far from support 1 should the diver stand to reduce that force to zero?

Fig. 12-75 Problem 77.

78 A makeshift swing is constructed by making a loop in one end of a rope and tying the other end to a tree limb. A child is sitting in the loop with the rope hanging vertically when the child's father pulls on the child with a horizontal force and displaces the child to one side. Just before the child is released from rest, the rope makes an angle of 15° with the vertical and the tension in the rope is 280 N. (a) How much does the child weigh? (b) What is the magnitude of the (horizontal) force of the father on the child just before the child is released? (c) If the maximum horizontal force the father can exert on the child is 93 N, what is the maximum angle with the vertical the rope can make while the father is pulling horizontally?

79 Four bricks of length L, identical and uniform, are stacked on a table in two ways, as shown in Fig. 12-76 (compare with Problem 53). We seek to maximize the overhang distance h in both arrangements. Find the optimum distances a_1, a_2, b_1, and b_2, and calculate h for the two arrangements.

80 Here is a way to move a heavy log in a tropical forest. Find a young tree that is located in the general direction you want to move the log and has a vine hanging from the top of the tree down to ground level. Wrap the vine around a limb on the log, pull hard enough on the vine to bend the tree over, and then tie off the vine on the limb. Repeat this procedure with several trees; eventually the net force of the vines moves the log forward. Although tedious, this technique let workers move logs long before modern machinery was available. Figure 12-77 shows the arrangement in which a single vine is attached to a branch at one end of a uniform log of mass M. The coefficient of static friction between log and ground is 0.80. If the log is on the verge of sliding, with the left end raised slightly by the vine, what are the (a) angle θ and (b) magnitude T of the force on the log from the vine?

(a)

(b)

Fig. 12-76 Problem 79.

Fig. 12-77 Problem 80.

81 The leaning Tower of Pisa (Fig. 12-78) is 55 m high and 7.0 m in diameter. The top of the tower is displaced 4.5 m from the vertical. Treat the tower as a uniform, circular cylinder. (a) What additional displacement, measured at the top, would bring the tower to the verge of toppling? (b) What angle would the tower then make with the vertical?

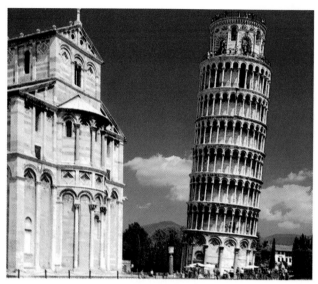

Fig. 12-78 Problem 81.

82 You are to build a large indoor sand mound and must be careful about the stress on the floor. The greatest stress occurs not directly beneath the apex (top) but at points a distance r_m from that central point (Fig. 12-79a). This outward displacement of the maximum stress is due to the sand grains forming arches within the mound. For mound height $H = 3.00$ m and angle $\theta = 33.0°$, Fig. 12-79b gives the stress σ as a function of radius r from the central point of the

(a)

(b)

Fig. 12-79 Problem 82.

mound's base: $\sigma_0 = 40\,000$ N/m^2, $\sigma_m = 40\,024$ N/m^2, $r_m = 1.82$ m, and 1.00 m^3 of sand has mass 1800 kg. (a) What is the volume of sand contained in the mound for $r \leq r_m/2$? (*Hint:* The volume is that of a vertical cylinder plus a cone on top of the cylinder. The volume of the cone is $\pi R^2 h/3$, where R is the cone's radius and h is the cone's height.) (b) What is the weight W of that volume of sand? (c) Use Fig. 12-79b to write an expression for the stress σ on the floor as a function of radius r, for $r \leq r_m$. (d) On the floor, what is the area dA of a thin ring of radius r centered on the mound's central vertical axis and with radial width dr? (e) What is the magnitude dF of the downward force on the ring due to the sand? (f) What is the magnitude F of the net downward force on the floor due to all the sand contained in the mound for $r \leq r_m/2$? (*Hint:* Integrate the expression of (e) from $r = 0$ to $r = r_m/2$.) Now note the surprise: This force magnitude F on the floor is less than the weight W of the sand above the floor, as found in (b). (g) By what fraction is F reduced from W; that is, what is $(F - W)/W$?

On-Line Simulation Problems

The website http://www.wiley.com/college/halliday has simulation problems about this chapter.

13 _Gravitation_

This is an image of the stars near our Milky Way galaxy's center, which is marked with a small cross. Note that nothing shows up exactly at the center, but slightly off center (at the 8:00 position) there is a small circle. That circle is the image of a star known as S2. The other circles are also images of stars (the halos around them are artificially produced by the method of processing the images). Most stars in our galaxy move so slowly that we cannot actually see them move relative to one another, not even over a lifetime of observations. However, S2 is very different—we can see it move. In fact, it is moving so rapidly that it makes a complete trip around the Galaxy's center in only 15.2 years. There must be something huge at the center, yet we see nothing there.

What monster lies at our galaxy's center?

The answer is in this chapter.

13-1 What Is Physics?

One of the long-standing goals of physics is to understand the gravitational force—the force that holds you to Earth, holds the Moon in orbit around Earth, and holds Earth in orbit around the Sun. It also reaches out through the whole of our Milky Way galaxy, holding together the billions and billions of stars in the Galaxy and the countless molecules and dust particles between stars. We are located somewhat near the edge of this disk-shaped collection of stars and other matter, 2.6×10^4 light-years (2.5×10^{20} m) from the galactic center, around which we slowly revolve.

The gravitational force also reaches across intergalactic space, holding together the Local Group of galaxies, which includes, in addition to the Milky Way, the Andromeda Galaxy (Fig. 13-1) at a distance of 2.3×10^6 light-years away from Earth, plus several closer dwarf galaxies, such as the Large Magellanic Cloud. The Local Group is part of the Local Supercluster of galaxies that is being drawn by the gravitational force toward an exceptionally massive region of space called the Great Attractor. This region appears to be about 3.0×10^8 light-years from Earth, on the opposite side of the Milky Way. And the gravitational force is even more far-reaching because it attempts to hold together the entire universe, which is expanding.

This force is also responsible for some of the most mysterious structures in the universe: *black holes.* When a star considerably larger than our Sun burns out, the gravitational force between all its particles can cause the star to collapse in on itself and thereby to form a black hole. The gravitational force at the surface of such a collapsed star is so strong that neither particles nor light can escape from the surface (thus the term "black hole"). Any star coming too near a black hole can be ripped apart by the strong gravitational force and pulled into the hole. Enough captures like this yields a *supermassive black hole.* Such mysterious monsters appear to be common in the universe.

Although the gravitational force is still not fully understood, the starting point in our understanding of it lies in the *law of gravitation* of Isaac Newton.

Fig. 13-1 The Andromeda Galaxy. Located 2.3×10^6 light-years from us, and faintly visible to the naked eye, it is very similar to our home galaxy, the Milky Way.

13-2 Newton's Law of Gravitation

Physicists like to study seemingly unrelated phenomena to show that a relationship can be found if the phenomena are examined closely enough. This search for unification has been going on for centuries. In 1665, the 23-year-old Isaac Newton made a basic contribution to physics when he showed that the force that holds the Moon in its orbit is the same force that makes an apple fall. We take this knowledge so much for granted now that it is not easy for us to comprehend the ancient belief that the motions of earthbound bodies and heavenly bodies were different in kind and were governed by different laws.

Newton concluded not only that Earth attracts both apples and the Moon but also that every body in the universe attracts every other body; this tendency of bodies to move toward each other is called **gravitation.** Newton's conclusion takes a little getting used to, because the familiar attraction of Earth for earthbound bodies is so great that it overwhelms the attraction that earthbound bodies have for each other. For example, Earth attracts an apple with a force magnitude of about 0.8 N. You also attract a nearby apple (and it attracts you), but the force of attraction has less magnitude than the weight of a speck of dust.

Newton proposed a *force law* that we call **Newton's law of gravitation:** Every particle attracts any other particle with a **gravitational force** of magnitude

$$F = G \frac{m_1 m_2}{r^2} \qquad \text{(Newton's law of gravitation).} \qquad (13\text{-}1)$$

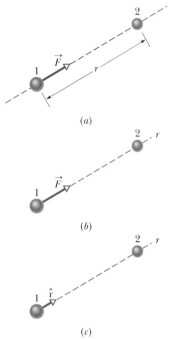

Fig. 13-2 (*a*) The gravitational force $\vec{F}$ on particle 1 due to particle 2 is an attractive force because particle 1 is attracted to particle 2. (*b*) Force $\vec{F}$ is directed along a radial coordinate axis *r* extending from particle 1 through particle 2. (*c*) $\vec{F}$ is in the direction of a unit vector $\hat{r}$ along the *r* axis.

Here m_1 and m_2 are the masses of the particles, *r* is the distance between them, and *G* is the **gravitational constant,** with a value that is now known to be

$$G = 6.67 \times 10^{-11} \; \text{N} \cdot \text{m}^2/\text{kg}^2$$
$$= 6.67 \times 10^{-11} \; \text{m}^3/\text{kg} \cdot \text{s}^2. \quad (13\text{-}2)$$

In Fig. 13-2*a*, $\vec{F}$ is the gravitational force acting on particle 1 (mass m_1) due to particle 2 (mass m_2). The force is directed toward particle 2 and is said to be an *attractive force* because particle 1 is attracted toward particle 2. The magnitude of the force is given by Eq. 13-1.

We can describe $\vec{F}$ as being in the positive direction of an *r* axis extending radially from particle 1 through particle 2 (Fig. 13-2*b*). We can also describe $\vec{F}$ by using a radial unit vector $\hat{r}$ (a dimensionless vector of magnitude 1) that is directed away from particle 1 along the *r* axis (Fig. 13-2*c*). From Eq. 13-1, the force on particle 1 is then

$$\vec{F} = G\frac{m_1 m_2}{r^2}\,\hat{r}. \quad (13\text{-}3)$$

The gravitational force on particle 2 due to particle 1 has the same magnitude as the force on particle 1 but the opposite direction. These two forces form a third-law force pair, and we can speak of the gravitational force *between* the two particles as having a magnitude given by Eq. 13-1. This force between two particles is not altered by other objects, even if they are located between the particles. Put another way, no object can shield either particle from the gravitational force due to the other particle.

The strength of the gravitational force—that is, how strongly two particles with given masses at a given separation attract each other—depends on the value of the gravitational constant *G*. If *G*—by some miracle—were suddenly multiplied by a factor of 10, you would be crushed to the floor by Earth's attraction. If *G* were divided by this factor, Earth's attraction would be so weak that you could jump over a building.

Although Newton's law of gravitation applies strictly to particles, we can also apply it to real objects as long as the sizes of the objects are small relative to the distance between them. The Moon and Earth are far enough apart so that, to a good approximation, we can treat them both as particles—but what about an apple and Earth? From the point of view of the apple, the broad and level Earth, stretching out to the horizon beneath the apple, certainly does not look like a particle.

Newton solved the apple–Earth problem by proving an important theorem called the *shell theorem:*

A uniform spherical shell of matter attracts a particle that is outside the shell as if all the shell's mass were concentrated at its center.

Earth can be thought of as a nest of such shells, one within another and each shell attracting a particle outside Earth's surface as if the mass of that shell were at the center of the shell. Thus, from the apple's point of view, Earth *does* behave like a particle, one that is located at the center of Earth and has a mass equal to that of Earth.

Suppose that, as in Fig. 13-3, Earth pulls down on an apple with a force of magnitude 0.80 N. The apple must then pull up on Earth with a force of magnitude 0.80 N, which we take to act at the center of Earth. Although the forces are matched in magnitude, they produce different accelerations when the apple is released. The acceleration of the apple is about 9.8 m/s^2, the familiar acceleration of a falling body near Earth's surface. The acceleration of Earth, however, measured in a reference frame attached to the center of mass of the apple–Earth system, is only about 1×10^{-25} m/s^2.

Fig. 13-3 The apple pulls up on Earth just as hard as Earth pulls down on the apple.

13-3 *Gravitation and the Principle of Superposition*

Given a group of particles, we find the net (or resultant) gravitational force on any one of them from the others by using the **principle of superposition.** This is a general principle that says a net effect is the sum of the individual effects. Here, the principle means that we first compute the individual gravitational forces that act on our selected particle due to each of the other particles. We then find the net force by adding these forces vectorially, as usual.

For n interacting particles, we can write the principle of superposition for the gravitational forces on particle 1 as

$$\vec{F}_{1,\text{net}} = \vec{F}_{12} + \vec{F}_{13} + \vec{F}_{14} + \vec{F}_{15} + \cdots + \vec{F}_{1n}. \tag{13-4}$$

Here $\vec{F}_{1,\text{net}}$ is the net force on particle 1 and, for example, $\vec{F}_{13}$ is the force on particle 1 from particle 3. We can express this equation more compactly as a vector sum:

$$\vec{F}_{1,\text{net}} = \sum_{i=2}^{n} \vec{F}_{1i}. \tag{13-5}$$

What about the gravitational force on a particle from a real (extended) object? This force is found by dividing the object into parts small enough to treat as particles and then using Eq. 13-5 to find the vector sum of the forces on the particle from all the parts. In the limiting case, we can divide the extended object into differential parts each of mass dm and each producing a differential force $d\vec{F}$ on the particle. In this limit, the sum of Eq. 13-5 becomes an integral and we have

$$\vec{F}_1 = \int d\vec{F}, \tag{13-6}$$

in which the integral is taken over the entire extended object and we drop the subscript "net." If the extended object is a uniform sphere or a spherical shell, we can avoid the integration of Eq. 13-6 by assuming that the object's mass is concentrated at the object's center and using Eq. 13-1.

Sample Problem 13-1

Figure 13-4a shows an arrangement of three particles, particle 1 of mass $m_1 = 6.0$ kg and particles 2 and 3 of mass $m_2 =$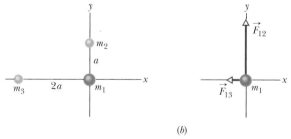

(a)

(b)

Fig. 13-4   (a) An arrangement of three particles. (b) The forces acting on the particle of mass m_1 due to the other particles.

$m_3 = 4.0$ kg, and distance $a = 2.0$ cm. What is the net gravitational force $\vec{F}_1$ on particle 1 due to the other particles?

Solution: One **Key Idea** here is that, because we have particles, the magnitude of the gravitational force on particle 1 due to either of the other particles is given by Eq. 13-1 ($F = Gm_1m_2/r^2$). Thus, the magnitude of the force $\vec{F}_{12}$ on particle 1 from particle 2 is

$$F_{12} = \frac{Gm_1m_2}{a^2}$$

$$= \frac{(6.67 \times 10^{-11}\ \text{m}^3/\text{kg} \cdot \text{s}^2)(6.0\ \text{kg})(4.0\ \text{kg})}{(0.020\ \text{m})^2}$$

$$= 4.00 \times 10^{-6}\ \text{N}.$$

Similarly, the magnitude of force $\vec{F}_{13}$ on particle 1 from particle 3 is

$$F_{13} = \frac{Gm_1m_3}{(2a)^2}$$

$$= \frac{(6.67 \times 10^{-11} \text{ m}^3/\text{kg} \cdot \text{s}^2)(6.0 \text{ kg})(4.0 \text{ kg})}{(0.040 \text{ m})^2}$$

$$= 1.00 \times 10^{-6} \text{ N}.$$

To determine the directions of $\vec{F}_{12}$ and $\vec{F}_{13}$, we use this **Key Idea:** Each force on particle 1 is directed toward the particle responsible for that force. Thus, $\vec{F}_{12}$ is directed in the positive direction of the y axis (Fig. 13-4b) and has only the y component F_{12}. Similarly, $\vec{F}_{13}$ is directed in the negative direction of the x axis and has only the x component $-F_{13}$.

To find the net force $\vec{F}_{1,\text{net}}$ on particle 1, we first use this very important **Key Idea:** Because the forces are not directed along the same line, we *cannot* simply add or subtract their magnitudes or their components to get their net force. Instead, we must add them as vectors.

We can do so on a vector-capable calculator. However, here we note that $-F_{13}$ and F_{12} are actually the x and y components of $\vec{F}_{1,\text{net}}$. Therefore, we can use Eq. 3-6 to find first the magnitude and then the direction of $\vec{F}_{1,\text{net}}$. The magnitude is

$$F_{1,\text{net}} = \sqrt{(F_{12})^2 + (-F_{13})^2}$$

$$= \sqrt{(4.00 \times 10^{-6} \text{ N})^2 + (-1.00 \times 10^{-6} \text{ N})^2}$$

$$= 4.1 \times 10^{-6} \text{ N}. \qquad \text{(Answer)}$$

Relative to the positive direction of the x axis, Eq. 3-6 gives the direction of $\vec{F}_{1,\text{net}}$ as

$$\theta = \tan^{-1}\frac{F_{12}}{-F_{13}} = \tan^{-1}\frac{4.00 \times 10^{-6} \text{ N}}{-1.00 \times 10^{-6} \text{ N}} = -76°.$$

Is this a reasonable direction? No, because the direction of $\vec{F}_{1,\text{net}}$ must be between the directions of $\vec{F}_{12}$ and $\vec{F}_{13}$. Recall from Chapter 3 (Problem-Solving Tactic 3) that a calculator displays only one of the two possible answers to a $\tan^{-1}$ function. We find the other answer by adding 180°:

$$-76° + 180° = 104°, \qquad \text{(Answer)}$$

which *is* a reasonable direction for $\vec{F}_{1,\text{net}}$.

✓**CHECKPOINT 2**   The figure shows four arrangements of three particles of equal masses. (a) Rank the arrangements according to the magnitude of the net gravitational force on the particle labeled m, greatest first. (b) In arrangement 2, is the direction of the net force closer to the line of length d or to the line of length D?

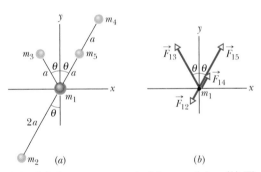

Sample Problem 13-2

Figure 13-5a shows an arrangement of five particles, with masses $m_1 = 8.0$ kg, $m_2 = m_3 = m_4 = m_5 = 2.0$ kg, and with $a = 2.0$ cm and $\theta = 30°$. What is the net gravitational force $\vec{F}_{1,\text{net}}$ on particle 1 due to the other particles?

Solution: Our **Key Ideas** are the same as in Sample Problem 13-1. However, this problem has a lot of symmetry that can help simplify the solution.

For the magnitudes of the forces on particle 1, first note that particles 2 and 4 have equal masses and equal distances of $r = 2a$ from particle 1. Thus, from Eq. 13-1, we find

$$F_{12} = F_{14} = \frac{Gm_1m_2}{(2a)^2}. \qquad (13\text{-}7)$$

Similarly, since particles 3 and 5 have equal masses and are both distance $r = a$ from particle 1, we find

$$F_{13} = F_{15} = \frac{Gm_1m_3}{a^2}. \qquad (13\text{-}8)$$

We could now substitute known data into these two equations to evaluate the magnitudes of the forces, indicate the directions of the forces on the free-body diagram of Fig. 13-5b, and then find the net force either (1) by resolving the vectors into x and y components, finding the net x and net y components, and then vectorially combining them or (2) by adding the vectors directly on a vector-capable calculator.

Fig. 13-5   (a) An arrangement of five particles. (b) The forces acting on the particle of mass m_1 due to the other four particles.

Instead, however, we shall make further use of the symmetry of the problem. First, we note that $\vec{F}_{12}$ and $\vec{F}_{14}$ are equal in magnitude but opposite in direction; thus, those forces *cancel*. Inspection of Fig. 13-5b and Eq. 13-8 reveals that the x components of $\vec{F}_{13}$ and $\vec{F}_{15}$ also *cancel*, and that their y components are identical in magnitude and both act in the positive direction of the y axis. Thus, $\vec{F}_{1,\text{net}}$ acts in that same direction, and its magnitude is twice the y component of $\vec{F}_{13}$:

$$F_{1,\text{net}} = 2F_{13} \cos \theta = 2\frac{Gm_1m_3}{a^2} \cos \theta$$

$$= 2\frac{(6.67 \times 10^{-11} \text{ m}^3/\text{kg} \cdot \text{s}^2)(8.0 \text{ kg})(2.0 \text{ kg})}{(0.020 \text{ m})^2} \cos 30°$$

$$= 4.6 \times 10^{-6} \text{ N}. \qquad \text{(Answer)}$$

Note that the presence of particle 5 along the line between particles 1 and 4 does not alter the gravitational force on particle 1 from particle 4.

CHECKPOINT 3 In the figure here, what is the direction of the net gravitational force on the particle of mass m_1 due to the other particles, each of mass m and arranged symmetrically relative to the y axis?

PROBLEM-SOLVING TACTICS

TACTIC 1: *Drawing Gravitational Force Vectors*

When you are given a diagram of particles, such as Fig. 13-4a, and asked to find the net gravitational force on one of them, you should usually draw a free-body diagram showing only the particle of concern and the forces on *it alone*, as in Fig. 13-4b. If, instead, you choose to superimpose the force vectors on the given diagram, be sure to draw the vectors with either all tails (preferably) or all heads on the particle experiencing those forces. If you draw the vectors elsewhere, you invite confusion—and confusion is guaranteed if you draw the vectors on the particles *causing* the forces.

TACTIC 2: *Simplifying a Sum of Forces with Symmetry*

In Sample Problem 13-2 we used the symmetry of the situation: By realizing that particles 2 and 4 are positioned symmetrically about particle 1, and thus that $\vec{F}_{12}$ and $\vec{F}_{14}$ cancel, we avoided calculating either force. By realizing that the x components of $\vec{F}_{13}$ and $\vec{F}_{15}$ cancel and that their y components are identical and add, we saved even more effort.

In problems with symmetry, you can save much effort and reduce the chance of error by identifying which calculations are not needed because of the symmetry. Such identification is a skill acquired only by doing many homework problems.

13-4 Gravitation Near Earth's Surface

Let us assume that Earth is a uniform sphere of mass M. The magnitude of the gravitational force from Earth on a particle of mass m, located outside Earth a distance r from Earth's center, is then given by Eq. 13-1 as

$$F = G\frac{Mm}{r^2}. \qquad (13\text{-}9)$$

If the particle is released, it will fall toward the center of Earth, as a result of the gravitational force $\vec{F}$, with an acceleration we shall call the **gravitational acceleration** $\vec{a}_g$. Newton's second law tells us that magnitudes F and a_g are related by

$$F = ma_g. \qquad (13\text{-}10)$$

Now, substituting F from Eq. 13-9 into Eq. 13-10 and solving for a_g, we find

$$a_g = \frac{GM}{r^2}. \qquad (13\text{-}11)$$

Table 13-1 shows values of a_g computed for various altitudes above Earth's surface. Notice a_g is significant even at 400 km.

Since Section 5-4, we have assumed that Earth is an inertial frame by neglecting its rotation. This simplification has allowed us to assume that the free-fall acceleration g of a particle is the same as the particle's gravitational acceleration (which we now call a_g). Furthermore, we assumed that g has the constant value 9.8 m/s² any place on Earth's surface. However, any g value measured at a given location will differ from the a_g value calculated with Eq. 13-11 for that location for three reasons: (1) Earth's mass is not distributed uniformly, (2) Earth is not a perfect sphere, and (3) Earth rotates. Moreover, because g differs from a_g, the same three reasons mean that the measured weight mg of a particle differs from the magnitude of the gravitational force on the particle as given by Eq. 13-9. Let us now examine those reasons.

TABLE 13-1
Variation of a_g with Altitude

Altitude (km)	a_g (m/s²)	Altitude Example
0	9.83	Mean Earth surface
8.8	9.80	Mt. Everest
36.6	9.71	Highest crewed balloon
400	8.70	Space shuttle orbit
35 700	0.225	Communications satellite

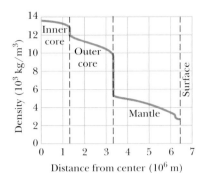

Fig. 13-6 The density of Earth as a function of distance from the center. The limits of the solid inner core, the largely liquid outer core, and the solid mantle are shown, but the crust of Earth is too thin to show clearly on this plot.

1. ***Earth's mass is not uniformly distributed.*** The density (mass per unit volume) of Earth varies radially as shown in Fig. 13-6, and the density of the crust (outer section) varies from region to region over Earth's surface. Thus, g varies from region to region over the surface.

2. ***Earth is not a sphere.*** Earth is approximately an ellipsoid, flattened at the poles and bulging at the equator. Its equatorial radius is greater than its polar radius by 21 km. Thus, a point at the poles is closer to the dense core of Earth than is a point on the equator. This is one reason the free-fall acceleration g increases as one proceeds, at sea level, from the equator toward either pole.

3. ***Earth is rotating.*** The rotation axis runs through the north and south poles of Earth. An object located on Earth's surface anywhere except at those poles must rotate in a circle about the rotation axis and thus must have a centripetal acceleration directed toward the center of the circle. This centripetal acceleration requires a centripetal net force that is also directed toward that center.

To see how Earth's rotation causes g to differ from a_g, let us analyze a simple situation in which a crate of mass m is on a scale at the equator. Figure 13-7a shows this situation as viewed from a point in space above the north pole.

Figure 13-7b, a free-body diagram for the crate, shows the two forces on the crate, both acting along a radial r axis that extends from Earth's center. The normal force $\vec{F}_N$ on the crate from the scale is directed outward, in the positive direction of the r axis. The gravitational force, represented with its equivalent $m\vec{a}_g$, is directed inward. Because it travels in a circle about the center of Earth as Earth turns, the crate has a centripetal acceleration $\vec{a}$ directed toward Earth's center. From Eq. 10-23 ($a_r = \omega^2 r$), we know this acceleration is equal to $\omega^2 R$, where ω is Earth's angular speed and R is the circle's radius (approximately Earth's radius). Thus, we can write Newton's second law for forces along the r axis ($F_{\text{net},r} = ma_r$) as

$$F_N - ma_g = m(-\omega^2 R). \qquad (13\text{-}12)$$

The magnitude F_N of the normal force is equal to the weight mg read on the scale. With mg substituted for F_N, Eq. 13-12 gives us

$$mg = ma_g - m(\omega^2 R), \qquad (13\text{-}13)$$

which says

$$\begin{pmatrix}\text{measured} \\ \text{weight}\end{pmatrix} = \begin{pmatrix}\text{magnitude of} \\ \text{gravitational force}\end{pmatrix} - \begin{pmatrix}\text{mass times} \\ \text{centripetal acceleration}\end{pmatrix}.$$

Thus, the measured weight is less than the magnitude of the gravitational force on the crate, because of Earth's rotation.

To find a corresponding expression for g and a_g, we cancel m from Eq. 13-13 to write

$$g = a_g - \omega^2 R, \qquad (13\text{-}14)$$

which says

$$\begin{pmatrix}\text{free-fall} \\ \text{acceleration}\end{pmatrix} = \begin{pmatrix}\text{gravitational} \\ \text{acceleration}\end{pmatrix} - \begin{pmatrix}\text{centripetal} \\ \text{acceleration}\end{pmatrix}.$$

Thus, the measured free-fall acceleration is less than the gravitational acceleration because of Earth's rotation.

The difference between accelerations g and a_g is equal to $\omega^2 R$ and is greatest on the equator (for one reason, the radius of the circle traveled by the crate is

(a)

(b)

Fig. 13-7 (a) A crate sitting on a scale at Earth's equator, as seen by an observer positioned on Earth's rotation axis at some point above the north pole. (b) A free-body diagram for the crate, with a radial r axis extending from Earth's center. The gravitational force on the crate is represented with its equivalent $m\vec{a}_g$. The normal force on the crate from the scale is $\vec{F}_N$. Because of Earth's rotation, the crate has a centripetal acceleration $\vec{a}$ that is directed toward Earth's center.

greatest there). To find the difference, we can use Eq. 10-5 ($\omega = \Delta\theta/\Delta t$) and Earth's radius $R = 6.37 \times 10^6$ m. For one rotation of Earth, θ is 2π rad and the time period Δt is about 24 h. Using these values (and converting hours to seconds), we find that g is less than a_g by only about 0.034 m/s² (small compared to 9.8 m/s²). Therefore, neglecting the difference in accelerations g and a_g is often justified. Similarly, neglecting the difference between weight and the magnitude of the gravitational force is also often justified.

Sample Problem 13-3

(a) An astronaut whose height h is 1.70 m floats "feet down" in an orbiting space shuttle at distance $r = 6.77 \times 10^6$ m away from the center of Earth. What is the difference between the gravitational acceleration at her feet and at her head?

Solution: One **Key Idea** here is that we can approximate Earth as a uniform sphere of mass M_E. Then, from Eq. 13-11, the gravitational acceleration at any distance r from the center of Earth is

$$a_g = \frac{GM_E}{r^2}. \tag{13-15}$$

We might simply apply this equation twice, first with $r = 6.77 \times 10^6$ m for the feet and then with $r = 6.77 \times 10^6$ m + 1.70 m for the head. However, a calculator may give us the same value for a_g twice, and thus a difference of zero, because h is so much smaller than r. A second **Key Idea** helps here: Because we have a differential change dr in r between the astronaut's feet and head, let us differentiate Eq. 13-15 with respect to r. That gives us

$$da_g = -2\frac{GM_E}{r^3}\,dr, \tag{13-16}$$

where da_g is the differential change in the gravitational acceleration due to the differential change dr in r. For the astronaut, $dr = h$ and $r = 6.77 \times 10^6$ m. Substituting data into Eq. 13-16, we find

$$da_g = -2\frac{(6.67 \times 10^{-11}\ \mathrm{m^3/kg \cdot s^2})(5.98 \times 10^{24}\ \mathrm{kg})}{(6.77 \times 10^6\ \mathrm{m})^3}(1.70\ \mathrm{m})$$
$$= -4.37 \times 10^{-6}\ \mathrm{m/s^2}, \qquad \text{(Answer)}$$

where the M_E value is taken from Appendix C. This result means that the gravitational acceleration of the astronaut's feet toward Earth is slightly greater than the gravitational acceleration of her head toward Earth. This difference in acceleration tends to stretch her body, but the difference is so small that the stretching is unnoticeable.

(b) If the astronaut is now "feet down" at the same orbital radius $r = 6.77 \times 10^6$ m about a black hole of mass $M_h = 1.99 \times 10^{31}$ kg (10 times our Sun's mass), what is the difference between the gravitational acceleration at her feet and at her head? The black hole has a mathematical surface (*event horizon*) of radius $R_h = 2.95 \times 10^4$ m. Nothing, not even light, can escape from that surface or anywhere inside it. Note that the astronaut is well outside the surface (at $r = 229R_h$).

Solution: The **Key Idea** here is that we again have a differential change dr in r between the astronaut's feet and head, so we can again use Eq. 13-16. However, now we substitute $M_h = 1.99 \times 10^{31}$ kg for M_E. We find

$$da_g = -2\frac{(6.67 \times 10^{-11}\ \mathrm{m^3/kg \cdot s^2})(1.99 \times 10^{31}\ \mathrm{kg})}{(6.77 \times 10^6\ \mathrm{m})^3}(1.70\ \mathrm{m})$$
$$= -14.5\ \mathrm{m/s^2}. \qquad \text{(Answer)}$$

This means that the gravitational acceleration of the astronaut's feet toward the black hole is noticeably larger than that of her head. The resulting tendency to stretch her body would be bearable but quite painful. If she drifted closer to the black hole, the stretching tendency would increase drastically.

13-5 *Gravitation Inside Earth*

Newton's shell theorem can also be applied to a situation in which a particle is located *inside* a uniform shell, to show the following:

> A uniform shell of matter exerts no *net* gravitational force on a particle located inside it.

Caution: This statement does *not* mean that the gravitational forces on the particle from the various elements of the shell magically disappear. Rather, it means that the *sum* of the force vectors on the particle from all the elements is zero.

If Earth's mass were uniformly distributed, the gravitational force acting on a particle would be a maximum at Earth's surface and would decrease as the particle moved outward, away from the planet. If the particle were to move inward, perhaps down a deep mine shaft, the gravitational force would change for two reasons. (1) It would tend to increase because the particle would be moving closer to the center of Earth. (2) It would tend to decrease because the thickening shell of material lying outside the particle's radial position would not exert any net force on the particle.

For a uniform Earth, the second influence would prevail and the force on the particle would steadily decrease to zero as the particle approached the center of Earth. However, for the real (nonuniform) Earth, the force on the particle actually increases as the particle begins to descend. The force reaches a maximum at a certain depth and then decreases as the particle descends farther.

Sample Problem 13-4

In *Pole to Pole*, an early science fiction story by George Griffith, three explorers attempt to travel by capsule through a naturally formed (and, of course, fictional) tunnel directly from the south pole to the north pole (Fig. 13-8). According to the story, as the capsule approaches Earth's center, the gravitational force on the explorers becomes alarmingly large and then, exactly at the center, it suddenly but only momentarily disappears. Then the capsule travels through the second half of the tunnel, to the north pole.

Check Griffith's description by finding the gravitational force on the capsule of mass m when it reaches a distance r from Earth's center. Assume that Earth is a sphere of uniform density ρ (mass per unit volume).

Solution: Newton's shell theorem gives us three **Key Ideas** here:

1. When the capsule is at radius r from Earth's center, the portion of Earth that lies outside a sphere of radius r does *not* produce a net gravitational force on the capsule.

2. The portion of Earth that lies inside that sphere *does* produce a net gravitational force on the capsule.

Fig. 13-8 A capsule of mass m falls from rest through a tunnel that connects Earth's south and north poles. When the capsule is at distance r from Earth's center, the portion of Earth's mass that is contained in a sphere of that radius is M_{ins}.

3. We can treat the mass M_{ins} of that inside portion of Earth as being the mass of a particle located at Earth's center.

All three ideas tell us that we can write Eq. 13-1, for the magnitude of the gravitational force on the capsule, as

$$F = \frac{GmM_{ins}}{r^2}. \tag{13-17}$$

To write the mass M_{ins} in terms of the radius r, we note that the volume V_{ins} containing this mass is $\frac{4}{3}\pi r^3$. Also, because we're assuming an Earth of uniform desnity, the density $\rho_{ins} = M_{ins}/V_{ins}$ is Earth's density ρ. Thus, we have

$$M_{ins} = \rho V_{ins} = \rho\,\frac{4\pi r^3}{3}. \tag{13-18}$$

Then, after substituting this expression into Eq. 13-17 and canceling, we have

$$F = \frac{4\pi Gm\rho}{3}\,r. \quad \text{(Answer)} \quad \text{(13-19)}$$

This equation tells us that the force magnitude F depends linearly on the capsule's distance r from Earth's center. Thus, as r decreases, F also decreases (opposite of Griffith's description), until it is zero at Earth's center. At least Griffith got the zero-at-the-center detail correct.

Equation 13-19 can also be written in terms of the force vector $\vec{F}$ and the capsule's position vector $\vec{r}$ along a radial axis extending from Earth's center. Let K represent the collection of constants $4\pi Gm\rho/3$. Then, Eq. 13-19 becomes

$$\vec{F} = -K\vec{r}, \tag{13-20}$$

in which we have inserted a minus sign to indicate that $\vec{F}$ and $\vec{r}$ have opposite directions. Equation 13-20 has the form of Hooke's law (Eq. 7-20, $\vec{F} = -k\vec{d}$). Thus, under the idealized conditions of the story, the capsule would oscillate like a block on a spring, with the center of the oscillation at Earth's center. After the capsule had fallen from the south pole to Earth's center, it would travel from the center to the north pole (as Griffith said) and then back again, repeating the cycle forever.

13-6 Gravitational Potential Energy

In Section 8-4, we discussed the gravitational potential energy of a particle–Earth system. We were careful to keep the particle near Earth's surface, so that we could regard the gravitational force as constant. We then chose some reference configuration of the system as having a gravitational potential energy of zero. Often, in this configuration the particle was on Earth's surface. For particles not on Earth's surface, the gravitational potential energy decreased when the separation between the particle and Earth decreased.

Here, we broaden our view and consider the gravitational potential energy U of two particles, of masses m and M, separated by a distance r. We again choose a reference configuration with U equal to zero. However, to simplify the equations, the separation distance r in the reference configuration is now large enough to be approximated as *infinite*. As before, the gravitational potential energy decreases when the separation decreases. Since $U = 0$ for $r = \infty$, the potential energy is negative for any finite separation and becomes progressively more negative as the particles move closer together.

With these facts in mind and as we shall justify next, we take the gravitational potential energy of the two-particle system to be

$$U = -\frac{GMm}{r} \quad \text{(gravitational potential energy).} \quad (13\text{-}21)$$

Note that $U(r)$ approaches zero as r approaches infinity and that for any finite value of r, the value of $U(r)$ is negative.

The potential energy given by Eq. 13-21 is a property of the system of two particles rather than of either particle alone. There is no way to divide this energy and say that so much belongs to one particle and so much to the other. However, if $M \gg m$, as is true for Earth (mass M) and a baseball (mass m), we often speak of "the potential energy of the baseball." We can get away with this because, when a baseball moves in the vicinity of Earth, changes in the potential energy of the baseball–Earth system appear almost entirely as changes in the kinetic energy of the baseball, since changes in the kinetic energy of Earth are too small to be measured. Similarly, in Section 13-8 we shall speak of "the potential energy of an artificial satellite" orbiting Earth, because the satellite's mass is so much smaller than Earth's mass. When we speak of the potential energy of bodies of comparable mass, however, we have to be careful to treat them as a system.

If our system contains more than two particles, we consider each pair of particles in turn, calculate the gravitational potential energy of that pair with Eq. 13-21 as if the other particles were not there, and then algebraically sum the results. Applying Eq. 13-21 to each of the three pairs of Fig. 13-9, for example, gives the potential energy of the system as

$$U = -\left(\frac{Gm_1m_2}{r_{12}} + \frac{Gm_1m_3}{r_{13}} + \frac{Gm_2m_3}{r_{23}}\right). \quad (13\text{-}22)$$

Proof of Equation 13-21

Let us shoot a baseball directly away from Earth along the path in Fig. 13-10. We want to find an expression for the gravitational potential energy U of the ball at point P along its path, at radial distance R from Earth's center. To do so, we first find the work W done on the ball by the gravitational force as the ball travels from point P to a great (infinite) distance from Earth. Because the gravitational force $\vec{F}(r)$ is a variable force (its magnitude depends on r), we must use the techniques of Section 7-8 to find the work. In vector notation, we can write

$$W = \int_R^\infty \vec{F}(r) \cdot d\vec{r}. \quad (13\text{-}23)$$

The integral contains the scalar (or dot) product of the force $\vec{F}(r)$ and the differential displacement vector $d\vec{r}$ along the ball's path. We can expand that product as

$$\vec{F}(r) \cdot d\vec{r} = F(r)\, dr \cos\phi, \quad (13\text{-}24)$$

where ϕ is the angle between the directions of $\vec{F}(r)$ and $d\vec{r}$. When we substitute

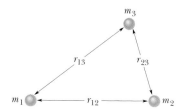

Fig. 13-9 A system consisting of three particles. The gravitational potential energy *of the system* is the sum of the gravitational potential energies of all three pairs of particles.

Fig. 13-10 A baseball is shot directly away from Earth, through point P at radial distance R from Earth's center. The gravitational force $\vec{F}$ on the ball and a differential displacement vector $d\vec{r}$ are shown, both directed along a radial r axis.

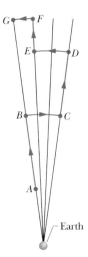

Fig. 13-11 Near Earth, a baseball is moved from point A to point G along a path consisting of radial lengths and circular arcs.

180° for ϕ and Eq. 13-1 for $F(r)$, Eq. 13-24 becomes

$$\vec{F}(r) \cdot d\vec{r} = -\frac{GMm}{r^2}\, dr,$$

where M is Earth's mass and m is the mass of the ball.

Substituting this into Eq. 13-23 and integrating gives us

$$W = -GMm \int_R^\infty \frac{1}{r^2}\, dr = \left[\frac{GMm}{r}\right]_R^\infty$$

$$= 0 - \frac{GMm}{R} = -\frac{GMm}{R}, \tag{13-25}$$

where W is the work required to move the ball from point P (at distance R) to infinity. Equation 8-1 ($\Delta U = -W$) tells us that we can also write that work in terms of potential energies as

$$U_\infty - U = -W.$$

Because the potential energy U_∞ at infinity is zero, U is the potential energy at P, and W is given by Eq. 13-25, this equation becomes

$$U = W = -\frac{GMm}{R}.$$

Switching R to r gives us Eq. 13-21, which we set out to prove.

Path Independence

In Fig. 13-11, we move a baseball from point A to point G along a path consisting of three radial lengths and three circular arcs (centered on Earth). We are interested in the total work W done by Earth's gravitational force $\vec{F}$ on the ball as it moves from A to G. The work done along each circular arc is zero, because the direction of $\vec{F}$ is perpendicular to the arc at every point. Thus, W is the sum of only the works done by $\vec{F}$ along the three radial lengths.

Now, suppose we mentally shrink the arcs to zero. We would then be moving the ball directly from A to G along a single radial length. Does that change W? No. Because no work was done along the arcs, eliminating them does not change the work. The path taken from A to G now is clearly different, but the work done by $\vec{F}$ is the same.

We discussed such a result in a general way in Section 8-3. Here is the point: The gravitational force is a conservative force. Thus, the work done by the gravitational force on a particle moving from an initial point i to a final point f is independent of the path taken between the points. From Eq. 8-1, the change ΔU in the gravitational potential energy from point i to point f is given by

$$\Delta U = U_f - U_i = -W. \tag{13-26}$$

Since the work W done by a conservative force is independent of the actual path taken, the change ΔU in gravitational potential energy is *also independent* of the path taken.

Potential Energy and Force

In the proof of Eq. 13-21, we derived the potential energy function $U(r)$ from the force function $\vec{F}(r)$. We should be able to go the other way—that is, to start from the potential energy function and derive the force function. Guided by Eq. 8-22 ($F(x) = -dU(x)/dx$), we can write

$$F = -\frac{dU}{dr} = -\frac{d}{dr}\left(-\frac{GMm}{r}\right)$$

$$= -\frac{GMm}{r^2}. \tag{13-27}$$

This is Newton's law of gravitation (Eq. 13-1). The minus sign indicates that the force on mass m points radially inward, toward mass M.

Escape Speed

If you fire a projectile upward, usually it will slow, stop momentarily, and return to Earth. There is, however, a certain minimum initial speed that will cause it to move upward forever, theoretically coming to rest only at infinity. This minimum initial speed is called the (Earth) **escape speed.**

Consider a projectile of mass m, leaving the surface of a planet (or some other astronomical body or system) with escape speed v. The projectile has a kinetic energy K given by $\frac{1}{2}mv^2$ and a potential energy U given by Eq. 13-21:

$$U = -\frac{GMm}{R},$$

in which M is the mass of the planet and R is its radius.

When the projectile reaches infinity, it stops and thus has no kinetic energy. It also has no potential energy because an infinite separation between two bodies is our zero-potential-energy configuration. Its total energy at infinity is therefore zero. From the principle of conservation of energy, its total energy at the planet's surface must also have been zero, and so

$$K + U = \tfrac{1}{2}mv^2 + \left(-\frac{GMm}{R}\right) = 0.$$

This yields
$$v = \sqrt{\frac{2GM}{R}}. \qquad (13\text{-}28)$$

Note that v does not depend on the direction in which a projectile is fired from a planet. However, attaining that speed is easier if the projectile is fired in the direction the launch site is moving as the planet rotates about its axis. For example, rockets are launched eastward at Cape Canaveral to take advantage of the Cape's eastward speed of 1500 km/h due to Earth's rotation.

Equation 13-28 can be applied to find the escape speed of a projectile from any astronomical body, provided we substitute the mass of the body for M and the radius of the body for R. Table 13-2 shows some escape speeds.

CHECKPOINT 4 You move a ball of mass m away from a sphere of mass M. (a) Does the gravitational potential energy of the ball–sphere system increase or decrease? (b) Is positive or negative work done by the gravitational force between the ball and the sphere?

TABLE 13-2
Some Escape Speeds

Body	Mass (kg)	Radius (m)	Escape Speed (km/s)
Ceres[a]	1.17×10^{21}	3.8×10^5	0.64
Earth's moon[a]	7.36×10^{22}	1.74×10^6	2.38
Earth	5.98×10^{24}	6.37×10^6	11.2
Jupiter	1.90×10^{27}	7.15×10^7	59.5
Sun	1.99×10^{30}	6.96×10^8	618
Sirius B[b]	2×10^{30}	1×10^7	5200
Neutron star[c]	2×10^{30}	1×10^4	2×10^5

[a] The most massive of the asteroids.

[b] A *white dwarf* (a star in a final stage of evolution) that is a companion of the bright star Sirius.

[c] The collapsed core of a star that remains after that star has exploded in a *supernova* event.

Sample Problem 13-5

An asteroid, headed directly toward Earth, has a speed of 12 km/s relative to the planet when the asteroid is 10 Earth radii from Earth's center. Neglecting the effects of Earth's atmosphere on the asteroid, find the asteroid's speed v_f when it reaches Earth's surface.

Solution: One Key Idea is that, because we are to neglect the effects of the atmosphere on the asteroid, the mechanical energy of the asteroid–Earth system is conserved during the fall. Thus, the final mechanical energy (when the asteroid reaches Earth's surface) is equal to the initial mechanical energy. With kinetic energy K and gravitational potential energy U, we can write this as

$$K_f + U_f = K_i + U_i. \qquad (13\text{-}29)$$

A second Key Idea is that, if we assume the system is isolated, the system's linear momentum must be conserved during the fall. Therefore, the momentum change of the asteroid and that of Earth must be equal in magnitude and opposite in sign. However, because Earth's mass is so much greater than the asteroid's mass, the change in Earth's speed is negligible relative to the change in the asteroid's speed. So, the change in Earth's kinetic energy is also negligible. Thus, we can assume that the kinetic energies in Eq. 13-29 are those of the asteroid alone.

Let m represent the asteroid's mass and M represent Earth's mass (5.98×10^{24} kg). The asteroid is initially at distance $10R_E$ and finally at distance R_E, where R_E is Earth's radius (6.37×10^6 m). Substituting Eq. 13-21 for U and $\frac{1}{2}mv^2$ for K, we rewrite Eq. 13-29 as

$$\tfrac{1}{2}mv_f^2 - \frac{GMm}{R_E} = \tfrac{1}{2}mv_i^2 - \frac{GMm}{10R_E}.$$

Rearranging and substituting known values, we find

$$
\begin{aligned}
v_f^2 &= v_i^2 + \frac{2GM}{R_E}\left(1 - \frac{1}{10}\right) \\
&= (12 \times 10^3 \text{ m/s})^2 \\
&\quad + \frac{2(6.67 \times 10^{-11} \text{ m}^3/\text{kg} \cdot \text{s}^2)(5.98 \times 10^{24} \text{ kg})}{6.37 \times 10^6 \text{ m}}\,0.9 \\
&= 2.567 \times 10^8 \text{ m}^2/\text{s}^2,
\end{aligned}
$$

and

$$v_f = 1.60 \times 10^4 \text{ m/s} = 16 \text{ km/s}. \qquad \text{(Answer)}$$

At this speed, the asteroid would not have to be particularly large to do considerable damage at impact. If it were only 5 m across, the impact could release about as much energy as the nuclear explosion at Hiroshima. Alarmingly, about 500 million asteroids of this size are near Earth's orbit, and in 1994 one of them apparently penetrated Earth's atmosphere and exploded 20 km above the South Pacific (setting off nuclear-explosion warnings on six military satellites). The impact of an asteroid 500 m across (there may be a million of them near Earth's orbit) could end modern civilization and almost eliminate humans worldwide.

13-7 Planets and Satellites: Kepler's Laws

The motions of the planets, as they seemingly wander against the background of the stars, have been a puzzle since the dawn of history. The "loop-the-loop" motion of Mars, shown in Fig. 13-12, was particularly baffling. Johannes Kepler (1571–1630), after a lifetime of study, worked out the empirical laws that govern these motions. Tycho Brahe (1546–1601), the last of the great astronomers to make observations without the help of a telescope, compiled the extensive data from which Kepler was able to derive the three laws of planetary motion that now bear Kepler's name. Later, Newton (1642–1727) showed that his law of gravitation leads to Kepler's laws.

In this section we discuss each of Kepler's three laws. Although here we apply the laws to planets orbiting the Sun, they hold equally well for satellites, either natural or artificial, orbiting Earth or any other massive central body.

▶ **1. THE LAW OF ORBITS:** All planets move in elliptical orbits, with the Sun at one focus.

Fig. 13-12 The path seen from Earth for the planet Mars as it moved against a background of the constellation Capricorn during 1971. The planet's position on four days is marked. Both Mars and Earth are moving in orbits around the Sun so that we see the position of Mars relative to us; this relative motion sometimes results in an apparent loop in the path of Mars.

Figure 13-13 shows a planet of mass m moving in such an orbit around the Sun, whose mass is M. We assume that $M \gg m$, so that the center of mass of the planet–Sun system is approximately at the center of the Sun.

The orbit in Fig. 13-13 is described by giving its **semimajor axis** a and its **eccentricity** e, the latter defined so that ea is the distance from the center of the ellipse to either focus F or F'. *An eccentricity of zero corresponds to a circle,* in which the two foci merge to a single central point. The eccentricities of the planetary orbits are not large; so if the orbits are drawn to scale, they look circular. The eccentricity of the ellipse of Fig. 13-13, which has been exaggerated for clarity, is 0.74. The eccentricity of Earth's orbit is only 0.0167.

2. THE LAW OF AREAS: A line that connects a planet to the Sun sweeps out equal areas in the plane of the planet's orbit in equal time intervals; that is, the rate dA/dt at which it sweeps out area A is constant.

Qualitatively, this second law tells us that the planet will move most slowly when it is farthest from the Sun and most rapidly when it is nearest to the Sun. As it turns out, Kepler's second law is totally equivalent to the law of conservation of angular momentum. Let us prove it.

The area of the shaded wedge in Fig. 13-14a closely approximates the area swept out in time Δt by a line connecting the Sun and the planet, which are separated by distance r. The area ΔA of the wedge is approximately the area of a triangle with base $r\,\Delta\theta$ and height r. Since the area of a triangle is one-half of the base times the height, $\Delta A \approx \frac{1}{2}r^2\,\Delta\theta$. This expression for ΔA becomes more exact as Δt (hence $\Delta\theta$) approaches zero. The instantaneous rate at which area is being swept out is then

$$\frac{dA}{dt} = \frac{1}{2}r^2\frac{d\theta}{dt} = \frac{1}{2}r^2\omega, \qquad (13\text{-}30)$$

in which ω is the angular speed of the rotating line connecting Sun and planet.

Figure 13-14b shows the linear momentum $\vec{p}$ of the planet, along with the radial and perpendicular components of $\vec{p}$. From Eq. 11-20 ($L = rp_\perp$), the magnitude of the angular momentum $\vec{L}$ of the planet about the Sun is given by the product of r and $p_\perp$, the component of $\vec{p}$ perpendicular to r. Here, for a planet of mass m,

$$L = rp_\perp = (r)(mv_\perp) = (r)(m\omega r)$$
$$= mr^2\omega, \qquad (13\text{-}31)$$

where we have replaced $v_\perp$ with its equivalent ωr (Eq. 10-18). Eliminating $r^2\omega$ between Eqs. 13-30 and 13-31 leads to

$$\frac{dA}{dt} = \frac{L}{2m}. \qquad (13\text{-}32)$$

If dA/dt is constant, as Kepler said it is, then Eq. 13-32 means that L must also be constant—angular momentum is conserved. Kepler's second law is indeed equivalent to the law of conservation of angular momentum.

3. THE LAW OF PERIODS: The square of the period of any planet is proportional to the cube of the semimajor axis of its orbit.

To see this, consider the circular orbit of Fig. 13-15, with radius r (the radius of a circle is equivalent to the semimajor axis of an ellipse). Applying Newton's second law ($F = ma$) to the orbiting planet in Fig. 13-15 yields

$$\frac{GMm}{r^2} = (m)(\omega^2 r). \qquad (13\text{-}33)$$

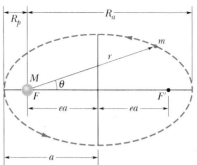

Fig. 13-13   A planet of mass m moving in an elliptical orbit around the Sun. The Sun, of mass M, is at one focus F of the ellipse. The other focus is F', which is located in empty space. Each focus is a distance ea from the ellipse's center, with e being the eccentricity of the ellipse. The semimajor axis a of the ellipse, the perihelion (nearest the Sun) distance R_p, and the aphelion (farthest from the Sun) distance R_a are also shown.

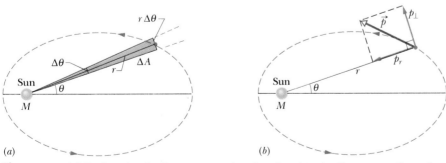

(a)

(b)

Fig. 13-14   (a) In time Δt, the line r connecting the planet to the Sun moves through an angle $\Delta\theta$, sweeping out an area ΔA (shaded). (b) The linear momentum $\vec{p}$ of the planet and the components of $\vec{p}$.

Fig. 13-15   A planet of mass m moving around the Sun in a circular orbit of radius r.

TABLE 13-3
Kepler's Law of Periods for the Solar System

Planet	Semimajor Axis a (10^{10} m)	Period T (y)	T^2/a^3 (10^{-34} y^2/m^3)
Mercury	5.79	0.241	2.99
Venus	10.8	0.615	3.00
Earth	15.0	1.00	2.96
Mars	22.8	1.88	2.98
Jupiter	77.8	11.9	3.01
Saturn	143	29.5	2.98
Uranus	287	84.0	2.98
Neptune	450	165	2.99
Pluto	590	248	2.99

Here we have substituted from Eq. 13-1 for the force magnitude F and used Eq. 10-23 to substitute $\omega^2 r$ for the centripetal acceleration. If we now use Eq. 10-20 to replace ω with $2\pi/T$, where T is the period of the motion, we obtain Kepler's third law:

$$T^2 = \left(\frac{4\pi^2}{GM}\right) r^3 \quad \text{(law of periods).} \tag{13-34}$$

The quantity in parentheses is a constant that depends only on the mass M of the central body about which the planet orbits.

Equation 13-34 holds also for elliptical orbits, provided we replace r with a, the semimajor axis of the ellipse. This law predicts that the ratio T^2/a^3 has essentially the same value for every planetary orbit around a given massive body. Table 13-3 shows how well it holds for the orbits of the planets of the solar system.

CHECKPOINT 5   Satellite 1 is in a certain circular orbit around a planet, while satellite 2 is in a larger circular orbit. Which satellite has (a) the longer period and (b) the greater speed?

Sample Problem 13-6

Comet Halley orbits the Sun with a period of 76 years and, in 1986, had a distance of closest approach to the Sun, its *perihelion distance* R_p, of 8.9×10^{10} m. Table 13-3 shows that this is between the orbits of Mercury and Venus.

(a) What is the comet's farthest distance from the Sun, which is called its *aphelion distance* R_a?

Solution: One **Key Idea** comes from Fig. 13-13, in which we see that $R_a + R_p = 2a$, where a is the semimajor axis of the orbit. Thus, we can find R_a if we first find a. A second **Key Idea** is that we can relate a to the given period via the law of periods (Eq. 13-34) if we simply substitute the semimajor axis a for r. Doing so and then solving for a, we have

$$a = \left(\frac{GMT^2}{4\pi^2}\right)^{1/3}. \tag{13-35}$$

If we substitute the mass M of the Sun, 1.99×10^{30} kg, and the period T of the comet, 76 years or 2.4×10^9 s, into Eq. 13-35, we find that $a = 2.7 \times 10^{12}$ m. Now we have

$$R_a = 2a - R_p$$
$$= (2)(2.7 \times 10^{12} \text{ m}) - 8.9 \times 10^{10} \text{ m}$$
$$= 5.3 \times 10^{12} \text{ m.} \quad \text{(Answer)}$$

Table 13-3 shows that this is a little less than the semimajor axis of the orbit of Pluto. Thus, the comet does not get farther from the Sun than Pluto.

(b) What is the eccentricity e of the orbit of comet Halley?

Solution: The **Key Idea** here is that we can relate e, a, and R_p via Fig. 13-13. We see there that $ea = a - R_p$, or

$$e = \frac{a - R_p}{a} = 1 - \frac{R_p}{a}$$
$$= 1 - \frac{8.9 \times 10^{10} \text{ m}}{2.7 \times 10^{12} \text{ m}} = 0.97. \quad \text{(Answer)}$$

This tells us that, with an eccentricity approaching unity, this orbit must be a long thin ellipse.

Sample Problem 13-7

Let's return to the story that opens this chapter. Figure 13-16 shows the observed orbit of the star S2 as the star moves around a mysterious and unobserved object called Sagittarius A* (pronounced "A star"), which is at the center of the Milky Way galaxy. S2 orbits Sagittarius A* with a period of $T = 15.2$ y and with a semimajor axis of $a = 5.50$ light-days ($= 1.42 \times 10^{14}$ m). What is the mass M of Sagittarius A*? What is Sagittarius A*?

Solution: The **Key Idea** here is that the period T and the semimajor axis a of the orbit are related to the mass M of Sagittarius A* according to Kepler's law of periods. From Eq. 13-34, with a replacing the radius r of a circular orbit, we have

$$T^2 = \left(\frac{4\pi^2}{GM}\right)a^3. \tag{13-36}$$

Solving for M and substituting the given data lead us to

$$M = \frac{4\pi^2 a^3}{GT^2}$$

$$= \frac{4\pi^2(1.42 \times 10^{14} \text{ m})^3}{(6.67 \times 10^{-11} \text{ N} \cdot \text{m}^2/\text{kg}^2)[(15.2 \text{ y})(3.16 \times 10^7 \text{ s/y})]^2}$$

$$= 7.35 \times 10^{36} \text{ kg.} \tag{Answer}$$

To figure out what Sagittarius A* might be, let's divide this mass by the mass of our Sun ($M_{Sun} = 1.99 \times 10^{30}$ kg) to find that

$$M = (3.7 \times 10^6)M_{Sun}.$$

Sagittarius A* has a mass of 3.7 million Suns! However, it cannot be seen. Thus, it is an extremely compact object. Such a huge mass in such a small object leads to the reasonable

Fig. 13-16 The orbit of star S2 about Sagittarius A* (Sgr A*). The elliptical orbit appears skewed because we do not see it from directly above the orbital plane. The observation dates are given in decimal fractions of a year. Uncertainties in the location of S2 are indicated by the cross bars. Uncertainty in the location of Sgr A* is indicated by the encircled cross bars.

conclusion that this object is a *supermassive* black hole. In fact, evidence is mounting that a supermassive black hole lurks at the center of most galaxies. (Movies of the stars orbiting Sagittarius A* are available on the Web; search under "black hole galactic center.")

13-8 Satellites: Orbits and Energy

As a satellite orbits Earth in an elliptical path, both its speed, which fixes its kinetic energy K, and its distance from the center of Earth, which fixes its gravitational potential energy U, fluctuate with fixed periods. However, the mechanical energy E of the satellite remains constant. (Since the satellite's mass is so much smaller than Earth's mass, we assign U and E for the Earth–satellite system to the satellite alone.)

The potential energy of the system is given by Eq. 13-21:

$$U = -\frac{GMm}{r}$$

(with $U = 0$ for infinite separation). Here r is the radius of the satellite's orbit, assumed for the time being to be circular, and M and m are the masses of Earth and the satellite, respectively.

To find the kinetic energy of a satellite in a circular orbit, we write Newton's second law ($F = ma$) as

$$\frac{GMm}{r^2} = m\frac{v^2}{r}, \tag{13-37}$$

where v^2/r is the centripetal acceleration of the satellite. Then, from Eq. 13-37, the kinetic energy is

$$K = \tfrac{1}{2}mv^2 = \frac{GMm}{2r}, \tag{13-38}$$

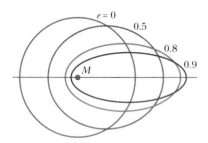

Fig. 13-17 Four orbits with different eccentricities *e* about an object of mass *M*. All four orbits have the same semimajor axis *a* and thus correspond to the same total mechanical energy *E*.

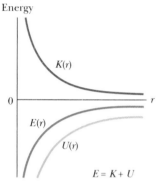

Fig. 13-18 The variation of kinetic energy *K*, potential energy *U*, and total energy *E* with radius *r* for a satellite in a circular orbit. For any value of *r*, the values of *U* and *E* are negative, the value of *K* is positive, and $E = -K$. As $r \rightarrow \infty$, all three energy curves approach a value of zero.

which shows us that for a satellite in a circular orbit,

$$K = -\frac{U}{2} \quad \text{(circular orbit).} \tag{13-39}$$

The total mechanical energy of the orbiting satellite is

$$E = K + U = \frac{GMm}{2r} - \frac{GMm}{r}$$

or

$$E = -\frac{GMm}{2r} \quad \text{(circular orbit).} \tag{13-40}$$

This tells us that for a satellite in a circular orbit, the total energy *E* is the negative of the kinetic energy *K*:

$$E = -K \quad \text{(circular orbit).} \tag{13-41}$$

For a satellite in an elliptical orbit of semimajor axis *a*, we can substitute *a* for *r* in Eq. 13-40 to find the mechanical energy:

$$E = -\frac{GMm}{2a} \quad \text{(elliptical orbit).} \tag{13-42}$$

Equation 13-42 tells us that the total energy of an orbiting satellite depends only on the semimajor axis of its orbit and not on its eccentricity *e*. For example, four orbits with the same semimajor axis are shown in Fig. 13-17; the same satellite would have the same total mechanical energy *E* in all four orbits. Figure 13-18 shows the variation of *K*, *U*, and *E* with *r* for a satellite moving in a circular orbit about a massive central body.

CHECKPOINT 6   In the figure here, a space shuttle is initially in a circular orbit of radius *r* about Earth. At point *P*, the pilot briefly fires a forward-pointing thruster to decrease the shuttle's kinetic energy *K* and mechanical energy *E*. (a) Which of the dashed elliptical orbits shown in the figure will the shuttle then take? (b) Is the orbital period *T* of the shuttle (the time to return to *P*) then greater than, less than, or the same as in the circular orbit?

Sample Problem 13-8

A playful astronaut releases a bowling ball, of mass $m = 7.20$ kg, into circular orbit about Earth at an altitude *h* of 350 km.

(a) What is the mechanical energy *E* of the ball in its orbit?

Solution: The **Key Idea** here is that we can get *E* from the orbital energy, given by Eq. 13-40 ($E = -GMm/2r$), if we first find the orbital radius *r*. That radius must be

$$r = R + h = 6370 \text{ km} + 350 \text{ km} = 6.72 \times 10^6 \text{ m,}$$

in which *R* is the radius of Earth. Then, from Eq. 13-40, the mechanical energy is

$$E = -\frac{GMm}{2r}$$

$$= -\frac{(6.67 \times 10^{-11} \text{ N} \cdot \text{m}^2/\text{kg}^2)(5.98 \times 10^{24} \text{ kg})(7.20 \text{ kg})}{(2)(6.72 \times 10^6 \text{ m})}$$

$$= -2.14 \times 10^8 \text{ J} = -214 \text{ MJ.} \tag{Answer}$$

(b) What is the mechanical energy E_0 of the ball on the launchpad at Cape Canaveral? From there to the orbit, what is the change ΔE in the ball's mechanical energy?

Solution: The **Key Idea** here is that, on the launchpad, the ball is *not* in orbit and thus Eq. 13-40 does *not* apply. Instead, we must find $E_0 = K_0 + U_0$, where K_0 is the ball's kinetic energy and U_0 is the gravitational potential energy of the ball–Earth system. To find U_0, we use Eq. 13-21 to write

$$U_0 = -\frac{GMm}{R}$$

$$= -\frac{(6.67 \times 10^{-11} \text{ N} \cdot \text{m}^2/\text{kg}^2)(5.98 \times 10^{24} \text{ kg})(7.20 \text{ kg})}{6.37 \times 10^6 \text{ m}}$$

$$= -4.51 \times 10^8 \text{ J} = -451 \text{ MJ.}$$

The kinetic energy K_0 of the ball is due to the ball's motion with Earth's rotation. You can show that K_0 is less than 1 MJ, which is negligible relative to U_0. Thus, the mechanical energy of the ball on the launchpad is

$$E_0 = K_0 + U_0 \approx 0 - 451 \text{ MJ} = -451 \text{ MJ}. \quad \text{(Answer)}$$

The *increase* in the mechanical energy of the ball from launchpad to orbit is

$$\Delta E = E - E_0 = (-214 \text{ MJ}) - (-451 \text{ MJ})$$
$$= 237 \text{ MJ}. \quad \text{(Answer)}$$

This is worth a few dollars at your utility company. Obviously the high cost of placing objects into orbit is not due to their required mechanical energy.

13-9 Einstein and Gravitation

Principle of Equivalence

Albert Einstein once said: "I was . . . in the patent office at Bern when all of a sudden a thought occurred to me: 'If a person falls freely, he will not feel his own weight.' I was startled. This simple thought made a deep impression on me. It impelled me toward a theory of gravitation."

Thus Einstein tells us how he began to form his **general theory of relativity.** The fundamental postulate of this theory about gravitation (the gravitating of objects toward each other) is called the **principle of equivalence,** which says that gravitation and acceleration are equivalent. If a physicist were locked up in a small box as in Fig. 13-19, he would not be able to tell whether the box was at rest on Earth (and subject only to Earth's gravitational force), as in Fig. 13-19*a*, or accelerating through interstellar space at 9.8 m/s² (and subject only to the force producing that acceleration), as in Fig. 13-19*b*. In both situations he would feel the same and would read the same value for his weight on a scale. Moreover, if he watched an object fall past him, the object would have the same acceleration relative to him in both situations.

Curvature of Space

We have thus far explained gravitation as due to a force between masses. Einstein showed that, instead, gravitation is due to a curvature of space that is caused by the masses. (As is discussed later in this book, space and time are entangled, so the curvature of which Einstein spoke is really a curvature of *spacetime,* the combined four dimensions of our universe.)

Picturing how space (such as vacuum) can have curvature is difficult. An analogy might help: Suppose that from orbit we watch a race in which two boats begin on Earth's equator with a separation of 20 km and head due south (Fig. 13-20*a*). To the sailors, the boats travel along flat, parallel paths. However, with time the boats draw together until, nearer the south pole, they touch. The sailors

Fig. 13-19 (*a*) A physicist in a box resting on Earth sees a cantaloupe falling with acceleration $a = 9.8$ m/s². (*b*) If he and the box accelerate in deep space at 9.8 m/s², the cantaloupe has the same acceleration relative to him. It is not possible, by doing experiments within the box, for the physicist to tell which situation he is in. For example, the platform scale on which he stands reads the same weight in both situations.

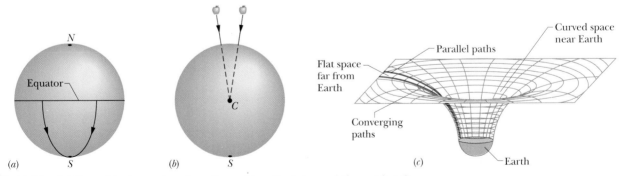

Fig. 13-20 (*a*) Two objects moving along lines of longitude toward the south pole converge because of the curvature of Earth's surface. (*b*) Two objects falling freely near Earth move along lines that converge toward the center of Earth because of the curvature of space near Earth. (*c*) Far from Earth (and other masses), space is flat and parallel paths remain parallel. Close to Earth, the parallel paths begin to converge because space is curved by Earth's mass.

in the boats can interpret this drawing together in terms of a force acting on the boats. Looking on from space, however, we can see that the boats draw together simply because of the curvature of Earth's surface. We can see this because we are viewing the race from "outside" that surface.

Figure 13-20b shows a similar race: Two horizontally separated apples are dropped from the same height above Earth. Although the apples may appear to travel along parallel paths, they actually move toward each other because they both fall toward Earth's center. We can interpret the motion of the apples in terms of the gravitational force on the apples from Earth. We can also interpret the motion in terms of a curvature of the space near Earth, a curvature due to the presence of Earth's mass. This time we cannot see the curvature because we cannot get "outside" the curved space, as we got "outside" the curved Earth in the boat example. However, we can depict the curvature with a drawing like Fig. 13-20c; there the apples would move along a surface that curves toward Earth because of Earth's mass.

When light passes near Earth, the path of the light bends slightly because of the curvature of space there, an effect called *gravitational lensing*. When light passes a more massive structure, like a galaxy or a black hole having large mass, its path can be bent more. If such a massive structure is between us and a quasar (an extremely bright, extremely distant source of light), the light from the quasar can bend around the massive structure and toward us (Fig. 13-21a). Then, because the light seems to be coming to us from a number of slightly different directions in the sky, we see the same quasar in all those different directions. In some situations, the quasars we see blend together to form a giant luminous arc, which is called an *Einstein ring* (Fig. 13-21b).

Should we attribute gravitation to the curvature of spacetime due to the presence of masses or to a force between masses? Or should we attribute it to the actions of a type of fundamental particle called a *graviton*, as conjectured in some modern physics theories? We just don't know.

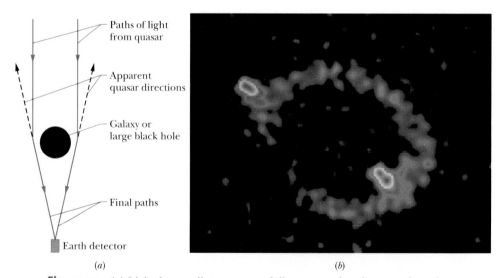

(a) *(b)*

Fig. 13-21 (a) Light from a distant quasar follows curved paths around a galaxy or a large black hole because the mass of the galaxy or black hole has curved the adjacent space. If the light is detected, it appears to have originated along the backward extensions of the final paths (dashed lines). (b) The Einstein ring known as MG1131+0456 on the computer screen of a telescope. The source of the light (actually, radio waves, which are a form of invisible light) is far behind the large, unseen galaxy that produces the ring; a portion of the source appears as the two bright spots seen along the ring.

Review & Summary

The Law of Gravitation Any particle in the universe attracts any other particle with a **gravitational force** whose magnitude is

$$F = G \frac{m_1 m_2}{r^2} \qquad \text{(Newton's law of gravitation)}, \quad (13\text{-}1)$$

where m_1 and m_2 are the masses of the particles, r is their separation, and G ($= 6.67 \times 10^{-11}$ N $\cdot$ m^2/kg^2) is the *gravitational constant*.

Gravitational Behavior of Uniform Spherical Shells Equation 13-1 holds only for particles. The gravitational force between extended bodies must generally be found by adding (integrating) the individual forces on individual particles within the bodies. However, if either of the bodies is a uniform spherical shell or a spherically symmetric solid, the net gravitational force it exerts on an *external* object may be computed as if all the mass of the shell or body were located at its center.

Superposition Gravitational forces obey the **principle of superposition;** that is, if n particles interact, the net force $\vec{F}_{1,\text{net}}$ on a particle labeled particle 1 is the sum of the forces on it from all the other particles taken one at a time:

$$\vec{F}_{1,\text{net}} = \sum_{i=2}^{n} \vec{F}_{1i}, \quad (13\text{-}5)$$

in which the sum is a vector sum of the forces $\vec{F}_{1i}$ on particle 1 from particles 2, 3, $\cdots$, n. The gravitational force $\vec{F}_1$ on a particle from an extended body is found by dividing the body into units of differential mass dm, each of which produces a differential force $d\vec{F}$ on the particle, and then integrating to find the sum of those forces:

$$\vec{F}_1 = \int d\vec{F}. \quad (13\text{-}6)$$

Gravitational Acceleration The *gravitational acceleration* a_g of a particle (of mass m) is due solely to the gravitational force acting on it. When the particle is at distance r from the center of a uniform, spherical body of mass M, the magnitude F of the gravitational force on the particle is given by Eq. 13-1. Thus, by Newton's second law,

$$F = ma_g, \quad (13\text{-}10)$$

which gives

$$a_g = \frac{GM}{r^2}. \quad (13\text{-}11)$$

Free-Fall Acceleration and Weight Because Earth's mass is not distributed uniformly, because the planet is not perfectly spherical, and because it rotates, the actual free-fall acceleration $\vec{g}$ of a particle near Earth differs slightly from the gravitational acceleration $\vec{a}_g$, and the particle's weight (equal to mg) differs from the magnitude of the gravitational force acting on the particle as computed with Eq. 13-1.

Gravitation Within a Spherical Shell A uniform shell of matter exerts no net gravitational force on a particle located inside it. This means that if a particle is located inside a uniform solid sphere at distance r from its center, the gravitational force exerted on the particle is due only to the mass M_{ins} that lies inside a sphere of radius r. This mass is given by

$$M_{\text{ins}} = \rho \frac{4\pi r^3}{3}, \quad (13\text{-}18)$$

where ρ is the density of the sphere.

Gravitational Potential Energy The gravitational potential energy $U(r)$ of a system of two particles, with masses M and m and separated by a distance r, is the negative of the work that would be done by the gravitational force of either particle acting on the other if the separation between the particles were changed from infinite (very large) to r. This energy is

$$U = -\frac{GMm}{r} \qquad \text{(gravitational potential energy)}. \quad (13\text{-}21)$$

Potential Energy of a System If a system contains more than two particles, its total gravitational potential energy U is the sum of terms representing the potential energies of all the pairs. As an example, for three particles, of masses m_1, m_2, and m^3,

$$U = -\left(\frac{Gm_1 m_2}{r_{12}} + \frac{Gm_1 m_3}{r_{13}} + \frac{Gm_2 m_3}{r_{23}} \right). \quad (13\text{-}22)$$

Escape Speed An object will escape the gravitational pull of an astronomical body of mass M and radius R (that is, it will reach an infinite distance) if the object's speed near the body's surface is at least equal to the **escape speed,** given by

$$v = \sqrt{\frac{2GM}{R}}. \quad (13\text{-}28)$$

Kepler's Laws Gravitational attraction holds the solar system together and makes possible orbiting Earth satellites, both natural and artificial. Such motions are governed by Kepler's three laws of planetary motion, all of which are direct consequences of Newton's laws of motion and gravitation:

1. **The law of orbits.** All planets move in elliptical orbits with the Sun at one focus.
2. **The law of areas.** A line joining any planet to the Sun sweeps out equal areas in equal time intervals. (This statement is equivalent to conservation of angular momentum.)
3. **The law of periods.** The square of the period T of any planet is proportional to the cube of the semimajor axis a of its orbit. For circular orbits with radius r,

$$T^2 = \left(\frac{4\pi^2}{GM} \right) r^3 \qquad \text{(law of periods)}, \quad (13\text{-}34)$$

where M is the mass of the attracting body—the Sun in the case of the solar system. For elliptical planetary orbits, the semimajor axis a is substituted for r.

Energy in Planetary Motion When a planet or satellite with mass m moves in a circular orbit with radius r, its potential energy U and kinetic energy K are given by

$$U = -\frac{GMm}{r} \quad \text{and} \quad K = \frac{GMm}{2r}. \quad (13\text{-}21, 13\text{-}38)$$

The mechanical energy $E = K + U$ is then

$$E = -\frac{GMm}{2r}. \quad (13\text{-}40)$$

For an elliptical orbit of semimajor axis a,

$$E = -\frac{GMm}{2a}. \quad (13\text{-}42)$$

Einstein's View of Gravitation Einstein pointed out that gravitation and acceleration are equivalent. This **principle of equivalence** led him to a theory of gravitation (the **general theory of relativity**) that explains gravitational effects in terms of a curvature of space.

Questions

1 In Fig. 13-22, two particles, of masses m and $2m$, are fixed in place on an axis. (a) Where on the axis can a third particle of mass $3m$ be placed (other than at infinity) so that the net gravitational force on it from the first two particles is zero: to the left of the first two particles, to their right, between them but closer to the more massive particle, or between them but closer to the less massive particle? (b) Does the answer change if the third particle has, instead, a mass of $16m$? (c) Is there a point off the axis (other than infinity) at which the net force on the third particle would be zero?

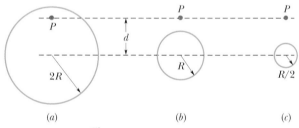

Fig. 13-22 Question 1.

2 Figure 13-23 shows three situations involving a point particle P with mass m and a spherical shell with a uniformly distributed mass M. The radii of the shells are given. Rank the situations according to the magnitude of the gravitational force on particle P due to the shell, greatest first.

Fig. 13-23 Question 2.

3 In Fig. 13-24, a central particle is surrounded by two circular rings of particles, at radii r and R, with $R > r$. All the particles have mass m. What are the magnitude and direction of the net gravitational force on the central particle due to the particles in the rings?

4 In Fig. 13-25, a central particle of mass M is surrounded by a square array of other particles, separated by either distance d or distance $d/2$ along the perimeter of the square. What are the magnitude and direction of the net gravitational force on the central particle due to the other particles?

Fig. 13-24 Question 3.

Fig. 13-25 Question 4.

5 Figure 13-26 shows three arrangements of the same identical particles, with three of them placed on a circle of radius 0.20 m and the fourth one placed at the center of the circle. (a) Rank the arrangements according to the magnitude of the net gravitational force on the central particle due to the other three particles, greatest first. (b) Rank them according to the gravitational potential energy of the four-particle system, least negative first.

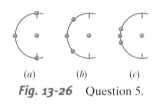

Fig. 13-26 Question 5.

6 Figure 13-27 shows three particles initially fixed in place, with B and C identical and positioned symmetrically about the y axis, at distance d from A. (a) In what direction is the net gravitational force $\vec{F}_{\text{net}}$ on A? (b) If we move C directly away from the origin, does $\vec{F}_{\text{net}}$ change in direction? If so, how and what is the limit of the change?

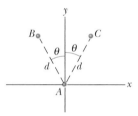

Fig. 13-27 Question 6.

7 In Fig. 13-28, three particles are fixed in place. The mass of B is greater than the mass of C. Can a fourth particle (particle D) be placed somewhere so that the net gravitational force on particle A from particles B, C, and D is zero? If so, in which quadrant should it be placed and which axis should it be near?

Fig. 13-28 Question 7.

8 Figure 13-29 gives the gravitational acceleration a_g for four planets as a function of the radial distance r from the center

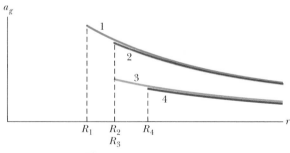

Fig. 13-29 Question 8.

of the planet, starting at the surface of the planet (at radius R_1, R_2, R_3, or R_4). Plots 1 and 2 coincide for $r \geq R_2$; plots 3 and 4 coincide for $r \geq R_4$. Rank the four planets according to (a) mass and (b) mass per unit volume, greatest first.

9 Figure 13-30 shows three uniform spherical planets that are identical in size and mass. The periods of rotation T for the planets are given, and six lettered points are indicated— three points are on the equators of the planets and three points are on the north poles. Rank the points according to the value of the free-fall acceleration g at them, greatest first.

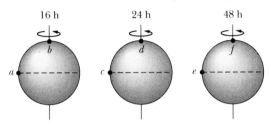

Fig. 13-30 Question 9.

10 Rank the four systems of equal-mass particles in Checkpoint 2 according to the absolute value of the gravitational potential energy of the system, greatest first.

11 Figure 13-31 shows six paths by which a rocket orbiting a moon might move from point a to point b. Rank the paths according to (a) the corresponding change in the gravitational potential energy of the rocket–moon system and (b) the net work done on the rocket by the gravitational force from the moon, greatest first.

Fig. 13-31 Question 11.

12 In Fig. 13-32, a particle of mass m (not shown) is to be moved from an infinite distance to one of the three possible locations a, b, and c. Two other particles, of masses m and $2m$, are fixed in place. Rank the three possible locations according to the work done by the net gravitational force on the moving particle due to the fixed particles, greatest first.

Fig. 13-32 Question 12.

Problems

sec. 13-2 Newton's Law of Gravitation

•1 What must the separation be between a 5.2 kg particle and a 2.4 kg particle for their gravitational attraction to have a magnitude of 2.3×10^{-12} N? SSM

•2 The Sun and Earth each exert a gravitational force on the Moon. What is the ratio F_{Sun}/F_{Earth} of these two forces? (The average Sun–Moon distance is equal to the Sun–Earth distance.)

••3 A mass M is split into two parts, m and $M - m$, which are then separated by a certain distance. What ratio m/M maximizes the magnitude of the gravitational force between the parts? ILW

sec. 13-3 Gravitation and the Principle of Superposition

•4 In Fig. 13-33, three 5.00 kg spheres are located at distances $d_1 = 0.300$ m and $d_2 = 0.400$ m. What are the (a) magnitude and (b) direction (relative to the positive direction of the x axis) of the net gravitational force on sphere B due to spheres A and C?

Fig. 13-33 Problem 4.

•5 How far from Earth must a space probe be along a line toward the Sun so that the Sun's gravitational pull on the probe balances Earth's pull? SSM WWW

•6 In Fig. 13-34, a square of edge length 20.0 cm is formed by four spheres of masses $m_1 = 5.00$ g, $m_2 = 3.00$ g, $m_3 = 1.00$ g, and $m_4 = 5.00$ g. In unit-vector notation, what is the net gravitational force from them on a central sphere with mass $m_5 = 2.50$ g?

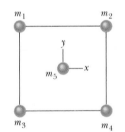

Fig. 13-34 Problem 6.

•7 *One dimension.* In Fig. 13-35, two point particles are fixed on an x axis separated by distance d. Particle A has mass m_A and particle B has mass $3.00m_A$. A third particle C, of mass $75.0m_A$, is to be placed on the x axis and near particles A and B. In terms of distance d, at what x coordinate should C be placed so that the net gravitational force on particle A from particles B and C is zero? ILW

Fig. 13-35 Problem 7.

••8 In Fig. 13-36a, particle A is fixed in place at $x = -0.20$ m on the x axis and particle B, with a mass of 1.0 kg, is

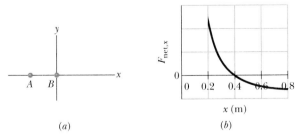

(a)   (b)

Fig. 13-36 Problem 8.

fixed in place at the origin. Particle C (not shown) can be moved along the x axis, between particle B and x = ∞. Figure 13-36b shows the x component $F_{net,x}$ of the net gravitational force on particle B due to particles A and C, as a function of position x of particle C. The plot actually extends to the right, approaching an asymptote of -4.17×10^{-10} N as $x \to \infty$. What are the masses of (a) particle A and (b) particle C?

••9 As seen in Fig. 13-37, two spheres of mass m and a third sphere of mass M form an equilateral triangle, and a fourth sphere of mass m_4 is at the center of the triangle. The net gravitational force on that central sphere from the three other spheres is zero. (a) What is M in terms of m? (b) If we double the value of m_4, what then is the magnitude of the net gravitational force on the central sphere?

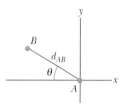

Fig. 13-37 Problem 9.

••10 Three point particles are fixed in position in an xy plane. Two of them, particle A of mass 6.00 g and particle B of mass 12.0 g, are shown in Fig. 13-38, with a separation of $d_{AB} = 0.500$ m at angle $\theta = 30°$. Particle C, with mass 8.00 g, is not shown. The net gravitational force acting on particle A due to particles B and C is 2.77×10^{-14} N at an angle of $-163.8°$ from the positive direction of the x axis. What are (a) the x coordinate and (b) the y coordinate of particle C?

Fig. 13-38 Problem 10.

••11 Figure 13-39 shows a spherical hollow inside a lead sphere of radius R = 4.00 cm; the surface of the hollow passes through the center of the sphere and "touches" the right side of the sphere. The mass of the sphere before hollowing was M = 2.95 kg. With what gravitational force does the hollowed-out lead sphere attract a small sphere of mass m = 0.431 kg that lies at a distance d = 9.00 cm from the center of the lead sphere, on the straight line connecting the centers of the spheres and of the hollow? **SSM**

Fig. 13-39 Problem 11.

••12 *Two dimensions.* In Fig. 13-40, three point particles are fixed in place in an xy plane. Particle A has mass m_A, particle B has mass $2.00m_A$, and particle C has mass $3.00m_A$. A fourth particle D, with mass $4.00m_A$, is to be placed near the other three particles. In terms of distance d, at what (a) x coordinate and (b) y coordinate should particle D be placed so that the net gravitational force on particle A from particles B, C, and D is zero?

Fig. 13-40 Problem 12.

•••13 *Three dimensions.* Three point particles are fixed in place in an xyz coordinate system. Particle A, at the origin, has mass m_A. Particle B, at xyz coordinates (2.00d, 1.00d, 2.00d), has mass $2.00m_A$, and particle C, at coordinates (−1.00d, 2.00d, −3.00d), has mass $3.00m_A$. A fourth particle D, with mass $4.00m_A$, is to be placed near the other particles. In terms of distance d, at what (a) x, (b) y, and (c) z coordinate should D be placed so that the net gravitational force on A from B, C, and D is zero?

sec. 13-4 Gravitation Near Earth's Surface

•14 In 1956, Frank Lloyd Wright proposed the construction of a mile-high building in Chicago. Suppose the building had been constructed. Ignoring Earth's rotation, find the change in your weight if you were to ride an elevator from the street level, where you weigh 600 N, to the top of the building.

•15 At what altitude above Earth's surface would the gravitational acceleration be 4.9 m/s²? **SSM**

•16 (a) What will an object weigh on the Moon's surface if it weighs 100 N on Earth's surface? (b) How many Earth radii must this same object be from the center of Earth if it is to weigh the same as it does on the Moon?

••17 One model for a certain planet has a core of radius R and mass M surrounded by an outer shell of inner radius R, outer radius 2R, and mass 4M. If $M = 4.1 \times 10^{24}$ kg and R = 6.0×10^6 m, what is the gravitational acceleration of a particle at points (a) R and (b) 3R from the center of the planet?

••18 The radius R_h and mass M_h of a black hole are related by $R_h = 2GM_h/c^2$, where c is the speed of light. Assume that the gravitational acceleration a_g of an object at a distance $r_o = 1.001R_h$ from the center of a black hole is given by Eq. 13-11 (it is, for large black holes). (a) In terms of M_h, find a_g at r_o. (b) Does a_g at r_o increase or decrease as M_h increases? (c) What is a_g at r_o for a very large black hole whose mass is 1.55×10^{12} times the solar mass of 1.99×10^{30} kg? (d) If the astronaut of Sample Problem 13-3 is at r_o with her feet toward this black hole, what is the difference in gravitational acceleration between her head and her feet? (e) Is the tendency to stretch the astronaut severe?

••19 Certain neutron stars (extremely dense stars) are believed to be rotating at about 1 rev/s. If such a star has a radius of 20 km, what must be its minimum mass so that material on its surface remains in place during the rapid rotation? **ILW**

sec. 13-5 Gravitation Inside Earth

•20 Two concentric spherical shells with uniformly distributed masses M_1 and M_2 are situated as shown in Fig. 13-41. Find the magnitude of the net gravitational force on a particle of mass m, due to the shells, when the particle is located at radial distance (a) a, (b) b, and (c) c.

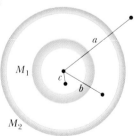

Fig. 13-41 Problem 20.

••21 A solid uniform sphere has a mass of 1.0×10^4 kg and a radius of 1.0 m. What is the magnitude of the gravitational force due to the sphere on a particle of mass m located at a distance of (a) 1.5 m and (b) 0.50 m from the center of the sphere? (c) Write a general expression for the magnitude of the gravitational force on the particle at a distance $r \leq 1.0$ m from the center of the sphere.

••22   Assume a planet is a uniform sphere of radius R that (somehow) has a narrow radial tunnel through its center (Fig. 13-8). Also assume we can position an apple anywhere along the tunnel or outside the sphere. Let F_R be the magnitude of the gravitational force on the apple when it is located at the planet's surface. How far from the surface is there a point where the magnitude of the gravitational force on the apple is $\frac{1}{2}F_R$ if we move the apple (a) away from the planet and (b) into the tunnel?

••23   Figure 13-42 shows, not to scale, a cross section through the interior of Earth. Rather than being uniform throughout, Earth is divided into three zones: an outer *crust*, a *mantle*, and an inner *core*. The dimensions of these zones and the masses contained within them are shown on the figure. Earth has a total mass of 5.98×10^{24} kg and a radius of 6370 km. Ignore rotation and assume that Earth is spherical. (a) Calculate a_g at the surface. (b) Suppose that a bore hole (the *Mohole*) is driven to the crust–mantle interface at a depth of 25.0 km; what would be the value of a_g at the bottom of the hole? (c) Suppose that Earth were a uniform sphere with the same total mass and size. What would be the value of a_g at a depth of 25.0 km? (Precise measurements of a_g are sensitive probes of the interior structure of Earth, although results can be clouded by local variations in mass distribution.) SSM WWW

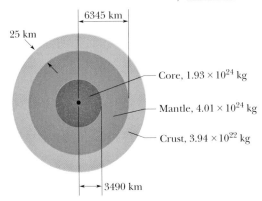

Fig. 13-42   Problem 23.

sec. 13-6   Gravitational Potential Energy

•24   (a) What is the gravitational potential energy of the two-particle system in Problem 1? If you triple the separation between the particles, how much work is done (b) by the gravitational force between the particles and (c) by you?

•25   The mean diameters of Mars and Earth are 6.9×10^3 km and 1.3×10^4 km, respectively. The mass of Mars is 0.11 times Earth's mass. (a) What is the ratio of the mean density (mass per unit volume) of Mars to that of Earth? (b) What is the value of the gravitational acceleration on Mars? (c) What is the escape speed on Mars? SSM

•26   In Problem 3, what ratio m/M gives the least gravitational potential energy for the system?

•27   What multiple of the energy needed to escape from Earth gives the energy needed to escape from (a) the Moon and (b) Jupiter?

•28   Figure 13-43 gives the potential energy function $U(r)$ of a projectile, plotted outward from the surface of a planet of radius R_s. If the projectile is launched radially outward from the surface with a mechanical energy of -2.0×10^9 J, what are (a) its kinetic energy at radius $r = 1.25R_s$ and (b) its *turning point* (see Section 8-6) in terms of R_s?

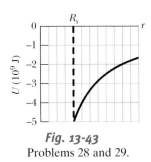

Fig. 13-43
Problems 28 and 29.

•29   Figure 13-43 gives the potential energy function $U(r)$ of a projectile, plotted outward from the surface of a planet of radius R_s. What least kinetic energy is required of a projectile launched at the surface if the projectile is to "escape" the planet?

••30   A projectile is shot directly away from Earth's surface. Neglect the rotation of Earth. What multiple of Earth's radius R_E gives the radial distance a projectile reaches if (a) its initial speed is 0.500 of the escape speed from Earth and (b) its initial kinetic energy is 0.500 of the kinetic energy required to escape Earth? (c) What is the least initial mechanical energy required at launch if the projectile is to escape Earth?

••31   The three spheres in Fig. 13-44, with masses $m_A = 80$ g, $m_B = 10$ g, and $m_C = 20$ g, have their centers on a common line, with $L = 12$ cm and $d = 4.0$ cm. You move sphere B along the line until its center-to-center separation from C is $d = 4.0$ cm. How much work is done on sphere B (a) by you and (b) by the net gravitational force on B due to spheres A and C? SSM

Fig. 13-44   Problem 31.

••32   Zero, a hypothetical planet, has a mass of 5.0×10^{23} kg, a radius of 3.0×10^6 m, and no atmosphere. A 10 kg space probe is to be launched vertically from its surface. (a) If the probe is launched with an initial energy of 5.0×10^7 J, what will be its kinetic energy when it is 4.0×10^6 m from the center of Zero? (b) If the probe is to achieve a maximum distance of 8.0×10^6 m from the center of Zero, with what initial kinetic energy must it be launched from the surface of Zero?

••33   (a) What is the escape speed on a spherical asteroid whose radius is 500 km and whose gravitational acceleration at the surface is 3.0 m/s²? (b) How far from the surface will a particle go if it leaves the asteroid's surface with a radial speed of 1000 m/s? (c) With what speed will an object hit the asteroid if it is dropped from 1000 km above the surface? SSM

••34   In deep space, sphere A of mass 20 kg is located at the origin of an x axis and sphere B of mass 10 kg is located on the axis at $x = 0.80$ m. Sphere B is released from rest while sphere A is held at the origin. (a) What is the gravitational potential energy of the two-sphere system just as B is released? (b) What is the kinetic energy of B when it has moved 0.20 m toward A?

••35 Two neutron stars are separated by a distance of 1.0×10^{10} m. They each have a mass of 1.0×10^{30} kg and a radius of 1.0×10^5 m. They are initially at rest with respect to each other. As measured from that rest frame, how fast are they moving when (a) their separation has decreased to one-half its initial value and (b) they are about to collide? SSM ILW

••36 Figure 13-45a shows a particle A that can be moved along a y axis from an infinite distance to the origin. That origin lies at the midpoint between particles B and C, which have identical masses, and the y axis is a perpendicular bisector between them. Distance D is 0.3057 m. Figure 13-45b shows the potential energy U of the three-particle system as a function of the position of particle A along the y axis. The curve actually extends rightward and approaches an asymptote of -2.7×10^{-11} J as $y \to \infty$. What are the masses of (a) particles B and C and (b) particle A?

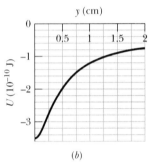

Fig. 13-45 Problem 36.

••37 Figure 13-46 shows four particles, each of mass 20.0 g, that form a square with an edge length of $d = 0.600$ m. If d is reduced to 0.200 m, what is the change in the gravitational potential energy of the four-particle system?

Fig. 13-46 Problem 37.

sec. 13-7 **Planets and Satellites: Kepler's Laws**

•38 The first known collision between space debris and a functioning satellite occurred in 1996: At an altitude of 700 km, a year-old French spy satellite was hit by a piece of an Ariane rocket that had been in orbit for 10 years. A stabilizing boom on the satellite was demolished, and the satellite was sent spinning out of control. Just before the collision and in kilometers per hour, what was the speed of the rocket piece relative to the satellite if both were in circular orbits and the collision was (a) head-on and (b) along perpendicular paths?

•39 The Martian satellite Phobos travels in an approximately circular orbit of radius 9.4×10^6 m with a period of 7 h 39 min. Calculate the mass of Mars from this information.

•40 The mean distance of Mars from the Sun is 1.52 times that of Earth from the Sun. From Kepler's law of periods, calculate the number of years required for Mars to make one revolution around the Sun; compare your answer with the value given in Appendix C.

•41 The Sun, which is 2.2×10^{20} m from the center of the Milky Way galaxy, revolves around that center once every

2.5×10^8 years. Assuming each star in the Galaxy has a mass equal to the Sun's mass of 2.0×10^{30} kg, the stars are distributed uniformly in a sphere about the galactic center, and the Sun is at the edge of that sphere, estimate the number of stars in the Galaxy. SSM WWW

•42 A satellite is put in a circular orbit about Earth with a radius equal to one-half the radius of the Moon's orbit. What is its period of revolution in lunar months? (A lunar month is the period of revolution of the Moon.)

•43 (a) What linear speed must an Earth satellite have to be in a circular orbit at an altitude of 160 km above Earth's surface? (b) What is the period of revolution?

•44 The Sun's center is at one focus of Earth's orbit. How far from this focus is the other focus, (a) in meters and (b) in terms of the solar radius, 6.96×10^8 m? The eccentricity of Earth's orbit is 0.0167, and the semimajor axis is 1.50×10^{11} m.

•45 A satellite, moving in an elliptical orbit, is 360 km above Earth's surface at its farthest point and 180 km above at its closest point. Calculate (a) the semimajor axis and (b) the eccentricity of the orbit. SSM

•46 An orbiting satellite stays over a certain spot on the equator of (rotating) Earth. What is the altitude of the orbit (called a *geosynchronous orbit*)?

•47 A comet that was seen in April 574 by Chinese astronomers on a day known by them as the Woo Woo day was spotted again in May 1994. Assume the time between observations is the period of the Woo Woo day comet and take its eccentricity as 0.11. What are (a) the semimajor axis of the comet's orbit and (b) its greatest distance from the Sun in terms of the mean orbital radius R_P of Pluto?

•48 In 1993 the spacecraft *Galileo* sent home an image (Fig. 13-47) of asteroid 243 Ida and a tiny orbiting moon (now known as Dactyl), the first confirmed example of an asteroid–moon system. In the image, the moon, which is 1.5 km wide, is 100 km from the center of the asteroid, which is 55 km long. The shape of the moon's orbit is not well known; assume it is circular with a period of 27 h. (a) What is the mass of the asteroid? (b) The volume of the asteroid, measured from the *Galileo* images, is 14 100 km³. What is the density (mass per unit volume) of the asteroid?

Fig. 13-47 Problem 48. A tiny moon (at right) orbits asteroid 243 Ida.

••49 In 1610, Galileo used his telescope to discover four prominent moons around Jupiter. Their mean orbital radii a and periods T are as follows:

Name	a (10^8 m)	T (days)
Io	4.22	1.77
Europa	6.71	3.55
Ganymede	10.7	7.16
Callisto	18.8	16.7

(a) Plot log a (y axis) against log T (x axis) and show that you get a straight line. (b) Measure the slope of the line and compare it with the value that you expect from Kepler's third law. (c) Find the mass of Jupiter from the intercept of this line with the y axis.

••**50** A 20 kg satellite has a circular orbit with a period of 2.4 h and a radius of 8.0×10^6 m around a planet of unknown mass. If the magnitude of the gravitational acceleration on the surface of the planet is 8.0 m/s², what is the radius of the planet?

••**51** In a certain binary-star system, each star has the same mass as our Sun, and they revolve about their center of mass. The distance between them is the same as the distance between Earth and the Sun. What is their period of revolution in years? **ILW**

•••**52** The presence of an unseen planet orbiting a distant star can sometimes be inferred from the motion of the star as we see it. As the star and planet orbit the center of mass of the star–planet system, the star moves toward and away from us with what is called the *line of sight velocity*, a motion that can be detected. Figure 13-48 shows a graph of the line of sight velocity versus time for the star 14 Herculis. The star's mass is believed to be 0.90 of the mass of our Sun. Assume that only one planet orbits the star and that our view is along the plane of the orbit. Then approximate (a) the planet's mass in terms of Jupiter's mass m_J and (b) the planet's orbital radius in terms of Earth's orbital radius r_E.

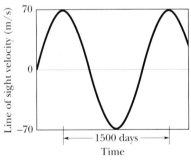

Fig. 13-48 Problem 52.

•••**53** Three identical stars of mass M form an equilateral triangle that rotates around the triangle's center as the stars move in a common circle about that center. The triangle has edge length L. What is the speed of the stars? **SSM**

sec. 13-8 Satellites: Orbits and Energy

•**54** A satellite orbits a planet of unknown mass in a circle of radius 2.0×10^7 m. The magnitude of the gravitational force on the satellite from the planet is $F = 80$ N. (a) What is the kinetic energy of the satellite in this orbit? (b) What would F be if the orbit radius were increased to 3.0×10^7 m?

•**55** An asteroid, whose mass is 2.0×10^{-4} times the mass of Earth, revolves in a circular orbit around the Sun at a distance that is twice Earth's distance from the Sun. (a) Calculate the period of revolution of the asteroid in years. (b) What is the ratio of the kinetic energy of the asteroid to the kinetic energy of Earth? **SSM WWW**

•**56** Two Earth satellites, A and B, each of mass m, are to be launched into circular orbits about Earth's center. Satellite A is to orbit at an altitude of 6370 km. Satellite B is to orbit at an altitude of 19 110 km. The radius of Earth R_E is 6370 km. (a) What is the ratio of the potential energy of satellite B to that of satellite A, in orbit? (b) What is the ratio of the kinetic energy of satellite B to that of satellite A, in orbit? (c) Which satellite has the greater total energy if each has a mass of 14.6 kg? (d) By how much?

•**57** (a) At what height above Earth's surface is the energy required to lift a satellite to that height equal to the kinetic energy required for the satellite to be in orbit at that height? (b) For greater heights, which is greater, the energy for lifting or the kinetic energy for orbiting?

•**58** In Fig. 13-49, two satellites, A and B, both of mass $m = 125$ kg, move in the same circular orbit of radius $r = 7.87 \times 10^6$ m around Earth but in opposite senses of rotation and therefore on a collision course. (a) Find the total mechanical energy $E_A + E_B$ of the *two satellites + Earth* system before the collision. (b) If the collision is completely inelastic so that the wreckage remains as one piece of tangled material (mass $= 2m$), find the total mechanical energy immediately after the collision. (c) Just after the collision, is the wreckage falling directly toward Earth's center or orbiting around Earth?

Fig. 13-49 Problem 58.

••**59** A satellite is in a circular Earth orbit of radius r. The area A enclosed by the orbit depends on r^2 because $A = \pi r^2$. Determine how the following properties of the satellite depend on r: (a) period, (b) kinetic energy, (c) angular momentum, and (d) speed.

••**60** One way to attack a satellite in Earth orbit is to launch a swarm of pellets in the same orbit as the satellite but in the opposite direction. Suppose a satellite in a circular orbit 500 km above Earth's surface collides with a pellet having mass 4.0 g. (a) What is the kinetic energy of the pellet in the reference frame of the satellite just before the collision? (b) What is the ratio of this kinetic energy to the kinetic energy of a 4.0 g bullet from a modern army rifle with a muzzle speed of 950 m/s?

•••**61** What are (a) the speed and (b) the period of a 220 kg satellite in an approximately circular orbit 640 km above the surface of Earth? Suppose the satellite loses mechanical energy at the average rate of 1.4×10^5 J per orbital revolution. Adopting the reasonable approximation that the satellite's orbit becomes a "circle of slowly diminishing radius," determine the satellite's (c) altitude, (d) speed, and (e) period at the end of its 1500th revolution. (f) What is the magnitude of the average retarding force on the satellite? Is angular momentum around Earth's center conserved for (g) the satellite and (h) the satellite–Earth system (assuming that system is isolated)? **SSM**

sec. 13-9 Einstein and Gravitation

•62 In Fig. 13-19b, the scale on which the 60 kg physicist stands reads 220 N. How long will the cantaloupe take to reach the floor if the physicist drops it (from rest relative to himself) at a height of 2.1 m above the floor?

Additional Problems

63 Figure 13-50 is a graph of the kinetic energy K of an asteroid versus its distance r from Earth's center, as the asteroid falls directly in toward that center. (a) What is the (approximate) mass of the asteroid? (b) What is its speed at $r = 1.945 \times 10^7$ m?

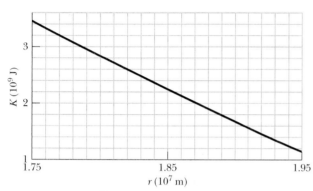

Fig. 13-50 Problem 63.

64 The radius R_h of a black hole is the radius of a mathematical sphere, called the event horizon, that is centered on the black hole. Information from events inside the event horizon cannot reach the outside world. According to Einstein's general theory of relativity, $R_h = 2GM/c^2$, where M is the mass of the black hole and c is the speed of light.

Suppose that you wish to study a black hole near it, at a radial distance of $50R_h$. However, you do not want the difference in gravitational acceleration between your feet and your head to exceed 10 m/s² when you are feet down (or head down) toward the black hole. (a) As a multiple of our Sun's mass M_S, approximately what is the limit to the mass of the black hole you can tolerate at the given radial distance? (You need to estimate your height.) (b) Is the limit an upper limit (you can tolerate smaller masses) or a lower limit (you can tolerate larger masses)?

65 Four identical 1.5 kg particles are placed at the corners of a square with sides equal to 20 cm. What is the magnitude of the net gravitational force on any one of the particles due to the others?

66 (a) If the legendary apple of Newton could be released from rest at a height of 2 m from the surface of a neutron star with a mass 1.5 times that of our Sun and a radius of 20 km, what would be the apple's speed when it reached the surface of the star? (b) If the apple could rest on the surface of the star, what would be the approximate difference between the gravitational acceleration at the top and at the bottom of the apple? (Choose a reasonable size for an apple; the answer indicates that an apple would never survive near a neutron star.)

67 Sphere A with mass 80 kg is located at the origin of an xy coordinate system; sphere B with mass 60 kg is located at

coordinates (0.25 m, 0); sphere C with mass 0.20 kg is located in the first quadrant 0.20 m from A and 0.15 m from B. In unit-vector notation, what is the gravitational force on C due to A and B?

68 A satellite is in elliptical orbit with a period of 8.00×10^4 s about a planet of mass 7.00×10^{24} kg. At aphelion, at radius 4.5×10^7 m, the satellite's angular speed is 7.158×10^{-5} rad/s. What is its angular speed at perihelion?

69 In a shuttle craft of mass $m = 3000$ kg, Captain Janeway orbits a planet of mass $M = 9.50 \times 10^{25}$ kg, in a circular orbit of radius $r = 4.20 \times 10^7$ m. What are (a) the period of the orbit and (b) the speed of the shuttle craft? Janeway briefly fires a forward-pointing thruster, reducing her speed by 2.00%. Just then, what are (c) the speed, (d) the kinetic energy, (e) the gravitational potential energy, and (f) the mechanical energy of the shuttle craft? (g) What is the semimajor axis of the elliptical orbit now taken by the craft? (h) What is the difference between the period of the original circular orbit and that of the new elliptical orbit? (i) Which orbit has the smaller period?

70 The mysterious visitor that appears in the enchanting story *The Little Prince* was said to come from a planet that "was scarcely any larger than a house!" Assume that the mass per unit volume of the planet is about that of Earth and that the planet does not appreciably spin. Approximate (a) the free-fall acceleration on the planet's surface and (b) the escape speed from the planet.

71 A typical neutron star may have a mass equal to that of the Sun but a radius of only 10 km. (a) What is the gravitational acceleration at the surface of such a star? (b) How fast would an object be moving if it fell from rest through a distance of 1.0 m on such a star? (Assume the star does not rotate.)

72 We watch two identical astronomical bodies A and B, each of mass m, fall toward each other from rest because of the gravitational force on each from the other. Their initial center-to-center separation is R_i. Assume that we are in an inertial reference frame that is stationary with respect to the center of mass of this two-body system. Use the principle of conservation of mechanical energy ($K_f + U_f = K_i + U_i$) to find the following when the center-to-center separation is $0.5R_i$: (a) the total kinetic energy of the system, (b) the kinetic energy of each body, (c) the speed of each body relative to us, and (d) the speed of body B relative to body A.

Next assume that we are in a reference frame attached to body A (we ride on the body). Now we see body B fall from rest toward us. From this reference frame, again use $K_f + U_f = K_i + U_i$ to find the following when the center-to-center separation is $0.5R_i$: (e) the kinetic energy of body B and (f) the speed of body B relative to body A. (g) Why are the answers to (d) and (f) different? Which answer is correct?

73 Four uniform spheres, with masses $m_A = 40$ kg, $m_B = 35$ kg, $m_C = 200$ kg, and $m_D = 50$ kg, have (x, y) coordinates of (0, 50 cm), (0, 0), (−80 cm, 0), and (40 cm, 0), respectively. In unit-vector notation, what is the net gravitational force on sphere B due to the other spheres?

74 (a) In Problem 73, remove sphere A and calculate the gravitational potential energy of the remaining three-particle system. (b) If A is then put back in place, is the potential

energy of the four-particle system more or less than that of the system in (a)? (c) In (a), is the work done by you to remove *A* positive or negative? (d) In (b), is the work done by you to replace *A* positive or negative?

75 A very early, simple satellite consisted of an inflated spherical aluminum balloon 30 m in diameter and of mass 20 kg. Suppose a meteor having a mass of 7.0 kg passes within 3.0 m of the surface of the satellite. What is the magnitude of the gravitational force on the meteor from the satellite at the closest approach?

76 A uniform solid sphere of radius *R* produces a gravitational acceleration of a_g on its surface. At what distance from the sphere's center are there points (a) inside and (b) outside the sphere where the gravitational acceleration is $a_g/3$?

77 A projectile is fired vertically from Earth's surface with an initial speed of 10 km/s. Neglecting air drag, how far above the surface of Earth will it go?

78 A 50 kg satellite circles planet Cruton every 6.0 h. The magnitude of the gravitational force exerted on the satellite by Cruton is 80 N. (a) What is the radius of the orbit? (b) What is the kinetic energy of the satellite? (c) What is the mass of planet Cruton?

79 In a double-star system, two stars of mass 3.0×10^{30} kg each rotate about the system's center of mass at radius 1.0×10^{11} m. (a) What is their common angular speed? (b) If a meteoroid passes through the system's center of mass perpendicular to their orbital plane, what minimum speed must it have at the center of mass if it is to escape to "infinity" from the two-star system?

80 An object lying on Earth's equator is accelerated (a) toward the center of Earth because Earth rotates, (b) toward the Sun because Earth revolves around the Sun in an almost circular orbit, and (c) toward the center of our galaxy because the Sun moves around the galactic center. For the latter, the period is 2.5×10^8 y and the radius is 2.2×10^{20} m. Calculate these three accelerations as multiples of $g = 9.8$ m/s².

81 The masses and coordinates of three spheres are as follows: 20 kg, *x* = 0.50 m, *y* = 1.0 m; 40 kg, *x* = −1.0 m, *y* = −1.0 m; 60 kg, *x* = 0 m, *y* = −0.50 m. What is the magnitude of the gravitational force on a 20 kg sphere located at the origin due to these three spheres?

82 In his 1865 science fiction novel *From the Earth to the Moon*, Jules Verne described how three astronauts are shot to the Moon by means of a huge gun. According to Verne, the aluminum capsule containing the astronauts is accelerated by ignition of nitrocellulose to a speed of 11 km/s along the gun barrel's length of 220 m. (a) In *g* units, what is the average acceleration of the capsule and astronauts in the gun barrel? (b) Is that acceleration tolerable or deadly to the astronauts?

A modern version of such gun-launched spacecraft (although without passengers) has been proposed. In this modern version, called the SHARP (Super High Altitude Research Project) gun, ignition of methane and air shoves a piston down the gun's tube, compressing hydrogen gas that then launches a rocket. During this launch, the rocket moves 3.5 km and reaches a speed of 7.0 km/s. Once launched, the rocket can be fired to gain additional speed. (c) In *g* units, what would be the average acceleration of the rocket within

the launcher? (d) How much additional speed is needed (via the rocket engine) if the rocket is to orbit Earth at an altitude of 700 km?

83 A 150.0 kg rocket moving radially outward from Earth has a speed of 3.70 km/s when its engine shuts off 200 km above Earth's surface. (a) Assuming negligible air drag, find the rocket's kinetic energy when the rocket is 1000 km above Earth's surface. (b) What maximum height above the surface is reached by the rocket?

84 Planet Roton, with a mass of 7.0×10^{24} kg and a radius of 1600 km, gravitationally attracts a meteorite that is initially at rest relative to the planet, at a distance great enough to take as infinite. The meteorite falls toward the planet. Assuming the planet is airless, find the speed of the meteorite when it reaches the planet's surface.

85 The orbit of Earth around the Sun is *almost* circular: The closest and farthest distances are 1.47×10^8 km and 1.52×10^8 km respectively. Determine the corresponding variations in (a) total energy, (b) gravitational potential energy, (c) kinetic energy, and (d) orbital speed. (*Hint:* Use conservation of energy and conservation of angular momentum.)

86 In your calculator, make a list of the periods *T* for the planets in Table 13-3 and a separate list of the semimajor axes *a*. Multiply the list of *T* values by an appropriate factor so that the unit of *T* is seconds. (a) Store values of T^2 and a^3 in two new lists. Have the calculator do a linear regression fit of T^2 versus a^3. From the parameters of the fit and using the known value of *G*, determine the mass of the Sun. (b) Store values of log *T* in one list and log *a* in another list. Have the calculator plot log *T* versus log *a* and then do a linear regression fit of the plot. From the parameters of this fit and using the known value of *G*, again determine the mass of the Sun.

87 A planet requires 300 (Earth) days to complete its circular orbit around its sun, which has a mass of 6.0×10^{30} kg. What are the planet's (a) orbital radius and (b) orbital speed?

88 Two 20 kg spheres are fixed in place on a *y* axis, one at *y* = 0.40 m and the other at *y* = −0.40 m. A 10 kg ball is then released from rest at a point on the *x* axis that is at a great distance (effectively infinite) from the spheres. If the only forces acting on the ball are the gravitational forces from the spheres, then when the ball reaches the (*x*, *y*) point (0.30 m, 0), what are (a) its kinetic energy and (b) the net force on it from the spheres, in unit-vector notation?

89 With what speed would mail pass through the center of Earth if falling in the tunnel of Sample Problem 13-4?

90 A satellite of mass 125 kg is in a circular orbit of radius 7.00×10^6 m around a planet. If the period is 8050 s, what is the mechanical energy of the satellite?

91 A particle of comet dust with mass *m* is a distance *R* from Earth's center and a distance *r* from the Moon's center. If Earth's mass is M_E and the Moon's mass is M_m, what is the sum of the gravitational potential energy of the particle–Earth system and the gravitational potential energy of the particle–Moon system?

92 A spaceship is on a straight-line path between Earth and the Moon. At what distance from Earth is the net gravitational force on the spaceship zero?

93 What is the percentage change in the acceleration of Earth toward the Sun when the alignment of Earth, Sun, and Moon changes from an eclipse of the Sun (with the Moon between Earth and Sun) to an eclipse of the Moon (Earth between Moon and Sun)?

94 The most valued orbit for a communications satellite is a *geosynchronous orbit,* in which the satellite remains above a certain point on Earth's equator as it orbits. Such a "geo satellite" would always be in the same position in the sky as seen from the transmitting or receiving equipment of, say, the BBC in London. Unfortunately, room along a geosynchronous orbit is limited because the satellites in such an orbit cannot be too close to one another or the signal quality is degraded. The least separation along the 360° of the circular orbit is 3.00°. What are (a) the maximum number of geo satellites the orbit should contain and (b) their orbital radius? (c) What then is the minimum separation d between any two geo satellites along the orbital path?

A magnetic storm can disrupt a satellite as well as move it, either toward or away from Earth. Suppose a magnetic storm turns off geo satellite *Hot Bird 2,* and when ground-based engineers turn it back on, it is east of its assigned orbital spot. (d) Is the period of *Hot Bird 2* now greater than or smaller than that of the other geo satellites? (e) Did the storm move the satellite toward or away from Earth?

95 An object of mass m is initially held in place at radial distance $r = 3R_E$ from the center of Earth, where R_E is the radius of Earth. Let M_E be the mass of Earth. A force is applied to the object to move it to a radial distance $r = 4R_E$, where it again is held in place. Calculate the work done by the applied force during the move by integrating the force magnitude.

96 Several planets (Jupiter, Saturn, Uranus) are encircled by rings, perhaps composed of material that failed to form a satellite. In addition, many galaxies contain ring-like structures. Consider a homogeneous thin ring of mass M and outer radius R (Fig. 13-51). (a)

Fig. 13-51 Problem 96.

What gravitational attraction does it exert on a particle of mass m located on the ring's central axis a distance x from the ring center? (b) Suppose the particle falls from rest as a result of the attraction of the ring of matter. What is the speed with which it passes through the center of the ring?

97 Show that, at the bottom of a vertical mine shaft dug to depth D, the gravitational acceleration a_g is

$$a_g = a_{gs}\left(1 - \frac{D}{R}\right),$$

where a_{gs} is the surface value. Assume that Earth is a uniform sphere of radius R.

98 The fastest possible rate of rotation of a planet is that for which the gravitational force on material at the equator just barely provides the centripetal force needed for the rotation. (Why?) (a) Show that the corresponding shortest period of rotation is

$$T = \sqrt{\frac{3\pi}{G\rho}},$$

where ρ is the uniform density (mass per unit volume) of the spherical planet. (b) Calculate the rotation period assuming a density of 3.0 g/cm³, typical of many planets, satellites, and asteroids. No astronomical object has ever been found to be spinning with a period shorter than that determined by this analysis.

99 The gravitational force between two particles with masses m and M, initially at rest at great separation, pulls them together. Show that at any instant the speed of either particle relative to the other is $\sqrt{2G(M + m)/d}$, where d is their separation at that instant. (*Hint:* Use the laws of conservation of energy and conservation of linear momentum.)

100 A projectile is launched directly away from the surface of a planet of mass M and radius R; the launch speed is $(GM/R)^{1/2}$. Use the principle of conservation of energy to determine the maximum distance from the center of the planet achieved by the projectile. Express your result in terms of R.

101 Show how, guided by Kepler's law of periods, Newton could deduce that the force holding the Moon in its orbit, assumed circular, depends on the inverse square of the Moon's distance from the center of Earth.

102 A sphere of matter, of mass M and radius a, has a concentric cavity of radius b, as shown in cross section in Fig. 13-52. (a) Sketch a curve of the magnitude of the gravitational force F from the sphere on a particle of mass m located a distance r from the center of the sphere, as a function of r in the range $0 \le r \le \infty$. Consider

Fig. 13-52 Problem 102.

$r = 0, b, a,$ and ∞ in particular. (b) Sketch the corresponding curve for the potential energy $U(r)$ of the system.

103 A certain triple-star system consists of two stars, each of mass m, revolving in the same circular orbit of radius r around a central star of mass M (Fig. 13-53). The two orbiting stars are always at opposite ends of a diameter of the orbit. Derive an expression for the period of revolution of the stars.

Fig. 13-53 Problem 103.

104 Some people believe that the positions of the planets at the time of birth influence the newborn. Others deride this belief and claim that the gravitational force exerted on a baby by the obstetrician is greater than the force exerted by the planets. To check this claim, calculate the magnitude of the gravitational force exerted on a 3 kg baby (a) by a 70 kg obstetrician who is 1 m away and roughly approximated as a point mass, (b) by the massive planet Jupiter ($m = 2 \times 10^{27}$ kg) at its closest approach to Earth ($= 6 \times 10^{11}$ m), and (c) by Jupiter at its greatest distance from Earth ($= 9 \times 10^{11}$ m). (d) Is the claim correct?

On-Line Simulation Problems

The website http://www.wiley.com/college/halliday has simulation problems about this chapter.

Fluids

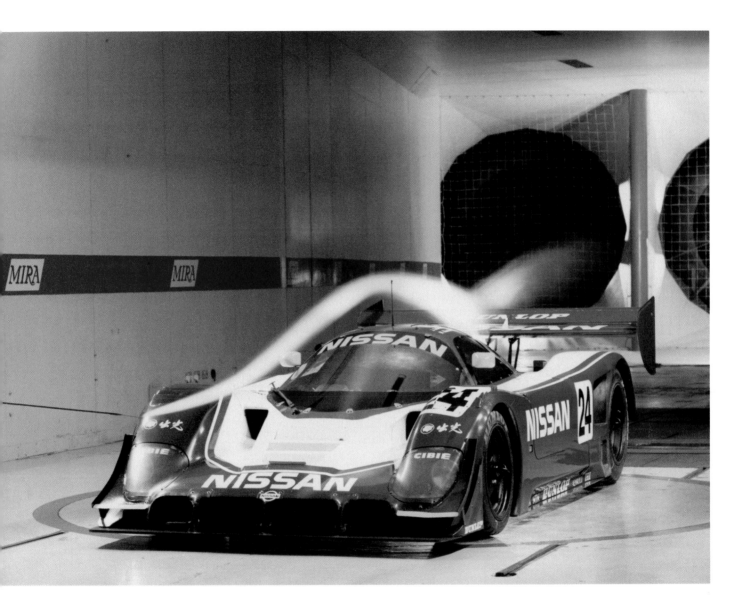

Many types of race cars depend on negative lift (also called downforce) to push them down against the track surface so they can take turns quickly without sliding out into the track wall. Part of the negative lift on a car is due to one or more wings on the car, and in the wind tunnel photograph here, smoke released at the front of the car reveals how the air flows over the car and past the wing at the rear. The rest of the negative lift is called ground effect and has to do with the airflow beneath the car. Ground effect is dependable when a car is the only one taking a turn as in, say, a time trial, but a skilled driver knows that ground effect can disappear during a race.

What causes ground effect and why can it disappear?

The answers are in this chapter.

14-1   *What Is Physics?*

The physics of fluids is the basis of hydraulic engineering, a branch of engineering that is applied in a great many fields. A nuclear engineer might study the fluid flow in the hydraulic system of an aging nuclear reactor, while a medical engineer might study the blood flow in the arteries of an aging patient. An environmental engineer might be concerned about the drainage from waste sites or the efficient irrigation of farmlands. A naval engineer might be concerned with the dangers faced by a deep-sea diver or with the possibility of a crew escaping from a downed submarine. An aeronautical engineer might design the hydraulic systems controlling the wing flaps that allow a jet airplane to land. Hydraulic engineering is also applied in many Broadway and Las Vegas shows, where huge sets are quickly put up and brought down by hydraulic systems.

Before we can study any such application of the physics of fluids, we must first answer the question "What is a fluid?"

14-2   *What Is a Fluid?*

A **fluid,** in contrast to a solid, is a substance that can flow. Fluids conform to the boundaries of any container in which we put them. They do so because a fluid cannot sustain a force that is tangential to its surface. (In the more formal language of Section 12-7, a fluid is a substance that flows because it cannot withstand a shearing stress. It can, however, exert a force in the direction perpendicular to its surface.) Some materials, such as pitch, take a long time to conform to the boundaries of a container, but they do so eventually; thus, we classify even those materials as fluids.

You may wonder why we lump liquids and gases together and call them fluids. After all (you may say), liquid water is as different from steam as it is from ice. Actually, it is not. Ice, like other crystalline solids, has its constituent atoms organized in a fairly rigid three-dimensional array called a crystalline lattice. In neither steam nor liquid water, however, is there any such orderly long-range arrangement.

14-3   *Density and Pressure*

When we discuss rigid bodies, we are concerned with particular lumps of matter, such as wooden blocks, baseballs, or metal rods. Physical quantities that we find useful, and in whose terms we express Newton's laws, are *mass* and *force*. We might speak, for example, of a 3.6 kg block acted on by a 25 N force.

With fluids, we are more interested in the extended substance and in properties that can vary from point to point in that substance. It is more useful to speak of **density** and **pressure** than of mass and force.

Density

To find the density ρ of a fluid at any point, we isolate a small volume element ΔV around that point and measure the mass Δm of the fluid contained within that element. The **density** is then

$$\rho = \frac{\Delta m}{\Delta V}. \tag{14-1}$$

In theory, the density at any point in a fluid is the limit of this ratio as the volume element ΔV at that point is made smaller and smaller. In practice, we assume that a fluid sample is large relative to atomic dimensions and thus is "smooth"

TABLE 14-1
Some Densities

Material or Object	Density (kg/m³)	Material or Object	Density (kg/m³)
Interstellar space	10^{-20}	Iron	7.9×10^3
Best laboratory vacuum	10^{-17}	Mercury (the metal, not the planet)	13.6×10^3
Air: 20°C and 1 atm pressure	1.21	Earth: average	5.5×10^3
20°C and 50 atm	60.5	core	9.5×10^3
Styrofoam	1×10^2	crust	2.8×10^3
Ice	0.917×10^3	Sun: average	1.4×10^3
Water: 20°C and 1 atm	0.998×10^3	core	1.6×10^5
20°C and 50 atm	1.000×10^3	White dwarf star (core)	10^{10}
Seawater: 20°C and 1 atm	1.024×10^3	Uranium nucleus	3×10^{17}
Whole blood	1.060×10^3	Neutron star (core)	10^{18}

(with uniform density), rather than "lumpy" with atoms. This assumption allows us to write Eq. 14-1 as

$$\rho = \frac{m}{V} \quad \text{(uniform density)}, \tag{14-2}$$

where m and V are the mass and volume of the sample.

Density is a scalar property; its SI unit is the kilogram per cubic meter. Table 14-1 shows the densities of some substances and the average densities of some objects. Note that the density of a gas (see Air in the table) varies considerably with pressure, but the density of a liquid (see Water) does not; that is, gases are readily *compressible* but liquids are not.

Pressure

Let a small pressure-sensing device be suspended inside a fluid-filled vessel, as in Fig. 14-1a. The sensor (Fig. 14-1b) consists of a piston of surface area ΔA riding in a close-fitting cylinder and resting against a spring. A readout arrangement allows us to record the amount by which the (calibrated) spring is compressed by the surrounding fluid, thus indicating the magnitude ΔF of the force that acts normal to the piston. We define the **pressure** on the piston from the fluid as

$$p = \frac{\Delta F}{\Delta A}. \tag{14-3}$$

In theory, the pressure at any point in the fluid is the limit of this ratio as the surface area ΔA of the piston, centered on that point, is made smaller and smaller. However, if the force is uniform over a flat area A, we can write Eq. 14-3 as

$$p = \frac{F}{A} \quad \text{(pressure of uniform force on flat area)}, \tag{14-4}$$

where F is the magnitude of the normal force on area A. (When we say a force is uniform over an area, we mean that the force is evenly distributed over every point of the area.)

We find by experiment that at a given point in a fluid at rest, the pressure p defined by Eq. 14-4 has the same value no matter how the pressure sensor is oriented. Pressure is a scalar, having no directional properties. It is true that the force acting on the piston of our pressure sensor is a vector quantity, but Eq. 14-4 involves only the *magnitude* of that force, a scalar quantity.

The SI unit of pressure is the newton per square meter, which is given a special name, the **pascal** (Pa). In metric countries, tire pressure gauges are calibrated in kilopascals. The pascal is related to some other common (non-SI) pres-

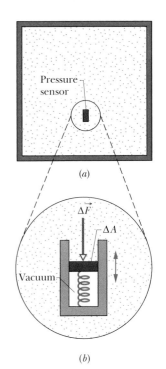

Fig. 14-1 (a) A fluid-filled vessel containing a small pressure sensor, shown in (b). The pressure is measured by the relative position of the movable piston in the sensor.

TABLE 14-2
Some Pressures

	Pressure (Pa)		Pressure (Pa)
Center of the Sun	2×10^{16}	Automobile tire[a]	2×10^5
Center of Earth	4×10^{11}	Atmosphere at sea level	1.0×10^5
Highest sustained laboratory pressure	1.5×10^{10}	Normal blood systolic pressure[a,b]	1.6×10^4
Deepest ocean trench (bottom)	1.1×10^8	Best laboratory vacuum	10^{-12}
Spike heels on a dance floor	10^6		

[a]Pressure in excess of atmospheric pressure.
[b]Equivalent to 120 torr on the physician's pressure gauge.

sure units as follows:

$$1 \text{ atm} = 1.01 \times 10^5 \text{ Pa} = 760 \text{ torr} = 14.7 \text{ lb/in.}^2.$$

The *atmosphere* (atm) is, as the name suggests, the approximate average pressure of the atmosphere at sea level. The *torr* (named for Evangelista Torricelli, who invented the mercury barometer in 1674) was formerly called the *millimeter of mercury* (mm Hg). The pound per square inch is often abbreviated psi. Table 14-2 shows some pressures.

Sample Problem 14-1

A living room has floor dimensions of 3.5 m and 4.2 m and a height of 2.4 m.

(a) What does the air in the room weigh when the air pressure is 1.0 atm?

Solution: The **Key Ideas** here are these: (1) The air's weight is equal to mg, where m is its mass. (2) Mass m is related to the air density ρ and the air volume V by Eq. 14-2 ($\rho = m/V$). Putting these two ideas together and taking the density of air at 1.0 atm from Table 14-1, we find

$$mg = (\rho V)g$$
$$= (1.21 \text{ kg/m}^3)(3.5 \text{ m} \times 4.2 \text{ m} \times 2.4 \text{ m})(9.8 \text{ m/s}^2)$$
$$= 418 \text{ N} \approx 420 \text{ N}. \qquad \text{(Answer)}$$

This is the weight of about 110 cans of Pepsi.

(b) What is the magnitude of the atmosphere's force on the floor of the room?

Solution: The **Key Idea** here is that the atmosphere pushes down on the floor with a force of magnitude F that is uniform over the floor. Thus, the atmosphere produces a pressure that is related to F and the flat area A of the floor by Eq. 14-4 ($p = F/A$), which gives us

$$F = pA = (1.0 \text{ atm})\left(\frac{1.01 \times 10^5 \text{ N/m}^2}{1.0 \text{ atm}}\right)(3.5 \text{ m})(4.2 \text{ m})$$
$$= 1.5 \times 10^6 \text{ N}. \qquad \text{(Answer)}$$

This enormous force is equal to the weight of the column of air that covers the floor and extends all the way to the top of the atmosphere.

14-4 Fluids at Rest

Figure 14-2a shows a tank of water—or other liquid—open to the atmosphere. As every diver knows, the pressure *increases* with depth below the air–water interface. The diver's depth gauge, in fact, is a pressure sensor much like that of Fig. 14-1b. As every mountaineer knows, the pressure *decreases* with altitude as one ascends into the atmosphere. The pressures encountered by the diver and the mountaineer are usually called *hydrostatic pressures,* because they are due to fluids that are static (at rest). Here we want to find an expression for hydrostatic pressure as a function of depth or altitude.

Let us look first at the increase in pressure with depth below the water's surface. We set up a vertical y axis in the tank, with its origin at the air–water interface and the positive direction upward. We next consider a water sample contained in an imaginary right circular cylinder of horizontal base (or face) area

A, such that y_1 and y_2 (both of which are *negative* numbers) are the depths below the surface of the upper and lower cylinder faces, respectively.

Figure 14-2*b* shows a free-body diagram for the water in the cylinder. The water is in *static equilibrium;* that is, it is stationary and the forces on it balance. Three forces act on it vertically: Force $\vec{F}_1$ acts at the top surface of the cylinder and is due to the water above the cylinder. Similarly, force $\vec{F}_2$ acts at the bottom surface of the cylinder and is due to the water below the cylinder. The gravitational force on the water in the cylinder is represented by $m\vec{g}$, where m is the mass of the water in the cylinder. The balance of these forces is written as

$$F_2 = F_1 + mg. \tag{14-5}$$

We want to transform Eq. 14-5 into an equation involving pressures. From Eq. 14-4, we know that

$$F_1 = p_1 A \quad \text{and} \quad F_2 = p_2 A. \tag{14-6}$$

The mass m of the water in the cylinder is, from Eq. 14-2, $m = \rho V$, where the cylinder's volume V is the product of its face area A and its height $y_1 - y_2$. Thus, m is equal to $\rho A(y_1 - y_2)$. Substituting this and Eq. 14-6 into Eq. 14-5, we find

$$p_2 A = p_1 A + \rho A g(y_1 - y_2)$$

or $$p_2 = p_1 + \rho g(y_1 - y_2). \tag{14-7}$$

This equation can be used to find pressure both in a liquid (as a function of depth) and in the atmosphere (as a function of altitude or height). For the former, suppose we seek the pressure p at a depth h below the liquid surface. Then we choose level 1 to be the surface, level 2 to be a distance h below it (as in Fig. 14-3), and p_0 to represent the atmospheric pressure on the surface. We then substitute

$$y_1 = 0, \quad p_1 = p_0 \quad \text{and} \quad y_2 = -h, \quad p_2 = p$$

into Eq. 14-7, which becomes

$$p = p_0 + \rho g h \quad \text{(pressure at depth } h\text{)}. \tag{14-8}$$

Note that the pressure at a given depth in the liquid depends on that depth but not on any horizontal dimension.

> The pressure at a point in a fluid in static equilibrium depends on the depth of that point but not on any horizontal dimension of the fluid or its container.

Thus, Eq. 14-8 holds no matter what the shape of the container. If the bottom surface of the container is at depth h, then Eq. 14-8 gives the pressure p there.

In Eq. 14-8, p is said to be the total pressure, or **absolute pressure,** at level 2. To see why, note in Fig. 14-3 that the pressure p at level 2 consists of two contributions: (1) p_0, the pressure due to the atmosphere, which bears down on the liquid, and (2) $\rho g h$, the pressure due to the liquid above level 2, which bears down on level 2. In general, the difference between an absolute pressure and an atmospheric pressure is called the **gauge pressure.** (The name comes from the use of a gauge to measure this difference in pressures.) For the situation of Fig. 14-3, the gauge pressure is $\rho g h$.

Equation 14-7 also holds above the liquid surface: It gives the atmospheric pressure at a given distance above level 1 in terms of the atmospheric pressure p_1 at level 1 (*assuming* that the atmospheric density is uniform over that distance). For example, to find the atmospheric pressure at a distance d above level 1 in Fig. 14-3, we substitute

$$y_1 = 0, \quad p_1 = p_0 \quad \text{and} \quad y_2 = d, \quad p_2 = p.$$

(b)

Fig. 14-2 (*a*) A tank of water in which a sample of water is contained in an imaginary cylinder of horizontal base area *A*. Force $\vec{F}_1$ acts at the top surface of the cylinder; force $\vec{F}_2$ acts at the bottom surface of the cylinder; the gravitational force on the water in the cylinder is represented by $m\vec{g}$. (*b*) A free-body diagram of the water sample.

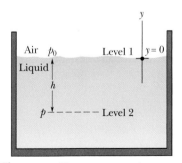

Fig. 14-3 The pressure p increases with depth h below the liquid surface according to Eq. 14-8.

Then with $\rho = \rho_{air}$, we obtain

$$p = p_0 - \rho_{air}gd.$$

✓CHECKPOINT 1   The figure shows four containers of olive oil. Rank them according to the pressure at depth h, greatest first.

(a)   (b)   (c)   (d)

Sample Problem 14-2

A novice scuba diver practicing in a swimming pool takes enough air from his tank to fully expand his lungs before abandoning the tank at depth L and swimming to the surface. He ignores instructions and fails to exhale during his ascent. When he reaches the surface, the difference between the external pressure on him and the air pressure in his lungs is 9.3 kPa. From what depth does he start? What potentially lethal danger does he face?

Solution: The **Key Idea** here is that when the diver fills his lungs at depth L, the external pressure on him (and thus the air pressure within his lungs) is greater than normal and given by Eq. 14-8 as

$$p = p_0 + \rho gL,$$

where p_0 is atmospheric pressure and ρ is the water's density (998 kg/m^3, from Table 14-1). As he ascends, the external pressure on him decreases, until it is atmospheric pressure p_0 at the surface. His blood pressure also decreases, until it is

normal. However, because he does not exhale, the air pressure in his lungs remains at the value it had at depth L. At the surface, the pressure difference between the higher pressure in his lungs and the lower pressure on his chest is

$$\Delta p = p - p_0 = \rho gL,$$

from which we find

$$L = \frac{\Delta p}{\rho g} = \frac{9300 \text{ Pa}}{(998 \text{ kg/m}^3)(9.8 \text{ m/s}^2)}$$
$$= 0.95 \text{ m.} \qquad \text{(Answer)}$$

This is not deep! Yet, the pressure difference of 9.3 kPa (about 9% of atmospheric pressure) is sufficient to rupture the diver's lungs and force air from them into the depressurized blood, which then carries the air to the heart, killing the diver. If the diver follows instructions and gradually exhales as he ascends, he allows the pressure in his lungs to equalize with the external pressure, and then there is no danger.

Sample Problem 14-3

The **U**-tube in Fig. 14-4 contains two liquids in static equilibrium: Water of density ρ_w ($= 998$ kg/m^3) is in the right arm, and oil of unknown density ρ_x is in the left. Measurement gives $l = 135$ mm and $d = 12.3$ mm. What is the density of the oil?

Solution: One **Key Idea** here is that the pressure p_{int} at the level of the oil–water interface in the left arm depends on the density ρ_x and height of the oil above the interface. A second **Key Idea** is that the water in the right arm *at the same level* must be at the same pressure p_{int}. The reason is that, because the water is in static equilibrium, pressures at points in the water at the same level must be the same even if the points are separated horizontally.

In the right arm, the interface is a distance l below the free surface of the *water*, and we have, from Eq. 14-8,

$$p_{int} = p_0 + \rho_w gl \qquad \text{(right arm)}.$$

In the left arm, the interface is a distance $l + d$ below the free surface of the *oil*, and we have, again from Eq. 14-8,

$$p_{int} = p_0 + \rho_x g(l + d) \qquad \text{(left arm)}.$$

Fig. 14-4   The oil in the left arm stands higher than the water in the right arm because the oil is less dense than the water. Both fluid columns produce the same pressure p_{int} at the level of the interface.

Equating these two expressions and solving for the unknown density yield

$$\rho_x = \rho_w \frac{l}{l + d} = (998 \text{ kg/m}^3) \frac{135 \text{ mm}}{135 \text{ mm} + 12.3 \text{ mm}}$$
$$= 915 \text{ kg/m}^3. \qquad \text{(Answer)}$$

Note that the answer does not depend on the atmospheric pressure p_0 or the free-fall acceleration g.

14-5 *Measuring Pressure*

The Mercury Barometer

Figure 14-5a shows a very basic *mercury barometer,* a device used to measure the pressure of the atmosphere. The long glass tube is filled with mercury and inverted with its open end in a dish of mercury, as the figure shows. The space above the mercury column contains only mercury vapor, whose pressure is so small at ordinary temperatures that it can be neglected.

We can use Eq. 14-7 to find the atmospheric pressure p_0 in terms of the height h of the mercury column. We choose level 1 of Fig. 14-2 to be that of the air–mercury interface and level 2 to be that of the top of the mercury column, as labeled in Fig. 14-5a. We then substitute

$$y_1 = 0, \quad p_1 = p_0 \quad \text{and} \quad y_2 = h, \quad p_2 = 0$$

into Eq. 14-7, finding that

$$p_0 = \rho gh, \tag{14-9}$$

where ρ is the density of the mercury.

For a given pressure, the height h of the mercury column does not depend on the cross-sectional area of the vertical tube. The fanciful mercury barometer of Fig. 14-5b gives the same reading as that of Fig. 14-5a; all that counts is the vertical distance h between the mercury levels.

Equation 14-9 shows that, for a given pressure, the height of the column of mercury depends on the value of g at the location of the barometer and on the density of mercury, which varies with temperature. The height of the column (in millimeters) is numerically equal to the pressure (in torr) *only* if the barometer is at a place where g has its accepted standard value of 9.80665 m/s^2 *and* the temperature of the mercury is 0°C. If these conditions do not prevail (and they rarely do), small corrections must be made before the height of the mercury column can be transformed into a pressure.

The Open-Tube Manometer

An *open-tube manometer* (Fig. 14-6) measures the gauge pressure p_g of a gas. It consists of a U-tube containing a liquid, with one end of the tube connected to the vessel whose gauge pressure we wish to measure and the other end open to the atmosphere. We can use Eq. 14-7 to find the gauge pressure in terms of the height h shown in Fig. 14-6. Let us choose levels 1 and 2 as shown in Fig. 14-6. We then substitute

$$y_1 = 0, \quad p_1 = p_0 \quad \text{and} \quad y_2 = -h, \quad p_2 = p$$

Fig. 14-5 (*a*) A mercury barometer. (*b*) Another mercury barometer. The distance h is the same in both cases.

Fig. 14-6 An open-tube manometer, connected to measure the gauge pressure of the gas in the tank on the left. The right arm of the U-tube is open to the atmosphere.

Fig. 14-7 Lead shot (small balls of lead) loaded onto the piston create a pressure p_{ext} at the top of the enclosed (incompressible) liquid. If p_{ext} is increased, by adding more lead shot, the pressure increases by the same amount at all points within the liquid.

into Eq. 14-7, finding that

$$p_g = p - p_0 = \rho gh, \quad (14\text{-}10)$$

where ρ is the density of the liquid in the tube. The gauge pressure p_g is directly proportional to h.

The gauge pressure can be positive or negative, depending on whether $p > p_0$ or $p < p_0$. In inflated tires or the human circulatory system, the (absolute) pressure is greater than atmospheric pressure, so the gauge pressure is a positive quantity, sometimes called the *overpressure*. If you suck on a straw to pull fluid up the straw, the (absolute) pressure in your lungs is actually less than atmospheric pressure. The gauge pressure in your lungs is then a negative quantity.

14-6 Pascal's Principle

When you squeeze one end of a tube to get toothpaste out the other end, you are watching **Pascal's principle** in action. This principle is also the basis for the Heimlich maneuver, in which a sharp pressure increase properly applied to the abdomen is transmitted to the throat, forcefully ejecting food lodged there. The principle was first stated clearly in 1652 by Blaise Pascal (for whom the unit of pressure is named):

 A change in the pressure applied to an enclosed incompressible fluid is transmitted undiminished to every portion of the fluid and to the walls of its container.

Demonstrating Pascal's Principle

Consider the case in which the incompressible fluid is a liquid contained in a tall cylinder, as in Fig. 14-7. The cylinder is fitted with a piston on which a container of lead shot rests. The atmosphere, container, and shot exert pressure p_{ext} on the piston and thus on the liquid. The pressure p at any point P in the liquid is then

$$p = p_{ext} + \rho gh. \quad (14\text{-}11)$$

Let us add a little more lead shot to the container to increase p_{ext} by an amount Δp_{ext}. The quantities ρ, g, and h in Eq. 14-11 are unchanged, so the pressure change at P is

$$\Delta p = \Delta p_{ext}. \quad (14\text{-}12)$$

This pressure change is independent of h, so it must hold for all points within the liquid, as Pascal's principle states.

Pascal's Principle and the Hydraulic Lever

Figure 14-8 shows how Pascal's principle can be made the basis of a hydraulic lever. In operation, let an external force of magnitude F_i be directed downward on the left-hand (or input) piston, whose surface area is A_i. An incompressible liquid in the device then produces an upward force of magnitude F_o on the right-hand (or output) piston, whose surface area is A_o. To keep the system in equilibrium, there must be a downward force of magnitude F_o on the output piston from an external load (not shown). The force $\vec{F}_i$ applied on the left and the downward force $\vec{F}_o$ from the load on the right produce a change Δp in the pressure of the liquid that is given by

$$\Delta p = \frac{F_i}{A_i} = \frac{F_o}{A_o},$$

so

$$F_o = F_i \frac{A_o}{A_i}. \quad (14\text{-}13)$$

Fig. 14-8 A hydraulic arrangement that can be used to magnify a force $\vec{F}_i$. The work done is, however, not magnified and is the same for both the input and output forces.

Equation 14-13 shows that the output force F_o on the load must be greater than the input force F_i if $A_o > A_i$, as is the case in Fig. 14-8.

If we move the input piston downward a distance d_i, the output piston moves upward a distance d_o, such that the same volume V of the incompressible liquid is displaced at both pistons. Then

$$V = A_i d_i = A_o d_o,$$

which we can write as

$$d_o = d_i \frac{A_i}{A_o}. \tag{14-14}$$

This shows that, if $A_o > A_i$ (as in Fig. 14-8), the output piston moves a smaller distance than the input piston moves.

From Eqs. 14-13 and 14-14 we can write the output work as

$$W = F_o d_o = \left(F_i \frac{A_o}{A_i} \right) \left(d_i \frac{A_i}{A_o} \right) = F_i d_i, \tag{14-15}$$

which shows that the work W done *on* the input piston by the applied force is equal to the work W done *by* the output piston in lifting the load placed on it.

The advantage of a hydraulic lever is this:

 With a hydraulic lever, a given force applied over a given distance can be transformed to a greater force applied over a smaller distance.

The product of force and distance remains unchanged so that the same work is done. However, there is often tremendous advantage in being able to exert the larger force. Most of us, for example, cannot lift an automobile directly but can with a hydraulic jack, even though we have to pump the handle farther than the automobile rises. In this device, the displacement d_i is accomplished not in a single stroke but over a series of small strokes.

14-7 *Archimedes' Principle*

Figure 14-9 shows a student in a swimming pool, manipulating a very thin plastic sack (of negligible mass) that is filled with water. She finds that the sack and its contained water are in static equilibrium, tending neither to rise nor to sink. The downward gravitational force $\vec{F}_g$ on the contained water must be balanced by a net upward force from the water surrounding the sack.

This net upward force is a **buoyant force** $\vec{F}_b$. It exists because the pressure in the surrounding water increases with depth below the surface. Thus, the pressure near the bottom of the sack is greater than the pressure near the top, which means the forces on the sack due to this pressure are greater in magnitude near the bottom of the sack than near the top. Some of the forces are represented in Fig. 14-10a, where the space occupied by the sack has been left empty. Note that the force vectors drawn near the bottom of that space (with upward components) have longer lengths than those drawn near the top of the sack (with downward components). If we vectorially add all the forces on the sack from the water, the horizontal components cancel and the vertical components add to yield the upward buoyant force $\vec{F}_b$ on the sack. (Force $\vec{F}_b$ is shown to the right of the pool in Fig. 14-10a.)

Because the sack of water is in static equilibrium, the magnitude of $\vec{F}_b$ is equal to the magnitude $m_f g$ of the gravitational force $\vec{F}_g$ on the sack of water: $F_b = m_f g$. (Subscript f refers to *fluid*, here the water.) In words, the magnitude of the buoyant force is equal to the weight of the water in the sack.

Fig. 14-9 A thin-walled plastic sack of water is in static equilibrium in the pool. The gravitational force on the sack must be balanced by a net upward force on it from the surrounding water.

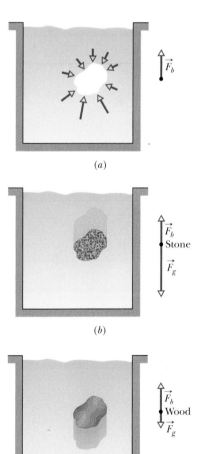

Fig. 14-10 (a) The water surrounding the hole in the water produces a net upward buoyant force on whatever fills the hole. (b) For a stone of the same volume as the hole, the gravitational force exceeds the buoyant force in magnitude. (c) For a lump of wood of the same volume, the gravitational force is less than the buoyant force in magnitude.

In Fig. 14-10b, we have replaced the sack of water with a stone that exactly fills the hole in Fig. 14-10a. The stone is said to *displace* the water, meaning that the stone occupies space that would otherwise be occupied by water. We have changed nothing about the shape of the hole, so the forces at the hole's surface must be the same as when the water-filled sack was in place. Thus, the same upward buoyant force that acted on the water-filled sack now acts on the stone; that is, the magnitude F_b of the buoyant force is equal to $m_f g$, the weight of the water displaced by the stone.

Unlike the water-filled sack, the stone is not in static equilibrium. The downward gravitational force $\vec{F}_g$ on the stone is greater in magnitude than the upward buoyant force, as is shown in the free-body diagram in Fig. 14-10b. The stone thus accelerates downward, sinking to the bottom of the pool.

Let us next exactly fill the hole in Fig. 14-10a with a block of lightweight wood, as in Fig. 14-10c. Again, nothing has changed about the forces at the hole's surface, so the magnitude F_b of the buoyant force is still equal to $m_f g$, the weight of the displaced water. Like the stone, the block is not in static equilibrium. However, this time the gravitational force $\vec{F}_g$ is lesser in magnitude than the buoyant force (as shown to the right of the pool), and so the block accelerates upward, rising to the top surface of the water.

Our results with the sack, stone, and block apply to all fluids and are summarized in **Archimedes' principle:**

> When a body is fully or partially submerged in a fluid, a buoyant force $\vec{F}_b$ from the surrounding fluid acts on the body. The force is directed upward and has a magnitude equal to the weight $m_f g$ of the fluid that has been displaced by the body.

The buoyant force on a body in a fluid has the magnitude

$$F_b = m_f g \qquad \text{(buoyant force)}, \qquad (14\text{-}16)$$

where m_f is the mass of the fluid that is displaced by the body.

Floating

When we release a block of lightweight wood just above the water in a pool, the block moves into the water because the gravitational force on it pulls it downward. As the block displaces more and more water, the magnitude F_b of the upward buoyant force acting on it increases. Eventually, F_b is large enough to equal the magnitude F_g of the downward gravitational force on the block, and the block comes to rest. The block is then in static equilibrium and is said to be *floating* in the water. In general,

> When a body floats in a fluid, the magnitude F_b of the buoyant force on the body is equal to the magnitude F_g of the gravitational force on the body.

We can write this statement as

$$F_b = F_g \qquad \text{(floating)}. \qquad (14\text{-}17)$$

From Eq. 14-16, we know that $F_b = m_f g$. Thus,

> When a body floats in a fluid, the magnitude F_g of the gravitational force on the body is equal to the weight $m_f g$ of the fluid that has been displaced by the body.

We can write this statement as

$$F_g = m_f g \qquad \text{(floating)}. \qquad (14\text{-}18)$$

In other words, a floating body displaces its own weight of fluid.

Apparent Weight in a Fluid

If we place a stone on a scale that is calibrated to measure weight, then the reading on the scale is the stone's weight. However, if we do this underwater, the upward buoyant force on the stone from the water decreases the reading. That reading is then an apparent weight. In general, an **apparent weight** is related to the actual weight of a body and the buoyant force on the body by

$$\begin{pmatrix} \text{apparent} \\ \text{weight} \end{pmatrix} = \begin{pmatrix} \text{actual} \\ \text{weight} \end{pmatrix} - \begin{pmatrix} \text{magnitude of} \\ \text{buoyant force} \end{pmatrix},$$

which we can write as

$$\text{weight}_{app} = \text{weight} - F_b \quad \text{(apparent weight).} \quad (14\text{-}19)$$

If, in some test of strength, you had to lift a heavy stone, you could do it more easily with the stone underwater. Then your applied force would need to exceed only the stone's apparent weight, not its larger actual weight, because the upward buoyant force would help you lift the stone.

The magnitude of the buoyant force on a floating body is equal to the body's weight. Equation 14-19 thus tells us that a floating body has an apparent weight of zero—the body would produce a reading of zero on a scale. (When astronauts prepare to perform a complex task in space, they practice the task floating underwater, where their apparent weight is zero as it is in space.)

✓**CHECKPOINT 2**   A penguin floats first in a fluid of density ρ_0, then in a fluid of density $0.95\rho_0$, and then in a fluid of density $1.1\rho_0$. (a) Rank the densities according to the magnitude of the buoyant force on the penguin, greatest first. (b) Rank the densities according to the amount of fluid displaced by the penguin, greatest first.

Sample Problem 14-4

What fraction of the volume of an iceberg floating in seawater is visible?

Solution: Let V_i be the total volume of the iceberg. The nonvisible portion is below water, and thus the volume of this portion is equal to the volume V_f of the fluid (the seawater) displaced by the iceberg. We seek the fraction (call it frac)

$$\text{frac} = \frac{V_i - V_f}{V_i} = 1 - \frac{V_f}{V_i}, \quad (14\text{-}20)$$

but we know neither volume. A **Key Idea** here is that, because the iceberg is floating, Eq. 14-18 ($F_g = m_f g$) applies. We can write that equation as

$$m_i g = m_f g,$$

from which we see that $m_i = m_f$. Thus, the mass of the iceberg is equal to the mass of the displaced fluid (seawater). Although we know neither mass, we can relate them to the densities of ice and seawater given in Table 14-1 by using Eq. 14-2 ($\rho = m/V$). Because $m_i = m_f$, we can write

$$\rho_i V_i = \rho_f V_f$$

or

$$\frac{V_f}{V_i} = \frac{\rho_i}{\rho_f}.$$

Substituting this into Eq. 14-20 and then using the known densities, we find

$$\text{frac} = 1 - \frac{\rho_i}{\rho_f} = 1 - \frac{917 \text{ kg/m}^3}{1024 \text{ kg/m}^3}$$

$$= 0.10 \text{ or } 10\%. \quad \text{(Answer)}$$

Sample Problem 14-5

A spherical, helium-filled balloon has a radius R of 12.0 m. The balloon, support cables, and basket have a mass m of 196 kg. What maximum load M can the balloon support while it floats at an altitude at which the helium density ρ_{He} is 0.160 kg/m^3 and the air density ρ_{air} is 1.25 kg/m^3? Assume that only the balloon displaces air.

Solution: The **Key Idea** here is that the balloon, cables, basket, load, *and* the contained helium form a floating body, with a total mass of $m + M + m_{He}$, where m_{He} is the mass of the

contained helium. Then the magnitude of the total gravitational force on this body must be equal to the weight of the air displaced by the body (the air is the fluid in which this body floats). Let m_{air} be the mass of the air displaced by the body. From Eq. 14-18 ($F_g = m_f g$), we have

$$(m + M + m_{He})g = m_{air} g$$

or

$$M = m_{air} - m_{He} - m. \quad (14\text{-}21)$$

We do not know m_{He} and m_{air} but we do know the corresponding densities; so we can use Eq. 14-2 ($\rho = m/V$) to rewrite Eq. 14-21 in terms of those densities. First we note that, because the load, support cables, and basket displace a negligible amount of air, the volume of the displaced air is equal to the volume $V\ (=\frac{4}{3}\pi R^3)$ of the spherical balloon. Then Eq. 14-21 becomes

$$
\begin{aligned}
M &= \rho_{\text{air}}V - \rho_{\text{He}}V - m \\
&= (\tfrac{4}{3}\pi R^3)(\rho_{\text{air}} - \rho_{\text{He}}) - m \\
&= (\tfrac{4}{3}\pi)(12.0\text{ m})^3(1.25\text{ kg/m}^3 - 0.160\text{ kg/m}^3) - 196\text{ kg} \\
&= 7694\text{ kg} \approx 7690\text{ kg}. \qquad\qquad\text{(Answer)}
\end{aligned}
$$

Fig. 14-11 At a certain point, the rising flow of smoke and heated gas changes from steady to turbulent.

14-8 Ideal Fluids in Motion

The motion of *real fluids* is very complicated and not yet fully understood. Instead, we shall discuss the motion of an **ideal fluid,** which is simpler to handle mathematically and yet provides useful results. Here are four assumptions that we make about our ideal fluid; they all are concerned with *flow:*

1. **Steady flow**   In *steady* (or *laminar*) flow, the velocity of the moving fluid at any fixed point does not change with time, either in magnitude or in direction. The gentle flow of water near the center of a quiet stream is steady; the flow in a chain of rapids is not. Figure 14-11 shows a transition from steady flow to *nonsteady* (or *nonlaminar* or *turbulent*) flow for a rising stream of smoke. The speed of the smoke particles increases as they rise and, at a certain critical speed, the flow changes from steady to nonsteady.

2. **Incompressible flow**   We assume, as for fluids at rest, that our ideal fluid is incompressible; that is, its density has a constant, uniform value.

3. **Nonviscous flow**   Roughly speaking, the viscosity of a fluid is a measure of how resistive the fluid is to flow. For example, thick honey is more resistive to flow than water, and so honey is said to be more viscous than water. Viscosity is the fluid analog of friction between solids; both are mechanisms by which the kinetic energy of moving objects can be transferred to thermal energy. In the absence of friction, a block could glide at constant speed along a horizontal surface. In the same way, an object moving through a nonviscous fluid would experience no *viscous drag force*—that is, no resistive force due to viscosity; it could move at constant speed through the fluid. The British scientist Lord Rayleigh noted that in an ideal fluid a ship's propeller would not work, but, on the other hand, in an ideal fluid a ship (once set into motion) would not need a propeller!

4. **Irrotational flow**   Although it need not concern us further, we also assume that the flow is *irrotational.* To test for this property, let a tiny grain of dust move with the fluid. Although this test body may (or may not) move in a circular path, in irrotational flow the test body will not rotate about an axis through its own center of mass. For a loose analogy, the motion of a Ferris wheel is rotational; that of its passengers is irrotational.

We can make the flow of a fluid visible by adding a *tracer.* This might be a dye injected into many points across a liquid stream (Fig. 14-12) or smoke par-

Fig. 14-12 The steady flow of a fluid around a cylinder, as revealed by a dye tracer that was injected into the fluid upstream of the cylinder.

ticles added to a gas flow (Fig. 14-11). Each bit of a tracer follows a *streamline,* which is the path that a tiny element of the fluid would take as the fluid flows. Recall from Chapter 4 that the velocity of a particle is always tangent to the path taken by the particle. Here the particle is the fluid element, and its velocity $\vec{v}$ is always tangent to a streamline (Fig. 14-13). For this reason, two streamlines can never intersect; if they did, then an element arriving at their intersection would have two different velocities simultaneously—an impossibility.

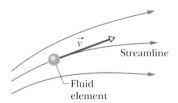

Fig. 14-13 A fluid element traces out a streamline as it moves. The velocity vector of the element is tangent to the streamline at every point.

14-9 *The Equation of Continuity*

You may have noticed that you can increase the speed of the water emerging from a garden hose by partially closing the hose opening with your thumb. Apparently the speed v of the water depends on the cross-sectional area A through which the water flows.

Here we wish to derive an expression that relates v and A for the steady flow of an ideal fluid through a tube with varying cross section, like that in Fig. 14-14. The flow there is toward the right, and the tube segment shown (part of a longer tube) has length L. The fluid has speeds v_1 at the left end of the segment and v_2 at the right end. The tube has cross-sectional areas A_1 at the left end and A_2 at the right end. Suppose that in a time interval Δt a volume ΔV of fluid enters the tube segment at its left end (that volume is colored purple in Fig. 14-14). Then, because the fluid is incompressible, an identical volume ΔV must emerge from the right end of the segment (it is colored green in Fig. 14-14).

We can use this common volume ΔV to relate the speeds and areas. To do so, we first consider Fig. 14-15, which shows a side view of a tube of *uniform* cross-sectional area A. In Fig. 14-15a, a fluid element e is about to pass through the dashed line drawn across the tube width. The element's speed is v, so during a time interval Δt, the element moves along the tube a distance $\Delta x = v\,\Delta t$. The volume ΔV of fluid that has passed through the dashed line in that time interval Δt is

$$\Delta V = A\,\Delta x = Av\,\Delta t. \tag{14-22}$$

Applying Eq. 14-22 to both the left and right ends of the tube segment in Fig. 14-14, we have

$$\Delta V = A_1 v_1\,\Delta t = A_2 v_2\,\Delta t$$

or

$$A_1 v_1 = A_2 v_2 \qquad \text{(equation of continuity)}. \tag{14-23}$$

This relation between speed and cross-sectional area is called the **equation of continuity** for the flow of an ideal fluid. It tells us that the flow speed increases when we decrease the cross-sectional area through which the fluid flows (as when we partially close off a garden hose with a thumb).

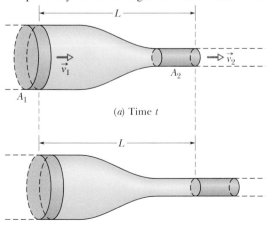

(a) Time t

(b) Time $t + \Delta t$

Fig. 14-14 Fluid flows from left to right at a steady rate through a tube segment of length L. The fluid's speed is v_1 at the left side and v_2 at the right side. The tube's cross-sectional area is A_1 at the left side and A_2 at the right side. From time t in (a) to time $t + \Delta t$ in (b), the amount of fluid shown in purple enters at the left side and the equal amount of fluid shown in green emerges at the right side.

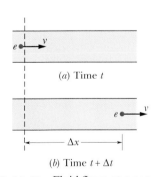

(a) Time t

(b) Time $t + \Delta t$

Fig. 14-15 Fluid flows at a constant speed v through a tube. (a) At time t, fluid element e is about to pass the dashed line. (b) At time $t + \Delta t$, element e is a distance $\Delta x = v\,\Delta t$ from the dashed line.

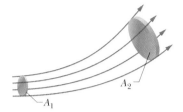

Fig. 14-16   A tube of flow is defined by the streamlines that form the boundary of the tube. The volume flow rate must be the same for all cross sections of the tube of flow.

Equation 14-23 applies not only to an actual tube but also to any so-called *tube of flow*, or imaginary tube whose boundary consists of streamlines. Such a tube acts like a real tube because no fluid element can cross a streamline; thus, all the fluid within a tube of flow must remain within its boundary. Figure 14-16 shows a tube of flow in which the cross-sectional area increases from area A_1 to area A_2 along the flow direction. From Eq. 14-23 we know that, with the increase in area, the speed must decrease, as is indicated by the greater spacing between streamlines at the right in Fig. 14-16. Similarly, you can see that in Fig. 14-12 the speed of the flow is greatest just above and just below the cylinder.

We can rewrite Eq. 14-23 as

$$R_V = Av = \text{a constant} \qquad \text{(volume flow rate, equation of continuity),} \qquad (14\text{-}24)$$

in which R_V is the **volume flow rate** of the fluid (volume past a given point per unit time). Its SI unit is the cubic meter per second (m^3/s). If the density ρ of the fluid is uniform, we can multiply Eq. 14-24 by that density to get the **mass flow rate** R_m (mass per unit time):

$$R_m = \rho R_V = \rho A v = \text{a constant} \qquad \text{(mass flow rate).} \qquad (14\text{-}25)$$

The SI unit of mass flow rate is the kilogram per second (kg/s). Equation 14-25 says that the mass that flows into the tube segment of Fig. 14-14 each second must be equal to the mass that flows out of that segment each second.

> ✓**CHECKPOINT 3**   The figure shows a pipe and gives the volume flow rate (in cm^3/s) and the direction of flow for all but one section. What are the volume flow rate and the direction of flow for that section?
>
>

Sample Problem 14-6

Figure 14-17 shows how the stream of water emerging from a faucet "necks down" as it falls. The indicated cross-sectional areas are $A_0 = 1.2 \text{ cm}^2$ and $A = 0.35 \text{ cm}^2$. The two levels are separated by a vertical distance $h = 45$ mm. What is the volume flow rate from the tap?

Solution: The **Key Idea** here is simply that the volume flow rate through the higher cross section must be the same as that through the lower cross section. Thus, from Eq. 14-24, we have

$$A_0 v_0 = Av, \qquad (14\text{-}26)$$

where v_0 and v are the water speeds at the levels corresponding to A_0 and A. From Eq. 2-16 we can also write, because the water is falling freely with acceleration g,

$$v^2 = v_0^2 + 2gh. \qquad (14\text{-}27)$$

Eliminating v between Eqs. 14-26 and 14-27 and solving for v_0, we obtain

$$v_0 = \sqrt{\frac{2ghA^2}{A_0^2 - A^2}}$$

$$= \sqrt{\frac{(2)(9.8 \text{ m/s}^2)(0.045 \text{ m})(0.35 \text{ cm}^2)^2}{(1.2 \text{ cm}^2)^2 - (0.35 \text{ cm}^2)^2}}$$

$$= 0.286 \text{ m/s} = 28.6 \text{ cm/s}.$$

Fig. 14-17   As water falls from a tap, its speed increases. Because the volume flow rate must be the same at all horizontal cross sections of the stream, the stream must "neck down" (narrow).

From Eq. 14-24, the volume flow rate R_V is then

$$R_V = A_0 v_0 = (1.2 \text{ cm}^2)(28.6 \text{ cm/s})$$

$$= 34 \text{ cm}^3/\text{s}. \qquad \text{(Answer)}$$

14-10 *Bernoulli's Equation*

Figure 14-18 represents a tube through which an ideal fluid is flowing at a steady rate. In a time interval Δt, suppose that a volume of fluid ΔV, colored purple in Fig. 14-18, enters the tube at the left (or input) end and an identical volume, colored green in Fig. 14-18, emerges at the right (or output) end. The emerging volume must be the same as the entering volume because the fluid is incompressible, with an assumed constant density ρ.

Let y_1, v_1, and p_1 be the elevation, speed, and pressure of the fluid entering at the left, and y_2, v_2, and p_2 be the corresponding quantities for the fluid emerging at the right. By applying the principle of conservation of energy to the fluid, we shall show that these quantities are related by

$$p_1 + \tfrac{1}{2}\rho v_1^2 + \rho g y_1 = p_2 + \tfrac{1}{2}\rho v_2^2 + \rho g y_2. \tag{14-28}$$

We can also write this equation as

$$p + \tfrac{1}{2}\rho v^2 + \rho g y = \text{a constant} \quad \text{(Bernoulli's equation).} \tag{14-29}$$

Equations 14-28 and 14-29 are equivalent forms of **Bernoulli's equation,** after Daniel Bernoulli, who studied fluid flow in the 1700s.* Like the equation of continuity (Eq. 14-24), Bernoulli's equation is not a new principle but simply the reformulation of a familiar principle in a form more suitable to fluid mechanics. As a check, let us apply Bernoulli's equation to fluids at rest, by putting $v_1 = v_2 = 0$ in Eq. 14-28. The result is

$$p_2 = p_1 + \rho g(y_1 - y_2),$$

which is Eq. 14-7.

A major prediction of Bernoulli's equation emerges if we take y to be a constant ($y = 0$, say) so that the fluid does not change elevation as it flows. Equation 14-28 then becomes

$$p_1 + \tfrac{1}{2}\rho v_1^2 = p_2 + \tfrac{1}{2}\rho v_2^2, \tag{14-30}$$

which tells us that:

> If the speed of a fluid element increases as the element travels along a horizontal streamline, the pressure of the fluid must decrease, and conversely.

Put another way, where the streamlines are relatively close together (where the velocity is relatively great), the pressure is relatively low, and conversely.

The link between a change in speed and a change in pressure makes sense if you consider a fluid element. When the element nears a narrow region, the higher pressure behind it accelerates it so that it then has a greater speed in the narrow region. When it nears a wide region, the higher pressure ahead of it decelerates it so that it then has a lesser speed in the wide region.

Bernoulli's equation is strictly valid only to the extent that the fluid is ideal. If viscous forces are present, thermal energy will be involved. We take no account of this in the derivation that follows.

Proof of Bernoulli's Equation

Let us take as our system the entire volume of the (ideal) fluid shown in Fig. 14-18. We shall apply the principle of conservation of energy to this system as it moves from its initial state (Fig. 14-18a) to its final state (Fig. 14-18b). The fluid lying between the two vertical planes separated by a distance L in Fig. 14-18 does not change its properties during this process; we need be concerned only with changes that take place at the input and output ends.

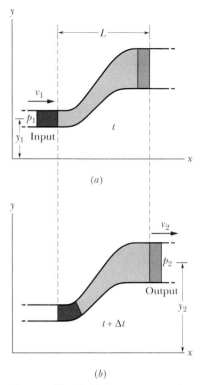

Fig. 14-18 Fluid flows at a steady rate through a length L of a tube, from the input end at the left to the output end at the right. From time t in (a) to time $t + \Delta t$ in (b), the amount of fluid shown in purple enters the input end and the equal amount shown in green emerges from the output end.

---

*For irrotational flow (which we assume), the constant in Eq. 14-29 has the same value for all points within the tube of flow; the points do not have to lie along the same streamline. Similarly, the points 1 and 2 in Eq. 14-28 can lie anywhere within the tube of flow.

First, we apply energy conservation in the form of the work–kinetic energy theorem,

$$W = \Delta K, \qquad (14\text{-}31)$$

which tells us that the change in the kinetic energy of our system must equal the net work done on the system. The change in kinetic energy results from the change in speed between the ends of the tube and is

$$\Delta K = \tfrac{1}{2}\Delta m \, v_2^2 - \tfrac{1}{2}\Delta m \, v_1^2$$
$$= \tfrac{1}{2}\rho \, \Delta V \, (v_2^2 - v_1^2), \qquad (14\text{-}32)$$

in which Δm $(= \rho \, \Delta V)$ is the mass of the fluid that enters at the input end and leaves at the output end during a small time interval Δt.

The work done on the system arises from two sources. The work W_g done by the gravitational force $(\Delta m \, \vec{g})$ on the fluid of mass Δm during the vertical lift of the mass from the input level to the output level is

$$W_g = -\Delta m \, g(y_2 - y_1)$$
$$= -\rho g \, \Delta V \, (y_2 - y_1). \qquad (14\text{-}33)$$

This work is negative because the upward displacement and the downward gravitational force have opposite directions.

Work must also be done *on* the system (at the input end) to push the entering fluid into the tube and *by* the system (at the output end) to push forward the fluid that is located ahead of the emerging fluid. In general, the work done by a force of magnitude F, acting on a fluid sample contained in a tube of area A to move the fluid through a distance Δx, is

$$F \, \Delta x = (pA)(\Delta x) = p(A \, \Delta x) = p \, \Delta V.$$

The work done on the system is then $p_1 \, \Delta V$, and the work done by the system is $-p_2 \, \Delta V$. Their sum W_p is

$$W_p = -p_2 \, \Delta V + p_1 \, \Delta V$$
$$= -(p_2 - p_1) \, \Delta V. \qquad (14\text{-}34)$$

The work–kinetic energy theorem of Eq. 14-31 now becomes

$$W = W_g + W_p = \Delta K.$$

Substituting from Eqs. 14-32, 14-33, and 14-34 yields

$$-\rho g \, \Delta V \, (y_2 - y_1) - \Delta V \, (p_2 - p_1) = \tfrac{1}{2}\rho \, \Delta V \, (v_2^2 - v_1^2).$$

This, after a slight rearrangement, matches Eq. 14-28, which we set out to prove.

CHECKPOINT 4 Water flows smoothly through the pipe shown in the figure, descending in the process. Rank the four numbered sections of pipe according to (a) the volume flow rate R_V through them, (b) the flow speed v through them, and (c) the water pressure p within them, greatest first.

Sample Problem 14-7

Ethanol of density $\rho = 791$ kg/m^3 flows smoothly through a horizontal pipe that tapers (as in Fig. 14-14) in cross-sectional area from $A_1 = 1.20 \times 10^{-3}$ m^2 to $A_2 = A_1/2$. The pressure difference between the wide and narrow sections of pipe is 4120 Pa. What is the volume flow rate R_V of the ethanol?

Solution: One **Key Idea** here is that, because the fluid flowing through the wide section of pipe must entirely pass

through the narrow section, the volume flow rate R_V must be the same in the two sections. Thus, from Eq. 14-24,

$$R_V = v_1 A_1 = v_2 A_2. \qquad (14\text{-}35)$$

However, with two unknown speeds, we cannot evaluate this equation for R_V.

A second **Key Idea** is that, because the flow is smooth, we can apply Bernoulli's equation. From Eq. 14-28, we can write

$$p_1 + \tfrac{1}{2}\rho v_1^2 + \rho g y = p_2 + \tfrac{1}{2}\rho v_2^2 + \rho g y, \quad (14\text{-}36)$$

where subscripts 1 and 2 refer to the wide and narrow sections of pipe, respectively, and y is their common elevation. This equation hardly seems to help because it does not contain the desired R_V and it contains the unknown speeds v_1 and v_2.

However, there is a neat way to make it work for us: First, we can use Eq. 14-35 and the fact that $A_2 = A_1/2$ to write

$$v_1 = \frac{R_V}{A_1} \quad \text{and} \quad v_2 = \frac{R_V}{A_2} = \frac{2R_V}{A_1}. \quad (14\text{-}37)$$

Then we can substitute these expressions into Eq. 14-36 to eliminate the unknown speeds and introduce the desired volume flow rate. Doing this and solving for R_V yield

$$R_V = A_1 \sqrt{\frac{2(p_1 - p_2)}{3\rho}}. \quad (14\text{-}38)$$

We still have a decision to make: We know that the pressure difference between the two sections is 4120 Pa, but does that mean that $p_1 - p_2$ is 4120 Pa or −4120 Pa? We could guess the former is true, or otherwise the square root in Eq. 14-38 would give us an imaginary number. Instead of guessing, however, let's try some reasoning. From Eq. 14-35 we see that speed v_2 in the narrow section (small A_2) must be greater than speed v_1 in the wider section (larger A_1). Recall that if the speed of a fluid increases as the fluid travels along a horizontal path (as here), the pressure of the fluid must decrease. Thus, p_1 is greater than p_2, and $p_1 - p_2 = 4120$ Pa. Inserting this and known data into Eq. 14-38 gives

$$R_V = 1.20 \times 10^{-3}\ \text{m}^2 \sqrt{\frac{(2)(4120\ \text{Pa})}{(3)(791\ \text{kg/m}^3)}}$$

$$= 2.24 \times 10^{-3}\ \text{m}^3/\text{s}. \quad \text{(Answer)}$$

Sample Problem 14-8

In the old West, a desperado fires a bullet into an open water tank (Fig. 14-19), creating a hole a distance h below the water surface. What is the speed v of the water exiting the tank?

Solution: A **Key Idea** is that this situation is essentially that of water moving (downward) with speed v_0 through a wide pipe (the tank) of cross-sectional area A and then moving (horizontally) with speed v through a narrow pipe (the hole) of cross-sectional area a. Another **Key Idea** is that, because the water flowing through the wide pipe must entirely pass through the narrow pipe, the volume flow rate R_V must be the same in the two "pipes." Then from Eq. 14-24,

$$R_V = av = Av_0$$

and thus

$$v_0 = \frac{a}{A} v.$$

Because $a \ll A$, we see that $v_0 \ll v$.

A third **Key Idea** is that we can also relate v to v_0 (and to h) through Bernoulli's equation (Eq. 14-28). We take the level of the hole as our reference level for measuring elevations (and thus gravitational potential energy). Noting that the pressure at the top of the tank and at the bullet hole is the atmospheric pressure p_0 (because both places are exposed to the atmosphere), we write Eq. 14-28 as

$$p_0 + \tfrac{1}{2}\rho v_0^2 + \rho g h = p_0 + \tfrac{1}{2}\rho v^2 + \rho g(0). \quad (14\text{-}39)$$

(Here the top of the tank is represented by the left side of the equation and the hole by the right side. The zero on the right indicates that the hole is at our reference level.) Before we

Fig. 14-19 Water pours through a hole in a water tank, at a distance h below the water surface. The pressure at the water surface and at the hole is atmospheric pressure p_0.

solve Eq. 14-39 for v, we can use our result that $v_0 \ll v$ to simplify it: We assume that v_0^2, and thus the term $\tfrac{1}{2}\rho v_0^2$ in Eq. 14-39, is negligible relative to the other terms, and we drop it. Solving the remaining equation for v then yields

$$v = \sqrt{2gh}. \quad \text{(Answer)}$$

This is the same speed that an object would have when falling a height h from rest.

Sample Problem 14-9

As a race car moves forward at 27.25 m/s, air is forced to flow over and under the car (Fig. 14-20a). The air forced to flow under the car enters through a vertical cross-sectional area $A_0 = 0.0330\ \text{m}^2$ at the front of the car (Fig. 14-20b) and then flows beneath the car where the vertical cross-sectional area is $A_1 = 0.0310\ \text{m}^2$. Treat this flow as steady flow through a stationary horizontal pipe that decreases in cross-sectional area from A_0 to A_1 (Fig. 14-20c).

(a) At the moment it passes through A_0, the air is at atmospheric pressure p_0. At what pressure p_1 is the air as it moves through A_1?

Solution: One **Key Idea** is that, because the flow is steady, we can apply Bernoulli's equation (Eq. 14-28) to the flow. To be consistent with the given subscripts, we write the equation as

$$p_0 + \tfrac{1}{2}\rho v_0^2 + \rho g y = p_1 + \tfrac{1}{2}\rho v_1^2 + \rho g y, \quad (14\text{-}40)$$

Fig. 14-20 (a) Air flows above and below a race car. (b) The flow beneath the car enters through vertical cross-sectional area A_0. (c) The flow is then constrained as in a pipe that narrows to vertical cross-sectional area A_1.

where ρ is the air density and y is the distance above the ground of the flowing air.

A second **Key Idea** is that because all the air entering through cross-sectional area A_0 flows through cross-sectional area A_1, the volume flow rate R_V through the two areas must be the same. Thus, from Eq. 14-24,

$$A_0 v_0 = A_1 v_1,$$

or

$$v_1 = v_0 \frac{A_0}{A_1}. \tag{14-41}$$

Substituting Eq. 14-41 into Eq. 14-40 and rearranging give us

$$p_1 = p_0 - \tfrac{1}{2}\rho v_0^2 \left(\frac{A_0^2}{A_1^2} - 1 \right). \tag{14-42}$$

The speed of the air as it enters A_0 at the front of the car is equal to 27.25 m/s, the speed of the car as it moves forward through the air. Substituting this speed, the air density $\rho = 1.21$ kg/m³, and the values for A_0 and A_1 into Eq. 14-42, we find

$$p_1 = p_0 - \tfrac{1}{2}(1.21 \text{ kg/m}^3)(27.25 \text{ m/s})^2 \left(\frac{(0.0330 \text{ m}^2)^2}{(0.0310 \text{ m}^2)^2} - 1 \right)$$

$$= p_0 - 59.838 \text{ Pa} \approx p_0 - 59.8 \text{ Pa}. \qquad \text{(Answer)}$$

Thus, the air pressure beneath the car is 59.8 Pa less than atmospheric pressure.

(b) If the horizontal cross-sectional area of the car is $A_h = 4.86$ m², what is the magnitude of the net vertical force $F_{\text{net},y}$ on the car due to the air pressures above and below the car?

Solution: The **Key Idea** is that the pressure on a surface is the force per unit area, as given by Eq. 14-4 ($p = F/A$). Here we are concerned with the top and bottom surfaces of the car, where we take both surfaces to have area A_h. Above the car the air is at atmospheric pressure p_0 and presses down on the car with a vertical component

$$F_{y,\text{above}} = -p_0 A_h.$$

Below the car, the air is at pressure $p_1 = p_0 - 59.838$ Pa and presses up on the car with a vertical component

$$F_{y,\text{below}} = (p_0 - 59.838 \text{ Pa}) A_h.$$

The net vertical force is then

$$\begin{aligned}
F_{\text{net},y} &= F_{y,\text{below}} + F_{y,\text{above}} \\
&= (p_0 - 59.838 \text{ Pa}) A_h - p_0 A_h \\
&= -(59.838 \text{ Pa})(4.86 \text{ m}^2) = -291 \text{ N}. \quad \text{(Answer)}
\end{aligned}$$

This net downward force, which is due to the reduced air pressure beneath the car, is the ground effect acting on the car. It is about 30% of the total negative lift that helps hold the car on the track. Without negative lift, a car must greatly slow for turns or else it will slide outward into the track wall. In a race, a driver can reduce the air drag on his car by closely following another car, a procedure known as *drafting*. However, the leading car disrupts the steady flow of air under the trailing car, eliminating the ground effect on the trailing car. If the trailing driver does not anticipate that elimination and slow accordingly, sliding into the track wall may be unavoidable.

Review & Summary

Density The **density** ρ of any material is defined as the material's mass per unit volume:

$$\rho = \frac{\Delta m}{\Delta V}. \tag{14-1}$$

Usually, where a material sample is much larger than atomic dimensions, we can write Eq. 14-1 as

$$\rho = \frac{m}{V}. \tag{14-2}$$

Fluid Pressure A **fluid** is a substance that can flow; it conforms to the boundaries of its container because it cannot withstand shearing stress. It can, however, exert a force perpendicular to its surface. That force is described in terms of **pressure** p:

$$p = \frac{\Delta F}{\Delta A}, \tag{14-3}$$

in which ΔF is the force acting on a surface element of area

ΔA. If the force is uniform over a flat area, Eq. 14-3 can be written as

$$p = \frac{F}{A}. \tag{14-4}$$

The force resulting from fluid pressure at a particular point in a fluid has the same magnitude in all directions. **Gauge pressure** is the difference between the actual pressure (or *absolute pressure*) at a point and the atmospheric pressure.

Pressure Variation with Height and Depth Pressure in a fluid at rest varies with vertical position y. For y measured positive upward,

$$p_2 = p_1 + \rho g(y_1 - y_2). \tag{14-7}$$

The pressure in a fluid is the same for all points at the same level. If h is the *depth* of a fluid sample below some reference level at which the pressure is p_0, Eq. 14-7 becomes

$$p = p_0 + \rho g h, \tag{14-8}$$

where p is the pressure in the sample.

Pascal's Principle A change in the pressure applied to an enclosed fluid is transmitted undiminished to every portion of the fluid and to the walls of the containing vessel.

Archimedes' Principle When a body is fully or partially submerged in a fluid, a buoyant force $\vec{F}_b$ from the surrounding fluid acts on the body. The force is directed upward and has a magnitude given by

$$F_b = m_f g, \qquad (14\text{-}16)$$

where m_f is the mass of the fluid that has been displaced by the body (that is, the fluid that has been pushed out of the way by the body).

When a body floats in a fluid, the magnitude F_b of the (upward) buoyant force on the body is equal to the magnitude F_g of the (downward) gravitational force on the body. The **apparent weight** of a body on which a buoyant force acts is related to its actual weight by

$$\text{weight}_{\text{app}} = \text{weight} - F_b. \qquad (14\text{-}19)$$

Flow of Ideal Fluids An **ideal fluid** is incompressible and lacks viscosity, and its flow is steady and irrotational. A *streamline* is the path followed by an individual fluid particle. A *tube of flow* is a bundle of streamlines. The flow within any tube of flow obeys the **equation of continuity:**

$$R_V = Av = \text{a constant}, \qquad (14\text{-}24)$$

in which R_V is the **volume flow rate**, A is the cross-sectional area of the tube of flow at any point, and v is the speed of the fluid at that point. The **mass flow rate** R_m is

$$R_m = \rho R_V = \rho Av = \text{a constant}. \qquad (14\text{-}25)$$

Bernoulli's Equation Applying the principle of conservation of mechanical energy to the flow of an ideal fluid leads to **Bernoulli's equation:**

$$p + \tfrac{1}{2}\rho v^2 + \rho gy = \text{a constant} \qquad (14\text{-}29)$$

along any tube of flow.

Questions

1 Figure 14-21 shows a tank filled with water. Five horizontal floors and ceilings are indicated; all have the same area and are located at distances L, $2L$, or $3L$ below the top of the tank. Rank them according to the force on them due to the water, greatest first.

Fig. 14-21 Question 1.

2 *The teapot effect:* Water poured slowly from a teapot spout can double back under the spout for a considerable distance before detaching and falling. (The water layer is held against the underside of the spout by atmospheric pressure.) In Fig. 14-22, in the water layer inside the spout, point a is at the top of the layer and point b is at the bottom of the layer; in the water layer outside the spout, point c is at the top of the layer and point d is at the bottom of the layer. Rank those four points according to the gauge pressure in the water there, most positive first.

Fig. 14-22 Question 2.

3 Figure 14-23 shows four situations in which a red liquid and a gray liquid are in a U-tube. In one situation the liquids

Fig. 14-23 Question 3.

cannot be in static equilibrium. (a) Which situation is that? (b) For the other three situations, assume static equilibrium. For each, is the density of the red liquid greater than, less than, or equal to the density of the gray liquid?

4 We fully submerge an irregular 3 kg lump of material in a certain fluid. The fluid that would have been in the space now occupied by the lump has a mass of 2 kg. (a) When we release the lump, does it move upward, move downward, or remain in place? (b) If we next fully submerge the lump in a less dense fluid and again release it, what does it do?

5 Figure 14-24 shows three identical open-top containers filled to the brim with water; toy ducks float in two of them. Rank the containers and contents according to their weight, greatest first.

(a)   (b)   (c)

Fig. 14-24 Question 5.

6 A boat with an anchor on board floats in a swimming pool that is somewhat wider than the boat. Does the pool water level move up, move down, or remain the same if the anchor is (a) dropped into the water or (b) thrown onto the surrounding ground? (c) Does the water level in the pool move upward, move downward, or remain the same if, instead, a cork is dropped from the boat into the water, where it floats?

7 The gauge pressure p_g versus depth h is plotted in Fig. 14-25 for three liquids. For a rigid plastic bead fully submerged in each liquid, rank the plots according to the magnitude of the buoyant force on the bead, greatest first.

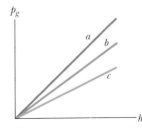

Fig. 14-25 Question 7.

8 A rectangular block is pushed face-down into three liquids, in turn. The apparent weight W_{app} of the block versus depth h in the three liquids is plotted in Fig. 14-26. Rank the liquids according to their weight per unit volume, greatest first.

Fig. 14-26 Question 8.

9 Water flows smoothly through a horizontal pipe. Figure 14-27 shows the kinetic energy K of a water element as it moves along an x axis that runs along the pipe. Rank the three lettered sections of the pipe according to the pipe radius, greatest first.

Fig. 14-27 Question 9.

10 Figure 14-28 shows four arrangements of pipes through which water flows smoothly toward the right. The radii of the pipe sections are indicated. In which arrangements is the net work done on a unit volume of water moving from the leftmost section to the rightmost section (a) zero, (b) positive, and (c) negative?

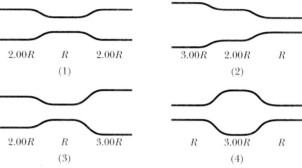

2.00R	R	2.00R
(1)		

3.00R	2.00R	R
(2)		

2.00R	R	3.00R
(3)		

R	3.00R	R
(4)		

Fig. 14-28 Question 10.

Problems

sec. 14-3 Density and Pressure

•**1** An office window has dimensions 3.4 m by 2.1 m. As a result of the passage of a storm, the outside air pressure drops to 0.96 atm, but inside the pressure is held at 1.0 atm. What net force pushes out on the window? **SSM**

•**2** Three liquids that will not mix are poured into a cylindrical container. The volumes and densities of the liquids are 0.50 L, 2.6 g/cm³; 0.25 L, 1.0 g/cm³; and 0.40 L, 0.80 g/cm³. What is the force on the bottom of the container due to these liquids? One liter = 1 L = 1000 cm³. (Ignore the contribution due to the atmosphere.)

•**3** Find the pressure increase in the fluid in a syringe when a nurse applies a force of 42 N to the syringe's circular piston, which has a radius of 1.1 cm. **SSM**

•**4** A partially evacuated airtight container has a tight-fitting lid of surface area 77 m² and negligible mass. If the force required to remove the lid is 480 N and the atmospheric pressure is 1.0×10^5 Pa, what is the air pressure in the container before it is opened?

•**5** A fish maintains its depth in fresh water by adjusting the air content of porous bone or air sacs to make its average density the same as that of the water. Suppose that with its air sacs collapsed, a fish has a density of 1.08 g/cm³. To what fraction of its expanded body volume must the fish inflate the air sacs to reduce its density to that of water? **ILW**

•**6** You inflate the front tires on your car to 28 psi. Later, you measure your blood pressure, obtaining a reading of 120/80, the readings being in mm Hg. In metric countries (which is to say, most of the world), these pressures are customarily reported in kilopascals (kPa). In kilopascals, what are (a) your tire pressure and (b) your blood pressure?

••**7** In 1654 Otto von Guericke, inventor of the air pump, gave a demonstration before the noblemen of the Holy Roman Empire in which two teams of eight horses could not pull apart two evacuated brass hemispheres. (a) Assuming the hemispheres have (strong) thin walls, so that R in Fig. 14-29 may be considered both the inside and outside radius, show that the force $\vec{F}$ required to pull apart the hemispheres has magnitude $F = \pi R^2 \, \Delta p$, where Δp is the difference between the pressures outside and inside the sphere. (b) Taking R as 30 cm, the inside pressure as 0.10 atm, and the outside pressure as 1.00 atm, find the force magnitude the teams of horses would have had to exert to pull apart the hemispheres. (c) Explain why one team of horses could have proved the point just as well if the hemispheres were attached to a sturdy wall. **SSM WWW**

Fig. 14-29 Problem 7.

sec. 14-4 Fluids at Rest

•**8** The maximum depth d_{max} that a diver can snorkel is set by the density of the water and the fact that human lungs can function against a maximum pressure difference (between inside and outside the chest cavity) of 0.050 atm. What is the difference in d_{max} for fresh water and the water of the Dead Sea (the saltiest natural water in the world, with a density of 1.5×10^3 kg/m³)?

•**9** Calculate the hydrostatic difference in blood pressure between the brain and the foot in a person of height 1.83 m. The density of blood is 1.06×10^3 kg/m³.

•**10** At a depth of 10.9 km, the Challenger Deep in the Marianas Trench of the Pacific Ocean is the deepest site in any ocean. Yet, in 1960, Donald Walsh and Jacques Piccard reached the Challenger Deep in the bathyscaph *Trieste*. Assuming that seawater has a uniform density of 1024 kg/m³, approximate the hydrostatic pressure (in atmospheres) that the *Trieste* had to withstand.

•11   Crew members attempt to escape from a damaged submarine 100 m below the surface. What force must be applied to a pop-out hatch, which is 1.2 m by 0.60 m, to push it out at that depth? Assume that the density of the ocean water is 1024 kg/m³. **SSM**

•12   What gauge pressure must a machine produce in order to suck mud of density 1800 kg/m³ up a tube by a height of 1.5 m?

•13   The plastic tube in Fig. 14-30 has a cross-sectional area of 5.00 cm². The tube is filled with water until the short arm (of length $d = 0.800$ m) is full. Then the short arm is sealed and more water is gradually poured into the long arm. If the seal will pop off when the force on it exceeds 9.80 N, what total height of water in the long arm will put the seal on the verge of popping?

Fig. 14-30   Problems 13 and 67.

•14   In Fig. 14-31, an open tube of length $L = 1.8$ m and cross-sectional area $A = 4.6$ cm² is fixed to the top of a cylindrical barrel of diameter $D = 1.2$ m and height $H = 1.8$ m. The barrel and tube are filled with water (to the top of the tube). Calculate the ratio of the hydrostatic force on the bottom of the barrel to the gravitational force on the water contained in the barrel. Why is that ratio not equal to 1.0? (You need not consider the atmospheric pressure.)

Fig. 14-31   Problem 14.

••15   Two identical cylindrical vessels with their bases at the same level each contain a liquid of density 1.30×10^3 kg/m³. The area of each base is 4.00 cm², but in one vessel the liquid height is 0.854 m and in the other it is 1.560 m. Find the work done by the gravitational force in equalizing the levels when the two vessels are connected. **SSM ILW**

••16   In analyzing certain geological features, it is often ap-

Fig. 14-32   Problem 16.

propriate to assume that the pressure at some horizontal *level of compensation*, deep inside Earth, is the same over a large region and is equal to the pressure due to the gravitational force on the overlying material. Thus, the pressure on the level of compensation is given by the fluid pressure formula. This model requires, for one thing, that mountains have *roots* of continental rock extending into the denser mantle (Fig. 14-32). Consider a mountain of height $H = 6.0$ km on a continent of thickness $T = 32$ km. The continental rock has a density of 2.9 g/cm³, and beneath this rock the mantle has a density of 3.3 g/cm³. Calculate the depth D of the root. (*Hint:* Set the pressure at points *a* and *b* equal; the depth *y* of the level of compensation will cancel out.)

••17   A large aquarium of height 5.00 m is filled with fresh water to a depth of 2.00 m. One wall of the aquarium consists of thick plastic 8.00 m wide. By how much does the total force on that wall increase if the aquarium is next filled to a depth of 4.00 m?

••18   The **L**-shaped tank shown in Fig. 14-33 is filled with water and is open at the top. If $d = 5.0$ m, what is the force due to the water (a) on face A and (b) on face B?

••19   In Fig. 14-34, water stands at depth $D = 35.0$ m behind the vertical upstream face of a dam of width $W = 314$ m. Find (a) the net horizontal force on the dam from the gauge pressure of the water and (b) the net torque due to that force about a line through O parallel to the width of the dam. (c) Find the moment arm of this torque. **SSM WWW**

Fig. 14-33   Problem 18.

sec. 14-5   Measuring Pressure

•20   To suck lemonade of density 1000 kg/m³ up a straw to a maximum height of 4.0 cm, what minimum gauge pressure (in atmospheres) must you produce in your lungs?

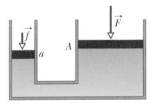

Fig. 14-34   Problem 19.

••21   What would be the height of the atmosphere if the air density (a) were uniform and (b) decreased linearly to zero with height? Assume that at sea level the air pressure is 1.0 atm and the air density is 1.3 kg/m³. **SSM**

sec. 14-6   Pascal's Principle

•22   A piston of cross-sectional area a is used in a hydraulic press to exert a small force of magnitude f on the enclosed liquid. A connecting pipe leads to a larger piston of cross-sectional area A (Fig. 14-35). (a) What force magnitude F will the larger piston sustain without moving? (b) If the piston diameters are 3.80 cm and 53.0 cm, what force magnitude on the small piston will balance a 20.0 kN force on the large piston?

Fig. 14-35   Problems 22 and 75.

••23 In Fig. 14-36, a spring of spring constant 3.00×10^4 N/m is between a rigid beam and the output piston of a hydraulic lever. An empty container with negligible mass sits on the input piston. The input piston has area A_i, and the output piston has area $18.0A_i$. Initially the spring is at its rest length. How many kilograms of sand must be (slowly) poured into the container to compress the spring by 5.00 cm?

Fig. 14-36 Problem 23.

sec. 14-7 Archimedes' Principle

•24 A boat floating in fresh water displaces water weighing 35.6 kN. (a) What is the weight of the water this boat displaces when floating in salt water of density 1.10×10^3 kg/m³? (b) What is the difference between the volume of fresh water displaced and the volume of salt water displaced?

•25 An iron anchor of density 7870 kg/m³ appears 200 N lighter in water than in air. (a) What is the volume of the anchor? (b) How much does it weigh in air? **SSM**

•26 In Fig. 14-37, a cube of edge length $L = 0.600$ m and mass 450 kg is suspended by a rope in an open tank of liquid of density 1030 kg/m³. Find (a) the magnitude of the total downward force on the top of the cube from the liquid and the atmosphere, assuming atmospheric pressure is 1.00 atm, (b) the magnitude of the total upward force on the bottom of the cube, and (c) the tension in the rope. (d) Calculate the magnitude of the buoyant force on the cube using Archimedes' principle. What relation exists among all these quantities?

Fig. 14-37 Problem 26.

•27 Three children, each of weight 356 N, make a log raft by lashing together logs of diameter 0.30 m and length 1.80 m. How many logs will be needed to keep them afloat in fresh water? Take the density of the logs to be 800 kg/m³.

•28 A flotation device is in the shape of a right cylinder, with a height of 0.500 m and a face area of 4.00 m² on top and bottom, and its density is 0.400 times that of fresh water. It is initially held fully submerged in fresh water, with its top face at the water surface. Then it is allowed to ascend gradually until it begins to float. How much work does the buoyant force do on the device during the ascent?

•29 A block of wood floats in fresh water with two-thirds of its volume V submerged and in oil with $0.90V$ submerged. Find the density of (a) the wood and (b) the oil. **SSM**

•30 A 5.00 kg object is released from rest while fully submerged in a liquid. The liquid displaced by the submerged object has a mass of 3.00 kg. How far and in what direction does the object move in 0.200 s, assuming that it moves freely and that the drag force on it from the liquid is negligible?

••31 A hollow sphere of inner radius 8.0 cm and outer radius 9.0 cm floats half-submerged in a liquid of density 800 kg/m³. (a) What is the mass of the sphere? (b) Calculate the density of the material of which it is made. **SSM WWW**

••32 A small solid ball is released from rest while fully submerged in a liquid and then its kinetic energy is measured when it has moved 4.0 cm in the liquid. Figure 14-38 gives the results after many liquids are used: The kinetic energy K is plotted versus the liquid density ρ_{liq}. What are (a) the density and (b) the volume of the ball?

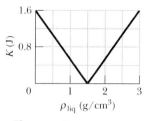

Fig. 14-38 Problem 32.

••33 A hollow spherical iron shell floats almost completely submerged in water. The outer diameter is 60.0 cm, and the density of iron is 7.87 g/cm³. Find the inner diameter. **ILW**

••34 In Fig. 14-39a, a rectangular block is gradually pushed face-down into a liquid. The block has height d; on the bottom and top the face area is $A = 5.67$ cm². Figure 14-39b gives the apparent weight W_{app} of the block as a function of the depth h of its lower face. What is the density of the liquid?

Fig. 14-39 Problem 34.

••35 An iron casting containing a number of cavities weighs 6000 N in air and 4000 N in water. What is the total volume of all the cavities in the casting? The density of iron (that is, a sample with no cavities) is 7.87 g/cm³. **SSM**

••36 Suppose that you release a small ball from rest at a depth of 0.600 m below the surface in a pool of water. If the density of the ball is 0.300 that of water and if the drag force on the ball from the water is negligible, how high above the water surface will the ball shoot as it emerges from the water? (Neglect any transfer of energy to the splashing and waves produced by the emerging ball.)

••37 The volume of air space in the passenger compartment of an 1800 kg car is 5.00 m³. The volume of the motor and front wheels is 0.750 m³, and the volume of the rear wheels, gas tank, and trunk is 0.800 m³; water cannot enter these two regions. The car rolls into a lake. (a) At first, no water enters the passenger compartment. How much of the car, in cubic meters, is below the water surface with the car floating (Fig. 14-40)? (b) As water slowly enters, the car sinks. How many cubic meters of water are in the car as it disappears below the water surface? (The car, with a heavy load in the trunk, remains horizontal.)

Fig. 14-40 Problem 37.

••38 A block of wood has a mass of 3.67 kg and a density of 600 kg/m³. It is to be loaded with lead (1.13×10^4 kg/m³) so that it will float in water with 0.900 of its volume submerged. What mass of lead is needed if the lead is attached to (a) the top of the wood and (b) the bottom of the wood?

••39 When researchers find a reasonably complete fossil of a dinosaur, they can determine the mass and weight of the living dinosaur with a scale model sculpted from plastic and based on the dimensions of the fossil bones. The scale of the model is 1/20; that is, lengths are 1/20 actual length, areas are $(1/20)^2$ actual areas, and

Fig. 14-41 Problem 39.

volumes are $(1/20)^3$ actual volumes. First, the model is suspended from one arm of a balance and weights are added to the other arm until equilibrium is reached. Then the model is fully submerged in water and enough weights are removed from the second arm to reestablish equilibrium (Fig. 14-41). For a model of a particular *T. rex* fossil, 637.76 g had to be removed to reestablish equilibrium. What was the volume of (a) the model and (b) the actual *T. rex*? (c) If the density of *T. rex* was approximately the density of water, what was its mass?

•••40 Figure 14-42 shows an iron ball suspended by thread of negligible mass from an upright cylinder that floats partially submerged in water. The cylinder has a height of 6.00

Fig. 14-42 Problem 40.

cm, a face area of 12.0 cm² on the top and bottom, and a density of 0.30 g/cm³, and 2.00 cm of its height is above the water surface. What is the radius of the iron ball?

sec. 14-9 The Equation of Continuity

•41 A garden hose with an internal diameter of 1.9 cm is connected to a (stationary) lawn sprinkler that consists merely of a container with 24 holes, each 0.13 cm in diameter. If the water in the hose has a speed of 0.91 m/s, at what speed does it leave the sprinkler holes? **SSM**

•42 Figure 14-43 shows the merging of two streams to form

Fig. 14-43 Problem 42.

a river. One stream has a width of 8.2 m, depth of 3.4 m, and current speed of 2.3 m/s. The other stream is 6.8 m wide and 3.2 m deep, and flows at 2.6 m/s. If the river has width 10.5 m and speed 2.9 m/s, what is its depth?

••43 Water is pumped steadily out of a flooded basement at a speed of 5.0 m/s through a uniform hose of radius 1.0 cm. The hose passes out through a window 3.0 m above the waterline. What is the power of the pump? **SSM**

••44 The water flowing through a 1.9 cm (inside diameter) pipe flows out through three 1.3 cm pipes. (a) If the flow rates in the three smaller pipes are 26, 19, and 11 L/min, what is the flow rate in the 1.9 cm pipe? (b) What is the ratio of the speed in the 1.9 cm pipe to that in the pipe carrying 26 L/min?

sec. 14-10 Bernoulli's Equation

•45 Water is moving with a speed of 5.0 m/s through a pipe with a cross-sectional area of 4.0 cm². The water gradually descends 10 m as the pipe cross-sectional area increases to 8.0 cm². (a) What is the speed at the lower level? (b) If the pressure at the upper level is 1.5×10^5 Pa, what is the pressure at the lower level? **SSM**

•46 The intake in Fig. 14-44 has cross-sectional area of 0.74 m² and water flow at 0.40 m/s. At the outlet, distance D = 180 m below the intake, the cross-sectional area is smaller than at the intake and the water flows out at 9.5 m/s.

Fig. 14-44 Problem 46.

What is the pressure difference between inlet and outlet?

•47 A water pipe having a 2.5 cm inside diameter carries water into the basement of a house at a speed of 0.90 m/s and a pressure of 170 kPa. If the pipe tapers to 1.2 cm and rises to the second floor 7.6 m above the input point, what are the (a) speed and (b) water pressure at the second floor? **ILW**

•48 Models of torpedoes are sometimes tested in a horizontal pipe of flowing water, much as a wind tunnel is used to test model airplanes. Consider a circular pipe of internal diameter 25.0 cm and a torpedo model aligned along the long axis of the pipe. The model has a 5.00 cm diameter and is to be tested with water flowing past it at 2.50 m/s. (a) With what speed must the water flow in the part of the pipe that is unconstricted by the model? (b) What will the pressure difference be between the constricted and unconstricted parts of the pipe?

•49 A cylindrical tank with a large diameter is filled with water to a depth D = 0.30 m. A hole of cross-sectional area A = 6.5 cm² in the bottom of the tank allows water to drain out. (a) What is the rate at which water flows out, in cubic meters per second? (b) At what distance below the bottom of the tank is the cross-sectional area of the stream equal to one-half the area of the hole? **SSM**

•50 Figure 14-45 shows two sections of an old pipe system

Fig. 14-45 Problem 50.

that runs through a hill, with distances $d_A = d_B = 30$ m and $D = 110$ m. On each side of the hill, the pipe radius is 2.00 cm. However, the radius of the pipe inside the hill is no longer known. To determine it, hydraulic engineers first establish that water flows through the left and right sections at 2.50 m/s. Then they release a dye in the water at point A and find that it takes 88.8 s to reach point B. What is the average radius of the pipe within the hill?

•51   How much work is done by pressure in forcing 1.4 m³ of water through a pipe having an internal diameter of 13 mm if the difference in pressure at the two ends of the pipe is 1.0 atm?

•52   Suppose that two tanks, 1 and 2, each with a large opening at the top, contain different liquids. A small hole is made in the side of each tank at the same depth h below the liquid surface, but the hole in tank 1 has half the cross-sectional area of the hole in tank 2. (a) What is the ratio ρ_1/ρ_2 of the densities of the liquids if the mass flow rate is the same for the two holes? (b) What is the ratio R_{V1}/R_{V2} of the volume flow rates from the two tanks? (c) At one instant, the liquid in tank 1 is 12.0 cm above the hole. If the tanks are to have *equal* volume flow rates, what height above the hole must the liquid in tank 2 be just then?

••53   In Fig. 14-46, the fresh water behind a reservoir dam has depth $D = 15$ m. A horizontal pipe 4.0 cm in diameter passes through the dam at depth $d = 6.0$ m. A plug secures the pipe opening. (a) Find the magnitude of the frictional force between plug and pipe wall. (b) The plug is removed. What water volume exits the pipe in 3.0 h? ILW

Fig. 14-46   Problem 53.

••54   In Fig. 14-47, water flows through a horizontal pipe and then out into the atmosphere at a speed $v_1 = 15$ m/s.

Fig. 14-47   Problem 54.

The diameters of the left and right sections of the pipe are 5.0 cm and 3.0 cm. (a) What volume of water flows into the atmosphere during a 10 min period? In the left section of the pipe, what are (b) the speed v_2 and (c) the gauge pressure?

••55   Fig. 14-48 shows a stream of water flowing through a hole at depth $h = 10$ cm in a tank holding water to height $H = 40$ cm. (a) At what distance x does the stream strike the floor? (b) At what depth should a second hole be made to give the same value of x? (c) At what depth should a hole be made to maximize x?

Fig. 14-48   Problem 55.

••56   Fresh water flows horizontally from pipe section 1 of cross-sectional area A_1 into pipe section 2 of cross-sectional area A_2. Figure 14-49 gives a plot of the pressure difference $p_2 - p_1$ versus the inverse area squared A_1^{-2} that would be expected for a volume flow rate of a certain value if the water

flow were laminar under all circumstances. For the conditions of the figure, what are the values of (a) A_2 and (b) the volume flow rate?

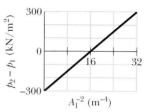

Fig. 14-49   Problem 56.

••57   A liquid of density 900 kg/m³ flows through a horizontal pipe that has a cross-sectional area of 1.90×10^{-2} m² in region A and a cross-sectional area of 9.50×10^{-2} m² in region B. The pressure difference between the two regions is 7.20×10^3 Pa. What are (a) the volume flow rate and (b) the mass flow rate?

••58   In Fig. 14-50, water flows steadily from the left pipe section (radius $r_1 = 2.00R$), through the middle section (radius R), and into the right section (radius $r_3 = 3.00R$). The speed of the water in the middle section is 0.500 m/s. What is the net work done on 0.400 m³ of the water as it moves from the left section to the right section?

Fig. 14-50   Problem 58.

••59   A *venturi meter* is used to measure the flow speed of a fluid in a pipe. The meter is connected between two sections of the pipe (Fig. 14-51); the cross-sectional area A of the entrance and exit of the meter matches the pipe's cross-sectional area. Between the entrance and exit, the fluid flows from the pipe with speed V and then through a narrow "throat" of cross-sectional area a with speed v. A manometer connects the wider portion of the meter to the narrower portion. The change in the fluid's speed is accompanied by a change Δp in the fluid's pressure, which causes a height difference h of the liquid in the two arms of the manometer. (Here Δp means pressure in the throat minus pressure in the pipe.) (a) By applying Bernoulli's equation and the equation of continuity to points 1 and 2 in Fig. 14-51, show that

$$V = \sqrt{\frac{2a^2\,\Delta p}{\rho(a^2 - A^2)}},$$

where ρ is the density of the fluid. (b) Suppose that the fluid is fresh water, that the cross-sectional areas are 64 cm² in the pipe and 32 cm² in the throat, and that the pressure is 55 kPa in the pipe and 41 kPa in the throat. What is the rate of water flow in cubic meters per second? SSM WWW

Fig. 14-51   Problems 59 and 60.

••60 Consider the venturi tube of Problem 59 and Fig. 14-51 without the manometer. Let A equal $5a$. Suppose the pressure p_1 at A is 2.0 atm. Compute the values of (a) the speed V at A and (b) the speed v at a that make the pressure p_2 at a equal to zero. (c) Compute the corresponding volume flow rate if the diameter at A is 5.0 cm. The phenomenon that occurs at a when p_2 falls to nearly zero is known as cavitation. The water vaporizes into small bubbles.

••61 A pitot tube (Fig. 14-52) is used to determine the airspeed of an airplane. It consists of an outer tube with a number of small holes B (four are shown) that allow air into the tube; that tube is connected to one arm of a U-tube. The other arm of the U-tube is connected to hole A at the front end of the device, which points in the direction the plane is headed. At A the air becomes stagnant so that $v_A = 0$. At B, however, the speed of the air presumably equals the airspeed v of the plane. (a) Use Bernoulli's equation to show that

$$v = \sqrt{\frac{2\rho g h}{\rho_{air}}},$$

where ρ is the density of the liquid in the U-tube and h is the difference in the liquid levels in that tube. (b) Suppose that the tube contains alcohol and the level difference h is 26.0 cm. What is the plane's speed relative to the air? The density of the air is 1.03 kg/m³ and that of alcohol is 810 kg/m³.

Fig. 14-52 Problems 61 and 62.

••62 A pitot tube (see Problem 61) on a high-altitude aircraft measures a differential pressure of 180 Pa. What is the aircraft's airspeed if the density of the air is 0.031 kg/m³?

•••63 A very simplified schematic of the rain drainage system for a home is shown in Fig. 14-53. Rain falling on the slanted roof runs off into gutters around the roof edge; it then

Fig. 14-53 Problem 63.

drains through downspouts (only one is shown) into a main drainage pipe M below the basement, which carries the water to an even larger pipe below the street. In Fig. 14-53, a floor drain in the basement is also connected to drainage pipe M. Suppose the following apply:

1. the downspouts have height $h_1 = 11$ m,
2. the floor drain has height $h_2 = 1.2$ m,
3. pipe M has radius 3.0 cm,
4. the house has side width $w = 30$ m and front length $L = 60$ m,
5. all the water striking the roof goes through pipe M,
6. the initial speed of the water in a downspout is negligible,
7. the wind speed is negligible (the rain falls vertically).

At what rainfall rate, in centimeters per hour, will water from pipe M reach the height of the floor drain and threaten to flood the basement?

Additional Problems

64 When you cough, you expel air at high speed through the trachea and upper bronchi so that the air will remove excess mucus lining the pathway. You produce the high speed by this procedure: You breathe in a large amount of air, trap it by closing the glottis (the narrow opening in the larynx), increase the air pressure by contracting the lungs, partially collapse the trachea and upper bronchi to narrow the pathway, and then expel the air through the pathway by suddenly reopening the glottis. Assume that during the expulsion the volume flow rate is 7.0×10^{-3} m³/s. What multiple of the speed of sound v_s ($= 343$ m/s) is the airspeed through the trachea if the trachea diameter (a) remains its normal value of 14 mm and (b) contracts to 5.2 mm?

65 When a rectangular block of height 8.00 cm floats in liquid 1, 6.00 cm of that height is above the liquid surface. What height of the block will be above the liquid surface when the block floats in liquid 2, which has 0.500 times the density of liquid 1?

66 A glass ball of radius 2.00 cm sits at the bottom of a container of milk that has a density of 1.03 g/cm³. The normal force on the ball from the container's lower surface has magnitude 9.48×10^{-2} N. What is the mass of the ball?

67 Figure 14-30 shows a modified U-tube: the right arm is shorter than the left arm. The open end of the right arm is height $d = 10.0$ cm above the laboratory bench. The radius throughout the tube is 1.50 cm. Water is gradually poured into the open end of the left arm until the water begins to flow out the open end of the right arm. Then a liquid of density 0.80 g/cm³ is gradually added to the left arm until its height in that arm is 8.0 cm (it does not mix with the water). How much water flows out of the right arm?

68 Caught in an avalanche, a skier is fully submerged in flowing snow of density 96 kg/m³. Assume that the average density of the skier, clothing, and skiing equipment is 1020 kg/m³. What percentage of the gravitational force on the skier is offset by the buoyant force from the snow?

69 Figure 14-54 shows a *siphon*, which is a device for removing liquid from a container. Tube ABC must initially be filled, but once this has been done, liquid will flow through the tube until the liquid surface in the container is level with the tube opening at A. The liquid has density 1000 kg/m^3 and negligible viscosity. The distances shown are $h_1 = 25$ cm, $d = 12$ cm, and $h_2 = 40$ cm. (a) With what speed does the liquid emerge from the tube at C? (b) If the atmospheric pressure is 1.0×10^5 Pa, what is the pressure in the liquid at the topmost point B? (c) Theoretically, what is the greatest possible height h_1 that a siphon can lift water?

Fig. 14-54   Problem 69.

70   Suppose that your body has a uniform density of 0.95 times that of water. (a) If you float in a swimming pool, what fraction of your body's volume is above the water surface?

Quicksand is a fluid produced when water is forced up into sand, moving the sand grains away from one another so they are no longer locked together by friction. Pools of quicksand can form when water drains underground from hills into valleys where there are sand pockets. (b) If you step into a deep pool of quicksand that has a density 1.6 times that of water, what fraction of your body's volume is above the quicksand surface? (c) In particular, are you submerged enough to be unable to breathe?

71   The dinosaur *Diplodocus* was enormous, with long neck and tail and a mass that was great enough to test its leg strength (Fig. 14-55). According to conjecture, *Diplodocus* waded in water, perhaps up to its head, so that buoyancy could offset its weight and lighten the load on its legs. To check the conjecture, take the density of *Diplodocus* to be

Fig. 14-55   Problem 71.

0.90 that of water, and assume that its mass was the published estimate of 1.85×10^4 kg. (a) What then would be its actual weight? Find its apparent weight when it had the following fractions of its volume submerged: (b) 0.50, (c) 0.80, and (d) 0.90. When submerged with only its head above water, its lungs would have been about 8.0 m below the water surface. (e) At that depth, what would be the difference between the (external) water pressure and the pressure of the air in its lungs? For the dinosaur to breathe in, its lung muscles would have had to expand its lungs against this pressure difference. It probably could not do so against a pressure difference of more than 8 kPa. (f) Did *Diplodocus* wade as conjectured?

72   If a bubble in sparkling water accelerates upward at the rate of 0.225 m/s^2 and has a radius of 0.500 mm, what is its mass? Assume that the drag force on the bubble is negligible.

73   An opening of area 0.25 cm^2 in an otherwise closed beverage keg is 50 cm below the level of the liquid (of density 1.0 g/cm^3) in the keg. What is the speed of the liquid flowing through the opening if the gauge pressure in the air space above the liquid is (a) zero and (b) 0.40 atm?

74   The cross-sectional area of the aorta (the major blood vessel emerging from the heart) of a normal resting person is 3 cm^2, and the speed of the blood through it is 30 cm/s. A typical capillary (diameter $\approx$ 6 μm) has a cross-sectional area of 3×10^{-7} cm^2 and a flow speed of 0.05 cm/s. How many capillaries does such a person have?

75   In the hydraulic press of Problem 22, through what distance must the large piston be moved to raise the small piston a distance of 0.85 m?

76   What is the acceleration of a rising hot-air balloon if the ratio of the air density outside the balloon to that inside is 1.39? Neglect the mass of the balloon fabric and the basket.

77   Water flows through a horizontal pipe that widens at two points; initial radius = 0.200 m, intermediate radius = 0.400 m, and final radius = 0.600 m. If the initial speed of the water's flow is 9.00×10^{-2} m/s, what is the final speed of the flow?

78   A lead sinker of volume 0.40 cm^3 and density 11.4 g/cm^3 is used in fishing. The sinker is suspended from a vertical string whose other end is attached to the bottom of a spherical cork (of density 0.20 g/cm^3) that is floating on the surface of a lake. Neglecting the effects of the line, hook, and bait, determine what the radius of the cork must be if it is to float with half its volume submerged.

79   A surrealistic plumber has constructed the pipe system shown in Fig. 14-56, where the radii of the pipe sections are given in terms of $R_B = 2.0$ cm. The pipe sections have vertical *standpipes*, which are tubes that are too tall for water to be pushed out of them by the pressure in the pipe system. The tubes are open at the top to an air pressure of 1.0 atm. (a) If the water extends up into the standpipe for section D by 0.50 m, and the volume flow rate through the pipe sections is 2.0×10^{-3} m^3/s, then how high does the water extend up into the standpipe for section B? (b) Without further written calculations, rank the standpipes according to how far water extends up into them, greatest first.

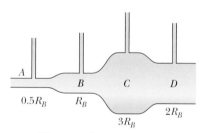

Fig. 14-56   Problem 79.

80   Find the absolute pressure, in pascals, at a depth of 150 m in the ocean. The density of seawater is 1.03 g/cm^3, and the atmospheric pressure at sea level is 1.01×10^5 Pa.

81   A tin can has a total volume of 1200 cm^3 and a mass of 130 g. How many grams of lead shot of density 11.4 g/cm^3 could it carry without sinking in water?

82   A simple open U-tube contains mercury. When 11.2 cm of water is poured into the right arm of the tube, how high above its initial level does the mercury rise in the left arm?

83   An object hangs from a spring balance. The balance registers 30 N in air, 20 N when this object is immersed in water,

and 24 N when the object is immersed in another liquid of unknown density. What is the density of that other liquid?

84 A 7.00 kg sphere of radius 5.00 cm is at a depth of 1.20 km in seawater that has an average density of 1025 kg/m³. What are the (a) gauge pressure, (b) total pressure, and (c) corresponding total force compressing the sphere's surface? What are (d) the magnitude of the buoyant force on the sphere and (e) the magnitude and (f) the direction of the sphere's acceleration if it is free to move?

85 A river 20 m wide and 4.0 m deep drains a 3000 km² land area in which the average precipitation is 48 cm/y. One-fourth of this rainfall returns to the atmosphere by evaporation, but the remainder ultimately drains into the river. What is the average speed of the river current?

86 In an experiment, a rectangular block with height h is allowed to float in four separate liquids. In the first liquid, which is water, it floats fully submerged. In liquids A, B, and C, it floats with heights $h/2$, $2h/3$, and $h/4$ above the liquid surface, respectively. What are the *relative densities* (the densities relative to that of water) of (a) A, (b) B, and (c) C?

87 According to one government specification, the "black box" recorders on airplanes must be able to withstand water pressures down to depths of 6.0 km so that they can be recovered in case of a crash over water. Assuming that seawater has a uniform density of 1024 kg/m³, find the pressure corresponding to that government specification.

88 Partially fill a tall drinking glass with water to depth h. Cut a square of sturdy paper somewhat wider than the mouth of the glass. Place the paper over the mouth (Fig. 14-57a). Spread the fingers of your left hand over the paper, pressing it against the mouth of

Fig. 14-57 Problem 88.

the glass. Grab the glass with your right hand and then as rapidly as you can, invert it with your left hand still pressing the paper against the rim. Chances are you can then remove your left hand without the water pouring out (Fig. 14-57b). The paper bulges downward slightly but stays against the rim. If $h = 11.0$ cm, what is the gauge pressure of the air now trapped in the glass above the water?

89 You have been asked to review plans for a 10.0 m by 10.0 m swimming pool in a new hotel. The water is to be supplied by a horizontal pipe of radius $R_1 = 6.00$ cm, with water under pressure of 2.00 atm. A vertical pipe of radius $R_2 = 1.00$ cm is to carry the water to a height of 9.40 m, where it is to pour out freely into the pool until the water depth is 2.00 m. (a) How much time will be required to fill the pool? (b) If more than a few days is considered unacceptable and less than a few hours is considered dangerous, is the filling time acceptable and safe?

90 A vertical cylinder holds liquid 1 (density 1.20 g/cm³, height 8.00 cm) and liquid 2 (density 2.00 g/cm³, height 4.00 cm), initially separated, with liquid 1 above liquid 2, but then well stirred. Assuming that liquids 1 and 2 are uniformly mixed and that the total volume does not change, what is the density of the mixture?

91 About one-third of the body of a person floating in the Dead Sea will be above the waterline. Assuming that the human body density is 0.98 g/cm³, find the density of the water in the Dead Sea. (Why is it so much greater than 1.0 g/cm³?)

92 (a) What is the minimum area of the top surface of an ice slab 0.30 m thick floating on fresh water that will hold up an 1100 kg automobile? (b) Does it matter where the car is placed on the slab?

93 (a) For seawater of density 1.03 g/cm³, find the weight of water on top of a submarine at a depth of 200 m if the horizontal cross-sectional hull area is 3000 m². (b) In atmospheres, what water pressure would a diver experience at this depth? (c) Could occupants of a damaged submarine at this depth escape without special equipment?

94 A uniform block of length 5.0 cm, width 4.0 cm, and height 2.0 cm floats in seawater of density 1025 kg/m³, with one broad face 1.5 cm below the surface. What is the block's mass?

95 A cube with a total surface area of 24 m² floats upright in water. If the density of water is 4.00 times the density of the cube, how far does the cube sink into the water?

96 You place a glass beaker, partially filled with water, in a sink (Fig. 14-58). The beaker has a mass of 390 g and an interior volume of 500 cm³. You now start to fill the sink with water and find that if the beaker is less than half full, it floats, but if it is more than half

Fig. 14-58 Problem 96.

full, it remains on the bottom of the sink as the water level in the sink rises to the rim of the beaker. What is the density of the glass of which the beaker is made?

97 Figure 14-59 shows a dam and part of the fresh-water reservoir behind it. The dam is made of concrete of density 3.2 g/cm³; the distances shown are $d_1 = 180$ m, $d_2 = 71$ m, $d_3 = 24$ m, and $d_4 = 48$ m. The water pushes horizontally on the dam and is resisted by the force of static friction between the dam and the bedrock on which it rests. The coefficient of static friction is 0.47. (a) Calculate the safety factor against slid-

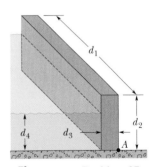

Fig. 14-59 Problem 97.

ing—that is, the ratio of the magnitudes of the maximum friction force to the force of the water. The water also tries to rotate the dam about a line running along the base through point A. The torque due to the gravitational force on the dam acts in the opposite sense about that line. (b) Calculate the safety factor against rotation—that is, the ratio of the magnitudes of the torque due to the gravitational force to the torque due to the net force from the water and the air.

98 Assume the density of brass weights to be 8.0 g/cm³ and that of air to be 0.0012 g/cm³. What percent error arises from neglecting the buoyancy of air in weighing an object of mass m and density ρ on a beam balance, as in Fig. 5-6?

15 _Oscillations_

The springboard used in modern diving competitions is a complex mechanical system that a skilled diver must master. In a running dive, for example, a skilled diver knows how to first take three quick steps along the board to start the board oscillating and then leap to the free end of the board so as to be catapulted high into the air. A novice diver can imitate that procedure but fail miserably to be catapulted upward and might even be knocked off the board.

What is the "secret" of a skilled diver's high catapult?

The answer is in this chapter.

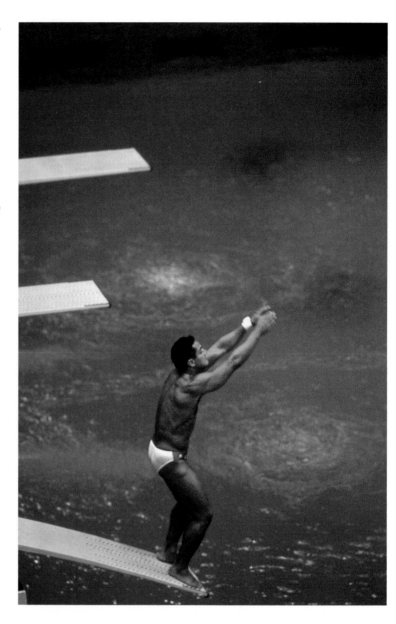

15-1 *What Is Physics?*

Our world is filled with oscillations in which objects move back and forth repeatedly. Many oscillations are merely amusing or annoying, but many others are financially important or dangerous. Here are a few examples: When a bat hits a baseball, the bat may oscillate enough to sting the batter's hands or even to break apart. When wind blows past a power line, the line may oscillate ("gallop" in electrical engineering terms) so severely that it rips apart, shutting off the power supply to a community. When an airplane is in flight, the turbulence of the air flowing past the wings makes them oscillate, eventually leading to metal fatigue and even failure. When a train travels around a curve, its wheels oscillate horizontally ("hunt" in mechanical engineering terms) as they are forced to turn in new directions (you can hear the oscillations).

When an earthquake occurs near a city, buildings may be set oscillating so severely that they are shaken apart. When an arrow is shot from a bow, the feathers at the end of the arrow manage to snake around the bow staff without hitting it because the arrow oscillates. When a coin drops into a metal collection plate, the coin oscillates with such a familiar ring that the coin's denomination can be determined from the sound. When a rodeo cowboy rides a bull, the cowboy oscillates wildly as the bull jumps and turns (at least the cowboy hopes to be oscillating).

The study and control of oscillations are two of the primary goals of both physics and engineering. In this book we discuss a basic type of oscillation called *simple harmonic motion.*

15-2 *Simple Harmonic Motion*

Figure 15-1*a* shows a sequence of "snapshots" of a simple oscillating system, a particle moving repeatedly back and forth about the origin of an *x* axis. In this section we simply describe the motion. Later, we shall discuss how to attain such motion.

One important property of oscillatory motion is its **frequency,** or number of oscillations that are completed each second. The symbol for frequency is *f*, and its SI unit is the **hertz** (abbreviated Hz), where

$$1 \text{ hertz} = 1 \text{ Hz} = 1 \text{ oscillation per second} = 1 \text{ s}^{-1}. \quad (15\text{-}1)$$

(b)

(a)

Fig. 15-1 (*a*) A sequence of "snapshots" (taken at equal time intervals) showing the position of a particle as it oscillates back and forth about the origin of an *x* axis, between the limits $+x_m$ and $-x_m$. The vector arrows are scaled to indicate the speed of the particle. The speed is maximum when the particle is at the origin and zero when it is at $\pm x_m$. If the time *t* is chosen to be zero when the particle is at $+x_m$, then the particle returns to $+x_m$ at $t = T$, where *T* is the period of the motion. The motion is then repeated. (*b*) A graph of *x* as a function of time for the motion of (*a*).

388   *Chapter 15*   Oscillations

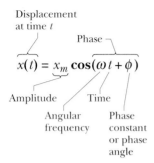

$$x(t) = x_m \cos(\omega t + \phi)$$

Displacement at time *t*

Phase

Amplitude

Time

Angular frequency

Phase constant or phase angle

Fig. 15-2 A handy reference to the quantities in Eq. 15-3 for simple harmonic motion.

Related to the frequency is the **period** *T* of the motion, which is the time for one complete oscillation (or **cycle**); that is,

$$T = \frac{1}{f}. \qquad (15\text{-}2)$$

Any motion that repeats itself at regular intervals is called **periodic motion** or **harmonic motion**. We are interested here in motion that repeats itself in a particular way—namely, like that in Fig. 15-1*a*. For such motion the displacement *x* of the particle from the origin is given as a function of time by

$$x(t) = x_m \cos(\omega t + \phi) \quad \text{(displacement)}, \qquad (15\text{-}3)$$

in which x_m, ω, and ϕ are constants. This motion is called **simple harmonic motion** (SHM), a term that means the periodic motion is a sinusoidal function of time. Equation 15-3, in which the sinusoidal function is a cosine function, is graphed in Fig. 15-1*b*. (You can get that graph by rotating Fig. 15-1*a* counterclockwise by 90° and then connecting the successive locations of the particle with a curve.) The quantities that determine the shape of the graph are displayed in Fig. 15-2 with their names. We now shall define those quantities.

The quantity x_m, called the **amplitude** of the motion, is a positive constant whose value depends on how the motion was started. The subscript *m* stands for *maximum* because the amplitude is the magnitude of the maximum displacement of the particle in either direction. The cosine function in Eq. 15-3 varies between the limits ± 1; so the displacement $x(t)$ varies between the limits $\pm x_m$.

The time-varying quantity $(\omega t + \phi)$ in Eq. 15-3 is called the **phase** of the motion, and the constant ϕ is called the **phase constant** (or **phase angle**). The value of ϕ depends on the displacement and velocity of the particle at time *t* = 0. For the $x(t)$ plots of Fig. 15-3*a*, the phase constant ϕ is zero.

To interpret the constant ω, called the **angular frequency** of the motion, we first note that the displacement $x(t)$ must return to its initial value after one period *T* of the motion; that is, $x(t)$ must equal $x(t + T)$ for all *t*. To simplify this analysis, let us put $\phi = 0$ in Eq. 15-3. From that equation we then can write

$$x_m \cos \omega t = x_m \cos \omega(t + T). \qquad (15\text{-}4)$$

The cosine function first repeats itself when its argument (the phase) has increased by 2π rad; so Eq. 15-4 gives us

$$\omega(t + T) = \omega t + 2\pi$$

or

$$\omega T = 2\pi.$$

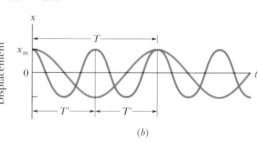

Fig. 15-3 In all three cases, the blue curve is obtained from Eq. 15-3 with $\phi = 0$. (*a*) The red curve differs from the blue curve *only* in that the red-curve amplitude x'_m is greater (the red-curve extremes of displacement are higher and lower). (*b*) The red curve differs from the blue curve *only* in that the red-curve period is $T' = T/2$ (the red curve is compressed horizontally). (*c*) The red curve differs from the blue curve *only* in that for the red-curve $\phi = -\pi/4$ rad rather than zero (the negative value of ϕ shifts the red curve to the right).

Thus, from Eq. 15-2 the angular frequency is

$$\omega = \frac{2\pi}{T} = 2\pi f. \qquad (15\text{-}5)$$

The SI unit of angular frequency is the radian per second. (To be consistent, then, ϕ must be in radians.) Figure 15-3 compares $x(t)$ for two simple harmonic motions that differ either in amplitude, in period (and thus in frequency and angular frequency), or in phase constant.

✔ **CHECKPOINT 1**   A particle undergoing simple harmonic oscillation of period T (like that in Fig. 15-1) is at $-x_m$ at time $t = 0$. Is it at $-x_m$, at $+x_m$, at 0, between $-x_m$ and 0, or between 0 and $+x_m$ when (a) $t = 2.00T$, (b) $t = 3.50T$, and (c) $t = 5.25T$?

The Velocity of SHM

By differentiating Eq. 15-3, we can find an expression for the velocity of a particle moving with simple harmonic motion; that is,

$$v(t) = \frac{dx(t)}{dt} = \frac{d}{dt}[x_m \cos(\omega t + \phi)]$$

or $\qquad\qquad v(t) = -\omega x_m \sin(\omega t + \phi) \qquad \text{(velocity).} \qquad (15\text{-}6)$

Figure 15-4a is a plot of Eq. 15-3 with $\phi = 0$. Figure 15-4b shows Eq. 15-6, also with $\phi = 0$. Analogous to the amplitude x_m in Eq. 15-3, the positive quantity ωx_m in Eq. 15-6 is called the **velocity amplitude** v_m. As you can see in Fig. 15-4b, the velocity of the oscillating particle varies between the limits $\pm v_m = \pm \omega x_m$. Note also in that figure that the curve of $v(t)$ is *shifted* (to the left) from the curve of $x(t)$ by one-quarter period; when the magnitude of the displacement is greatest (that is, $x(t) = x_m$), the magnitude of the velocity is least (that is, $v(t) = 0$). When the magnitude of the displacement is least (that is, zero), the magnitude of the velocity is greatest (that is, $v_m = \omega x_m$).

The Acceleration of SHM

Knowing the velocity $v(t)$ for simple harmonic motion, we can find an expression for the acceleration of the oscillating particle by differentiating once more. Thus, we have, from Eq. 15-6,

$$a(t) = \frac{dv(t)}{dt} = \frac{d}{dt}[-\omega x_m \sin(\omega t + \phi)]$$

or $\qquad\qquad a(t) = -\omega^2 x_m \cos(\omega t + \phi) \qquad \text{(acceleration).} \qquad (15\text{-}7)$

Figure 15-4c is a plot of Eq. 15-7 for the case $\phi = 0$. The positive quantity $\omega^2 x_m$ in Eq. 15-7 is called the **acceleration amplitude** a_m; that is, the acceleration of the particle varies between the limits $\pm a_m = \pm \omega^2 x_m$, as Fig. 15-4c shows. Note also that the acceleration curve $a(t)$ is shifted (to the left) by $\frac{1}{4}T$ relative to the velocity curve $v(t)$.

We can combine Eqs. 15-3 and 15-7 to yield

$$a(t) = -\omega^2 x(t), \qquad (15\text{-}8)$$

which is the hallmark of simple harmonic motion:

➤   In SHM, the acceleration is proportional to the displacement but opposite in sign, and the two quantities are related by the square of the angular frequency.

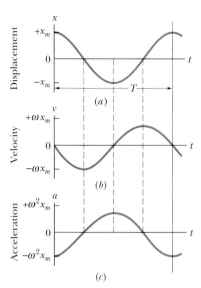

Fig. 15-4   (*a*) The displacement $x(t)$ of a particle oscillating in SHM with phase angle ϕ equal to zero. The period T marks one complete oscillation. (*b*) The velocity $v(t)$ of the particle. (*c*) The acceleration $a(t)$ of the particle.

Thus, as Fig. 15-4 shows, when the displacement has its greatest positive value, the acceleration has its greatest negative value, and conversely. When the displacement is zero, the acceleration is also zero.

PROBLEM-SOLVING TACTICS

TACTIC 1: *Phase Angles*

Note the effect of the phase angle ϕ on a plot of $x(t)$. When $\phi = 0$, $x(t)$ has a graph like that in Fig. 15-4a, a typical cosine curve. Increasing ϕ shifts the curve leftward along the t axis. (You might remember this with the symbol $\dashv\phi$, where the up arrow indicates an increase in ϕ and the left arrow indicates the resulting shift in the curve.) Decreasing ϕ shifts the curve rightward, as in Fig. 15-3c for $\phi = -\pi/4$.

Two plots of SHM with different phase angles are said to have a *phase difference*; each is said to be *phase-shifted* from the other, or *out of phase* with the other. The curves in Fig. 15-3c, for example, have a phase difference of $\pi/4$ rad.

Because SHM repeats after each period T and the cosine function repeats after each 2π rad, one period T represents a phase difference of 2π rad. In Fig. 15-4, $x(t)$ is phase-shifted to the right from $v(t)$ by one-quarter period, or $-\pi/2$ rad; it is shifted to the right from $a(t)$ by one-half period, or $-\pi$ rad. A phase shift of 2π rad causes a curve of SHM to coincide with itself; that is, it looks unchanged.

15-3 The Force Law for Simple Harmonic Motion

Once we know how the acceleration of a particle varies with time, we can use Newton's second law to learn what force must act on the particle to give it that acceleration. If we combine Newton's second law and Eq. 15-8, we find, for simple harmonic motion,

$$F = ma = -(m\omega^2)x. \qquad (15\text{-}9)$$

This result—a restoring force that is proportional to the displacement but opposite in sign—is familiar. It is Hooke's law,

$$F = -kx, \qquad (15\text{-}10)$$

for a spring, the spring constant here being

$$k = m\omega^2. \qquad (15\text{-}11)$$

We can in fact take Eq. 15-10 as an alternative definition of simple harmonic motion. It says:

> Simple harmonic motion is the motion executed by a particle subject to a force that is proportional to the displacement of the particle but opposite in sign.

The block–spring system of Fig. 15-5 forms a **linear simple harmonic oscillator** (linear oscillator, for short), where "linear" indicates that F is proportional to x rather than to some other power of x. The angular frequency ω of the simple harmonic motion of the block is related to the spring constant k and the mass m of the block by Eq. 15-11, which yields

$$\omega = \sqrt{\frac{k}{m}} \qquad \text{(angular frequency).} \qquad (15\text{-}12)$$

By combining Eqs. 15-5 and 15-12, we can write, for the **period** of the linear oscillator of Fig. 15-5,

$$T = 2\pi\sqrt{\frac{m}{k}} \qquad \text{(period).} \qquad (15\text{-}13)$$

Equations 15-12 and 15-13 tell us that a large angular frequency (and thus a small period) goes with a stiff spring (large k) and a light block (small m).

Fig. 15-5 A linear simple harmonic oscillator. The surface is frictionless. Like the particle of Fig. 15-1, the block moves in simple harmonic motion once it has been either pulled or pushed away from the $x = 0$ position and released. Its displacement is then given by Eq. 15-3.

Every oscillating system, be it a diving board or a violin string, has some element of "springiness" and some element of "inertia" or mass, and thus resembles a linear oscillator. In the linear oscillator of Fig. 15-5, these elements are located in separate parts of the system: The springiness is entirely in the spring, which we assume to be massless, and the inertia is entirely in the block, which we assume to be rigid. In a violin string, however, the two elements are both within the string, as you will see in Chapter 16.

✓**CHECKPOINT 2**   Which of the following relationships between the force F on a particle and the particle's position x implies simple harmonic oscillation: (a) $F = -5x$, (b) $F = -400x^2$, (c) $F = 10x$, (d) $F = 3x^2$?

Sample Problem 15-1

A block whose mass m is 680 g is fastened to a spring whose spring constant k is 65 N/m. The block is pulled a distance $x = 11$ cm from its equilibrium position at $x = 0$ on a frictionless surface and released from rest at $t = 0$.

(a) What are the angular frequency, the frequency, and the period of the resulting motion?

Solution: The **Key Idea** here is that the block–spring system forms a linear simple harmonic oscillator, with the block undergoing SHM. Then the angular frequency is given by Eq. 15-12:

$$\omega = \sqrt{\frac{k}{m}} = \sqrt{\frac{65 \text{ N/m}}{0.68 \text{ kg}}} = 9.78 \text{ rad/s}$$

$$\approx 9.8 \text{ rad/s.} \qquad \text{(Answer)}$$

The frequency follows from Eq. 15-5, which yields

$$f = \frac{\omega}{2\pi} = \frac{9.78 \text{ rad/s}}{2\pi \text{ rad}} = 1.56 \text{ Hz} \approx 1.6 \text{ Hz.} \quad \text{(Answer)}$$

The period follows from Eq. 15-2, which yields

$$T = \frac{1}{f} = \frac{1}{1.56 \text{ Hz}} = 0.64 \text{ s} = 640 \text{ ms.} \qquad \text{(Answer)}$$

(b) What is the amplitude of the oscillation?

Solution: The **Key Idea** here is that, with no friction involved, the mechanical energy of the spring–block system is conserved. The block is released from rest 11 cm from its equilibrium position, with zero kinetic energy and the elastic potential energy of the system at a maximum. Thus, the block will have zero kinetic energy whenever it is again 11 cm from its equilibrium position, which means it will never be farther than 11 cm from that position. Its maximum displacement is 11 cm:

$$x_m = 11 \text{ cm.} \qquad \text{(Answer)}$$

(c) What is the maximum speed v_m of the oscillating block, and where is the block when it has this speed?

Solution: The **Key Idea** here is that the maximum speed v_m is the velocity amplitude ωx_m in Eq. 15-6; that is,

$$v_m = \omega x_m = (9.78 \text{ rad/s})(0.11 \text{ m})$$

$$= 1.1 \text{ m/s.} \qquad \text{(Answer)}$$

This maximum speed occurs when the oscillating block is rushing through the origin; compare Figs. 15-4a and 15-4b, where you can see that the speed is a maximum whenever $x = 0$.

(d) What is the magnitude a_m of the maximum acceleration of the block?

Solution: The **Key Idea** this time is that the magnitude a_m of the maximum acceleration is the acceleration amplitude $\omega^2 x_m$ in Eq. 15-7; that is,

$$a_m = \omega^2 x_m = (9.78 \text{ rad/s})^2(0.11 \text{ m})$$

$$= 11 \text{ m/s}^2. \qquad \text{(Answer)}$$

This maximum acceleration occurs when the block is at the ends of its path. At those points, the force acting on the block has its maximum magnitude; compare Figs. 15-4a and 15-4c, where you can see that the magnitudes of the displacement and acceleration are maximum at the same times.

(e) What is the phase constant ϕ for the motion?

Solution: Here the **Key Idea** is that Eq. 15-3 gives the displacement of the block as a function of time. We know that at time $t = 0$, the block is located at $x = x_m$. Substituting these *initial conditions*, as they are called, into Eq. 15-3 and canceling x_m give us

$$1 = \cos \phi. \qquad (15\text{-}14)$$

Taking the inverse cosine then yields

$$\phi = 0 \text{ rad.} \qquad \text{(Answer)}$$

(Any angle that is an integer multiple of 2π rad also satisfies Eq. 15-14; we chose the smallest angle.)

(f) What is the displacement function $x(t)$ for the spring–block system?

Solution: The **Key Idea** here is that $x(t)$ is given in general form by Eq. 15-3. Substituting known quantities into that equation gives us

$$x(t) = x_m \cos(\omega t + \phi)$$

$$= (0.11 \text{ m}) \cos[(9.8 \text{ rad/s})t + 0]$$

$$= 0.11 \cos(9.8t), \qquad \text{(Answer)}$$

where x is in meters and t is in seconds.

Sample Problem 15-2

At $t = 0$, the displacement $x(0)$ of the block in a linear oscillator like that of Fig. 15-5 is -8.50 cm. (Read $x(0)$ as "x at time zero.") The block's velocity $v(0)$ then is -0.920 m/s, and its acceleration $a(0)$ is $+47.0$ m/s^2.

(a) What is the angular frequency ω of this system?

Solution: A **Key Idea** here is that, with the block in SHM, Eqs. 15-3, 15-6, and 15-7 give its displacement, velocity, and acceleration, respectively, and each contains ω. Let's substitute $t = 0$ into each to see whether we can solve any one of them for ω. We find

$$x(0) = x_m \cos \phi, \qquad (15\text{-}15)$$

$$v(0) = -\omega x_m \sin \phi, \qquad (15\text{-}16)$$

and

$$a(0) = -\omega^2 x_m \cos \phi. \qquad (15\text{-}17)$$

In Eq. 15-15, ω has disappeared. In Eqs. 15-16 and 15-17, we know values for the left sides, but we do not know x_m and ϕ. However, if we divide Eq. 15-17 by Eq. 15-15, we neatly eliminate both x_m and ϕ and can then solve for ω as

$$\omega = \sqrt{-\frac{a(0)}{x(0)}} = \sqrt{-\frac{47.0 \text{ m/s}^2}{-0.0850 \text{ m}}}$$

$$= 23.5 \text{ rad/s.} \qquad \text{(Answer)}$$

(b) What are the phase constant ϕ and amplitude x_m?

Solution: The same **Key Idea** as in part (a) also applies here, as do Eqs. 15-15 through 15-17. Now, however, we know ω and want ϕ and x_m. If we divide Eq. 15-16 by Eq. 15-15, we find

$$\frac{v(0)}{x(0)} = \frac{-\omega x_m \sin \phi}{x_m \cos \phi} = -\omega \tan \phi.$$

Solving for $\tan \phi$, we find

$$\tan \phi = -\frac{v(0)}{\omega x(0)} = -\frac{-0.920 \text{ m/s}}{(23.5 \text{ rad/s})(-0.0850 \text{ m})}$$

$$= -0.461.$$

This equation has two solutions:

$$\phi = -25° \quad \text{and} \quad \phi = 180° + (-25°) = 155°.$$

(Normally only the first solution here is displayed by a calculator.) A **Key Idea** in choosing the proper solution is to test them both by using them to compute values for the amplitude x_m. From Eq. 15-15, we find that if $\phi = -25°$, then

$$x_m = \frac{x(0)}{\cos \phi} = \frac{-0.0850 \text{ m}}{\cos(-25°)} = -0.094 \text{ m.}$$

We find similarly that if $\phi = 155°$, then $x_m = 0.094$ m. Because the amplitude of SHM must be a positive constant, the correct phase constant and amplitude here are

$$\phi = 155° \quad \text{and} \quad x_m = 0.094 \text{ m} = 9.4 \text{ cm.} \quad \text{(Answer)}$$

PROBLEM-SOLVING TACTICS

TACTIC 2: *Identifying SHM*

In linear SHM the acceleration a and displacement x of the system are related by an equation of the form

$$a = -(\text{a positive constant})x,$$

which says that the acceleration is proportional to the displacement from the equilibrium position but is in the opposite direction. Once you find such an expression for an oscillating system, you can immediately compare it with Eq. 15-8, identify the positive constant as being equal to ω^2, and so quickly get an expression for the angular frequency of the motion. With Eq. 15-5 you then can find the period T and the frequency f.

In some problems you might derive an expression for the force F as a function of displacement x. If the motion is linear SHM, the force and displacement are related by

$$F = -(\text{a positive constant})x,$$

which says that the force is proportional to the displacement but is in the opposite direction. Once you have found such an expression, you can immediately compare it with Eq. 15-10 and identify the positive constant as being k. If you know the mass that is involved, you can then use Eqs. 15-12, 15-13, and 15-5 to find the angular frequency ω, period T, and frequency f.

15-4 Energy in Simple Harmonic Motion

In Chapter 8 we saw that the energy of a linear oscillator transfers back and forth between kinetic energy and potential energy, while the sum of the two—the mechanical energy E of the oscillator—remains constant. We now consider this situation quantitatively.

The potential energy of a linear oscillator like that of Fig. 15-5 is associated entirely with the spring. Its value depends on how much the spring is stretched or compressed—that is, on $x(t)$. We can use Eqs. 8-11 and 15-3 to find

$$U(t) = \tfrac{1}{2}kx^2 = \tfrac{1}{2}kx_m^2 \cos^2(\omega t + \phi). \qquad (15\text{-}18)$$

Caution: A function written in the form $\cos^2 A$ (as here) means $(\cos A)^2$ and is *not* the same as one written $\cos A^2$, which means $\cos (A^2)$.

The kinetic energy of the system of Fig. 15-5 is associated entirely with the block. Its value depends on how fast the block is moving—that is, on $v(t)$. We can use Eq. 15-6 to find

$$K(t) = \tfrac{1}{2}mv^2 = \tfrac{1}{2}m\omega^2 x_m^2 \sin^2(\omega t + \phi). \qquad (15\text{-}19)$$

If we use Eq. 15-12 to substitute k/m for ω^2, we can write Eq. 15-19 as

$$K(t) = \tfrac{1}{2}mv^2 = \tfrac{1}{2}k x_m^2 \sin^2(\omega t + \phi). \qquad (15\text{-}20)$$

The mechanical energy follows from Eqs. 15-18 and 15-20 and is

$$
\begin{aligned}
E &= U + K \\
&= \tfrac{1}{2}k x_m^2 \cos^2(\omega t + \phi) + \tfrac{1}{2}k x_m^2 \sin^2(\omega t + \phi) \\
&= \tfrac{1}{2}k x_m^2 [\cos^2(\omega t + \phi) + \sin^2(\omega t + \phi)].
\end{aligned}
$$

For any angle α,

$$\cos^2 \alpha + \sin^2 \alpha = 1.$$

Thus, the quantity in the square brackets above is unity and we have

$$E = U + K = \tfrac{1}{2}k x_m^2. \qquad (15\text{-}21)$$

The mechanical energy of a linear oscillator is indeed constant and independent of time. The potential energy and kinetic energy of a linear oscillator are shown as functions of time t in Fig. 15-6a, and they are shown as functions of displacement x in Fig. 15-6b.

You might now understand why an oscillating system normally contains an element of springiness and an element of inertia: The former stores its potential energy and the latter stores its kinetic energy.

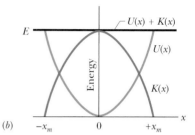

Fig. 15-6 (a) Potential energy $U(t)$, kinetic energy $K(t)$, and mechanical energy E as functions of time t for a linear harmonic oscillator. Note that all energies are positive and that the potential energy and the kinetic energy peak twice during every period. (b) Potential energy $U(x)$, kinetic energy $K(x)$, and mechanical energy E as functions of position x for a linear harmonic oscillator with amplitude x_m. For $x = 0$ the energy is all kinetic, and for $x = \pm x_m$ it is all potential.

✓CHECKPOINT 3   In Fig. 15-5, the block has a kinetic energy of 3 J and the spring has an elastic potential energy of 2 J when the block is at $x = +2.0$ cm. (a) What is the kinetic energy when the block is at $x = 0$? What is the elastic potential energy when the block is at (b) $x = -2.0$ cm and (c) $x = -x_m$?

Sample Problem 15-3

(a) What is the mechanical energy E of the linear oscillator of Sample Problem 15-1? (Initially, the block's position is $x = 11$ cm and its speed is $v = 0$. Spring constant k is 65 N/m.)

Solution: The **Key Idea** here is that the mechanical energy E (the sum of the kinetic energy $K = \tfrac{1}{2}mv^2$ of the block and the potential energy $U = \tfrac{1}{2}kx^2$ of the spring) is constant throughout the motion of the oscillator. Thus, we can evaluate E at any point during the motion. Because we are given the initial conditions of the oscillator as $x = 11$ cm and $v = 0$, let us evaluate E for those conditions. We find

$$E = K + U = \tfrac{1}{2}mv^2 + \tfrac{1}{2}kx^2 = 0 + \tfrac{1}{2}(65 \text{ N/m})(0.11 \text{ m})^2$$
$$= 0.393 \text{ J} \approx 0.39 \text{ J}. \qquad \text{(Answer)}$$

(b) What are the potential energy U and kinetic energy K of the oscillator when the block is at $x = \tfrac{1}{2}x_m$? What are they when the block is at $x = -\tfrac{1}{2}x_m$?

Solution: The **Key Idea** here is that, because we are given the location of the block, we can easily find the spring's potential energy with $U = \tfrac{1}{2}kx^2$. For $x = \tfrac{1}{2}x_m$, we have

$$U = \tfrac{1}{2}kx^2 = \tfrac{1}{2}k(\tfrac{1}{2}x_m)^2 = (\tfrac{1}{2})(\tfrac{1}{4})k x_m^2.$$

We can substitute for k and x_m, or we can use the **Key Idea** that the total mechanical energy, which we know from part (a), is $\tfrac{1}{2}k x_m^2$. That idea allows us to write, from the above equation,

$$U = \tfrac{1}{4}(\tfrac{1}{2}k x_m^2) = \tfrac{1}{4}E = \tfrac{1}{4}(0.393 \text{ J}) = 0.098 \text{ J}. \quad \text{(Answer)}$$

Now, using again the **Key Idea** that $E = K + U$, we can write

$$K = E - U = 0.393 \text{ J} - 0.098 \text{ J} \approx 0.30 \text{ J}. \quad \text{(Answer)}$$

By repeating these calculations for $x = -\tfrac{1}{2}x_m$, we would find the same answers for that displacement, consistent with the left–right symmetry of Fig. 15-6b.

15-5   *An Angular Simple Harmonic Oscillator*

Figure 15-7 shows an angular version of a simple harmonic oscillator; the element of springiness or elasticity is associated with the twisting of a suspension wire

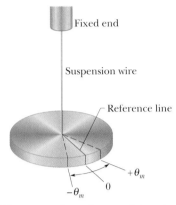

Fig. 15-7 A torsion pendulum is an angular version of a linear simple harmonic oscillator. The disk oscillates in a horizontal plane; the reference line oscillates with angular amplitude θ_m. The twist in the suspension wire stores potential energy as a spring does and provides the restoring torque.

rather than the extension and compression of a spring as we previously had. The device is called a **torsion pendulum,** with *torsion* referring to the twisting.

If we rotate the disk in Fig. 15-7 by some angular displacement θ from its rest position (where the reference line is at $\theta = 0$) and release it, it will oscillate about that position in **angular simple harmonic motion.** Rotating the disk through an angle θ in either direction introduces a restoring torque given by

$$\tau = -\kappa\theta. \qquad (15\text{-}22)$$

Here κ (Greek *kappa*) is a constant, called the **torsion constant,** that depends on the length, diameter, and material of the suspension wire.

Comparison of Eq. 15-22 with Eq. 15-10 leads us to suspect that Eq. 15-22 is the angular form of Hooke's law, and that we can transform Eq. 15-13, which gives the period of linear SHM, into an equation for the period of angular SHM: We replace the spring constant k in Eq. 15-13 with its equivalent, the constant κ of Eq. 15-22, and we replace the mass m in Eq. 15-13 with *its* equivalent, the rotational inertia I of the oscillating disk. These replacements lead to

$$T = 2\pi\sqrt{\frac{I}{\kappa}} \quad \text{(torsion pendulum)}, \qquad (15\text{-}23)$$

which is the correct equation for the period of an angular simple harmonic oscillator, or torsion pendulum.

PROBLEM-SOLVING TACTICS

TACTIC 3: *Identifying Angular SHM*

When a system undergoes angular simple harmonic motion, its angular acceleration α and angular displacement θ are related by an equation of the form

$$\alpha = -(\text{a positive constant})\theta.$$

This equation is the angular equivalent of Eq. 15-8 ($a = -\omega^2 x$). It says that the angular acceleration α is proportional to the angular displacement θ from the equilibrium position but is in the direction opposite the displacement. If you have an expression of this form, you can identify the positive constant as being ω^2, and then you can determine ω, f, and T.

You can also identify angular SHM if you have an expression for the torque τ in terms of the angular displacement, because that expression must be in the form of Eq. 15-22 ($\tau = -\kappa\theta$) or

$$\tau = -(\text{a positive constant})\theta.$$

This equation is the angular equivalent of Eq. 15-10 ($F = -kx$). It says that the torque τ is proportional to the angular displacement θ from the equilibrium position but tends to rotate the system in the opposite direction. If you have an expression of this form, then you can identify the positive constant as being the system's torsion constant κ. If you know the rotational inertia I of the system, you can then determine T.

Sample Problem 15-4

Figure 15-8*a* shows a thin rod whose length L is 12.4 cm and whose mass m is 135 g, suspended at its midpoint from a long wire. Its period T_a of angular SHM is measured to be 2.53 s. An irregularly shaped object, which we call object X, is then hung from the same wire, as in Fig. 15-8*b*, and its period T_b is found to be 4.76 s. What is the rotational inertia of object X about its suspension axis?

Solution: The **Key Idea** here is that the rotational inertia of either the rod or object X is related to the measured period by Eq. 15-23. In Table 10-2*e*, the rotational inertia of a thin rod about a perpendicular axis through its midpoint is given as $\frac{1}{12}mL^2$. Thus, we have, for the rod in Fig. 15-8*a*,

$$I_a = \tfrac{1}{12}mL^2 = (\tfrac{1}{12})(0.135 \text{ kg})(0.124 \text{ m})^2$$
$$= 1.73 \times 10^{-4} \text{ kg} \cdot \text{m}^2.$$

Now let us write Eq. 15-23 twice, once for the rod and once for object X:

$$T_a = 2\pi\sqrt{\frac{I_a}{\kappa}} \quad \text{and} \quad T_b = 2\pi\sqrt{\frac{I_b}{\kappa}}.$$

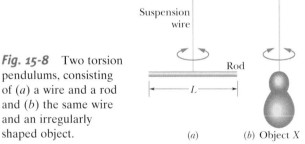

Fig. 15-8 Two torsion pendulums, consisting of (*a*) a wire and a rod and (*b*) the same wire and an irregularly shaped object.

The constant κ, which is a property of the wire, is the same for both figures; only the periods and the rotational inertias differ.

Let us square each of these equations, divide the second by the first, and solve the resulting equation for I_b. The result is

$$I_b = I_a \frac{T_b^2}{T_a^2} = (1.73 \times 10^{-4} \text{ kg} \cdot \text{m}^2)\frac{(4.76 \text{ s})^2}{(2.53 \text{ s})^2}$$
$$= 6.12 \times 10^{-4} \text{ kg} \cdot \text{m}^2. \qquad \text{(Answer)}$$

15-6 Pendulums

We turn now to a class of simple harmonic oscillators in which the springiness is associated with the gravitational force rather than with the elastic properties of a twisted wire or a compressed or stretched spring.

The Simple Pendulum

If you hang an apple at the end of a long thread fixed at its upper end and then set the apple swinging back and forth a small distance, you easily see that the apple's motion is periodic. Is it, in fact, simple harmonic motion? If so, what is the period T? To answer, we consider a **simple pendulum,** which consists of a particle of mass m (called the *bob* of the pendulum) suspended from one end of an unstretchable, massless string of length L that is fixed at the other end, as in Fig. 15-9a. The bob is free to swing back and forth in the plane of the page, to the left and right of a vertical line through the pendulum's pivot point.

The forces acting on the bob are the force $\vec{T}$ from the string and the gravitational force $\vec{F}_g$, as shown in Fig. 15-9b, where the string makes an angle θ with the vertical. We resolve $\vec{F}_g$ into a radial component $F_g \cos \theta$ and a component $F_g \sin \theta$ that is tangent to the path taken by the bob. This tangential component produces a restoring torque about the pendulum's pivot point because the component always acts opposite the displacement of the bob so as to bring the bob back toward its central location. That location is called the *equilibrium position* ($\theta = 0$) because the pendulum would be at rest there were it not swinging.

From Eq. 10-41 ($\tau = r_\perp F$), we can write this restoring torque as

$$\tau = -L(F_g \sin \theta), \qquad (15\text{-}24)$$

where the minus sign indicates that the torque acts to reduce θ and L is the moment arm of the force component $F_g \sin \theta$ about the pivot point. Substituting Eq. 15-24 into Eq. 10-44 ($\tau = I\alpha$) and then substituting mg as the magnitude of F_g, we obtain

$$-L(mg \sin \theta) = I\alpha, \qquad (15\text{-}25)$$

where I is the pendulum's rotational inertia about the pivot point and α is its angular acceleration about that point.

We can simplify Eq. 15-25 if we assume the angle θ is small, for then we can approximate $\sin \theta$ with θ (expressed in radian measure). (As an example, if $\theta = 5.00° = 0.0873$ rad, then $\sin \theta = 0.0872$, a difference of only about 0.1%.) With that approximation and some rearranging, we then have

$$\alpha = -\frac{mgL}{I}\,\theta. \qquad (15\text{-}26)$$

This equation is the angular equivalent of Eq. 15-8, the hallmark of SHM. It tells us that the angular acceleration α of the pendulum is proportional to the angular displacement θ but opposite in sign. Thus, as the pendulum bob moves to the right, as in Fig. 15-9a, its acceleration *to the left* increases until the bob stops and begins moving to the left. Then, when it is to the left of the equilibrium position, its acceleration to the right tends to return it to the right, and so on, as it swings back and forth in SHM. More precisely, the motion of a *simple pendulum swinging through only small angles* is approximately SHM. We can state this restriction to small angles another way: The **angular amplitude** θ_m of the motion (the maximum angle of swing) must be small.

Comparing Eqs. 15-26 and Eq. 15-8, we see that the angular frequency of the pendulum is $\omega = \sqrt{mgL/I}$. Next, if we substitute this expression for ω into Eq. 15-5 ($\omega = 2\pi/T$), we see that the period of the pendulum may be written as

$$T = 2\pi\sqrt{\frac{I}{mgL}}. \qquad (15\text{-}27)$$

All the mass of a simple pendulum is concentrated in the mass m of the particle-

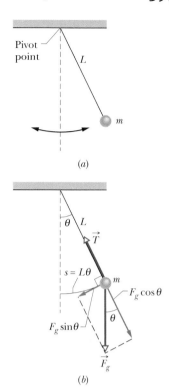

Fig. 15-9 (a) A simple pendulum. (b) The forces acting on the bob are the gravitational force $\vec{F}_g$ and the force $\vec{T}$ from the string. The tangential component $F_g \sin \theta$ of the gravitational force is a restoring force that tends to bring the pendulum back to its central position.

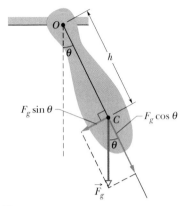

Fig. 15-10 A physical pendulum. The restoring torque is $hF_g \sin\theta$. When $\theta = 0$, center of mass C hangs directly below pivot point O.

like bob, which is at radius L from the pivot point. Thus, we can use Eq. 10-33 ($I = mr^2$) to write $I = mL^2$ for the rotational inertia of the pendulum. Substituting this into Eq. 15-27 and simplifying then yield

$$T = 2\pi\sqrt{\frac{L}{g}} \qquad \text{(simple pendulum, small amplitude)} \qquad (15\text{-}28)$$

as a simpler expression for the period of a simple pendulum swinging through only small angles. (We assume small-angle swinging in this chapter.)

The Physical Pendulum

A real pendulum, usually called a **physical pendulum,** can have a complicated distribution of mass, much different from that of a simple pendulum. Does a physical pendulum also undergo SHM? If so, what is its period?

Figure 15-10 shows an arbitrary physical pendulum displaced to one side by angle θ. The gravitational force $\vec{F}_g$ acts at its center of mass C, at a distance h from the pivot point O. Comparison of Figs. 15-10 and 15-9b reveals only one important difference between an arbitrary physical pendulum and a simple pendulum. For a physical pendulum the restoring component $F_g \sin\theta$ of the gravitational force has a moment arm of distance h about the pivot point, rather than of string length L. In all other respects, an analysis of the physical pendulum would duplicate our analysis of the simple pendulum up through Eq. 15-27. Again, (for small θ_m) we would find that the motion is approximately SHM.

If we replace L with h in Eq. 15-27, we can write the period as

$$T = 2\pi\sqrt{\frac{I}{mgh}} \qquad \text{(physical pendulum, small amplitude)}. \qquad (15\text{-}29)$$

As with the simple pendulum, I is the rotational inertia of the pendulum about O. However, now I is not simply mL^2 (it depends on the shape of the physical pendulum), but it is still proportional to m.

A physical pendulum will not swing if it pivots at its center of mass. Formally, this corresponds to putting $h = 0$ in Eq. 15-29. That equation then predicts $T \to \infty$, which implies that such a pendulum will never complete one swing.

Corresponding to any physical pendulum that oscillates about a given pivot point O with period T is a simple pendulum of length L_0 with the same period T. We can find L_0 with Eq. 15-28. The point along the physical pendulum at distance L_0 from point O is called the *center of oscillation* of the physical pendulum for the given suspension point.

Measuring g

We can use a physical pendulum to measure the free-fall acceleration g at a particular location on Earth's surface. (Countless thousands of such measurements have been made during geophysical prospecting.)

To analyze a simple case, take the pendulum to be a uniform rod of length L, suspended from one end. For such a pendulum, h in Eq. 15-29, the distance between the pivot point and the center of mass, is $\frac{1}{2}L$. Table 10-2e tells us that the rotational inertia of this pendulum about a perpendicular axis through its center of mass is $\frac{1}{12}mL^2$. From the parallel-axis theorem of Eq. 10-36 ($I = I_{\text{com}} + Mh^2$), we then find that the rotational inertia about a perpendicular axis through one end of the rod is

$$I = I_{\text{com}} + mh^2 = \tfrac{1}{12}mL^2 + m(\tfrac{1}{2}L)^2 = \tfrac{1}{3}mL^2. \qquad (15\text{-}30)$$

If we put $h = \frac{1}{2}L$ and $I = \frac{1}{3}mL^2$ in Eq. 15-29 and solve for g, we find

$$g = \frac{8\pi^2 L}{3T^2}. \qquad (15\text{-}31)$$

Thus, by measuring L and the period T, we can find the value of g at the pendulum's location. (If precise measurements are to be made, a number of refinements are needed, such as swinging the pendulum in an evacuated chamber.)

Sample Problem 15-5

In Fig. 15-11a, a meter stick swings about a pivot point at one end, at distance h from the stick's center of mass.

(a) What is the period of oscillation T?

Solution: One **Key Idea** here is that the stick is not a simple pendulum because its mass is not concentrated in a bob at the end opposite the pivot point—so the stick is a physical pendulum. Then its period is given by Eq. 15-29, for which we need the rotational inertia I of the stick about the pivot point. We can treat the stick as a uniform rod of length L and mass m. Then Eq. 15-30 tells us that $I = \frac{1}{3}mL^2$, and the distance h in Eq. 15-29 is $\frac{1}{2}L$. Substituting these quantities into Eq. 15-29, we find

$$T = 2\pi\sqrt{\frac{I}{mgh}} = 2\pi\sqrt{\frac{\frac{1}{3}mL^2}{mg(\frac{1}{2}L)}} = 2\pi\sqrt{\frac{2L}{3g}} \quad (15\text{-}32)$$

$$= 2\pi\sqrt{\frac{(2)(1.00 \text{ m})}{(3)(9.8 \text{ m/s}^2)}} = 1.64 \text{ s.} \quad \text{(Answer)}$$

Note the result is independent of the pendulum's mass m.

(b) What is the distance L_0 between the pivot point O of the stick and the center of oscillation of the stick?

Solution: The **Key Idea** here is that we want the length L_0 of the simple pendulum (drawn in Fig. 15-11b) that has the same period as the physical pendulum (the stick) of Fig.

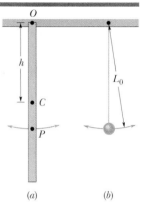

Fig. **15-11**   (a) A meter stick suspended from one end as a physical pendulum. (b) A simple pendulum whose length L_0 is chosen so that the periods of the two pendulums are equal. Point P on the pendulum of (a) marks the center of oscillation.

15-11a. Setting Eqs. 15-28 and 15-32 equal yields

$$T = 2\pi\sqrt{\frac{L_0}{g}} = 2\pi\sqrt{\frac{2L}{3g}}.$$

You can see by inspection that

$$L_0 = \tfrac{2}{3}L = (\tfrac{2}{3})(100 \text{ cm}) = 66.7 \text{ cm.} \quad \text{(Answer)}$$

In Fig. 15-11a, point P marks this distance from suspension point O. Thus, point P is the stick's center of oscillation for the given suspension point.

> **CHECKPOINT 4**   Three physical pendulums, of masses m_0, $2m_0$, and $3m_0$, have the same shape and size and are suspended at the same point. Rank the masses according to the periods of the pendulums, greatest first.

Sample Problem 15-6

A competition diving board sits on a fulcrum about one-third of the way out from the fixed end of the board (Fig. 15-12a). In a running dive, a diver takes three quick steps along the

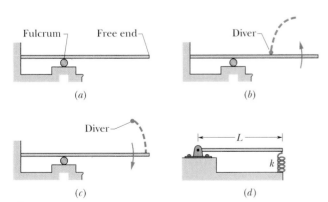

Fig. **15-12**   (a) A diving board. (b) The diver leaps upward and forward as the board moves through the horizontal. (c) The diver lands 2.5 oscillations later. (d) A spring-oscillator model of the oscillating board.

board, out past the fulcrum so as to rotate the board's free end downward. As the board rebounds back through the horizontal, the diver leaps upward and toward the board's free end (Fig. 15-12b). A skilled diver trains to land on the free end just as the board has completed 2.5 oscillations during the leap. With such timing, the diver lands as the free end is moving downward with greatest speed (Fig. 15-12c). The landing then drives the free end down substantially, and the rebound catapults the diver high into the air.

Figure 15-12d shows a simple but realistic model of a competition board. The board section beyond the fulcrum is treated as a stiff rod of length L that can rotate about a hinge at the fulcrum, compressing an (imaginary) spring under the board's free end. If the rod's mass is $m = 20.0$ kg and the diver's leap lasts $t_{fl} = 0.620$ s, what spring constant k is required of the spring for a proper landing?

Solution: Because a spring is involved, we might guess that the oscillations are SHM, but we shall not assume that. Instead, we use the following **Key Idea**: If the rod is in SHM, then the acceleration and displacement of the oscillating end of the rod must be related by an expression in the form of Eq. 15-8 ($a = -\omega^2 x$). If so, we shall be able to find ω and then the desired k from the expression. Let us check by finding the relation between the acceleration and displacement of the rod's right end.

Because the rod rotates about the hinge as the free end oscillates, we are concerned with a torque $\vec{\tau}$ on the rod about the hinge. That torque is due to the force $\vec{F}$ on the rod from the spring. Because $\vec{F}$ varies with time, $\vec{\tau}$ must also. However, at any given instant we can relate the magnitudes of $\vec{\tau}$ and $\vec{F}$ with Eq. 10-39 ($\tau = rF \sin \phi$). Here we have

$$\tau = LF \sin 90°, \qquad (15\text{-}33)$$

where L is the moment arm of force $\vec{F}$ and 90° is the angle between the moment arm and the force's line of action. Combining Eq. 15-33 with Eq. 10-44 ($\tau = I\alpha$) gives us

$$I\alpha = LF, \qquad (15\text{-}34)$$

where I is the rod's rotational inertia about the hinge and α is its angular acceleration about that point. From Eq. 15-30, the rod's rotational inertia I is $\frac{1}{3}mL^2$.

Now let us mentally erect a vertical x axis through the oscillating right end of the rod, with the positive direction upward. Then the force on the right end of the rod from the spring is $F = -kx$, where x is the vertical displacement of the right end.

Substituting these expressions for I and F into Eq. 15-34 gives us

$$\frac{mL^2\alpha}{3} = -Lkx. \qquad (15\text{-}35)$$

We now have a mixture of linear displacement x (vertically) and rotational acceleration α (about the hinge). We can replace α in Eq. 15-35 with the (linear) acceleration a along the x axis by substituting according to Eq. 10-22 ($a_t = \alpha r$) for tangential acceleration. Here the tangential acceleration is a and the radius of rotation r is L, so $\alpha = a/L$. With that sub-

stitution, Eq. 15-35 becomes

$$\frac{mL^2 a}{3L} = -Lkx,$$

which yields

$$a = -\frac{3k}{m}x. \qquad (15\text{-}36)$$

Equation 15-36 is, in fact, of the same form as Eq. 15-8 ($a = -\omega^2 x$). Therefore, the rod does indeed undergo SHM, and comparison of Eqs. 15-36 and 15-8 shows that

$$\omega^2 = \frac{3k}{m}.$$

Solving for k and substituting for ω from Eq. 15-5 ($\omega = 2\pi/T$) give us

$$k = \frac{m}{3}\left(\frac{2\pi}{T}\right)^2, \qquad (15\text{-}37)$$

where T is the period of the board's oscillation. We want the time of flight t_{fl} to last for 2.5 oscillations of the board and thus also for 2.5 oscillations of our rod. Thus we want $t_{fl} = 2.5T$. Substituting this and given data into Eq. 15-37 leads to

$$k = \frac{m}{3}\left(\frac{2\pi}{t_{fl}}2.5\right)^2 \qquad (15\text{-}38)$$

$$= \frac{(20.0 \text{ kg})}{3}\left(\frac{2\pi}{0.620 \text{ s}}2.5\right)^2$$

$$= 4.28 \times 10^3 \text{ N/m.} \qquad \text{(Answer)}$$

This is the effective spring constant k of the diving board.

From Eq. 15-38, we see that a diver with a longer flight time t_{fl} requires a smaller spring constant k in order to land at the proper instant at the end of the board. The value of k can be increased by moving the fulcrum toward the free end or decreased by moving it in the opposite direction. A skilled diver trains to leap with a certain flight time t_{fl} and knows how to set the fulcrum position accordingly.

15-7   Simple Harmonic Motion and Uniform Circular Motion

In 1610, Galileo, using his newly constructed telescope, discovered the four principal moons of Jupiter. Over weeks of observation, each moon seemed to him to be moving back and forth relative to the planet in what today we would call simple harmonic motion; the disk of the planet was the midpoint of the motion. The record of Galileo's observations, written in his own hand, is still available. A. P. French of MIT used Galileo's data to work out the position of the moon Callisto relative to Jupiter. In the results shown in Fig. 15-13, the circles are based on Galileo's observations and the curve is a best fit to the data. The curve strongly suggests Eq. 15-3, the displacement function for SHM. A period of about 16.8 days can be measured from the plot.

Actually, Callisto moves with essentially constant speed in an essentially circular orbit around Jupiter. Its true motion — far from being simple harmonic — is uniform circular motion. What Galileo saw — and what you can see with a good pair of binoculars and a little patience — is the projection of this uniform circular motion on a line in the plane of the motion. We are led by Galileo's remarkable observations to the conclusion that simple harmonic motion is uni-

(a)

(b)

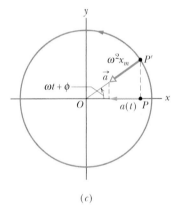

(c)

Fig. 15-13 The angle between Jupiter and its moon Callisto as seen from Earth. The circles are based on Galileo's 1610 measurements. The curve is a best fit, strongly suggesting simple harmonic motion. At Jupiter's mean distance from Earth, 10 minutes of arc corresponds to about 2×10^6 km. (Adapted from A. P. French, *Newtonian Mechanics*, W. W. Norton & Company, New York, 1971, p. 288.)

form circular motion viewed edge-on. In more formal language:

> Simple harmonic motion is the projection of uniform circular motion on a diameter of the circle in which the circular motion occurs.

Figure 15-14*a* gives an example. It shows a *reference particle P'* moving in uniform circular motion with (constant) angular speed ω in a *reference circle*. The radius x_m of the circle is the magnitude of the particle's position vector. At any time t, the angular position of the particle is $\omega t + \phi$, where ϕ is its angular position at $t = 0$.

The projection of particle P' onto the x axis is a point P, which we take to be a second particle. The projection of the position vector of particle P' onto the x axis gives the location $x(t)$ of P. Thus, we find

$$x(t) = x_m \cos(\omega t + \phi),$$

which is precisely Eq. 15-3. Our conclusion is correct. If reference particle P' moves in uniform circular motion, its projection particle P moves in simple harmonic motion along a diameter of the circle.

Figure 15-14*b* shows the velocity $\vec{v}$ of the reference particle. From Eq. 10-18 ($v = \omega r$), the magnitude of the velocity vector is ωx_m; its projection on the x axis is

$$v(t) = -\omega x_m \sin(\omega t + \phi),$$

which is exactly Eq. 15-6. The minus sign appears because the velocity component of P in Fig. 15-14*b* is directed to the left, in the negative direction of x.

Figure 15-14*c* shows the radial acceleration $\vec{a}$ of the reference particle. From Eq. 10-23 ($a_r = \omega^2 r$), the magnitude of the radial acceleration vector is $\omega^2 x_m$; its projection on the x axis is

$$a(t) = -\omega^2 x_m \cos(\omega t + \phi),$$

which is exactly Eq. 15-7. Thus, whether we look at the displacement, the velocity, or the acceleration, the projection of uniform circular motion is indeed simple harmonic motion.

Fig. 15-14 (*a*) A reference particle P' moving with uniform circular motion in a reference circle of radius x_m. Its projection P on the x axis executes simple harmonic motion. (*b*) The projection of the velocity $\vec{v}$ of the reference particle is the velocity of SHM. (*c*) The projection of the radial acceleration $\vec{a}$ of the reference particle is the acceleration of SHM.

15-8 Damped Simple Harmonic Motion

A pendulum will swing only briefly underwater, because the water exerts on the pendulum a drag force that quickly eliminates the motion. A pendulum swinging in air does better, but still the motion dies out eventually, because the air exerts

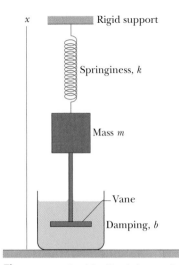

x Rigid support

Springiness, *k*

Mass *m*

Vane

Damping, *b*

Fig. 15-15 An idealized damped simple harmonic oscillator. A vane immersed in a liquid exerts a damping force on the block as the block oscillates parallel to the *x* axis.

a drag force on the pendulum (and friction acts at its support point), transferring energy from the pendulum's motion.

When the motion of an oscillator is reduced by an external force, the oscillator and its motion are said to be **damped.** An idealized example of a damped oscillator is shown in Fig. 15-15, where a block with mass *m* oscillates vertically on a spring with spring constant *k*. From the block, a rod extends to a vane (both assumed massless) that is submerged in a liquid. As the vane moves up and down, the liquid exerts an inhibiting drag force on it and thus on the entire oscillating system. With time, the mechanical energy of the block–spring system decreases, as energy is transferred to thermal energy of the liquid and vane.

Let us assume the liquid exerts a **damping force** $\vec{F}_d$ that is proportional to the velocity $\vec{v}$ of the vane and block (an assumption that is accurate if the vane moves slowly). Then, for components along the *x* axis in Fig. 15-15, we have

$$F_d = -bv, \tag{15-39}$$

where *b* is a **damping constant** that depends on the characteristics of both the vane and the liquid and has the SI unit of kilogram per second. The minus sign indicates that $\vec{F}_d$ opposes the motion.

The force on the block from the spring is $F_s = -kx$. Let us assume that the gravitational force on the block is negligible relative to F_d and F_s. Then we can write Newton's second law for components along the *x* axis ($F_{\text{net},x} = ma_x$) as

$$-bv - kx = ma. \tag{15-40}$$

Substituting dx/dt for *v* and d^2x/dt^2 for *a* and rearranging give us the differential equation

$$m\frac{d^2x}{dt^2} + b\frac{dx}{dt} + kx = 0. \tag{15-41}$$

The solution of this equation is

$$x(t) = x_m\, e^{-bt/2m} \cos(\omega' t + \phi), \tag{15-42}$$

where x_m is the amplitude and ω' is the angular frequency of the damped oscillator. This angular frequency is given by

$$\omega' = \sqrt{\frac{k}{m} - \frac{b^2}{4m^2}}. \tag{15-43}$$

If $b = 0$ (there is no damping), then Eq. 15-43 reduces to Eq. 15-12 ($\omega = \sqrt{k/m}$) for the angular frequency of an undamped oscillator, and Eq. 15-42 reduces to Eq. 15-3 for the displacement of an undamped oscillator. If the damping constant is small but not zero (so that $b \ll \sqrt{km}$), then $\omega' \approx \omega$.

We can regard Eq. 15-42 as a cosine function whose amplitude, which is $x_m e^{-bt/2m}$, gradually decreases with time, as Fig. 15-16 suggests. For an undamped oscillator, the mechanical energy is constant and is given by Eq. 15-21 ($E = \frac{1}{2}kx_m^2$). If the oscillator is damped, the mechanical energy is not constant but decreases with time. If the damping is small, we can find $E(t)$ by replacing x_m in Eq. 15-21 with $x_m e^{-bt/2m}$, the amplitude of the damped oscillations. By doing so,

Fig. 15-16 The displacement function $x(t)$ for the damped oscillator of Fig. 15-15, with values given in Sample Problem 15-7. The amplitude, which is $x_m e^{-bt/2m}$, decreases exponentially with time.

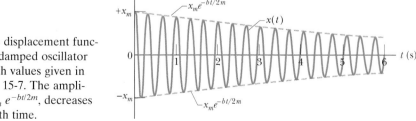

we find that

$$E(t) \approx \tfrac{1}{2}k x_m^2 \, e^{-bt/m}, \qquad (15\text{-}44)$$

which tells us that, like the amplitude, the mechanical energy decreases exponentially with time.

CHECKPOINT 5   Here are three sets of values for the spring constant, damping constant, and mass for the damped oscillator of Fig. 15-15. Rank the sets according to the time required for the mechanical energy to decrease to one-fourth of its initial value, greatest first.

Set 1	$2k_0$	b_0	m_0
Set 2	k_0	$6b_0$	$4m_0$
Set 3	$3k_0$	$3b_0$	m_0

Sample Problem 15-7

For the damped oscillator of Fig. 15-15, $m = 250$ g, $k = 85$ N/m, and $b = 70$ g/s.

(a) What is the period of the motion?

Solution: The **Key Idea** here is that because $b \ll \sqrt{km} = 4.6$ kg/s, the period is approximately that of the undamped oscillator. From Eq. 15-13, we then have

$$T = 2\pi\sqrt{\frac{m}{k}} = 2\pi\sqrt{\frac{0.25 \text{ kg}}{85 \text{ N/m}}} = 0.34 \text{ s}. \quad \text{(Answer)}$$

(b) How long does it take for the amplitude of the damped oscillations to drop to half its initial value?

Solution: Now the **Key Idea** is that the amplitude at time t is displayed in Eq. 15-42 as $x_m \, e^{-bt/2m}$. It has the value x_m at $t = 0$. Thus, we must find the value of t for which

$$x_m \, e^{-bt/2m} = \tfrac{1}{2}x_m.$$

Canceling x_m and taking the natural logarithm of the equation that remains, we have $\ln \tfrac{1}{2}$ on the right side and

$$\ln(e^{-bt/2m}) = -bt/2m$$

on the left side. Thus,

$$t = \frac{-2m \ln \tfrac{1}{2}}{b} = \frac{-(2)(0.25 \text{ kg})(\ln \tfrac{1}{2})}{0.070 \text{ kg/s}}$$

$$= 5.0 \text{ s}. \qquad \text{(Answer)}$$

Because $T = 0.34$ s, this is about 15 periods of oscillation.

(c) How long does it take for the mechanical energy to drop to one-half its initial value?

Solution: Here the **Key Idea** is that, from Eq. 15-44, the mechanical energy at time t is $\tfrac{1}{2}k x_m^2 \, e^{-bt/m}$. It has the value $\tfrac{1}{2}k x_m^2$ at $t = 0$. Thus, we must find the value of t for which

$$\tfrac{1}{2}k x_m^2 \, e^{-bt/m} = \tfrac{1}{2}(\tfrac{1}{2}k x_m^2).$$

If we divide both sides of this equation by $\tfrac{1}{2}k x_m^2$ and solve for t as we did above, we find

$$t = \frac{-m \ln \tfrac{1}{2}}{b} = \frac{-(0.25 \text{ kg})(\ln \tfrac{1}{2})}{0.070 \text{ kg/s}} = 2.5 \text{ s}. \quad \text{(Answer)}$$

This is exactly half the time we calculated in (b), or about 7.5 periods of oscillation. Figure 15-16 was drawn to illustrate this sample problem.

15-9   Forced Oscillations and Resonance

A person swinging in a swing without anyone pushing it is an example of *free oscillation*. However, if someone pushes the swing periodically, the swing has *forced*, or *driven*, *oscillations*. Two angular frequencies are associated with a system undergoing driven oscillations: (1) the *natural* angular frequency ω of the system, which is the angular frequency at which it would oscillate if it were suddenly disturbed and then left to oscillate freely, and (2) the angular frequency ω_d of the external driving force causing the driven oscillations.

We can use Fig. 15-15 to represent an idealized forced simple harmonic oscillator if we allow the structure marked "rigid support" to move up and down at a variable angular frequency ω_d. Such a forced oscillator oscillates at the angular frequency ω_d of the driving force, and its displacement $x(t)$ is given by

$$x(t) = x_m \cos(\omega_d t + \phi), \qquad (15\text{-}45)$$

where x_m is the amplitude of the oscillations.

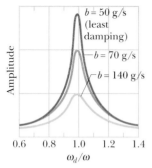

Fig. 15-17 The displacement amplitude x_m of a forced oscillator varies as the angular frequency ω_d of the driving force is varied. The curves here correspond to three values of the damping constant b.

How large the displacement amplitude x_m is depends on a complicated function of ω_d and ω. The velocity amplitude v_m of the oscillations is easier to describe: it is greatest when

$$\omega_d = \omega \quad \text{(resonance),} \tag{15-46}$$

a condition called **resonance**. Equation 15-46 is also *approximately* the condition at which the displacement amplitude x_m of the oscillations is greatest. Thus, if you push a swing at its natural angular frequency, the displacement and velocity amplitudes will increase to large values, a fact that children learn quickly by trial and error. If you push at other angular frequencies, either higher or lower, the displacement and velocity amplitudes will be smaller.

Figure 15-17 shows how the displacement amplitude of an oscillator depends on the angular frequency ω_d of the driving force, for three values of the damping coefficient b. Note that for all three the amplitude is approximately greatest when $\omega_d/\omega = 1$ (the resonance condition of Eq. 15-46). The curves of Fig. 15-17 show that less damping gives a taller and narrower *resonance peak*.

All mechanical structures have one or more natural angular frequencies, and if a structure is subjected to a strong external driving force that matches one of these angular frequencies, the resulting oscillations of the structure may rupture it. Thus, for example, aircraft designers must make sure that none of the natural angular frequencies at which a wing can oscillate matches the angular frequency of the engines in flight. A wing that flaps violently at certain engine speeds would obviously be dangerous.

Resonance appears to be one reason buildings in Mexico City collapsed in September 1985 when a major earthquake (8.1 on the Richter scale) occurred on the western coast of Mexico. The seismic waves from the earthquake should have been too weak to cause extensive damage when they reached Mexico City about 400 km away. However, Mexico City is largely built on an ancient lake bed, where the soil is still soft with water. Although the amplitude of the seismic waves was small in the firmer ground en route to Mexico City, their amplitude substantially increased in the loose soil of the city. Acceleration amplitudes of the waves were as much as $0.20g$, and the angular frequency was (surprisingly) concentrated around 3 rad/s. Not only was the ground severely oscillated, but many intermediate-height buildings had resonant angular frequencies of about 3 rad/s. Most of those buildings collapsed during the shaking (Fig. 15-18), while shorter buildings (with higher resonant angular frequencies) and taller buildings (with lower resonant angular frequencies) remained standing.

Fig. 15-18 In 1985, buildings of intermediate height collapsed in Mexico City as a result of an earthquake far from the city. Taller and shorter buildings remained standing.

Review & Summary

Frequency The *frequency* f of periodic, or oscillatory, motion is the number of oscillations per second. In the SI system, it is measured in hertz:

$$1 \text{ hertz} = 1 \text{ Hz} = 1 \text{ oscillation per second} = 1 \text{ s}^{-1}. \tag{15-1}$$

Period The *period* T is the time required for one complete oscillation, or **cycle**. It is related to the frequency by

$$T = \frac{1}{f}. \tag{15-2}$$

Simple Harmonic Motion In *simple harmonic motion*

(SHM), the displacement $x(t)$ of a particle from its equilibrium position is described by the equation

$$x = x_m \cos(\omega t + \phi) \quad \text{(displacement),} \tag{15-3}$$

in which x_m is the **amplitude** of the displacement, the quantity $(\omega t + \phi)$ is the **phase** of the motion, and ϕ is the **phase constant**. The **angular frequency** ω is related to the period and frequency of the motion by

$$\omega = \frac{2\pi}{T} = 2\pi f \quad \text{(angular frequency).} \tag{15-5}$$

Differentiating Eq. 15-3 leads to equations for the particle's

SHM velocity and acceleration as functions of time:

$$v = -\omega x_m \sin(\omega t + \phi) \quad \text{(velocity)} \quad (15\text{-}6)$$

and $$a = -\omega^2 x_m \cos(\omega t + \phi) \quad \text{(acceleration)}. \quad (15\text{-}7)$$

In Eq. 15-6, the positive quantity ωx_m is the **velocity amplitude** v_m of the motion. In Eq. 15-7, the positive quantity $\omega^2 x_m$ is the **acceleration amplitude** a_m of the motion.

The Linear Oscillator   A particle with mass m that moves under the influence of a Hooke's law restoring force given by $F = -kx$ exhibits simple harmonic motion with

$$\omega = \sqrt{\frac{k}{m}} \quad \text{(angular frequency)} \quad (15\text{-}12)$$

and $$T = 2\pi\sqrt{\frac{m}{k}} \quad \text{(period)}. \quad (15\text{-}13)$$

Such a system is called a **linear simple harmonic oscillator.**

Energy   A particle in simple harmonic motion has, at any time, kinetic energy $K = \frac{1}{2}mv^2$ and potential energy $U = \frac{1}{2}kx^2$. If no friction is present, the mechanical energy $E = K + U$ remains constant even though K and U change.

Pendulums   Examples of devices that undergo simple harmonic motion are the **torsion pendulum** of Fig. 15-7, the **simple pendulum** of Fig. 15-9, and the **physical pendulum** of Fig. 15-10. Their periods of oscillation for small oscillations are, respectively,

$$T = 2\pi\sqrt{I/\kappa} \quad \text{(torsion pendulum)}, \quad (15\text{-}23)$$

$$T = 2\pi\sqrt{L/g} \quad \text{(simple pendulum)}, \quad (15\text{-}28)$$

$$T = 2\pi\sqrt{I/mgh} \quad \text{(physical pendulum)}. \quad (15\text{-}29)$$

Simple Harmonic Motion and Uniform Circular Motion
Simple harmonic motion is the projection of uniform circular motion onto the diameter of the circle in which the circular motion occurs. Figure 15-14 shows that all parameters of circular motion (position, velocity, and acceleration) project to the corresponding values for simple harmonic motion.

Damped Harmonic Motion   The mechanical energy E in a real oscillating system decreases during the oscillations because external forces, such as a drag force, inhibit the oscillations and transfer mechanical energy to thermal energy. The real oscillator and its motion are then said to be **damped.** If the **damping force** is given by $\vec{F}_d = -b\vec{v}$, where $\vec{v}$ is the velocity of the oscillator and b is a **damping constant,** then the displacement of the oscillator is given by

$$x(t) = x_m e^{-bt/2m} \cos(\omega't + \phi), \quad (15\text{-}42)$$

where ω', the angular frequency of the damped oscillator, is given by

$$\omega' = \sqrt{\frac{k}{m} - \frac{b^2}{4m^2}}. \quad (15\text{-}43)$$

If the damping constant is small ($b \ll \sqrt{km}$), then $\omega' \approx \omega$, where ω is the angular frequency of the undamped oscillator. For small b, the mechanical energy E of the oscillator is given by

$$E(t) \approx \frac{1}{2}kx_m^2 e^{-bt/m}. \quad (15\text{-}44)$$

Forced Oscillations and Resonance   If an external driving force with angular frequency ω_d acts on an oscillating system with *natural* angular frequency ω, the system oscillates with angular frequency ω_d. The velocity amplitude v_m of the system is greatest when

$$\omega_d = \omega, \quad (15\text{-}46)$$

a condition called **resonance.** The amplitude x_m of the system is (approximately) greatest under the same condition.

Questions

1 Which of the following relationships between the acceleration a and the displacement x of a particle involve SHM: (a) $a = 0.5x$, (b) $a = 400x^2$, (c) $a = -20x$, (d) $a = -3x^2$?

2 The acceleration $a(t)$ of a particle undergoing SHM is graphed in Fig. 15-19. (a) Which of the labeled points corresponds to the particle at $-x_m$? (b) At point 4, is the velocity of the particle positive, negative, or zero? (c) At point 5, is the particle at $-x_m$, at $+x_m$, at 0, between $-x_m$ and 0, or between 0 and $+x_m$?

Fig. 15-19   Question 2.

3 Which of the following describe ϕ for the SHM of Fig. 15-20a:
(a) $-\pi < \phi < -\pi/2$,
(b) $\pi < \phi < 3\pi/2$,
(c) $-3\pi/2 < \phi < -\pi$?

Fig. 15-20   Questions 3 and 4.

4 The velocity $v(t)$ of a particle undergoing SHM is graphed in Fig. 15-20b. Is the particle momentarily stationary, headed toward $-x_m$, or headed toward $+x_m$ at (a) point A on the graph and (b) point B? Is the particle at $-x_m$, at $+x_m$, at 0, between $-x_m$ and 0, or between 0 and $+x_m$ when its velocity is represented by (c) point A and (d) point B? Is the speed of the particle increasing or decreasing at (e) point A and (f) point B?

5 Figure 15-21 gives, for three situations, the displacements $x(t)$ of a pair of simple harmonic oscillators (A and B) that are identical except for phase. For each pair, what phase shift (in radians and in degrees) is needed to shift the curve for *A* to coincide with the curve for *B*? Of the many possible answers, choose the shift with the smallest absolute magnitude.

Fig. 15-21 Question 5.

6 You are to complete Fig. 15-22*a* so that it is a plot of velocity *v* versus time *t* for the spring–block oscillator that is shown in Fig. 15-22*b* for *t* = 0. (a) In Fig. 15-22*a*, at which lettered point or in what region between the points should the (vertical) *v* axis intersect the *t* axis? (For example, should it intersect at point *A*, or maybe in the region between points *A* and *B*?) (b) If the velocity of the block is given by $v = -v_m \sin(\omega t + \phi)$, what is the value of ϕ? Make it positive, and if you cannot specify the value (such as $+\pi/2$ rad), then give a range of values (such as between 0 and $\pi/2$).

Fig. 15-22 Question 6.

7 You are to complete Fig. 15-23*a* so that it is a plot of acceleration *a* versus time *t* for the spring–block oscillator that is shown in Fig. 15-23*b* for *t* = 0. (a) In Fig. 15-23*a*, at which lettered point or in what region between the points should the (vertical) *a* axis intersect the *t* axis? (For example, should it intersect at point *A*, or maybe in the region between points *A* and *B*?) (b) If the block's acceleration is given by $a = -a_m \cos(\omega t + \phi)$, what is the value of ϕ? Make it positive, and if you cannot specify the value (such as $+\pi/2$ rad), then give a range of values (such as between 0 and $\pi/2$).

Fig. 15-23 Question 7.

8 Figure 15-24 shows the *x*(*t*) curves for three experiments

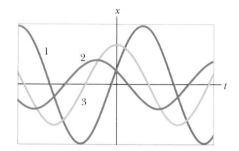

Fig. 15-24
Question 8.

involving a particular spring–box system oscillating in SHM. Rank the curves according to (a) the system's angular frequency, (b) the spring's potential energy at time *t* = 0, (c) the box's kinetic energy at *t* = 0, (d) the box's speed at *t* = 0, and (e) the box's maximum kinetic energy, greatest first.

9 Figure 15-25 shows plots of the kinetic energy *K* versus position *x* for three harmonic oscillators that have the same mass. Rank the plots according to (a) the corresponding spring constant and (b) the corresponding period of the oscillator, greatest first.

Fig. 15-25 Question 9.

10 In Fig. 15-26, a spring–block system is put into SHM in two experiments. In the first, the block is pulled from the equilibrium position through a displacement d_1 and then released. In the second, it is pulled from the equilibrium position through a greater displacement d_2 and then released. Are the (a) amplitude, (b) period, (c) frequency, (d) maximum kinetic energy, and (e) maximum potential energy in the second experiment greater than, less than, or the same as those in the first experiment?

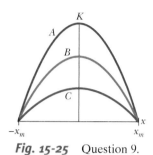

Fig. 15-26 Question 10.

11 Figure 15-27 shows three physical pendulums consisting of identical uniform spheres of the same mass that are rigidly connected by identical rods of negligible mass. Each pendulum is vertical and can pivot about suspension point *O*. Rank the pendulums according to period of oscillation, greatest first.

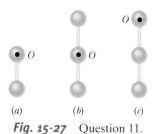

Fig. 15-27 Question 11.

12 You are to build the oscillation transfer device shown in Fig. 15-28. It consists of two spring–block systems hanging from a flexible rod. When the spring of system 1 is stretched and then released, the resulting SHM of system 1 at frequency f_1 oscillates the rod. The rod then exerts a driving force on system 2, at the same frequency f_1. You can choose from four springs with spring constants *k* of 1600, 1500, 1400, and 1200 N/m, and four blocks with masses *m* of 800, 500, 400, and 200 kg. Mentally determine which spring should go with which block in each of the two systems to maximize the amplitude of oscillations in system 2.

Fig. 15-28 Question 12.

Problems

sec. 15-3 The Force Law for Simple Harmonic Motion

•1 In an electric shaver, the blade moves back and forth over a distance of 2.0 mm in simple harmonic motion, with frequency 120 Hz. Find (a) the amplitude, (b) the maximum blade speed, and (c) the magnitude of the maximum blade acceleration. **SSM**

•2 A 0.12 kg body undergoes simple harmonic motion of amplitude 8.5 cm and period 0.20 s. (a) What is the magnitude of the maximum force acting on it? (b) If the oscillations are produced by a spring, what is the spring constant?

•3 A particle with a mass of 1.00×10^{-20} kg is oscillating with simple harmonic motion with a period of 1.00×10^{-5} s and a maximum speed of 1.00×10^3 m/s. Calculate (a) the angular frequency and (b) the maximum displacement of the particle.

•4 What is the maximum acceleration of a platform that oscillates at amplitude 2.20 cm and frequency 6.60 Hz?

•5 An oscillator consists of a block of mass 0.500 kg connected to a spring. When set into oscillation with amplitude 35.0 cm, the oscillator repeats its motion every 0.500 s. Find the (a) period, (b) frequency, (c) angular frequency, (d) spring constant, (e) maximum speed, and (f) magnitude of the maximum force on the block from the spring. **SSM**

•6 An oscillating block–spring system takes 0.75 s to begin repeating its motion. Find (a) the period, (b) the frequency in hertz, and (c) the angular frequency in radians per second.

•7 An object undergoing simple harmonic motion takes 0.25 s to travel from one point of zero velocity to the next such point. The distance between those points is 36 cm. Calculate the (a) period, (b) frequency, and (c) amplitude of the motion.

•8 An automobile can be considered to be mounted on four identical springs as far as vertical oscillations are concerned. The springs of a certain car are adjusted so that the oscillations have a frequency of 3.00 Hz. (a) What is the spring constant of each spring if the mass of the car is 1450 kg and the mass is evenly distributed over the springs? (b) What will be the oscillation frequency if five passengers, averaging 73.0 kg each, ride in the car with an even distribution of mass?

•9 The function $x = (6.0 \text{ m}) \cos[(3\pi \text{ rad/s})t + \pi/3 \text{ rad}]$ gives the simple harmonic motion of a body. At $t = 2.0$ s, what are the (a) displacement, (b) velocity, (c) acceleration, and (d) phase of the motion? Also, what are the (e) frequency and (f) period of the motion?

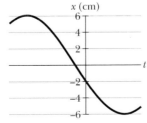

Fig. 15-29 Problem 10.

•10 What is the phase constant for the harmonic oscillator with the position function $x(t)$ given in Fig. 15-29 if the position function has the form $x = x_m \cos(\omega t + \phi)$?

•11 In Fig. 15-30, two identical springs of spring constant 7580 N/m are attached to a block of mass 0.245 kg. The block is set oscillating on the frictionless floor. What is the frequency of oscillation?

Fig. 15-30 Problems 11 and 24.

•12 What is the phase constant for the harmonic oscillator with the velocity function $v(t)$ given in Fig. 15-31 if the position function $x(t)$ has the form $x = x_m \cos(\omega t + \phi)$?

Fig. 15-31 Problem 12.

•13 A loudspeaker produces a musical sound by means of the oscillation of a diaphragm whose amplitude is limited to 1.00 μm. (a) At what frequency is the magnitude a of the diaphragm's acceleration equal to g? (b) For greater frequencies, is a greater than or less than g? **SSM**

••14 At a certain harbor, the tides cause the ocean surface to rise and fall a distance d (from highest level to lowest level) in simple harmonic motion, with a period of 12.5 h. How long does it take for the water to fall a distance $0.250d$ from its highest level?

••15 A block is on a horizontal surface (a shake table) that is moving back and forth horizontally with simple harmonic motion of frequency 2.0 Hz. The coefficient of static friction between block and surface is 0.50. How great can the amplitude of the SHM be if the block is not to slip along the surface? **SSM WWW**

••16 Figure 15-32 shows block 1 of mass 0.200 kg sliding to the right over a frictionless elevated surface at a speed of 8.00 m/s. The block undergoes an elastic collision with stationary block 2, which is attached to a spring of spring constant 1208.5 N/m. (Assume that the spring does not affect the collision.) After the collision, block 2 oscillates in SHM with a period of 0.140 s, and block 1 slides off the opposite end of the elevated surface, landing a distance d from the base of that surface after falling height $h = 4.90$ m. What is the value of d?

Fig. 15-32 Problem 16.

••17 An oscillator consists of a block attached to a spring ($k = 400$ N/m). At some time t, the position (measured from the system's equilibrium location), velocity, and acceleration of the block are $x = 0.100$ m, $v = -13.6$ m/s, and $a = -123$ m/s². Calculate (a) the frequency of oscillation, (b) the mass of the block, and (c) the amplitude of the motion. **ILW**

••**18** Figure 15-33*a* is a partial graph of the position function $x(t)$ for a simple harmonic oscillator with an angular frequency of 1.20 rad/s; Fig. 15-33*b* is a partial graph of the corresponding velocity function $v(t)$. What is the phase constant of the SHM if the position function $x(t)$ is given the form $x = x_m \cos(\omega t + \phi)$?

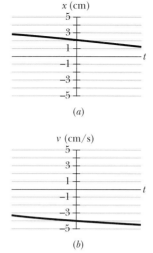

(a)

(b)

Fig. 15-33 Problem 18.

••**19** A simple harmonic oscillator consists of a block of mass 2.00 kg attached to a spring of spring constant 100 N/m. When $t = 1.00$ s, the position and velocity of the block are $x = 0.129$ m and $v = 3.415$ m/s. (a) What is the amplitude of the oscillations? What were the (b) position and (c) velocity of the block at $t = 0$ s?

••**20** A block rides on a piston that is moving vertically with simple harmonic motion. (a) If the SHM has period 1.0 s, at what amplitude of motion will the block and piston separate? (b) If the piston has an amplitude of 5.0 cm, what is the maximum frequency for which the block and piston will be in contact continuously?

••**21** Two particles oscillate in simple harmonic motion along a common straight-line segment of length A. Each particle has a period of 1.5 s, but they differ in phase by $\pi/6$ rad. (a) How far apart are they (in terms of A) 0.50 s after the lagging particle leaves one end of the path? (b) Are they then moving in the same direction, toward each other, or away from each other? **SSM ILW**

••**22** Two particles execute simple harmonic motion of the same amplitude and frequency along close parallel lines. They pass each other moving in opposite directions each time their displacement is half their amplitude. What is their phase difference?

••**23** A massless spring hangs from the ceiling with a small object attached to its lower end. The object is initially held at rest in a position y_i such that the spring is at its rest length. The object is then released from y_i and oscillates up and down, with its lowest position being 10 cm below y_i. (a) What is the frequency of the oscillation? (b) What is the speed of the object when it is 8.0 cm below the initial position? (c) An object of mass 300 g is attached to the first object, after which the system oscillates with half the original frequency. What is the mass of the first object? (d) How far below y_i is the new equilibrium (rest) position with both objects attached to the spring? **SSM**

••**24** In Fig. 15-30, two springs are attached to a block that can oscillate over a frictionless floor. If the left spring is removed, the block oscillates at a frequency of 30 Hz. If, instead, the spring on the right is removed, the block oscillates at a frequency of 45 Hz. At what frequency does the block oscillate with both springs attached?

•••**25** In Fig. 15-34, two blocks ($m = 1.0$ kg and $M = 10$ kg) and a spring ($k = 200$ N/m) are arranged on a horizontal, frictionless surface. The coefficient of static friction between the two blocks is 0.40. What amplitude of simple harmonic motion of the spring–blocks system puts the smaller block on the verge of slipping over the larger block?

Fig. 15-34 Problem 25.

•••**26** In Fig. 15-35, two springs are joined and connected to a block of mass 0.245 kg that is set oscillating over a frictionless floor. The springs each have spring constant $k = 6430$ N/m. What is the frequency of the oscillations?

Fig. 15-35 Problem 26.

•••**27** In Fig. 15-36, a block weighing 14.0 N, which can slide without friction on an incline at angle $\theta = 40.0°$, is connected to the top of the incline by a massless spring of unstretched length 0.450 m and spring constant 120 N/m. (a) How far from the top of the incline is the block's equilibrium point? (b) If the block is pulled slightly down the incline and released, what is the period of the resulting oscillations?

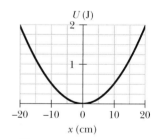

Fig. 15-36 Problem 27.

sec. 15-4 Energy in Simple Harmonic Motion

•**28** An oscillating block–spring system has a mechanical energy of 1.00 J, an amplitude of 10.0 cm, and a maximum speed of 1.20 m/s. Find (a) the spring constant, (b) the mass of the block, and (c) the frequency of oscillation.

•**29** Find the mechanical energy of a block–spring system having a spring constant of 1.3 N/cm and an oscillation amplitude of 2.4 cm. **SSM**

•**30** Figure 15-37 gives the one-dimensional potential energy well for a 2.0 kg particle (the function $U(x)$ has the form bx^2). (a) If the particle passes through the equilibrium position with a velocity of 85 cm/s, will it be turned back before it reaches $x = 15$ cm? (b) If yes, at what position, and if no, what is the speed of the particle at $x = 15$ cm?

Fig. 15-37 Problem 30.

•**31** When the displacement in SHM is one-half the amplitude x_m, what fraction of the total energy is (a) kinetic energy and (b) potential energy? (c) At what displacement, in terms of the amplitude, is the energy of the system half kinetic energy and half potential energy? **SSM**

•**32** Figure 15-38 shows the kinetic energy K of a simple

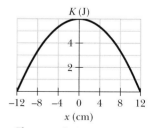

Fig. 15-38 Problem 32.

harmonic oscillator versus its position x. What is the spring constant?

•**33** A 5.00 kg object on a horizontal frictionless surface is attached to a spring with $k = 1000$ N/m. The object is displaced from equilibrium 50.0 cm horizontally and given an initial velocity of 10.0 m/s back toward the equilibrium position. What are (a) the motion's frequency, (b) the initial potential energy of the block–spring system, (c) the initial kinetic energy, and (d) the motion's amplitude? **ILW**

••**34** If the phase angle for a block–spring system in SHM is $\pi/6$ rad and the block's position is given by $x = x_m \cos(\omega t + \phi)$, what is the ratio of the kinetic energy to the potential energy at time $t = 0$?

••**35** A 10 g particle undergoes SHM with an amplitude of 2.0 mm, a maximum acceleration of magnitude 8.0×10^3 m/s², and an unknown phase constant ϕ. What are (a) the period of the motion, (b) the maximum speed of the particle, and (c) the total mechanical energy of the oscillator? What is the magnitude of the force on the particle when the particle is at (d) its maximum displacement and (e) half its maximum displacement?

••**36** In Fig. 15-39, block 2 of mass 2.0 kg oscillates on the end of a spring in SHM with a period of 20 ms. The position of the block is given by $x = $ (1.0 cm) $\cos(\omega t + \pi/2)$. Block 1 of mass 4.0 kg slides toward block 2 with a velocity of magnitude 6.0 m/s, directed along the spring's length. The two blocks undergo a completely inelastic collision at time $t = 5.0$ ms. (The duration of the collision is much less than the period of motion.) What is the amplitude of the SHM after the collision?

Fig. 15-39 Problem 36.

••**37** A block of mass $M = 5.4$ kg, at rest on a horizontal frictionless table, is attached to a rigid support by a spring of constant $k = 6000$ N/m. A bullet of mass $m = 9.5$ g and velocity $\vec{v}$ of magnitude 630 m/s strikes and is embedded in the block (Fig. 15-40). Assuming the compression of the spring is negligible until the bullet is embedded, determine (a) the speed of the block immediately after the collision and (b) the amplitude of the resulting simple harmonic motion.

Fig. 15-40 Problem 37.

sec. 15-5 An Angular Simple Harmonic Oscillator

•**38** A 95 kg solid sphere with a 15 cm radius is suspended by a vertical wire. A torque of 0.20 N · m is required to rotate the sphere through an angle of 0.85 rad and then maintain that orientation. What is the period of the oscillations that result when the sphere is then released?

••**39** The balance wheel of a watch oscillates with angular amplitude π rad and period 0.500 s. Find (a) the maximum angular speed of the wheel, (b) the angular speed at displacement $\pi/2$ rad, and (c) the magnitude of the angular acceleration at displacement $\pi/4$ rad. **SSM WWW**

sec. 15-6 Pendulums

•**40** (a) If the physical pendulum of Sample Problem 15-5 is inverted and suspended at point P, what is its period of oscil-

lation? (b) Is the period now greater than, less than, or equal to its previous value?

•**41** In Sample Problem 15-5, we saw that a physical pendulum has a center of oscillation at distance $2L/3$ from its point of suspension. Show that the distance between the point of suspension and the center of oscillation for a physical pendulum of any form is I/mh, where I and h have the meanings assigned to them in Eq. 15-29 and m is the mass of the pendulum.

•**42** Suppose that a simple pendulum consists of a small 60.0 g bob at the end of a cord of negligible mass. If the angle θ between the cord and the vertical is given by

$$\theta = (0.0800 \text{ rad}) \cos[(4.43 \text{ rad/s}) t + \phi],$$

what are (a) the pendulum's length and (b) its maximum kinetic energy?

•**43** In Fig. 15-41, the pendulum consists of a uniform disk with radius $r = 10.0$ cm and mass 500 g attached to a uniform rod with length $L = 500$ mm and mass 270 g. (a) Calculate the rotational inertia of the pendulum about the pivot point. (b) What is the distance between the pivot point and the center of mass of the pendulum? (c) Calculate the period of oscillation. **SSM**

Fig. 15-41 Problem 43.

•**44** In Fig. 15-42, a physical pendulum consists of a uniform solid disk (of radius $R = 2.35$ cm) supported in a vertical plane by a pivot located a distance $d = 1.75$ cm from the center of the disk. The disk is displaced by a small angle and released. What is the period of the resulting simple harmonic motion?

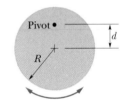

Fig. 15-42 Problem 44.

•**45** A physical pendulum consists of a meter stick that is pivoted at a small hole drilled through the stick a distance d from the 50 cm mark. The period of oscillation is 2.5 s. Find d. **ILW**

•**46** A performer seated on a trapeze is swinging back and forth with a period of 8.85 s. If she stands up, thus raising the center of mass of the *trapeze + performer* system by 35.0 cm, what will be the new period of the system? Treat *trapeze + performer* as a simple pendulum.

•**47** A physical pendulum consists of two meter-long sticks joined together as shown in Fig. 15-43. What is the pendulum's period of oscillation about a pin inserted through point A at the center of the horizontal stick?

Fig. 15-43 Problem 47.

••**48** A thin uniform rod (mass = 0.50 kg) swings about an axis that passes through one end of the rod and is perpendicular to the plane of the swing. The rod swings with a period of 1.5 s and an angular amplitude of 10°. (a) What is the length of the rod? (b) What is the maximum kinetic energy of the rod as it swings?

••49 In the overhead view of Fig. 15-44, a long uniform rod of mass 0.600 kg is free to rotate in a horizontal plane about a vertical axis through its center. A spring with force constant k = 1850 N/m is connected horizontally between one end of the rod and a fixed wall. When the rod is in equilibrium, it is parallel to the wall. What is the period of the small oscillations that result when the rod is rotated slightly and released? SSM ILW

Fig. 15-44 Problem 49.

••50 A rectangular block, with face lengths a = 35 cm and b = 45 cm, is to be suspended on a thin horizontal rod running through a narrow hole in the block. The block is then to be set swinging about the rod like a pendulum, through small angles so that it is in SHM. Figure 15-45 shows one possible position of the hole, at distance r from the block's center, along a line connecting the center with a corner. (a) Plot the period of the pendulum versus distance r along that line such that the minimum in the curve is apparent. (b) For what value of r does that minimum occur? There is actually a line of points around the block's center for which the period of swinging has the same minimum value. (c) What shape does that line make?

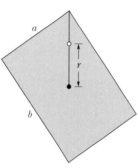

Fig. 15-45 Problem 50.

••51 In Fig. 15-46, a stick of length L = 1.85 m oscillates as a physical pendulum. (a) What value of distance x between the stick's center of mass and its pivot point O gives the least period? (b) What is that least period?

••52 The 3.00 kg cube in Fig. 15-47 has edge lengths d = 6.00 cm and is mounted on an axle through its center. A spring (k = 1200 N/m) connects the cube's upper corner to a rigid wall. Initially the spring is at its rest length. If the cube is rotated 3° and released, what is the period of the resulting SHM?

••53 The angle of the pendulum of Fig. 15-9b is given by $\theta = \theta_m \cos[(4.44 \text{ rad/s})t + \phi]$. If at t = 0, θ = 0.040 rad and $d\theta/dt$ = −0.200 rad/s, what are (a) the phase constant ϕ and (b) the maximum angle θ_m? (*Hint:* Don't confuse the rate $d\theta/dt$ at which θ changes with the ω of the SHM.)

Fig. 15-46 Problem 51.

Fig. 15-47 Problem 52.

••54 In Fig. 15-48a, a metal plate is mounted on an axle through its center of mass. A spring with k = 2000 N/m connects a wall with a point on the rim a distance r = 2.5 cm from the center of mass. Initially the spring is at its rest length. If the plate is rotated by 7° and released, it rotates about the axle in SHM, with its angular position given by Fig. 15-48b. What is the rotational inertia of the plate about its center of mass?

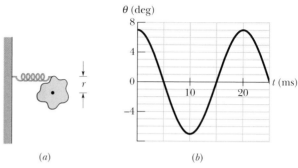

Fig. 15-48 Problem 54.

•••55 A pendulum is formed by pivoting a long thin rod about a point on the rod. In a series of experiments, the period is measured as a function of the distance d between the pivot point and the rod's center. (a) If the rod's length is L = 2.20 m and its mass is m = 22.1 g, what is the minimum period? (b) If d is chosen to minimize the period and then L is increased, does the period increase, decrease, or remain the same? (c) If, instead, m is increased without L increasing, does the period increase, decrease, or remain the same? SSM WWW

•••56 In Fig. 15-49, a 2.50 kg disk of diameter D = 42.0 cm is supported by a rod of length L = 76.0 cm and negligible mass that is pivoted at its end. (a) With the massless torsion spring unconnected, what is the period of oscillation? (b) With the torsion spring connected, the rod is vertical at equilibrium. What is the torsion constant of the spring if the period of oscillation has been decreased by 0.500 s?

Fig. 15-49 Problem 56.

sec. 15-8 Damped Simple Harmonic Motion

•57 In Sample Problem 15-7, what is the ratio of the amplitude of the damped oscillations to the initial amplitude at the end of 20 cycles?

•58 The amplitude of a lightly damped oscillator decreases by 3.0% during each cycle. What percentage of the mechanical energy of the oscillator is lost in each cycle?

•59 In Fig. 15-15, the block has a mass of 1.50 kg and the spring constant is 8.00 N/m. The damping force is given by $-b(dx/dt)$, where b = 230 g/s. The block is pulled down 12.0 cm and released. (a) Calculate the time required for the amplitude of the resulting oscillations to fall to one-third of its initial value. (b) How many oscillations are made by the block in this time? SSM

••60 The suspension system of a 2000 kg automobile "sags" 10 cm when the chassis is placed on it. Also, the oscillation amplitude decreases by 50% each cycle. Estimate the values of (a) the spring constant k and (b) the damping constant b for the spring and shock absorber system of one wheel, assuming each wheel supports 500 kg.

sec. 15-9 Forced Oscillations and Resonance

•61 Hanging from a horizontal beam are nine simple pendulums of the following lengths: (a) 0.10, (b) 0.30, (c) 0.40, (d) 0.80, (e) 1.2, (f) 2.8, (g) 3.5, (h) 5.0, and (i) 6.2 m. Suppose the beam undergoes horizontal oscillations with angular frequencies in the range from 2.00 rad/s to 4.00 rad/s. Which of the pendulums will be (strongly) set in motion?

•62 For Eq. 15-45, suppose the amplitude x_m is given by

$$x_m = \frac{F_m}{[m^2(\omega_d^2 - \omega^2)^2 + b^2\omega_d^2]^{1/2}},$$

where F_m is the (constant) amplitude of the external oscillating force exerted on the spring by the rigid support in Fig. 15-15. At resonance, what are the (a) amplitude and (b) velocity amplitude of the oscillating object?

••63 A 1000 kg car carrying four 82 kg people travels over a "washboard" dirt road with corrugations 4.0 m apart. The car bounces with maximum amplitude when its speed is 16 km/h. When the car stops, and the people get out, by how much does the car body rise on its suspension?

Additional Problems

64 What is the phase constant for the harmonic oscillator with the acceleration function $a(t)$ given in Fig. 15-50 if the position function $x(t)$ has the form $x = x_m \cos(\omega t + \phi)$?

65 Figure 15-51 gives the position of a 20 g block oscillating in SHM on the end of a spring. What are (a) the maximum kinetic energy of the block and (b) the number of times per second that maximum is reached? (*Hint:* Measuring a slope will probably not be very accurate. Can you think of another approach?)

Fig. 15-50 Problem 64.

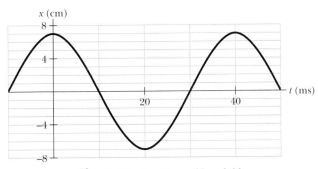

Fig. 15-51 Problems 65 and 66.

66 Figure 15-51 gives the position $x(t)$ of a block oscillating in SHM on the end of a spring. What are (a) the speed and (b) the magnitude of the radial acceleration of a particle in the corresponding uniform circular motion?

67 Figure 15-52 shows the kinetic energy K of a simple pendulum versus its angle θ from the vertical. The pendulum bob has mass 0.200 kg. What is the length of the pendulum?

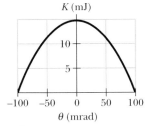

Fig. 15-52 Problem 67.

68 A block is in SHM on the end of a spring, with position given by $x = x_m \cos(\omega t + \phi)$. If $\phi = \pi/5$ rad, then at $t = 0$ what percentage of the total mechanical energy is potential energy?

69 A 55.0 g block oscillates in SHM on the end of a spring with $k = 1500$ N/m according to $x = x_m \cos(\omega t + \phi)$. How long does the block take to move from position $+0.800x_m$ to (a) position $+0.600x_m$ and (b) position $-0.800x_m$?

70 Although California is known for earthquakes, it has large regions dotted with precariously balanced rocks that would be easily toppled by even a mild earthquake. Evidence shows that the rocks have stood this way for thousands of years, suggesting that major earthquakes have not occurred in those regions during that time. If an earthquake were to put such a rock into sinusoidal oscillation (parallel to the ground) with a frequency of 2.2 Hz, an oscillation amplitude of 1.0 cm would cause the rock to topple. What would be the magnitude of the maximum acceleration of the oscillation, in terms of g?

71 A 4.00 kg block is suspended from a spring with $k = 500$ N/m. A 50.0 g bullet is fired into the block from directly below with a speed of 150 m/s and becomes embedded in the block. (a) Find the amplitude of the resulting simple harmonic motion. (b) What percentage of the original kinetic energy of the bullet is transferred to mechanical energy of the oscillator?

72 A wheel is free to rotate about its fixed axle. A spring is attached to one of its spokes a distance r from the axle, as shown in Fig. 15-53. (a) Assuming that the wheel is a hoop of mass m and radius R, what is the angular frequency ω of small oscillations of this system in terms of m, R, r, and the spring constant k? What is ω if (b) $r = R$ and (c) $r = 0$?

Fig. 15-53 Problem 72.

73 A torsion pendulum consists of a metal disk with a wire running through its center and soldered in place. The wire is mounted vertically on clamps and pulled taut. Figure 15-54a gives the magnitude τ of the torque needed to rotate the disk about its center (and thus twist the wire) versus the rotation angle θ. The disk is rotated to $\theta = 0.200$ rad and then released. Figure 15-54b shows the resulting oscillation in terms of angular position θ versus time t. (a) What is the rotational inertia of the disk about its

Fig. 15-54 Problem 73.

center? (b) What is the maximum angular speed $d\theta/dt$ of the disk? (*Caution:* Do not confuse the (constant) angular frequency of the SHM with the (varying) angular speed of the rotating disk, even though they usually have the same symbol ω. *Hint:* The potential energy U of a torsion pendulum is equal to $\frac{1}{2}\kappa\theta^2$, analogous to $U = \frac{1}{2}kx^2$ for a spring.)

74 (a) In Eq. 15-41, find the ratio of the maximum damping force $(-b\,dx/dt)$ to the maximum spring force $(-kx)$ during the first oscillation for a damped simple harmonic oscillator, using the data of Sample Problem 15-7. (b) Does this ratio change appreciably during later oscillations?

75 The scale of a spring balance that reads from 0 to 15.0 kg is 12.0 cm long. A package suspended from the balance is found to oscillate vertically with a frequency of 2.00 Hz. (a) What is the spring constant? (b) How much does the package weigh?

76 A loudspeaker diaphragm is oscillating in simple harmonic motion with a frequency of 440 Hz and a maximum displacement of 0.75 mm. What are the (a) angular frequency, (b) maximum speed, and (c) magnitude of the maximum acceleration?

77 The piston in the cylinder head of a locomotive has a stroke (twice the amplitude) of 0.76 m. If the piston moves with simple harmonic motion with an angular frequency of 180 rev/min, what is its maximum speed?

78 The tip of one prong of a tuning fork undergoes SHM of frequency 1000 Hz and amplitude 0.40 mm. For this tip, what is the magnitude of the (a) maximum acceleration, (b) maximum velocity, (c) acceleration at tip displacement 0.20 mm, and (d) velocity at tip displacement 0.20 mm?

79 In Fig. 15-55, three 10 000 kg ore cars are held at rest on a mine railway using a cable that is parallel to the rails, which are inclined at angle $\theta = 30°$. The cable stretches 15 cm just before the coupling between the two lower cars breaks, detaching the lowest car. Assuming that the cable obeys Hooke's law, find the (a) frequency and (b) amplitude of the resulting oscillations of the remaining two cars.

Fig. 15-55 Problem 79.

80 A uniform spring with $k = 8600$ N/m is cut into pieces 1 and 2 of unstretched lengths $L_1 = 7.0$ cm and $L_2 = 10$ cm. What are (a) k_1 and (b) k_2? A block attached to the original spring as in Fig. 15-5 oscillates at 200 Hz. What is the oscillation frequency of the block attached to (c) piece 1 and (d) piece 2?

81 A vertical spring stretches 9.6 cm when a 1.3 kg block is hung from its end. (a) Calculate the spring constant. This block is then displaced an additional 5.0 cm downward and released from rest. Find the (b) period, (c) frequency, (d) amplitude, and (e) maximum speed of the resulting SHM.

82 A flat uniform circular disk has a mass of 3.00 kg and a radius of 70.0 cm. It is suspended in a horizontal plane by a vertical wire attached to its center. If the disk is rotated 2.50

rad about the wire, a torque of 0.0600 N·m is required to maintain that orientation. Calculate (a) the rotational inertia of the disk about the wire, (b) the torsion constant, and (c) the angular frequency of this torsion pendulum when it is set oscillating.

83 A uniform circular disk whose radius R is 12.6 cm is suspended as a physical pendulum from a point on its rim. (a) What is its period? (b) At what radial distance $r < R$ is there a pivot point that gives the same period?

84 For a simple pendulum, find the angular amplitude θ_m at which the restoring torque required for simple harmonic motion deviates from the actual restoring torque by 1.0%. (See "Trigonometric Expansions" in Appendix E.)

85 What is the frequency of a simple pendulum 2.0 m long (a) in a room, (b) in an elevator accelerating upward at a rate of 2.0 m/s², and (c) in free fall?

86 A simple pendulum of length 20 cm and mass 5.0 g is suspended in a race car traveling with constant speed 70 m/s around a circle of radius 50 m. If the pendulum undergoes small oscillations in a radial direction about its equilibrium position, what is the frequency of oscillation?

87 A particle executes linear SHM with frequency 0.25 Hz about the point $x = 0$. At $t = 0$, it has displacement $x = 0.37$ cm and zero velocity. For the motion, determine the (a) period, (b) angular frequency, (c) amplitude, (d) displacement $x(t)$, (e) velocity $v(t)$, (f) maximum speed, (g) magnitude of the maximum acceleration, (h) displacement at $t = 3.0$ s, and (i) speed at $t = 3.0$ s.

88 A 50.0 g stone is attached to the bottom of a vertical spring and set vibrating. If the maximum speed of the stone is 15.0 cm/s and the period is 0.500 s, find the (a) spring constant of the spring, (b) amplitude of the motion, and (c) frequency of oscillation.

89 A 2.00 kg block hangs from a spring. A 300 g body hung below the block stretches the spring 2.00 cm farther. (a) What is the spring constant? (b) If the 300 g body is removed and the block is set into oscillation, find the period of the motion.

90 The end point of a spring oscillates with a period of 2.0 s when a block with mass m is attached to it. When this mass is increased by 2.0 kg, the period is found to be 3.0 s. Find m.

91 A 0.10 kg block oscillates back and forth along a straight line on a frictionless horizontal surface. Its displacement from the origin is given by

$$x = (10 \text{ cm}) \cos[(10 \text{ rad/s})t + \pi/2 \text{ rad}].$$

(a) What is the oscillation frequency? (b) What is the maximum speed acquired by the block? (c) At what value of x does this occur? (d) What is the magnitude of the maximum acceleration of the block? (e) At what value of x does this occur? (f) What force, applied to the block by the spring, results in the given oscillation?

92 A block weighing 20 N oscillates at one end of a vertical spring for which $k = 100$ N/m; the other end of the spring is attached to a ceiling. At a certain instant the spring is stretched 0.30 m beyond its relaxed length (the length when no object is attached) and the block has zero velocity. (a) What is the net force on the block at this instant? What are the (b) amplitude and (c) period of the resulting simple harmonic motion? (d) What is the maximum kinetic energy of the block as it oscillates?

93 A simple harmonic oscillator consists of a block attached to a spring with $k = 200$ N/m. The block slides on a frictionless surface, with equilibrium point $x = 0$ and amplitude 0.20 m. A graph of the block's velocity v as a function of time t is shown in Fig. 15-56. What are (a) the period of the SHM, (b) the block's mass, (c) its displacement at $t = 0$, (d) its acceleration at $t = 0.10$ s, and (e) its maximum kinetic energy?

Fig. 15-56 Problem 93.

94 A simple harmonic oscillator consists of a 0.50 kg block attached to a spring. The block slides back and forth along a straight line on a frictionless surface with equilibrium point $x = 0$. At $t = 0$ the block is at $x = 0$ and moving in the positive x direction. A graph of the magnitude of the net force $\vec{F}$ on the block as a function of its position is shown in Fig. 15-57. What are (a) the amplitude and (b) the period of the motion, (c) the magnitude of the maximum acceleration, and (d) the maximum kinetic energy?

Fig. 15-57 Problem 94.

95 A 3.0 kg particle is in simple harmonic motion in one dimension and moves according to the equation

$$x = (5.0 \text{ m}) \cos[(\pi/3 \text{ rad/s})t - \pi/4 \text{ rad}],$$

with t in seconds. (a) At what value of x is the potential energy of the particle equal to half the total energy? (b) How long does the particle take to move to this position x from the equilibrium position?

96 In Fig. 15-58, a solid cylinder attached to a horizontal spring ($k = 3.00$ N/m) rolls without slipping along a horizontal surface. If the system is released from rest when the spring is stretched by 0.250 m, find (a) the translational kinetic energy and (b) the rotational kinetic energy of the cylinder as it passes through the equilibrium position. (c) Show that under these conditions the cylinder's center of mass executes simple harmonic motion with period

Fig. 15-58 Problem 96.

$$T = 2\pi \sqrt{\frac{3M}{2k}},$$

where M is the cylinder mass. (*Hint:* Find the time derivative of the total mechanical energy.)

97 A 1.2 kg block sliding on a horizontal frictionless surface is attached to a horizontal spring with $k = 480$ N/m. Let x be the displacement of the block from the position at which the spring is unstretched. At $t = 0$ the block passes through $x = 0$ with a speed of 5.2 m/s in the positive x direction. What are the (a) frequency and (b) amplitude of the block's motion? (c) Write an expression for x as a function of time.

98 A massless spring with spring constant 19 N/m hangs vertically. A body of mass 0.20 kg is attached to its free end and then released. Assume that the spring was unstretched before the body was released. Find (a) how far below the initial position the body descends, and the (b) frequency and (c) amplitude of the resulting SHM.

99 An engineer has an odd-shaped 10 kg object and needs to find its rotational inertia about an axis through its center of mass. The object is supported on a wire stretched along the desired axis. The wire has a torsion constant $\kappa = 0.50$ N · m. If this torsion pendulum oscillates through 20 cycles in 50 s, what is the rotational inertia of the object?

100 A simple harmonic oscillator consists of an 0.80 kg block attached to a spring ($k = 200$ N/m). The block slides on a horizontal frictionless surface about the equilibrium point $x = 0$ with a total mechanical energy of 4.0 J. (a) What is the amplitude of the oscillation? (b) How many oscillations does the block complete in 10 s? (c) What is the maximum kinetic energy attained by the block? (d) What is the speed of the block at $x = 0.15$ m?

101 A block weighing 10.0 N is attached to the lower end of a vertical spring ($k = 200.0$ N/m), the other end of which is attached to a ceiling. The block oscillates vertically and has a kinetic energy of 2.00 J as it passes through the point at which the spring is unstretched. (a) What is the period of the oscillation? (b) Use the law of conservation of energy to determine the maximum distance the block moves both above and below the point at which the spring is unstretched. (These are not necessarily the same.) (c) What is the amplitude of the oscillation? (d) What is the maximum kinetic energy of the block as it oscillates?

102 A grandfather clock has a pendulum that consists of a thin brass disk of radius $r = 15.00$ cm and mass 1.000 kg that is attached to a long thin rod of negligible mass. The pendulum swings freely about an axis perpendicular to the rod and through the end of the rod opposite the disk, as shown in Fig. 15-59. If the pendulum is to have a period of 2.000 s for small oscillations at a place where $g = 9.800$ m/s², what must be the rod length L to the nearest tenth of a millimeter?

Fig. 15-59 Problem 102.

103 A 2.0 kg block executes SHM while attached to a horizontal spring of spring constant 200 N/m. The maximum speed of the block as it slides on a horizontal frictionless surface is 3.0 m/s. What are (a) the amplitude of the block's motion, (b) the magnitude of its maximum acceleration, and (c) the magnitude of its minimum acceleration? (d) How long does the block take to complete 7.0 cycles of its motion?

104 A block sliding on a horizontal frictionless surface is attached to a horizontal spring with a spring constant of 600 N/m. The block executes SHM about its equilibrium position with a period of 0.40 s and an amplitude of 0.20 m. As the block slides through its equilibrium position, a 0.50 kg putty wad is dropped vertically onto the block. If the putty wad sticks to the block, determine (a) the new period of the motion and (b) the new amplitude of the motion.

105 A 4.00 kg block hangs from a spring, extending it 16.0 cm from its unstretched position. (a) What is the spring constant? (b) The block is removed, and a 0.500 kg body is hung from the same spring. If the spring is then stretched and released, what is its period of oscillation?

106 The vibration frequencies of atoms in solids at normal temperatures are of the order of 10^{13} Hz. Imagine the atoms to be connected to one another by springs. Suppose that a single silver atom in a solid vibrates with this frequency and that all the other atoms are at rest. Compute the effective spring constant. One mole of silver (6.02×10^{23} atoms) has a mass of 108 g.

107 When a 20 N can is hung from the bottom of a vertical spring, it causes the spring to stretch 20 cm. (a) What is the spring constant? (b) This spring is now placed horizontally on a frictionless table. One end of it is held fixed, and the other end is attached to a 5.0 N can. The can is then moved (stretching the spring) and released from rest. What is the period of the resulting oscillation?

108 A damped harmonic oscillator consists of a block ($m = 2.00$ kg), a spring ($k = 10.0$ N/m), and a damping force ($F = -bv$). Initially, it oscillates with an amplitude of 25.0 cm; because of the damping, the amplitude falls to three-fourths of this initial value at the completion of four oscillations. (a) What is the value of b? (b) How much energy has been "lost" during these four oscillations?

109 In Fig. 15-60, a 2500 kg demolition ball swings from the end of a crane. The length of the swinging segment of cable is 17 m. (a) Find the period of the swinging, assuming that the system can be treated as a simple pendulum. (b) Does the period depend on the ball's mass?

Fig. 15-60 Problem 109.

110 The center of oscillation of a physical pendulum has this interesting property: If an impulse (assumed horizontal and in the plane of oscillation) acts at the center of oscillation, no oscillations are felt at the point of support. Baseball players (and players of many other sports) know that unless the ball hits the bat at this point (called the "sweet spot" by athletes), the oscillations due to the impact will sting their hands. To prove this property, let the stick in Fig. 15-11a simulate a baseball bat. Suppose that a horizontal force $\vec{F}$ (due to impact with

the ball) acts toward the right at P, the center of oscillation. The batter is assumed to hold the bat at O, the pivot point of the stick. (a) What acceleration does the point O undergo as a result of $\vec{F}$? (b) What angular acceleration is produced by $\vec{F}$ about the center of mass of the stick? (c) As a result of the angular acceleration in (b), what linear acceleration does point O undergo? (d) Considering the magnitudes and directions of the accelerations in (a) and (c), convince yourself that P is indeed the "sweet spot."

111 A common device for entertaining a toddler is a *jump seat* that hangs from the horizontal portion of a doorframe via elastic cords (Fig. 15-61). Assume that only one cord is on each side in spite of the more realistic arrangement shown. When a child is placed in the seat, they both descend by a distance d_s as the cords stretch (treat them as springs). Then the seat is pulled down an extra distance d_m and re-

Fig. 15-61 Problem 111.

leased, so that the child oscillates vertically, like a block on the end of a spring. Suppose you are the safety engineer for the manufacturer of the seat. You do not want the magnitude of the child's acceleration to exceed $0.20g$ for fear of hurting the child's neck. If $d_m = 10$ cm, what value of d_s corresponds to that acceleration magnitude?

112 A 2.0 kg block is attached to the end of a spring with a spring constant of 350 N/m and forced to oscillate by an applied force $F = (15 \text{ N}) \sin(\omega_d t)$, where $\omega_d = 35$ rad/s. The damping constant is $b = 15$ kg/s. At $t = 0$, the block is at rest with the spring at its rest length. (a) Use numerical integration to plot the displacement of the block for the first 1.0 s. Use the motion near the end of the 1.0 s interval to estimate the amplitude, period, and angular frequency. Repeat the calculation for (b) $\omega_d = \sqrt{k/m}$ and (c) $\omega_d = 20$ rad/s.

113 The physical pendulum in Fig. 15-62 has two possible pivot points A and B; point A has a fixed position, and B is adjustable along the length of the pendulum. The period of the pendulum when suspended from A is found to be T. The pendulum is then reversed and suspended from B, which is moved until the pendulum again has period T. Show that the free-fall acceleration g at the pendulum's location is given by

Fig. 15-62 Problem 113.

$$g = \frac{4\pi^2 L}{T^2},$$

in which L is the distance between A and B for equal periods T. (This method can be used to measure g without knowledge of the rotational inertia of the pendulum or any of its dimensions except L.)

On-Line Simulation Problems

The website http://www.wiley.com/college/halliday has simulation problems about this chapter.

Waves — I

In August 2000, as Russia's Northern Fleet conducted exercises in the Barents Sea north of Russia, the nuclear submarine Kursk mysteriously sank. As word of the loss spread, seismologists from around the Northern Hemisphere realized that, on the day the Kursk sank, they had recorded unusual seismic waves originating in the Barents Sea. Analysis of the data suggested the reason the submarine sank, and—more surprising— it also revealed its depth.

How could the submarine's depth be determined from measurements made very far away?

The answer is in this chapter.

Fig. 16-1   (a) A single pulse is sent along a stretched string. A typical string element (marked with a dot) moves up once and then down as the pulse passes. The element's motion is perpendicular to the wave's direction of travel, so the pulse is a *transverse wave*. (b) A sinusoidal wave is sent along the string. A typical string element moves up and down continuously as the wave passes. This too is a transverse wave.

16-1   What Is Physics?

One of the primary subjects of physics is waves. To see how important waves are in the modern world, just consider the music industry. Every piece of music you hear, from some retro-punk band playing in a campus dive to the most eloquent concerto playing on the Web, depends on performers producing waves and your detecting those waves. In between production and detection, the information carried by the waves might need to be transmitted (as in a live performance on the Web) or recorded and then reproduced (as with CDs, DVDs, or the other devices currently being developed in engineering labs worldwide). The financial importance of controlling music waves is staggering, and the rewards to engineers who develop new control techniques can be rich.

This chapter focuses on waves traveling along a stretched string, such as on a guitar. The next chapter focuses on sound waves, such as those produced by a guitar string being played. Before we do all this, though, our first job is to classify the countless waves of the everyday world into basic types.

16-2   Types of Waves

Waves are of three main types:

1. **Mechanical waves.**   These waves are most familiar because we encounter them almost constantly; common examples include water waves, sound waves, and seismic waves. All these waves have two central features: They are governed by Newton's laws, and they can exist only within a material medium, such as water, air, and rock.

2. **Electromagnetic waves.**   These waves are less familiar, but you use them constantly; common examples include visible and ultraviolet light, radio and television waves, microwaves, x rays, and radar waves. These waves require no material medium to exist. Light waves from stars, for example, travel through the vacuum of space to reach us. All electromagnetic waves travel through a vacuum at the same speed $c = 299\ 792\ 458$ m/s.

3. **Matter waves.**   Although these waves are commonly used in modern technology, they are probably very unfamiliar to you. These waves are associated with electrons, protons, and other fundamental particles, and even atoms and molecules. Because we commonly think of these particles as constituting matter, such waves are called matter waves.

Much of what we discuss in this chapter applies to waves of all kinds. However, for specific examples we shall refer to mechanical waves.

16-3   Transverse and Longitudinal Waves

A wave sent along a stretched, taut string is the simplest mechanical wave. If you give one end of a stretched string a single up-and-down jerk, a wave in the form of a single *pulse* travels along the string, as in Fig. 16-1a. This pulse and its motion can occur because the string is under tension. When you pull your end of the string upward, it begins to pull upward on the adjacent section of the string via tension between the two sections. As the adjacent section moves upward, it begins to pull the next section upward, and so on. Meanwhile, you have pulled down on your end of the string. As each section moves upward in turn, it begins to be pulled back downward by neighboring sections that are already on the way down. The net result is that a distortion in the string's shape (the pulse) moves along the string at some velocity $\vec{v}$.

If you move your hand up and down in continuous simple harmonic motion, a continuous wave travels along the string at velocity $\vec{v}$. Because the motion of your hand is a sinusoidal function of time, the wave has a sinusoidal shape at any given instant, as in Fig. 16-1*b*; that is, the wave has the shape of a sine curve or a cosine curve.

We consider here only an "ideal" string, in which no friction-like forces within the string cause the wave to die out as it travels along the string. In addition, we assume that the string is so long that we need not consider a wave rebounding from the far end.

One way to study the waves of Fig. 16-1 is to monitor the **wave forms** (shapes of the waves) as they move to the right. Alternatively, we could monitor the motion of an element of the string as the element oscillates up and down while a wave passes through it. We would find that the displacement of every such oscillating string element is *perpendicular* to the direction of travel of the wave, as indicated in Fig. 16-1*b*. This motion is said to be **transverse,** and the wave is said to be a **transverse wave.**

Figure 16-2 shows how a sound wave can be produced by a piston in a long, air-filled pipe. If you suddenly move the piston rightward and then leftward, you can send a pulse of sound along the pipe. The rightward motion of the piston moves the elements of air next to it rightward, changing the air pressure there. The increased air pressure then pushes rightward on the elements of air somewhat farther along the pipe. Moving the piston leftward reduces the air pressure next to it. As a result, first the elements nearest the piston and then farther elements move leftward. Thus, the motion of the air and the change in air pressure travel rightward along the pipe as a pulse.

If you push and pull on the piston in simple harmonic motion, as is being done in Fig. 16-2, a sinusoidal wave travels along the pipe. Because the motion of the elements of air is parallel to the direction of the wave's travel, the motion is said to be **longitudinal,** and the wave is said to be a **longitudinal wave.** In this chapter we focus on transverse waves, and string waves in particular; in Chapter 17 we focus on longitudinal waves, and sound waves in particular.

Both a transverse wave and a longitudinal wave are said to be **traveling waves** because they both travel from one point to another, as from one end of the string to the other end in Fig. 16-1 and from one end of the pipe to the other end in Fig. 16-2. Note that it is the wave that moves from end to end, not the material (string or air) through which the wave moves.

Fig. 16-2 A sound wave is set up in an air-filled pipe by moving a piston back and forth. Because the oscillations of an element of the air (represented by the dot) are parallel to the direction in which the wave travels, the wave is a *longitudinal wave.*

Sample Problem 16-1

Seismic waves are waves that travel either through Earth's interior or along the ground. Seismology stations are set up mainly to record seismic waves generated by earthquakes, but they also record seismic waves generated by any large release of energy near Earth's surface, such as an explosion. As the seismic waves travel past a station, they oscillate a recording pen and the pen traces out a graph. Figure 16-3*a* is one of the graphs created by seismic waves from the sinking of the submarine *Kursk*. The first oscillations of the pen are marked with an arrow and were of small amplitude. Much stronger oscillations began about 134 s later.

Fig. 16-3 (*a*) Graph made by a recording device as seismic waves from the *Kursk* passed the device. Amplitude is plotted vertically; time increases rightward. The first *Kursk* explosion produced the small-amplitude waves marked with an arrow. The next *Kursk* explosion produced waves of larger amplitude. (*b*) With the submarine sitting at depth *D*, the large explosion sent pulses both into the ground and up through the water.

(*a*)

(*b*)

From this, analysts concluded that the first seismic waves were generated by an onboard explosion, possibly a torpedo that failed to launch when fired. The explosion presumably breached the hull, started a fire, and sank the submarine. The later, much stronger seismic waves were generated after the submarine was sunk and were possibly generated when the fire caused several of the powerful missiles on board to explode simultaneously. These stronger waves arrived at seismology stations as pulses separated by a time interval Δt of about 0.11 s. To what depth D did the submarine sink?

Solution: Let us assume that the stronger explosion occurred after the *Kursk* was sitting on the ocean floor (the seabed). That explosion sent a pulse into the seabed and a pulse upward through the water (Fig. 16-3b). The pulse traveling through the water "bounced" several times between the water surface and the seabed. Each time it hit the seabed, it sent another pulse into the ground, and seismology stations detected those ground pulses as they arrived one after another.

Thus, the time Δt between any two detected ground pulses was equal to the round-trip time for the water pulse to travel up to the water surface and back to the seabed. From Eq. 2-2 ($v_{\text{avg}} = \Delta x/\Delta t$), we can relate the speed v of the water pulse to the round-trip distance $2D$ and the round-trip time Δt as

$$v = \frac{2D}{\Delta t},$$

or

$$D = \frac{v\,\Delta t}{2}. \qquad (16\text{-}1)$$

Waves travel through water at about 1500 m/s. Substituting this value and $\Delta t = 0.11$ s in Eq. 16-1, we find that the seismic data predicted the sunken *Kursk* was at a depth of about

$$D = \frac{(1500 \text{ m/s})(0.11 \text{ s})}{2}$$

$$= 82.5 \text{ m} \approx 83 \text{ m}. \qquad \text{(Answer)}$$

In fact, the wreck was discovered at a depth of about 115 m.

16-4 Wavelength and Frequency

To completely describe a wave on a string (and the motion of any element along its length), we need a function that gives the shape of the wave. This means that we need a relation in the form $y = h(x, t)$, in which y is the transverse displacement of any string element as a function h of the time t and the position x of the element along the string. In general, a sinusoidal shape like the wave in Fig. 16-1b can be described with h being either a sine or cosine function; both give the same general shape for the wave. In this chapter we use the sine function.

Imagine a sinusoidal wave like that of Fig. 16-1b traveling in the positive direction of an x axis. As the wave sweeps through succeeding elements (that is, very short sections) of the string, the elements oscillate parallel to the y axis. At time t, the displacement y of the element located at position x is given by

$$y(x, t) = y_m \sin(kx - \omega t). \qquad (16\text{-}2)$$

Because this equation is written in terms of position x, it can be used to find the displacements of all the elements of the string as a function of time. Thus, it can tell us the shape of the wave at any given time and how that shape changes as the wave moves along the string.

The names of the quantities in Eq. 16-2 are displayed in Fig. 16-4 and defined next. Before we discuss them, however, let us examine Fig. 16-5, which shows five "snapshots" of a sinusoidal wave traveling in the positive direction of an x axis. The movement of the wave is indicated by the rightward progress of the short arrow pointing to a high point of the wave. From snapshot to snapshot, the short arrow moves to the right with the wave shape, but the string moves *only* parallel to the y axis. To see that, let us follow the motion of the red-dyed string element at $x = 0$. In the first snapshot (Fig. 16-5a), this element is at displacement $y = 0$. In the next snapshot, it is at its extreme downward displacement because a *valley* (or extreme low point) of the wave is passing through it. It then moves back up through $y = 0$. In the fourth snapshot, it is at its extreme upward displacement because a *peak* (or extreme high point) of the wave is passing through it. In the fifth snapshot, it is again at $y = 0$, having completed one full oscillation.

Amplitude and Phase

The **amplitude** y_m of a wave, such as that in Fig. 16-5, is the magnitude of the maximum displacement of the elements from their equilibrium positions as the

Fig. 16-4 The names of the quantities in Eq. 16-2, for a transverse sinusoidal wave.

wave passes through them. (The subscript m stands for maximum.) Because y_m is a magnitude, it is always a positive quantity, even if it is measured downward instead of upward as drawn in Fig. 16-5a.

The **phase** of the wave is the *argument* $kx - \omega t$ of the sine in Eq. 16-2. As the wave sweeps through a string element at a particular position x, the phase changes linearly with time t. This means that the sine also changes, oscillating between $+1$ and -1. Its extreme positive value ($+1$) corresponds to a peak of the wave moving through the element; at that instant the value of y at position x is y_m. Its extreme negative value (-1) corresponds to a valley of the wave moving through the element; at that instant the value of y at position x is $-y_m$. Thus, the sine function and the time-dependent phase of a wave correspond to the oscillation of a string element, and the amplitude of the wave determines the extremes of the element's displacement.

Wavelength and Angular Wave Number

The **wavelength** λ of a wave is the distance (parallel to the direction of the wave's travel) between repetitions of the shape of the wave (or *wave shape*). A typical wavelength is marked in Fig. 16-5a, which is a snapshot of the wave at time $t = 0$. At that time, Eq. 16-2 gives, for the description of the wave shape,

$$y(x, 0) = y_m \sin kx. \tag{16-3}$$

By definition, the displacement y is the same at both ends of this wavelength—that is, at $x = x_1$ and $x = x_1 + \lambda$. Thus, by Eq. 16-3,

$$y_m \sin kx_1 = y_m \sin k(x_1 + \lambda)$$
$$= y_m \sin(kx_1 + k\lambda). \tag{16-4}$$

A sine function begins to repeat itself when its angle (or argument) is increased by 2π rad, so in Eq. 16-4 we must have $k\lambda = 2\pi$, or

$$k = \frac{2\pi}{\lambda} \quad \text{(angular wave number).} \tag{16-5}$$

We call k the **angular wave number** of the wave; its SI unit is the radian per meter, or the inverse meter. (Note that the symbol k here does *not* represent a spring constant as previously.)

Notice that the wave in Fig. 16-5 moves to the right by $\frac{1}{4}\lambda$ from one snapshot to the next. Thus, by the fifth snapshot, it has moved to the right by 1λ.

Period, Angular Frequency, and Frequency

Figure 16-6 shows a graph of the displacement y of Eq. 16-2 versus time t at a certain position along the string, taken to be $x = 0$. If you were to monitor the string, you would see that the single element of the string at that position moves up and down in simple harmonic motion given by Eq. 16-2 with $x = 0$:

$$y(0, t) = y_m \sin(-\omega t)$$
$$= -y_m \sin \omega t \quad (x = 0). \tag{16-6}$$

Here we have made use of the fact that $\sin(-\alpha) = -\sin \alpha$, where α is any angle. Figure 16-6 is a graph of this equation; it *does not* show the shape of the wave.

We define the **period** of oscillation T of a wave to be the time any string element takes to move through one full oscillation. A typical period is marked on the graph of Fig. 16-6. Applying Eq. 16-6 to both ends of this time interval and equating the results yield

$$-y_m \sin \omega t_1 = -y_m \sin \omega(t_1 + T)$$
$$= -y_m \sin(\omega t_1 + \omega T). \tag{16-7}$$

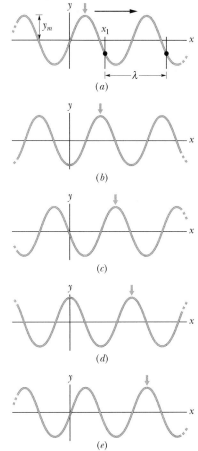

Fig. 16-5 Five "snapshots" of a string wave traveling in the positive direction of an x axis. The amplitude y_m is indicated. A typical wavelength λ, measured from an arbitrary position x_1, is also indicated.

Fig. 16-6 A graph of the displacement of the string element at $x = 0$ as a function of time, as the sinusoidal wave of Fig. 16-5 passes through the element. The amplitude y_m is indicated. A typical period T, measured from an arbitrary time t_1, is also indicated.

Fig. 16-7 A sinusoidal traveling wave at *t* = 0 with a phase constant *φ* of (*a*) 0 and (*b*) *π*/5 rad.

This can be true only if *ωT* = 2*π*, or if

$$\omega = \frac{2\pi}{T} \quad \text{(angular frequency).}\tag{16-8}$$

We call *ω* the **angular frequency** of the wave; its SI unit is the radian per second.

Look back at the five snapshots of a traveling wave in Fig. 16-5. The time between snapshots is $\frac{1}{4}T$. Thus, by the fifth snapshot, every string element has made one full oscillation.

The **frequency** *f* of a wave is defined as 1/*T* and is related to the angular frequency *ω* by

$$f = \frac{1}{T} = \frac{\omega}{2\pi} \quad \text{(frequency).}\tag{16-9}$$

Like the frequency of simple harmonic motion in Chapter 15, this frequency *f* is a number of oscillations per unit time—here, the number made by a string element as the wave moves through it. As in Chapter 15, *f* is usually measured in hertz or its multiples, such as kilohertz.

CHECKPOINT 1   The figure is a composite of three snapshots, each of a wave traveling along a particular string. The phases for the waves are given by (a) 2*x* − 4*t*, (b) 4*x* − 8*t*, and (c) 8*x* − 16*t*. Which phase corresponds to which wave in the figure?

Phase Constant

When a sinusoidal traveling wave is given by the wave function of Eq. 16-2, the wave near *x* = 0 looks like Fig. 16-7*a* when *t* = 0. Note that at *x* = 0, the displacement is *y* = 0 and the slope is at its maximum positive value. We can generalize Eq. 16-2 by inserting a **phase constant** *φ* in the wave function:

$$y = y_m \sin(kx - \omega t + \phi).\tag{16-10}$$

The value of *φ* can be chosen so that the function gives some other displacement and slope at *x* = 0 when *t* = 0. For example, a choice of *φ* = +*π*/5 rad gives the displacement and slope shown in Fig. 16-7*b* when *t* = 0. The wave is still sinusoidal with the same values of *y_m*, *k*, and *ω*, but it is now shifted from what you see in Fig. 16-7*a* (where *φ* = 0).

16-5 The Speed of a Traveling Wave

Figure 16-8 shows two snapshots of the wave of Eq. 16-2, taken a small time interval Δ*t* apart. The wave is traveling in the positive direction of *x* (to the right in Fig. 16-8), the entire wave pattern moving a distance Δ*x* in that direction during the interval Δ*t*. The ratio Δ*x*/Δ*t* (or, in the differential limit, *dx*/*dt*) is the **wave speed** *v*. How can we find its value?

As the wave in Fig. 16-8 moves, each point of the moving wave form, such as point *A* marked on a peak, retains its displacement *y*. (Points on the string do not retain their displacement, but points on the wave *form* do.) If point *A* retains its displacement as it moves, the phase in Eq. 16-2 giving it that displacement must remain a constant:

$$kx - \omega t = \text{a constant.}\tag{16-11}$$

Note that although this argument is constant, both *x* and *t* are changing. In fact, as *t* increases, *x* must also, to keep the argument constant. This confirms that the wave pattern is moving in the positive direction of *x*.

Fig. 16-8 Two snapshots of the wave of Fig. 16-5, at time *t* = 0 and then at time *t* = Δ*t*. As the wave moves to the right at velocity $\vec{v}$, the entire curve shifts a distance Δ*x* during Δ*t*. Point *A* "rides" with the wave form, but the string elements move only up and down.

To find the wave speed v, we take the derivative of Eq. 16-11, getting

$$k\frac{dx}{dt} - \omega = 0$$

or $$\frac{dx}{dt} = v = \frac{\omega}{k}. \tag{16-12}$$

Using Eq. 16-5 ($k = 2\pi/\lambda$) and Eq. 16-8 ($\omega = 2\pi/T$), we can rewrite the wave speed as

$$v = \frac{\omega}{k} = \frac{\lambda}{T} = \lambda f \quad \text{(wave speed)}. \tag{16-13}$$

The equation $v = \lambda/T$ tells us that the wave speed is one wavelength per period; the wave moves a distance of one wavelength in one period of oscillation.

Equation 16-2 describes a wave moving in the positive direction of x. We can find the equation of a wave traveling in the opposite direction by replacing t in Eq. 16-2 with $-t$. This corresponds to the condition

$$kx + \omega t = \text{a constant}, \tag{16-14}$$

which (compare Eq. 16-11) requires that x *decrease* with time. Thus, a wave traveling in the negative direction of x is described by the equation

$$y(x, t) = y_m \sin(kx + \omega t). \tag{16-15}$$

If you analyze the wave of Eq. 16-15 as we have just done for the wave of Eq. 16-2, you will find for its velocity

$$\frac{dx}{dt} = -\frac{\omega}{k}. \tag{16-16}$$

The minus sign (compare Eq. 16-12) verifies that the wave is indeed moving in the negative direction of x and justifies our switching the sign of the time variable.

Consider now a wave of arbitrary shape, given by

$$y(x, t) = h(kx \pm \omega t), \tag{16-17}$$

where h represents *any* function, the sine function being one possibility. Our previous analysis shows that all waves in which the variables x and t enter into the combination $kx \pm \omega t$ are traveling waves. Furthermore, all traveling waves *must* be of the form of Eq. 16-17. Thus, $y(x, t) = \sqrt{ax + bt}$ represents a possible (though perhaps physically a little bizarre) traveling wave. The function $y(x, t) = \sin(ax^2 - bt)$, on the other hand, does *not* represent a traveling wave.

Sample Problem 16-2

A wave traveling along a string is described by

$$y(x, t) = 0.00327 \sin(72.1x - 2.72t), \tag{16-18}$$

in which the numerical constants are in SI units (0.00327 m, 72.1 rad/m, and 2.72 rad/s).

(a) What is the amplitude of this wave?

Solution: The Key Idea is that Eq. 16-18 is of the same form as Eq. 16-2,

$$y = y_m \sin(kx - \omega t), \tag{16-19}$$

so we have a sinusoidal wave. By comparing the two equations, we see that the amplitude is

$$y_m = 0.00327 \text{ m} = 3.27 \text{ mm}. \quad \text{(Answer)}$$

(b) What are the wavelength, period, and frequency of this wave?

Solution: By comparing Eqs. 16-18 and 16-19, we see that the angular wave number and angular frequency are

$$k = 72.1 \text{ rad/m} \quad \text{and} \quad \omega = 2.72 \text{ rad/s}.$$

We then relate wavelength λ to k via Eq. 16-5:

$$\lambda = \frac{2\pi}{k} = \frac{2\pi \text{ rad}}{72.1 \text{ rad/m}}$$

$$= 0.0871 \text{ m} = 8.71 \text{ cm}. \quad \text{(Answer)}$$

Next, we relate T to ω with Eq. 16-8:

$$T = \frac{2\pi}{\omega} = \frac{2\pi \text{ rad}}{2.72 \text{ rad/s}} = 2.31 \text{ s}, \quad \text{(Answer)}$$

and from Eq. 16-9 we have

$$f = \frac{1}{T} = \frac{1}{2.31 \text{ s}} = 0.433 \text{ Hz.} \qquad \text{(Answer)}$$

(c) What is the velocity of this wave?

Solution: The speed of the wave is given by Eq. 16-13:

$$v = \frac{\omega}{k} = \frac{2.72 \text{ rad/s}}{72.1 \text{ rad/m}} = 0.0377 \text{ m/s}$$

$$= 3.77 \text{ cm/s.} \qquad \text{(Answer)}$$

Because the phase in Eq. 16-18 contains the position variable x, the wave is moving along the x axis. Also, because the wave equation is written in the form of Eq. 16-2, the *minus* sign in front of the ωt term indicates that the wave is moving in the

positive direction of the x axis. (Note that the quantities calculated in (b) and (c) are independent of the amplitude of the wave.)

(d) What is the displacement y at $x = 22.5$ cm and $t = 18.9$ s?

Solution: The **Key Idea** here is that Eq. 16-18 gives the displacement as a function of position x and time t. Substituting the given values into the equation yields

$$y = 0.00327 \sin(72.1 \times 0.225 - 2.72 \times 18.9)$$

$$= (0.00327 \text{ m}) \sin(-35.1855 \text{ rad})$$

$$= (0.00327 \text{ m})(0.588)$$

$$= 0.00192 \text{ m} = 1.92 \text{ mm.} \qquad \text{(Answer)}$$

Thus, the displacement is positive. (Be sure to change your calculator mode to radians before evaluating the sine.)

Sample Problem 16-3

In Sample Problem 16-2d, we showed that at $t = 18.9$ s the transverse displacement y of the element of the string at $x = 0.255$ m due to the wave of Eq. 16-18 is 1.92 mm.

(a) What is u, the transverse velocity of the same element of the string, at that time? (This velocity, which is associated with the transverse oscillation of an element of the string, is in the y direction. Do not confuse it with v, the constant velocity at which the *wave form* travels along the x axis.)

Solution: The **Key Idea** here is that the transverse velocity u is the rate at which the displacement y of the element is changing. In general, that displacement is given by

$$y(x, t) = y_m \sin(kx - \omega t). \qquad (16\text{-}20)$$

For an element at a certain location x, we find the rate of change of y by taking the derivative of Eq. 16-20 with respect to t while treating x as a constant. A derivative taken while one (or more) of the variables is treated as a constant is called a *partial derivative* and is represented by the symbol $\partial/\partial x$ rather than d/dx. Here we have

$$u = \frac{\partial y}{\partial t} = -\omega y_m \cos(kx - \omega t). \qquad (16\text{-}21)$$

Next, substituting numerical values from Sample Problem 16-2, we obtain

$$u = (-2.72 \text{ rad/s})(3.27 \text{ mm}) \cos(-35.1855 \text{ rad})$$

$$= 7.20 \text{ mm/s.} \qquad \text{(Answer)}$$

Thus, at $t = 18.9$ s, the element of the string at $x = 22.5$ cm is moving in the positive direction of y with a speed of 7.20 mm/s.

(b) What is the transverse acceleration a_y of the same element at that time?

Solution: The **Key Idea** here is that the transverse acceleration a_y is the rate at which the transverse velocity of the element is changing. From Eq. 16-21, again treating x as a constant but allowing t to vary, we find

$$a_y = \frac{\partial u}{\partial t} = -\omega^2 y_m \sin(kx - \omega t).$$

Comparison with Eq. 16-20 shows that we can write this as

$$a_y = -\omega^2 y.$$

We see that the transverse acceleration of an oscillating string element is proportional to its transverse displacement but opposite in sign. This is completely consistent with the action of the element itself—namely, that it is moving transversely in simple harmonic motion. Substituting numerical values yields

$$a_y = -(2.72 \text{ rad/s})^2(1.92 \text{ mm})$$

$$= -14.2 \text{ mm/s}^2. \qquad \text{(Answer)}$$

Thus, at $t = 18.9$ s, the element of string at $x = 22.5$ cm is displaced from its equilibrium position by 1.92 mm in the positive y direction and has an acceleration of magnitude 14.2 mm/s² in the negative y direction.

CHECKPOINT 2   Here are the equations of three waves:
(1) $y(x, t) = 2 \sin(4x - 2t)$,   (2) $y(x, t) = \sin(3x - 4t)$,
(3) $y(x, t) = 2 \sin(3x - 3t)$.
Rank the waves according to their (a) wave speed and (b) maximum transverse speed, greatest first.

PROBLEM-SOLVING TACTICS

TACTIC 1: *Evaluating Large Phases*

Sometimes, as in Sample Problems 16-2d and 16-3, an angle much greater than 2π rad (or 360°) crops up and you are asked to find its sine or cosine. Adding or subtracting an integral multiple of 2π rad to such an angle does not change the value of any of its trigonometric functions. In Sample Problem

16-2d, for example, we used the angle -35.1855 rad in a sine function. Adding $(6)(2\pi \text{ rad})$ to this angle yields

$$-35.1855 \text{ rad} + (6)(2\pi \text{ rad}) = 2.51361 \text{ rad},$$

an angle of less than 2π rad that has the same trigonometric

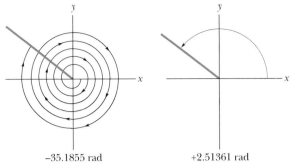

Fig. 16-9 These two angles are different, but all their trigonometric functions are identical.

−35.1855 rad   +2.51361 rad

functions as −35.1855 rad (Fig. 16-9). As an example, the sine of both 2.51361 rad and −35.1855 rad is 0.588.

    Your calculator will reduce such large angles for you automatically. *Caution:* Do not round off large angles if you intend to take their sines or cosines. In taking the sine of a very large angle, you are throwing away most of the angle and taking the sine of what is left over. If, for example, you were to round −35.1855 rad to −35 rad (a change of 0.5% and normally a reasonable step), you would be changing the sine of the angle by 27%. Also, if you change a large angle from degrees to radians, be sure to use an exact conversion factor (such as $180° = \pi$ rad) rather than an approximate one (such as $57.3° \approx 1$ rad).

16-6 *Wave Speed on a Stretched String*

The speed of a wave is related to the wave's wavelength and frequency by Eq. 16-13, but *it is set by the properties of the medium.* If a wave is to travel through a medium such as water, air, steel, or a stretched string, it must cause the particles of that medium to oscillate as it passes. For that to happen, the medium must possess both mass (so that there can be kinetic energy) and elasticity (so that there can be potential energy). Thus, the medium's mass and elasticity properties determine how fast the wave can travel in the medium. Conversely, it should be possible to calculate the speed of the wave through the medium in terms of these properties. We do so now for a stretched string, in two ways.

Dimensional Analysis

In dimensional analysis we carefully examine the dimensions of all the physical quantities that enter into a given situation to determine the quantities they produce. In this case, we examine mass and elasticity to find a speed v, which has the dimension of length divided by time, or LT^{-1}.

    For the mass, we use the mass of a string element, which is represented by the mass m of the string divided by the length l of the string. We call this ratio the *linear density* μ of the string. Thus, $\mu = m/l$, its dimension being mass divided by length, ML^{-1}.

    You cannot send a wave along a string unless the string is under tension, which means that it has been stretched and pulled taut by forces at its two ends. The tension τ in the string is equal to the common magnitude of those two forces. As a wave travels along the string, it displaces elements of the string by causing additional stretching, with adjacent sections of string pulling on each other because of the tension. Thus, we can associate the tension in the string with the stretching (elasticity) of the string. The tension and the stretching forces it produces have the dimension of a force—namely, MLT^{-2} (from $F = ma$).

    The goal here is to combine μ (dimension ML^{-1}) and τ (dimension MLT^{-2}) in such a way as to generate v (dimension LT^{-1}). A little juggling of various combinations suggests

$$v = C\sqrt{\frac{\tau}{\mu}}, \qquad (16\text{-}22)$$

in which C is a dimensionless constant that cannot be determined with dimensional analysis. In our second approach to determining wave speed, you will see that Eq. 16-22 is indeed correct and that $C = 1$.

Derivation from Newton's Second Law

Instead of the sinusoidal wave of Fig. 16-1*b*, let us consider a single symmetrical pulse such as that of Fig. 16-10, moving from left to right along a string with speed

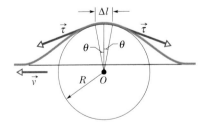

Fig. 16-10 A symmetrical pulse, viewed from a reference frame in which the pulse is stationary and the string appears to move right to left with speed v. We find speed v by applying Newton's second law to a string element of length Δl, located at the top of the pulse.

v. For convenience, we choose a reference frame in which the pulse remains stationary; that is, we run along with the pulse, keeping it constantly in view. In this frame, the string appears to move past us, from right to left in Fig. 16-10, with speed *v*.

Consider a small string element of length Δl within the pulse, an element that forms an arc of a circle of radius *R* and subtending an angle 2θ at the center of that circle. A force $\vec{\tau}$ with a magnitude equal to the tension in the string pulls tangentially on this element at each end. The horizontal components of these forces cancel, but the vertical components add to form a radial restoring force $\vec{F}$. In magnitude,

$$F = 2(\tau \sin \theta) \approx \tau(2\theta) = \tau \frac{\Delta l}{R} \qquad \text{(force)}, \qquad (16\text{-}23)$$

where we have approximated $\sin \theta$ as θ for the small angles θ in Fig. 16-10. From that figure, we have also used $2\theta = \Delta l/R$.

The mass of the element is given by

$$\Delta m = \mu \, \Delta l \qquad \text{(mass)}, \qquad (16\text{-}24)$$

where μ is the string's linear density.

At the moment shown in Fig. 16-10, the string element Δl is moving in an arc of a circle. Thus, it has a centripetal acceleration toward the center of that circle, given by

$$a = \frac{v^2}{R} \qquad \text{(acceleration)}. \qquad (16\text{-}25)$$

Equations 16-23, 16-24, and 16-25 contain the elements of Newton's second law. Combining them in the form

$$\text{force} = \text{mass} \times \text{acceleration}$$

gives $\qquad\qquad \dfrac{\tau \, \Delta l}{R} = (\mu \, \Delta l) \dfrac{v^2}{R}.$

Solving this equation for the speed *v* yields

$$v = \sqrt{\frac{\tau}{\mu}} \qquad \text{(speed)}, \qquad (16\text{-}26)$$

in exact agreement with Eq. 16-22 if the constant *C* in that equation is given the value unity. Equation 16-26 gives the speed of the pulse in Fig. 16-10 and the speed of *any* other wave on the same string under the same tension.

Equation 16-26 tells us:

> The speed of a wave along a stretched ideal string depends only on the tension and linear density of the string and not on the frequency of the wave.

The *frequency* of the wave is fixed entirely by whatever generates the wave (for example, the person in Fig. 16-1*b*). The *wavelength* of the wave is then fixed by Eq. 16-13 in the form $\lambda = v/f$.

CHECKPOINT 3 You send a traveling wave along a particular string by oscillating one end. If you increase the frequency of the oscillations, do (a) the speed of the wave and (b) the wavelength of the wave increase, decrease, or remain the same? If, instead, you increase the tension in the string, do (c) the speed of the wave and (d) the wavelength of the wave increase, decrease, or remain the same?

Sample Problem 16-4

In Fig. 16-11, two strings have been tied together with a knot and then stretched between two rigid supports. The strings have linear densities $\mu_1 = 1.4 \times 10^{-4}$ kg/m and $\mu_2 = 2.8 \times 10^{-4}$ kg/m. Their lengths are $L_1 = 3.0$ m and $L_2 = 2.0$ m, and string 1 is under a tension of 400 N. Simultaneously, on each string a pulse is sent from the rigid support end, toward the knot. Which pulse reaches the knot first?

Solution: We need several **Key Ideas** here:

1. The time t taken by a pulse to travel a length L is $t = L/v$, where v is the constant speed of the pulse.
2. The speed of a pulse on a stretched string depends on the

string's tension τ and linear density μ, and is given by Eq. 16-26 ($v = \sqrt{\tau/\mu}$).

3. Because the two strings are stretched together, they must both be under the same tension τ ($= 400$ N).

Putting these three ideas together gives us, as the time for the pulse on string 1 to reach the knot,

$$t_1 = \frac{L_1}{v_1} = L_1\sqrt{\frac{\mu_1}{\tau}} = (3.0 \text{ m})\sqrt{\frac{1.4 \times 10^{-4} \text{ kg/m}}{400 \text{ N}}}$$
$$= 1.77 \times 10^{-3} \text{ s}.$$

Similarly, the data for the pulse on string 2 give us

$$t_2 = L_2\sqrt{\frac{\mu_2}{\tau}} = 1.67 \times 10^{-3} \text{ s}.$$

Thus, the pulse on string 2 reaches the knot first.

Now look back at **Key Idea** 2. The linear density of string 2 is greater than that of string 1, so the pulse on string 2 must be slower than that on string 1. Could we have guessed the answer from that fact alone? No, because from the first **Key Idea** we see that the distance traveled by a pulse also matters.

Fig. 16-11   Two strings, of lengths L_1 and L_2, tied together with a knot and stretched between two rigid supports.

16-7   Energy and Power of a Wave Traveling Along a String

When we set up a wave on a stretched string, we provide energy for the motion of the string. As the wave moves away from us, it transports that energy as both kinetic energy and elastic potential energy. Let us consider each form in turn.

Kinetic Energy

A string element of mass dm, oscillating transversely in simple harmonic motion as the wave passes through it, has kinetic energy associated with its transverse velocity $\vec{u}$. When the element is rushing through its $y = 0$ position (element b in Fig. 16-12), its transverse velocity—and thus its kinetic energy—is a maximum. When the element is at its extreme position $y = y_m$ (as is element a), its transverse velocity—and thus its kinetic energy—is zero.

Elastic Potential Energy

To send a sinusoidal wave along a previously straight string, the wave must necessarily stretch the string. As a string element of length dx oscillates transversely, its length must increase and decrease in a periodic way if the string element is to fit the sinusoidal wave form. Elastic potential energy is associated with these length changes, just as for a spring.

When the string element is at its $y = y_m$ position (element a in Fig. 16-12), its length has its normal undisturbed value dx, so its elastic potential energy is zero. However, when the element is rushing through its $y = 0$ position, it has maximum stretch and thus maximum elastic potential energy.

Energy Transport

The oscillating string element thus has both its maximum kinetic energy and its maximum elastic potential energy at $y = 0$. In the snapshot of Fig. 16-12, the

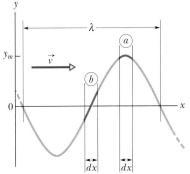

Fig. 16-12   A snapshot of a traveling wave on a string at time $t = 0$. String element a is at displacement $y = y_m$, and string element b is at displacement $y = 0$. The kinetic energy of the string element at each position depends on the transverse velocity of the element. The potential energy depends on the amount by which the string element is stretched as the wave passes through it.

regions of the string at maximum displacement have no energy, and the regions at zero displacement have maximum energy. As the wave travels along the string, forces due to the tension in the string continuously do work to transfer energy from regions with energy to regions with no energy.

Suppose we set up a wave on a string stretched along a horizontal x axis so that Eq. 16-2 describes the string's displacement. We might send a wave along the string by continuously oscillating one end of the string, as in Fig. 16-1b. In doing so, we continuously provide energy for the motion and stretching of the string—as the string sections oscillate perpendicularly to the x axis, they have kinetic energy and elastic potential energy. As the wave moves into sections that were previously at rest, energy is transferred into those new sections. Thus, we say that the wave *transports* the energy along the string.

The Rate of Energy Transmission

The kinetic energy dK associated with a string element of mass dm is given by

$$dK = \tfrac{1}{2}\, dm\, u^2, \tag{16-27}$$

where u is the transverse speed of the oscillating string element. To find u, we differentiate Eq. 16-2 with respect to time while holding x constant:

$$u = \frac{\partial y}{\partial t} = -\omega y_m \cos(kx - \omega t). \tag{16-28}$$

Using this relation and putting $dm = \mu\, dx$, we rewrite Eq. 16-27 as

$$dK = \tfrac{1}{2}(\mu\, dx)(-\omega y_m)^2 \cos^2(kx - \omega t). \tag{16-29}$$

Dividing Eq. 16-29 by dt gives the rate at which kinetic energy passes through a string element, and thus the rate at which kinetic energy is carried along by the wave. The ratio dx/dt that then appears on the right of Eq. 16-29 is the wave speed v, so we obtain

$$\frac{dK}{dt} = \tfrac{1}{2}\mu v \omega^2 y_m^2 \cos^2(kx - \omega t). \tag{16-30}$$

The *average* rate at which kinetic energy is transported is

$$\left(\frac{dK}{dt}\right)_{\text{avg}} = \tfrac{1}{2}\mu v \omega^2 y_m^2 [\cos^2(kx - \omega t)]_{\text{avg}}$$
$$= \tfrac{1}{4}\mu v \omega^2 y_m^2. \tag{16-31}$$

Here we have taken the average over an integer number of wavelengths and have used the fact that the average value of the square of a cosine function over an integer number of periods is $\tfrac{1}{2}$.

Elastic potential energy is also carried along with the wave, and at the same average rate given by Eq. 16-31. Although we shall not examine the proof, you should recall that, in an oscillating system such as a pendulum or a spring–block system, the average kinetic energy and the average potential energy are equal.

The **average power,** which is the average rate at which energy of both kinds is transmitted by the wave, is then

$$P_{\text{avg}} = 2\left(\frac{dK}{dt}\right)_{\text{avg}} \tag{16-32}$$

or, from Eq. 16-31,

$$P_{\text{avg}} = \tfrac{1}{2}\mu v \omega^2 y_m^2 \qquad \text{(average power).} \tag{16-33}$$

The factors μ and v in this equation depend on the material and tension of the string. The factors ω and y_m depend on the process that generates the wave. The dependence of the average power of a wave on the square of its amplitude and also on the square of its angular frequency is a general result, true for waves of all types.

Sample Problem 16-5

A stretched string has linear density $\mu = 525$ g/m and is under tension $\tau = 45$ N. We send a sinusoidal wave with frequency $f = 120$ Hz and amplitude $y_m = 8.5$ mm along the string. At what average rate does the wave transport energy?

Solution: The Key Idea here is that the average rate of energy transport is the average power P_{avg} as given by Eq. 16-33. To use that equation, however, we first must calculate angular frequency ω and wave speed v. From Eq. 16-9,

$$\omega = 2\pi f = (2\pi)(120 \text{ Hz}) = 754 \text{ rad/s}.$$

From Eq. 16-26 we have

$$v = \sqrt{\frac{\tau}{\mu}} = \sqrt{\frac{45 \text{ N}}{0.525 \text{ kg/m}}} = 9.26 \text{ m/s}.$$

Equation 16-33 then yields

$$
\begin{aligned}
P_{avg} &= \tfrac{1}{2}\mu v \omega^2 y_m^2 \\
&= (\tfrac{1}{2})(0.525 \text{ kg/m})(9.26 \text{ m/s})(754 \text{ rad/s})^2(0.0085 \text{ m})^2 \\
&\approx 100 \text{ W.} \qquad\qquad\qquad\qquad\qquad \text{(Answer)}
\end{aligned}
$$

CHECKPOINT 4   For the string and wave of this sample problem, we can adjust three parameters: the tension in the string, the frequency of the wave, and the amplitude of the wave. Does the average rate at which the wave transports energy along the string increase, decrease, or remain the same if we increase (a) the tension, (b) the frequency, and (c) the amplitude?

16-8   The Wave Equation

As a wave passes through any element on a stretched string, the element moves perpendicularly to the wave's direction of travel. By applying Newton's second law to the element's motion, we can derive a general differential equation, called the *wave equation*, that governs the travel of waves of any type.

Figure 16-13a shows a snapshot of a string element of mass dm and length ℓ as a wave travels along a string of linear density μ that is stretched along a horizontal x axis. Let us assume that the wave amplitude is small so that the element can be tilted only slightly from the x axis as the wave passes. The force $\vec{F}_2$ on the right end of the element has a magnitude equal to tension τ in the string and is directed slightly upward. The force $\vec{F}_1$ on the left end of the element also has a magnitude equal to the tension τ but is directed slightly downward. Because of the slight curvature of the element, these two forces produce a net force that causes the element to have an upward acceleration a_y. Newton's second law written for y components ($F_{net,y} = ma_y$) gives us

$$F_{2y} - F_{1y} = dm\, a_y. \qquad (16\text{-}34)$$

Let's analyze this equation in parts.

Mass. The element's mass dm can be written in terms of the string's linear density μ and the element's length ℓ as $dm = \mu\ell$. Because the element can have only a slight tilt, $\ell \approx dx$ (Fig. 16-13a) and we have the approximation

$$dm = \mu\, dx. \qquad (16\text{-}35)$$

Acceleration. The acceleration a_y in Eq. 16-34 is the second derivative of the displacement y with respect to time:

$$a_y = \frac{d^2 y}{dt^2}. \qquad (16\text{-}36)$$

Forces. Figure 16-13b shows that $\vec{F}_2$ is tangent to the string at the right end of the string element. Thus we can relate the components of the force to the string slope S_2 at the right end as

$$\frac{F_{2y}}{F_{2x}} = S_2. \qquad (16\text{-}37)$$

We can also relate the components to the magnitude $F_2 (= \tau)$ with

$$F_2 = \sqrt{F_{2x}^2 + F_{2y}^2}$$

or

$$\tau = \sqrt{F_{2x}^2 + F_{2y}^2}. \qquad (16\text{-}38)$$

(a)

(b)

Fig. 16-13   (a) A string element as a sinusoidal transverse wave travels on a stretched string. Forces $\vec{F}_1$ and $\vec{F}_2$ act at the left and right ends, producing acceleration $\vec{a}$ having a vertical component a_y. (b) The force at the element's right end is directed along a tangent to the element's right side.

However, because we assume that the element is only slightly tilted, $F_{2y} \ll F_{2x}$ and therefore we can rewrite Eq. 16-38 as

$$\tau = F_{2x}. \tag{16-39}$$

Substituting this into Eq. 16-37 and solving for F_{2y} yield

$$F_{2y} = \tau S_2. \tag{16-40}$$

Similar analysis at the left end of the string element gives us

$$F_{1y} = \tau S_1. \tag{16-41}$$

We can now substitute Eqs. 16-35, 16-36, 16-40, and 16-41 into Eq. 16-34 to write

$$\tau S_2 - \tau S_1 = (\mu \, dx) \frac{d^2 y}{dt^2},$$

or

$$\frac{S_2 - S_1}{dx} = \frac{\mu}{\tau} \frac{d^2 y}{dt^2}. \tag{16-42}$$

Because the string element is short, slopes S_2 and S_1 differ by only a differential amount dS, where S is the slope at any point:

$$S = \frac{dy}{dx}. \tag{16-43}$$

First replacing $S_2 - S_1$ in Eq. 16-42 with dS and then using Eq. 16-43 to substitute dy/dx for S, we find

$$\frac{dS}{dx} = \frac{\mu}{\tau} \frac{d^2 y}{dt^2},$$

$$\frac{d(dy/dx)}{dx} = \frac{\mu}{\tau} \frac{d^2 y}{dt^2},$$

and

$$\frac{\partial^2 y}{\partial x^2} = \frac{\mu}{\tau} \frac{\partial^2 y}{\partial t^2}. \tag{16-44}$$

In the last step, we switched to the notation of partial derivatives because on the left we differentiate only with respect to x and on the right we differentiate only with respect to t. Finally, substituting from Eq. 16-26 ($v = \sqrt{\tau/\mu}$), we find

$$\frac{\partial^2 y}{\partial x^2} = \frac{1}{v^2} \frac{\partial^2 y}{\partial t^2} \quad \text{(wave equation).} \tag{16-45}$$

This is the general differential equation that governs the travel of waves of all types.

16-9 The Principle of Superposition for Waves

It often happens that two or more waves pass simultaneously through the same region. When we listen to a concert, for example, sound waves from many instruments fall simultaneously on our eardrums. The electrons in the antennas of our radio and television receivers are set in motion by the net effect of many electromagnetic waves from many different broadcasting centers. The water of a lake or harbor may be churned up by waves in the wakes of many boats.

Suppose that two waves travel simultaneously along the same stretched string. Let $y_1(x, t)$ and $y_2(x, t)$ be the displacements that the string would experience if each wave traveled alone. The displacement of the string when the waves overlap is then the algebraic sum

$$y'(x, t) = y_1(x, t) + y_2(x, t). \tag{16-46}$$

This summation of displacements along the string means that

 Overlapping waves algebraically add to produce a **resultant wave** (or **net wave**).

This is another example of the **principle of superposition,** which says that when several effects occur simultaneously, their net effect is the sum of the individual effects.

Figure 16-14 shows a sequence of snapshots of two pulses traveling in opposite directions on the same stretched string. When the pulses overlap, the resultant pulse is their sum. Moreover, each pulse moves through the other as if the other were not present:

 Overlapping waves do not in any way alter the travel of each other.

16-10 *Interference of Waves*

Suppose we send two sinusoidal waves of the same wavelength and amplitude in the same direction along a stretched string. The superposition principle applies. What resultant wave does it predict for the string?

The resultant wave depends on the extent to which the waves are *in phase* (in step) with respect to each other — that is, how much one wave form is shifted from the other wave form. If the waves are exactly in phase (so that the peaks and valleys of one are exactly aligned with those of the other), they combine to double the displacement of either wave acting alone. If they are exactly out of phase (the peaks of one are exactly aligned with the valleys of the other), they combine to cancel everywhere, and the string remains straight. We call this phenomenon of combining waves **interference,** and the waves are said to **interfere.** (These terms refer only to the wave displacements; the travel of the waves is unaffected.)

Let one wave traveling along a stretched string be given by

$$y_1(x, t) = y_m \sin(kx - \omega t) \tag{16-47}$$

and another, shifted from the first, by

$$y_2(x, t) = y_m \sin(kx - \omega t + \phi). \tag{16-48}$$

These waves have the same angular frequency ω (and thus the same frequency f), the same angular wave number k (and thus the same wavelength λ), and the same amplitude y_m. They both travel in the positive direction of the x axis, with the same speed, given by Eq. 16-26. They differ only by a constant angle ϕ, the phase constant. These waves are said to be *out of phase* by ϕ or to have a *phase difference* of ϕ, or one wave is said to be *phase-shifted* from the other by ϕ.

From the principle of superposition (Eq. 16-46), the resultant wave is the algebraic sum of the two interfering waves and has displacement

$$y'(x, t) = y_1(x, t) + y_2(x, t)$$
$$= y_m \sin(kx - \omega t) + y_m \sin(kx - \omega t + \phi). \tag{16-49}$$

In Appendix E we see that we can write the sum of the sines of two angles α and β as

$$\sin \alpha + \sin \beta = 2 \sin \tfrac{1}{2}(\alpha + \beta) \cos \tfrac{1}{2}(\alpha - \beta). \tag{16-50}$$

Applying this relation to Eq. 16-49 leads to

$$y'(x, t) = [2y_m \cos \tfrac{1}{2}\phi] \sin(kx - \omega t + \tfrac{1}{2}\phi). \tag{16-51}$$

As Fig. 16-15 shows, the resultant wave is also a sinusoidal wave traveling in the direction of increasing x. It is the only wave you would actually see on the string (you would *not* see the two interfering waves of Eqs. 16-47 and 16-48).

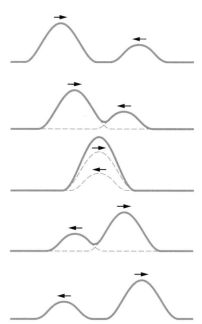

Fig. 16-14 A series of snapshots that show two pulses traveling in opposite directions along a stretched string. The superposition principle applies as the pulses move through each other.

Displacement

$$y'(x,t) = \underbrace{[2y_m \cos \tfrac{1}{2}\phi]}_{\substack{\text{Magnitude} \\ \text{gives} \\ \text{amplitude}}} \underbrace{\sin(kx - \omega t + \tfrac{1}{2}\phi)}_{\substack{\text{Oscillating} \\ \text{term}}}$$

Fig. 16-15 The resultant wave of Eq. 16-51, due to the interference of two sinusoidal transverse waves, is also a sinusoidal transverse wave, with an amplitude and an oscillating term.

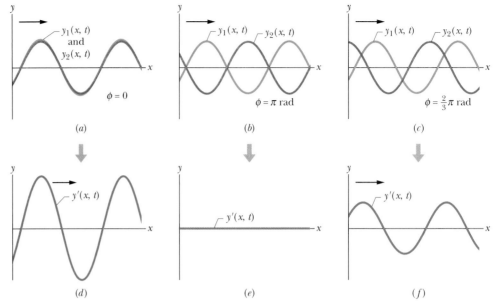

Fig. 16-16 Two identical sinusoidal waves, $y_1(x, t)$ and $y_2(x, t)$, travel along a string in the positive direction of an x axis. They interfere to give a resultant wave $y'(x, t)$. The resultant wave is what is actually seen on the string. The phase difference ϕ between the two interfering waves is (a) 0 rad or 0°, (b) π rad or 180°, and (c) $\frac{2}{3}\pi$ rad or 120°. The corresponding resultant waves are shown in (d), (e), and (f).

> If two sinusoidal waves of the same amplitude and wavelength travel in the *same* direction along a stretched string, they interfere to produce a resultant sinusoidal wave traveling in that direction.

The resultant wave differs from the interfering waves in two respects: (1) its phase constant is $\frac{1}{2}\phi$, and (2) its amplitude y'_m is the magnitude of the quantity in the brackets in Eq. 16-51:

$$y'_m = |2y_m \cos \tfrac{1}{2}\phi| \qquad \text{(amplitude)}. \tag{16-52}$$

If $\phi = 0$ rad (or 0°), the two interfering waves are exactly in phase, as in Fig. 16-16a. Then Eq. 16-51 reduces to

$$y'(x, t) = 2y_m \sin(kx - \omega t) \qquad (\phi = 0). \tag{16-53}$$

This resultant wave is plotted in Fig. 16-16d. Note from both that figure and Eq. 16-53 that the amplitude of the resultant wave is twice the amplitude of either interfering wave. That is the greatest amplitude the resultant wave can have, because the cosine term in Eqs. 16-51 and 16-52 has its greatest value (unity) when $\phi = 0$. Interference that produces the greatest possible amplitude is called *fully constructive interference.*

If $\phi = \pi$ rad (or 180°), the interfering waves are exactly out of phase as in Fig. 16-16b. Then $\cos \frac{1}{2}\phi$ becomes $\cos \pi/2 = 0$, and the amplitude of the resultant wave as given by Eq. 16-52 is zero. We then have, for all values of x and t,

$$y'(x, t) = 0 \qquad (\phi = \pi \text{ rad}). \tag{16-54}$$

The resultant wave is plotted in Fig. 16-16e. Although we sent two waves along the string, we see no motion of the string. This type of interference is called *fully destructive interference.*

Because a sinusoidal wave repeats its shape every 2π rad, a phase difference $\phi = 2\pi$ rad (or 360°) corresponds to a shift of one wave relative to the other wave by a distance equivalent to one wavelength. Thus, phase differences can be

TABLE 16-1
Phase Differences and Resulting Interference Types[a]

Phase Difference, in			Amplitude of Resultant Wave	Type of Interference
Degrees	Radians	Wavelengths		
0	0	0	$2y_m$	Fully constructive
120	$\frac{2}{3}\pi$	0.33	y_m	Intermediate
180	π	0.50	0	Fully destructive
240	$\frac{4}{3}\pi$	0.67	y_m	Intermediate
360	2π	1.00	$2y_m$	Fully constructive
865	15.1	2.40	$0.60y_m$	Intermediate

[a]The phase difference is between two otherwise identical waves, with amplitude y_m, moving in the same direction.

described in terms of wavelengths as well as angles. For example, in Fig. 16-16b the waves may be said to be 0.50 wavelength out of phase. Table 16-1 shows some other examples of phase differences and the interference they produce. Note that when interference is neither fully constructive nor fully destructive, it is called *intermediate interference*. The amplitude of the resultant wave is then intermediate between 0 and $2y_m$. For example, from Table 16-1, if the interfering waves have a phase difference of 120° ($\phi = \frac{2}{3}\pi$ rad = 0.33 wavelength), then the resultant wave has an amplitude of y_m, the same as that of the interfering waves (see Figs. 16-16c and f).

Two waves with the same wavelength are in phase if their phase difference is zero or any integer number of wavelengths. Thus, the integer part of any phase difference *expressed in wavelengths* may be discarded. For example, a phase difference of 0.40 wavelength is equivalent in every way to one of 2.40 wavelengths, and so the simpler of the two numbers can be used in computations.

Sample Problem 16-6

Two identical sinusoidal waves, moving in the same direction along a stretched string, interfere with each other. The amplitude y_m of each wave is 9.8 mm, and the phase difference ϕ between them is 100°.

(a) What is the amplitude y'_m of the resultant wave due to the interference, and what is the type of this interference?

Solution: The **Key Idea** here is that these are identical sinusoidal waves traveling in the *same direction* along a string, so they interfere to produce a sinusoidal traveling wave. Because they are identical, they have the *same amplitude*. Thus, the amplitude y'_m of the resultant wave is given by Eq. 16-52:

$$y'_m = |2y_m \cos \tfrac{1}{2}\phi| = |(2)(9.8 \text{ mm}) \cos(100°/2)|$$
$$= 13 \text{ mm.} \qquad \text{(Answer)}$$

We can tell that the interference is *intermediate* in two ways. The phase difference is between 0 and 180°, and, correspondingly, the amplitude y'_m is between 0 and $2y_m$ (= 19.6 mm).

(b) What phase difference, in radians and wavelengths, will give the resultant wave an amplitude of 4.9 mm?

Solution: The same **Key Idea** applies here as in part (a), but

now we are given y'_m and seek ϕ. From Eq. 16-52,

$$y'_m = |2y_m \cos \tfrac{1}{2}\phi|,$$

we now have

$$4.9 \text{ mm} = (2)(9.8 \text{ mm}) \cos \tfrac{1}{2}\phi,$$

which gives us (with a calculator in the radian mode)

$$\phi = 2 \cos^{-1} \frac{4.9 \text{ mm}}{(2)(9.8 \text{ mm})}$$
$$= \pm 2.636 \text{ rad} \approx \pm 2.6 \text{ rad.} \qquad \text{(Answer)}$$

There are two solutions because we can obtain the same resultant wave by letting the first wave *lead* (travel ahead of) or *lag* (travel behind) the second wave by 2.6 rad. In wavelengths, the phase difference is

$$\frac{\phi}{2\pi \text{ rad/wavelength}} = \frac{\pm 2.636 \text{ rad}}{2\pi \text{ rad/wavelength}}$$
$$= \pm 0.42 \text{ wavelength.} \qquad \text{(Answer)}$$

✔**CHECKPOINT 5**   Here are four other possible phase differences between the two waves of this sample problem, expressed in wavelengths: 0.20, 0.45, 0.60, and 0.80. Rank them according to the amplitude of the resultant wave, greatest first.

16-11 Phasors

We can represent a string wave (or any other type of wave) vectorially with a **phasor.** In essence, a phasor is a vector that has a magnitude equal to the amplitude of the wave and that rotates around an origin; the angular speed of the phasor is equal to the angular frequency ω of the wave. For example, the wave

$$y_1(x, t) = y_{m1} \sin(kx - \omega t) \qquad (16\text{-}55)$$

is represented by the phasor shown in Fig. 16-17a. The magnitude of the phasor is the amplitude y_{m1} of the wave. As the phasor rotates around the origin at angular speed ω, its projection y_1 on the vertical axis varies sinusoidally, from a maximum of y_{m1} through zero to a minimum of $-y_{m1}$ and then back to y_{m1}. This variation corresponds to the sinusoidal variation in the displacement y_1 of any point along the string as the wave passes through that point.

When two waves travel along the same string in the same direction, we can represent them and their resultant wave in a *phasor diagram*. The phasors in Fig. 16-17b represent the wave of Eq. 16-55 and a second wave given by

$$y_2(x, t) = y_{m2} \sin(kx - \omega t + \phi). \qquad (16\text{-}56)$$

This second wave is phase-shifted from the first wave by phase constant ϕ. Because the phasors rotate at the same angular speed ω, the angle between the two phasors is always ϕ. If ϕ is a *positive* quantity, then the phasor for wave 2 *lags* the phasor for wave 1 as they rotate, as drawn in Fig. 16-17b. If ϕ is a negative quantity, then the phasor for wave 2 *leads* the phasor for wave 1.

Because waves y_1 and y_2 have the same angular wave number k and angular frequency ω, we know from Eqs. 16-51 and 16-52 that their resultant is of the form

$$y'(x, t) = y'_m \sin(kx - \omega t + \beta), \qquad (16\text{-}57)$$

where y'_m is the amplitude of the resultant wave and β is its phase constant. To find the values of y'_m and β, we would have to sum the two combining waves, as we did to obtain Eq. 16-51. To do this on a phasor diagram, we vectorially add the two phasors at any instant during their rotation, as in Fig. 16-17c where phasor y_{m2} has been shifted to the head of phasor y_{m1}. The magnitude of the vector sum equals the amplitude y'_m in Eq. 16-57. The angle between the vector sum and the phasor for y_1 equals the phase constant β in Eq. 16-57.

Note that, in contrast to the method of Section 16-10:

We can use phasors to combine waves *even if their amplitudes are different.*

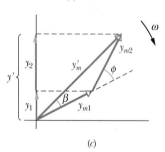

(a)

(b)

(c)

Fig. 16-17   (a) A phasor of magnitude y_{m1} rotating about an origin at angular speed ω represents a sinusoidal wave. The phasor's projection y_1 on the vertical axis represents the displacement of a point through which the wave passes. (b) A second phasor, also of angular speed ω but of magnitude y_{m2} and rotating at a constant angle ϕ from the first phasor, represents a second wave, with a phase constant ϕ. (c) The resultant wave is represented by the vector sum y'_m of the two phasors.

Sample Problem 16-7

Two sinusoidal waves $y_1(x, t)$ and $y_2(x, t)$ have the same wavelength and travel together in the same direction along a string. Their amplitudes are $y_{m1} = 4.0$ mm and $y_{m2} = 3.0$ mm, and their phase constants are 0 and $\pi/3$ rad, respectively. What are the amplitude y'_m and phase constant β of the resultant wave? Write the resultant wave in the form of Eq. 16-57.

Solution: One **Key Idea** here is that the two waves have a number of properties in common: Because they travel along the same string, they must have the same speed v, as set by the tension and linear density of the string according to Eq. 16-26. With the same wavelength λ, they must have the same angular wave number k ($= 2\pi/\lambda$). Also, with the same wave

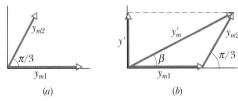

Fig. 16-18 (*a*) Two phasors of magnitudes y_{m1} and y_{m2} and with phase difference $\pi/3$. (*b*) Vector addition of these phasors at any instant during their rotation gives the magnitude y'_m of the phasor for the resultant wave.

number k and speed v, they must have the same angular frequency ω ($= kv$).

A second **Key Idea** is that the waves (call them waves 1 and 2) can be represented by phasors rotating at the same angular speed ω about an origin. Because the phase constant for wave 2 is *greater* than that for wave 1 by $\pi/3$, phasor 2 must *lag* phasor 1 by $\pi/3$ rad in their clockwise rotation, as shown in Fig. 16-18*a*. The resultant wave due to the interference of waves 1 and 2 can then be represented by a phasor that is the vector sum of phasors 1 and 2.

To simplify the vector summation, we drew phasors 1 and 2 in Fig. 16-18*a* at the instant when phasor 1 lies along the horizontal axis. We then drew lagging phasor 2 at positive angle $\pi/3$ rad. In Fig. 16-18*b* we shifted phasor 2 so its tail is at the head of phasor 1. Then we can draw the phasor y'_m of the resultant wave from the tail of phasor 1 to the head of

phasor 2. The phase constant β is the angle phasor y'_m makes with phasor 1.

To find values for y'_m and β, we can sum phasors 1 and 2 directly on a vector-capable calculator (by adding a vector of magnitude 4.0 and angle 0 rad to a vector of magnitude 3.0 and angle $\pi/3$ rad) or we can add the vectors by components. For the horizontal components we have

$$y'_{mh} = y_{m1} \cos 0 + y_{m2} \cos \pi/3$$
$$= 4.0 \text{ mm} + (3.0 \text{ mm}) \cos \pi/3 = 5.50 \text{ mm}.$$

For the vertical components we have

$$y'_{mv} = y_{m1} \sin 0 + y_{m2} \sin \pi/3$$
$$= 0 + (3.0 \text{ mm}) \sin \pi/3 = 2.60 \text{ mm}.$$

Thus, the resultant wave has an amplitude of

$$y'_m = \sqrt{(5.50 \text{ mm})^2 + (2.60 \text{ mm})^2}$$
$$= 6.1 \text{ mm} \qquad \text{(Answer)}$$

and a phase constant of

$$\beta = \tan^{-1} \frac{2.60 \text{ mm}}{5.50 \text{ mm}} = 0.44 \text{ rad.} \qquad \text{(Answer)}$$

From Fig. 16-18*b*, phase constant β is a *positive* angle relative to phasor 1. Thus, the resultant wave *lags* wave 1 in their travel by phase constant $\beta = +0.44$ rad. From Eq. 16-57, we can write the resultant wave as

$$y'(x, t) = (6.1 \text{ mm}) \sin(kx - \omega t + 0.44 \text{ rad}). \quad \text{(Answer)}$$

16-12 *Standing Waves*

In Section 16-10, we discussed two sinusoidal waves of the same wavelength and amplitude traveling *in the same direction* along a stretched string. What if they travel in opposite directions? We can again find the resultant wave by applying the superposition principle.

Figure 16-19 suggests the situation graphically. It shows the two combining waves, one traveling to the left in Fig. 16-19*a*, the other to the right in Fig. 16-19*b*. Figure 16-19*c* shows their sum, obtained by applying the superposition principle graphically. The outstanding feature of the resultant wave is that

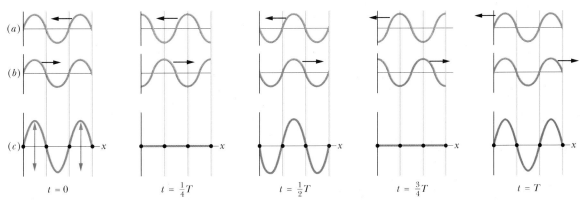

Fig. 16-19 (*a*) Five snapshots of a wave traveling to the left, at the times t indicated below part (*c*) (T is the period of oscillation). (*b*) Five snapshots of a wave identical to that in (*a*) but traveling to the right, at the same times t. (*c*) Corresponding snapshots for the superposition of the two waves on the same string. At $t = 0, \frac{1}{2}T$, and T, fully constructive interference occurs because of the alignment of peaks with peaks and valleys with valleys. At $t = \frac{1}{4}T$ and $\frac{3}{4}T$, fully destructive interference occurs because of the alignment of peaks with valleys. Some points (the nodes, marked with dots) never oscillate; some points (the antinodes) oscillate the most.

Displacement

$$y'(x,t) = [2y_m \sin kx]\cos \omega t$$

Magnitude   Oscillating
gives   term
amplitude
at position x

Fig. 16-20 The resultant wave of Eq. 16-60 is a standing wave and is due to the interference of two sinusoidal waves of the same amplitude and wavelength that travel in opposite directions.

there are places along the string, called **nodes**, where the string never moves. Four such nodes are marked by dots in Fig. 16-19c. Halfway between adjacent nodes are **antinodes,** where the amplitude of the resultant wave is a maximum. Wave patterns such as that of Fig. 16-19c are called **standing waves** because the wave patterns do not move left or right; the locations of the maxima and minima do not change.

⟶ If two sinusoidal waves of the same amplitude and wavelength travel in *opposite* directions along a stretched string, their interference with each other produces a standing wave.

To analyze a standing wave, we represent the two combining waves with the equations

$$y_1(x, t) = y_m \sin(kx - \omega t) \tag{16-58}$$

and

$$y_2(x, t) = y_m \sin(kx + \omega t). \tag{16-59}$$

The principle of superposition gives, for the combined wave,

$$y'(x, t) = y_1(x, t) + y_2(x, t) = y_m \sin(kx - \omega t) + y_m \sin(kx + \omega t).$$

Applying the trigonometric relation of Eq. 16-50 leads to

$$y'(x, t) = [2y_m \sin kx] \cos \omega t, \tag{16-60}$$

which is displayed in Fig. 16-20. This equation does not describe a traveling wave because it is not of the form of Eq. 16-17. Instead, it describes a standing wave.

The quantity $2y_m \sin kx$ in the brackets of Eq. 16-60 can be viewed as the amplitude of oscillation of the string element that is located at position x. However, since an amplitude is always positive and $\sin kx$ can be negative, we take the absolute value of the quantity $2y_m \sin kx$ to be the amplitude at x.

In a traveling sinusoidal wave, the amplitude of the wave is the same for all string elements. That is not true for a standing wave, in which the amplitude *varies with position.* In the standing wave of Eq. 16-60, for example, the amplitude is zero for values of kx that give $\sin kx = 0$. Those values are

$$kx = n\pi, \qquad \text{for } n = 0, 1, 2, \ldots. \tag{16-61}$$

Substituting $k = 2\pi/\lambda$ in this equation and rearranging, we get

$$x = n\frac{\lambda}{2}, \qquad \text{for } n = 0, 1, 2, \ldots \quad \text{(nodes),} \tag{16-62}$$

as the positions of zero amplitude—the nodes—for the standing wave of Eq. 16-60. Note that adjacent nodes are separated by $\lambda/2$, half a wavelength.

The amplitude of the standing wave of Eq. 16-60 has a maximum value of $2y$, which occurs for values of kx that give $|\sin kx| = 1$. Those values are

$$kx = \tfrac{1}{2}\pi, \tfrac{3}{2}\pi, \tfrac{5}{2}\pi, \ldots$$
$$= (n + \tfrac{1}{2})\pi, \qquad \text{for } n = 0, 1, 2, \ldots. \tag{16-63}$$

Substituting $k = 2\pi/\lambda$ in Eq. 16-63 and rearranging, we get

$$x = \left(n + \frac{1}{2}\right)\frac{\lambda}{2}, \qquad \text{for } n = 0, 1, 2, \ldots \quad \text{(antinodes),} \tag{16-64}$$

as the positions of maximum amplitude—the antinodes—of the standing wave of Eq. 16-60. The antinodes are separated by $\lambda/2$ and are located halfway between pairs of nodes.

Reflections at a Boundary

We can set up a standing wave in a stretched string by allowing a traveling wave to be reflected from the far end of the string so that the wave travels back through

Fig. 16-21 (*a*) A pulse incident from the right is reflected at the left end of the string, which is tied to a wall. Note that the reflected pulse is inverted from the incident pulse. (*b*) Here the left end of the string is tied to a ring that can slide without friction up and down the rod. Now the pulse is not inverted by the reflection.

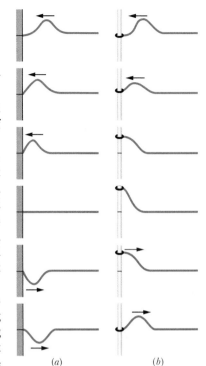

itself. The incident (original) wave and the reflected wave can then be described by Eqs. 16-58 and 16-59, respectively, and they can combine to form a pattern of standing waves.

In Fig. 16-21, we use a single pulse to show how such reflections take place. In Fig. 16-21*a*, the string is fixed at its left end. When the pulse arrives at that end, it exerts an upward force on the support (the wall). By Newton's third law, the support exerts an opposite force of equal magnitude on the string. This second force generates a pulse at the support, which travels back along the string in the direction opposite that of the incident pulse. In a "hard" reflection of this kind, there must be a node at the support because the string is fixed there. The reflected and incident pulses must have opposite signs, so as to cancel each other at that point.

In Fig. 16-21*b*, the left end of the string is fastened to a light ring that is free to slide without friction along a rod. When the incident pulse arrives, the ring moves up the rod. As the ring moves, it pulls on the string, stretching the string and producing a reflected pulse with the same sign and amplitude as the incident pulse. Thus, in such a "soft" reflection, the incident and reflected pulses reinforce each other, creating an antinode at the end of the string; the maximum displacement of the ring is twice the amplitude of either of these pulses.

> **CHECKPOINT 6** Two waves with the same amplitude and wavelength interfere in three different situations to produce resultant waves with the following equations:
> (1) $y'(x, t) = 4 \sin(5x - 4t)$
> (2) $y'(x, t) = 4 \sin(5x) \cos(4t)$
> (3) $y'(x, t) = 4 \sin(5x + 4t)$
> In which situation are the two combining waves traveling (a) toward positive x, (b) toward negative x, and (c) in opposite directions?

16-13 Standing Waves and Resonance

Consider a string, such as a guitar string, that is stretched between two clamps. Suppose we send a continuous sinusoidal wave of a certain frequency along the string, say, toward the right. When the wave reaches the right end, it reflects and begins to travel back to the left. That left-going wave then overlaps the wave that is still traveling to the right. When the left-going wave reaches the left end, it reflects again and the newly reflected wave begins to travel to the right, overlapping the left-going and right-going waves. In short, we very soon have many overlapping traveling waves, which interfere with one another.

For certain frequencies, the interference produces a standing wave pattern (or **oscillation mode**) with nodes and large antinodes like those in Fig. 16-22.

Fig. 16-22 Stroboscopic photographs reveal (imperfect) standing wave patterns on a string being made to oscillate by an oscillator at the left end. The patterns occur at certain frequencies of oscillation.

Fig. 16-23 A string, stretched between two clamps, is made to oscillate in standing wave patterns. (a) The simplest possible pattern consists of one *loop,* which refers to the composite shape formed by the string in its extreme displacements (the solid and dashed lines). (b) The next simplest pattern has two loops. (c) The next has three loops.

Fig. 16-24 One of many possible standing wave patterns for a kettledrum head, made visible by dark powder sprinkled on the drumhead. As the head is set into oscillation at a single frequency by a mechanical oscillator at the upper left of the photograph, the powder collects at the nodes, which are circles and straight lines in this two-dimensional example.

Such a standing wave is said to be produced at **resonance,** and the string is said to *resonate* at these certain frequencies, called **resonant frequencies.** If the string is oscillated at some frequency other than a resonant frequency, a standing wave is not set up. Then the interference of the right-going and left-going traveling waves results in only small (perhaps imperceptible) oscillations of the string.

Let a string be stretched between two clamps separated by a fixed distance L. To find expressions for the resonant frequencies of the string, we note that a node must exist at each of its ends, because each end is fixed and cannot oscillate. The simplest pattern that meets this key requirement is that in Fig. 16-23a, which shows the string at both its extreme displacements (one solid and one dashed, together forming a single "loop"). There is only one antinode, which is at the center of the string. Note that half a wavelength spans the length L, which we take to be the string's length. Thus, for this pattern, $\lambda/2 = L$. This condition tells us that if the left-going and right-going traveling waves are to set up this pattern by their interference, they must have the wavelength $\lambda = 2L$.

A second simple pattern meeting the requirement of nodes at the fixed ends is shown in Fig. 16-23b. This pattern has three nodes and two antinodes and is said to be a two-loop pattern. For the left-going and right-going waves to set it up, they must have a wavelength $\lambda = L$. A third pattern is shown in Fig. 16-23c. It has four nodes, three antinodes, and three loops, and the wavelength is $\lambda = \frac{2}{3}L$. We could continue this progression by drawing increasingly more complicated patterns. In each step of the progression, the pattern would have one more node and one more antinode than the preceding step, and an additional $\lambda/2$ would be fitted into the distance L.

Thus, a standing wave can be set up on a string of length L by a wave with a wavelength equal to one of the values

$$\lambda = \frac{2L}{n}, \qquad \text{for } n = 1, 2, 3, \dots . \tag{16-65}$$

The resonant frequencies that correspond to these wavelengths follow from Eq. 16-13:

$$f = \frac{v}{\lambda} = n\frac{v}{2L}, \qquad \text{for } n = 1, 2, 3, \dots . \tag{16-66}$$

Here v is the speed of traveling waves on the string.

Equation 16-66 tells us that the resonant frequencies are integer multiples of the lowest resonant frequency, $f = v/2L$, which corresponds to $n = 1$. The oscillation mode with that lowest frequency is called the *fundamental mode* or the *first harmonic.* The *second harmonic* is the oscillation mode with $n = 2$, the *third harmonic* is that with $n = 3$, and so on. The frequencies associated with these modes are often labeled f_1, f_2, f_3, and so on. The collection of all possible oscillation modes is called the **harmonic series,** and n is called the **harmonic number** of the nth harmonic.

The phenomenon of resonance is common to all oscillating systems and can occur in two and three dimensions. For example, Fig. 16-24 shows a two-dimensional standing wave pattern on the oscillating head of a kettledrum.

CHECKPOINT 7 In the following series of resonant frequencies, one frequency (lower than 400 Hz) is missing: 150, 225, 300, 375 Hz. (a) What is the missing frequency? (b) What is the frequency of the seventh harmonic?

Sample Problem 16-8

In Fig. 16-25, a string, tied to a sinusoidal oscillator at P and running over a support at Q, is stretched by a block of mass m. The separation L between P and Q is 1.2 m, the linear density of the string is 1.6 g/m, and the frequency f of the

Fig. 16-25 A string under tension connected to an oscillator. For a fixed oscillator frequency, standing wave patterns occur for certain values of the string tension.

oscillator is fixed at 120 Hz. The amplitude of the motion at P is small enough for that point to be considered a node. A node also exists at Q.

(a) What mass m allows the oscillator to set up the fourth harmonic on the string?

Solution: One **Key Idea** here is that the string will resonate at only certain frequencies, determined by the wave speed v on the string and the length L of the string. From Eq. 16-66, these resonant frequencies are

$$f = n\frac{v}{2L}, \qquad \text{for } n = 1, 2, 3, \dots . \qquad (16\text{-}67)$$

To set up the fourth harmonic (for which $n = 4$), we need to adjust the right side of this equation, with $n = 4$, so that the left side equals the frequency of the oscillator (120 Hz).

We cannot adjust L in Eq. 16-67; it is set. However, a second **Key Idea** is that we *can* adjust v, because it depends on how much mass m we hang on the string. According to Eq. 16-26, wave speed $v = \sqrt{\tau/\mu}$. Here the tension τ in the string is equal to the weight mg of the block. Thus,

$$v = \sqrt{\frac{\tau}{\mu}} = \sqrt{\frac{mg}{\mu}}. \qquad (16\text{-}68)$$

Substituting v from Eq. 16-68 into Eq. 16-67, setting $n = 4$ for the fourth harmonic, and solving for m give us

$$m = \frac{4L^2 f^2 \mu}{n^2 g} \qquad (16\text{-}69)$$

$$= \frac{(4)(1.2 \text{ m})^2(120 \text{ Hz})^2(0.0016 \text{ kg/m})}{(4)^2(9.8 \text{ m/s}^2)}$$

$$= 0.846 \text{ kg} \approx 0.85 \text{ kg}. \qquad \text{(Answer)}$$

(b) What standing wave mode is set up if $m = 1.00$ kg?

Solution: If we insert this value of m into Eq. 16-69 and solve for n, we find that $n = 3.7$. A **Key Idea** here is that n must be an integer, so $n = 3.7$ is impossible. Thus, with $m = 1.00$ kg, the oscillator cannot set up a standing wave on the string, and any oscillation of the string will be small, perhaps even imperceptible.

PROBLEM-SOLVING TACTICS

TACTIC 2: *Harmonics on a String*

When you need to obtain information about a certain harmonic on a stretched string of given length L, first draw that harmonic (as in Fig. 16-23). If you are asked about, say, the fifth harmonic, you need to draw five loops between the fixed support points. That would mean that five loops, each of length $\lambda/2$, occupy the length L of the string. Thus, $5(\lambda/2) =$

L, and $\lambda = 2L/5$. You can then use Eq. 16-13 ($f = v/\lambda$) to find the frequency of the harmonic.

Keep in mind that the wavelength of a harmonic is set only by the length L of the string, but the frequency depends also on the wave speed v, which is set by the tension and linear density of the string via Eq. 16-26.

Review & Summary

Transverse and Longitudinal Waves Mechanical waves can exist only in material media and are governed by Newton's laws. **Transverse** mechanical waves, like those on a stretched string, are waves in which the particles of the medium oscillate perpendicular to the wave's direction of travel. Waves in which the particles of the medium oscillate parallel to the wave's direction of travel are **longitudinal** waves.

Sinusoidal Waves A sinusoidal wave moving in the positive direction of an x axis has the mathematical form

$$y(x, t) = y_m \sin(kx - \omega t), \qquad (16\text{-}2)$$

where y_m is the **amplitude** of the wave, k is the **angular wave number**, ω is the **angular frequency**, and $kx - \omega t$ is the **phase**. The **wavelength** λ is related to k by

$$k = \frac{2\pi}{\lambda}. \qquad (16\text{-}5)$$

The **period** T and **frequency** f of the wave are related to ω by

$$\frac{\omega}{2\pi} = f = \frac{1}{T}. \qquad (16\text{-}9)$$

Finally, the **wave speed** v is related to these other parameters by

$$v = \frac{\omega}{k} = \frac{\lambda}{T} = \lambda f. \qquad (16\text{-}13)$$

Equation of a Traveling Wave Any function of the form

$$y(x, t) = h(kx \pm \omega t) \qquad (16\text{-}17)$$

can represent a **traveling wave** with a wave speed given by Eq. 16-13 and a wave shape given by the mathematical form of h. The plus sign denotes a wave traveling in the negative direction of the x axis, and the minus sign a wave traveling in the positive direction.

Wave Speed on Stretched String   The speed of a wave on a stretched string is set by properties of the string. The speed on a string with tension τ and linear density μ is

$$v = \sqrt{\frac{\tau}{\mu}}. \qquad (16\text{-}26)$$

Power   The **average power** of, or average rate at which energy is transmitted by, a sinusoidal wave on a stretched string is given by

$$P_{\text{avg}} = \tfrac{1}{2}\mu v\omega^2 y_m^2. \qquad (16\text{-}33)$$

Superposition of Waves   When two or more waves traverse the same medium, the displacement of any particle of the medium is the sum of the displacements that the individual waves would give it.

Interference of Waves   Two sinusoidal waves on the same string exhibit **interference,** adding or canceling according to the principle of superposition. If the two are traveling in the same direction and have the same amplitude y_m and frequency (hence the same wavelength) but differ in phase by a **phase constant** ϕ, the result is a single wave with this same frequency:

$$y'(x, t) = [2y_m \cos \tfrac{1}{2}\phi] \sin(kx - \omega t + \tfrac{1}{2}\phi). \qquad (16\text{-}51)$$

If $\phi = 0$, the waves are exactly in phase and their interference is fully constructive; if $\phi = \pi$ rad, they are exactly out of phase and their interference is fully destructive.

Phasors   A wave $y(x, t)$ can be represented with a **phasor.** This is a vector that has a magnitude equal to the amplitude y_m of the wave and that rotates about an origin with an angular speed equal to the angular frequency ω of the wave. The projection of the rotating phasor on a vertical axis gives the displacement y of a point along the wave's travel.

Standing Waves   The interference of two identical sinusoidal waves moving in opposite directions produces **standing waves.** For a string with fixed ends, the standing wave is given by

$$y'(x, t) = [2y_m \sin kx] \cos \omega t. \qquad (16\text{-}60)$$

Standing waves are characterized by fixed locations of zero displacement called **nodes** and fixed locations of maximum displacement called **antinodes.**

Resonance   Standing waves on a string can be set up by reflection of traveling waves from the ends of the string. If an end is fixed, it must be the position of a node. This limits the frequencies at which standing waves will occur on a given string. Each possible frequency is a **resonant frequency,** and the corresponding standing wave pattern is an **oscillation mode.** For a stretched string of length L with fixed ends, the resonant frequencies are

$$f = \frac{v}{\lambda} = n\frac{v}{2L}, \qquad \text{for } n = 1, 2, 3, \ldots. \qquad (16\text{-}66)$$

The oscillation mode corresponding to $n = 1$ is called the *fundamental mode* or the *first harmonic*; the mode corresponding to $n = 2$ is the *second harmonic*; and so on.

Questions

1   Figure 16-26 shows three waves that are *separately* sent along a string that is stretched under a certain tension along an x axis. Rank the waves according to their (a) wavelengths, (b) speeds, and (c) angular frequencies, greatest first.

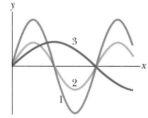

Fig. 16-26   Question 1.

2   Figure 16-27*a* gives a snapshot of a wave traveling in the direction of positive x along a string under tension. Four string elements are indicated by the lettered points. For each of those elements, determine whether, at the instant of the snapshot, the element is moving upward or downward or is momentarily at rest. (*Hint:* Imagine the wave as it moves through the four string elements.)

Figure 16-27*b* gives the displacement of a string element located at, say, $x = 0$ as a function of time. At the lettered times, is the element moving upward or downward or is it momentarily at rest?

3   The following four waves are sent along strings with the same linear densities (x is in meters and t is in seconds). Rank the waves according to (a) their wave speed and (b) the tension in the strings along which they travel, greatest first:

(1) $y_1 = (3 \text{ mm}) \sin(x - 3t)$,   (3) $y_3 = (1 \text{ mm}) \sin(4x - t)$,
(2) $y_2 = (6 \text{ mm}) \sin(2x - t)$,   (4) $y_4 = (2 \text{ mm}) \sin(x - 2t)$.

4   In Fig. 16-28, wave 1 consists of a rectangular peak of height 4 units and width d, and a rectangular valley of depth 2 units and width d. The wave travels rightward along an x

Fig. 16-27   Question 2.

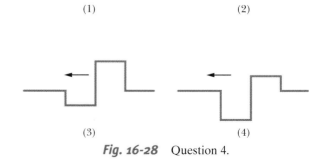

Fig. 16-28   Question 4.

axis. Choices 2, 3, and 4 are similar waves, with the same heights, depths, and widths, that will travel leftward along that axis and through wave 1. Right-going wave 1 and one of the left-going waves will interfere as they pass through each other. With which left-going wave will the interference give, for an instant, (a) the deepest valley, (b) a flat line, and (c) a flat peak $2d$ wide?

5 If you start with two sinusoidal waves of the same amplitude traveling in phase on a string and then somehow phase-shift one of them by 5.4 wavelengths, what type of interference will occur on the string?

6 A sinusoidal wave is sent along a cord under tension, transporting energy at the average rate of $P_{avg,1}$. Two waves, identical to that first one, are then to be sent along the cord with a phase difference ϕ of either 0, 0.2 wavelength, or 0.5 wavelength. (a) With only mental calculation, rank those choices of ϕ according to the average rate at which the waves will transport energy, greatest first. (b) For the first choice of ϕ, what is the average rate in terms of $P_{avg,1}$?

7 The amplitudes and phase differences for four pairs of waves of equal wavelengths are (a) 2 mm, 6 mm, and π rad; (b) 3 mm, 5 mm, and π rad; (c) 7 mm, 9 mm, and π rad; (d) 2 mm, 2 mm, and 0 rad. Each pair travels in the same direction along the same string. Without written calculation, rank the four pairs according to the amplitude of their resultant wave, greatest first. (*Hint:* Construct phasor diagrams.)

8 Figure 16-29 shows phasor diagrams for three situations in which two waves travel along the same string. All six waves have the same amplitude. Rank the situations according to the amplitude of the net wave on the string, greatest first.

(a)   (b)   (c)

Fig. 16-29 Question 8.

9 If you set up the seventh harmonic on a string, (a) how many nodes are present, and (b) is there a node, antinode, or some intermediate state at the midpoint? If you next set up the sixth harmonic, (c) is its resonant wavelength longer or shorter than that for the seventh harmonic, and (d) is the resonant frequency higher or lower?

10 Strings A and B have identical lengths and linear densities, but string B is under greater tension than string A. Figure 16-30 shows four situations, (*a*) through (*d*), in which standing wave patterns exist on the two strings. In which situations is there the possibility that strings A and B are oscillating at the same resonant frequency?

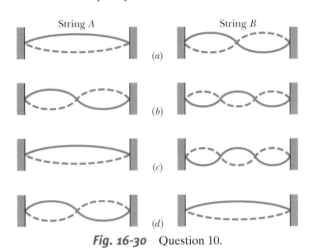

Fig. 16-30 Question 10.

11 (a) If a standing wave on a string is given by

$$y'(t) = (3 \text{ mm}) \sin(5x) \cos(4t),$$

is there a node or an antinode of the oscillations of the string at $x = 0$? (b) If the standing wave is given by

$$y'(t) = (3 \text{ mm}) \sin(5x + \pi/2) \cos(4t),$$

is there a node or an antinode at $x = 0$?

Problems

SSM	Solution is in the Student Solutions Manual.
WWW	Solution is at http://www.wiley.com/college/halliday
ILW	Interactive LearningWare solution is at
	http://www.wiley.com/college/halliday
• – •••	Number of dots indicates level of problem difficulty.

sec. 16-5 The Speed of a Traveling Wave

•**1** A sinusoidal wave travels along a string. The time for a particular point to move from maximum displacement to zero is 0.170 s. What are the (a) period and (b) frequency? (c) The wavelength is 1.40 m; what is the wave speed?

•**2** A wave has an angular frequency of 110 rad/s and a wavelength of 1.80 m. Calculate (a) the angular wave number and (b) the speed of the wave.

•**3** If $y(x, t) = (6.0 \text{ mm}) \sin(kx + (600 \text{ rad/s})t + \phi)$ describes a wave traveling along a string, how much time does any given point on the string take to move between displacements $y = +2.0$ mm and $y = -2.0$ mm?

••**4** Figure 16-31 shows the transverse velocity u versus time t of the point on a string at $x = 0$, as a wave passes through it. The wave has form $y(x, t) = y_m \sin(kx - \omega t + \phi)$. What is ϕ? (*Caution:* A calculator does not always give the proper inverse trig function, so check your answer by substituting it and an assumed value of ω into $y(x, t)$ and then plotting the function.)

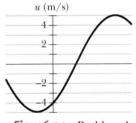

Fig. 16-31 Problem 4.

••**5** A sinusoidal wave of frequency 500 Hz has a speed of 350 m/s. (a) How far apart are two points that differ in phase by $\pi/3$ rad? (b) What is the phase difference between two displacements at a certain point at times 1.00 ms apart? **ILW**

••**6** The equation of a transverse wave traveling along a very long string is $y = 6.0 \sin(0.020\pi x + 4.0\pi t)$, where x and y are expressed in centimeters and t is in seconds. Determine (a) the amplitude, (b) the wavelength, (c) the frequency, (d) the speed, (e) the direction of propagation of the wave, and (f) the maximum transverse speed of a particle in the string. (g) What is the transverse displacement at $x = 3.5$ cm when $t = 0.26$ s?

••7   A transverse sinusoidal wave is moving along a string in the positive direction of an x axis with a speed of 80 m/s. At $t = 0$, the string particle at $x = 0$ has a transverse displacement of 4.0 cm from its equilibrium position and is not moving. The maximum transverse speed of the string particle at $x = 0$ is 16 m/s. (a) What is the frequency of the wave? (b) What is the wavelength of the wave? If the wave equation is of the form $y(x, t) = y_m \sin(kx \pm \omega t + \phi)$, what are (c) y_m, (d) k, (e) ω, (f) ϕ, and (g) the correct choice of sign in front of ω?

••8   The function $y(x, t) = (15.0 \text{ cm}) \cos(\pi x - 15\pi t)$, with x in meters and t in seconds, describes a wave on a taut string. What is the transverse speed for a point on the string at an instant when that point has the displacement $y = +12.0$ cm?

••9   A sinusoidal wave moving along a string is shown twice in Fig. 16-32, as crest A travels in the positive direction of an x axis by distance $d = 6.0$ cm in 4.0 ms. The tick marks along the axis are separated by 10 cm. If the wave equation is of the form $y(x, t) = y_m \sin(kx \pm \omega t)$, what are (a) y_m, (b) k, (c) ω, and (d) the correct choice of sign in front of ω?

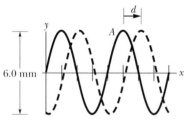

Fig. 16-32   Problem 9.

••10   A sinusoidal wave travels along a string under tension. Figure 16-33 gives the slopes along the string at time $t = 0$. What is the amplitude of the wave?

••11   A sinusoidal transverse wave of wavelength 20 cm travels along a string in the positive direction of an x axis. The displacement y of the string particle at $x = 0$ is given in Fig. 16-34 as a function of time t. The wave equation is to be in the form of $y(x, t) = y_m \sin(kx \pm \omega t + \phi)$. (a) At $t = 0$, is a plot of y versus x in the shape of a positive sine function or a negative sine function? What are (b) y_m, (c) k, (d) ω, (e) ϕ, (f) the sign in front of ω, and (g) the speed of the wave? (h) What is the transverse velocity of the particle at $x = 0$ when $t = 5.0$ s?

Fig. 16-33   Problem 10.

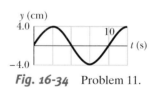

Fig. 16-34   Problem 11.

sec. 16-6   Wave Speed on a Stretched String

•12   The heaviest and lightest strings on a certain violin have linear densities of 3.0 and 0.29 g/m. What is the ratio of the diameter of the heaviest string to that of the lightest string, assuming that the strings are of the same material?

•13   What is the speed of a transverse wave in a rope of length 2.00 m and mass 60.0 g under a tension of 500 N? **SSM**

•14   The tension in a wire clamped at both ends is doubled without appreciably changing the wire's length between the clamps. What is the ratio of the new to the old wave speed for transverse waves traveling along this wire?

•15   The linear density of a string is 1.6×10^{-4} kg/m. A transverse wave on the string is described by the equation

$$y = (0.021 \text{ m}) \sin[(2.0 \text{ m}^{-1})x + (30 \text{ s}^{-1})t].$$

What are (a) the wave speed and (b) the tension in the string?

•16   The speed of a transverse wave on a string is 170 m/s when the string tension is 120 N. To what value must the tension be changed to raise the wave speed to 180 m/s?

•17   A stretched string has a mass per unit length of 5.00 g/cm and a tension of 10.0 N. A sinusoidal wave on this string has an amplitude of 0.12 mm and a frequency of 100 Hz and is traveling in the negative direction of an x axis. If the wave equation is of the form $y(x, t) = y_m \sin(kx \pm \omega t)$, what are (a) y_m, (b) k, (c) ω, and (d) the correct choice of sign in front of ω? **SSM WWW**

•18   The equation of a transverse wave on a string is

$$y = (2.0 \text{ mm}) \sin[(20 \text{ m}^{-1})x - (600 \text{ s}^{-1})t].$$

The tension in the string is 15 N. (a) What is the wave speed? (b) Find the linear density of this string in grams per meter.

••19   A sinusoidal transverse wave is traveling along a string in the negative direction of an x axis. Figure 16-35 shows a plot of the displacement as a function of position at time $t = 0$; the y intercept is 4.0 cm. The string tension is 3.6 N, and its linear density is 25 g/m. Find the (a) amplitude, (b) wavelength, (c) wave speed, and (d) period of the wave. (e) Find the maximum transverse speed of a particle in the string. If the wave is of the form $y(x, t) = y_m \sin(kx \pm \omega t + \phi)$, what are (f) k, (g) ω, (h) ϕ, and (i) the correct choice of sign in front of ω? **SSM ILW**

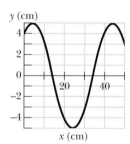

Fig. 16-35   Problem 19.

••20   A sinusoidal wave is traveling on a string with speed 40 cm/s. The displacement of the particles of the string at $x = 10$ cm is found to vary with time according to the equation $y = (5.0 \text{ cm}) \sin[1.0 - (4.0 \text{ s}^{-1})t]$. The linear density of the string is 4.0 g/cm. What are (a) the frequency and (b) the wavelength of the wave? If the wave equation is of the form $y(x, t) = y_m \sin(kx \pm \omega t)$, what are (c) y_m, (d) k, (e) ω, and (f) the correct choice of sign in front of ω? (g) What is the tension in the string?

••21   A 100 g wire is held under a tension of 250 N with one end at $x = 0$ and the other at $x = 10.0$ m. At time $t = 0$, pulse 1 is sent along the wire from the end at $x = 10.0$ m. At time $t = 30.0$ ms, pulse 2 is sent along the wire from the end at $x = 0$. At what position x do the pulses begin to meet? **ILW**

•••22   In Fig. 16-36a, string 1 has a linear density of 3.00 g/m, and string 2 has a linear density of 5.00 g/m. They are under tension due to the hanging block of mass $M = 500$ g. Calculate the wave speed on (a) string 1 and (b) string 2. (*Hint:* When a string loops halfway around a pulley, it pulls on the pulley with a net force that is twice the tension in the string.) Next the block is divided into two blocks (with $M_1 + M_2 = M$) and the apparatus is rearranged as shown in Fig. 16-36b. Find (c) M_1 and (d) M_2 such that the wave speeds in the two strings are equal.

(a)

(b)

Fig. 16-36   Problem 22.

•••23   A uniform rope of mass m and length L hangs from a ceiling. (a) Show that the speed of a transverse wave on the rope is a function of y, the distance from the lower end, and is given by $v = \sqrt{gy}$. (b) Show that the time a transverse wave takes to travel the length of the rope is given by $t = 2\sqrt{L/g}$.   SSM WWW

sec. 16-7   Energy and Power of a Wave Traveling Along a String

•24   A string along which waves can travel is 2.70 m long and has a mass of 260 g. The tension in the string is 36.0 N. What must be the frequency of traveling waves of amplitude 7.70 mm for the average power to be 85.0 W?

••25   A sinusoidal wave is sent along a string with a linear density of 2.0 g/m. As it travels, the kinetic energies of the mass elements along the string vary. Figure 16-37a gives the rate dK/dt at which kinetic energy passes through the string elements at a particular instant, plotted as a function of distance x along the string. Figure 16-37b is similar except that it gives the rate at which kinetic energy passes through a particular mass element (at a particular location), plotted as a function of time t. What is the amplitude of the wave?

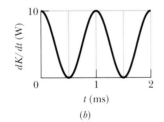

(a)

(b)

Fig. 16-37   Problem 25.

sec. 16-8   The Wave Equation

•26   Use the wave equation to find the speed of a wave given by

$$y(x, t) = (3.00 \text{ mm}) \sin[(4.00 \text{ m}^{-1})x - (7.00 \text{ s}^{-1})t].$$

••27   Use the wave equation to find the speed of a wave given by

$$y(x, t) = (2.00 \text{ mm})[(20 \text{ m}^{-1})x - (4.0 \text{ s}^{-1})t]^{0.5}.$$

•••28   Use the wave equation to find the speed of a wave given in terms of the general function $h(x, t)$:

$$y(x, t) = (4.00 \text{ mm}) h[(30 \text{ m}^{-1})x + (6.0 \text{ s}^{-1})t].$$

sec. 16-10   Interference of Waves

•29   Two identical traveling waves, moving in the same direction, are out of phase by $\pi/2$ rad. What is the amplitude of the resultant wave in terms of the common amplitude y_m of the two combining waves?   SSM

•30   What phase difference between two otherwise identical traveling waves, moving in the same direction along a stretched string, will result in the combined wave having an amplitude 1.50 times that of the common amplitude of the two combining waves? Express your answer in (a) degrees, (b) radians, and (c) wavelengths.

••31   Two sinusoidal waves with the same amplitude of 9.00 mm and the same wavelength travel together along a string that is stretched along an x axis. Their resultant wave is shown twice in Fig. 16-38, as valley A travels in the negative direction of the x axis by distance $d = 56.0$ cm in 8.0 ms. The tick marks along the axis are separated by 10 cm. Let the equation for one wave be of the form $y(x, t) = y_m \sin(kx \pm \omega t + \phi_1)$, where $\phi_1 = 0$ and you must choose the correct sign in front of ω. For the equation for the other wave, what are (a) y_m, (b) k, (c) ω, (d) ϕ_2, and (e) the sign in front of ω?

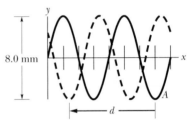

Fig. 16-38   Problem 31.

•••32   A sinusoidal wave of angular frequency 1200 rad/s and amplitude 3.00 mm is sent along a cord with linear density 2.00 g/m and tension 1200 N. (a) What is the average rate at which energy is transported by the wave to the opposite end of the cord? (b) If, simultaneously, an identical wave travels along an adjacent, identical cord, what is the total average rate at which energy is transported to the opposite ends of the two cords by the waves? If, instead, those two waves are sent along the *same* cord simultaneously, what is the total average rate at which they transport energy when their phase difference is (c) 0, (d) 0.4π rad, and (e) π rad?

sec. 16-11   Phasors

•33   Two sinusoidal waves of the same frequency travel in the same direction along a string. If $y_{m1} = 3.0$ cm, $y_{m2} = 4.0$ cm, $\phi_1 = 0$, and $\phi_2 = \pi/2$ rad, what is the amplitude of the resultant wave?   SSM

••34   Two sinusoidal waves of the same period, with amplitudes of 5.0 and 7.0 mm, travel in the same direction along a stretched string; they produce a resultant wave with an amplitude of 9.0 mm. The phase constant of the 5.0 mm wave is 0. What is the phase constant of the 7.0 mm wave?

••35   Two sinusoidal waves of the same frequency are to be sent in the same direction along a taut string. One wave has an amplitude of 5.0 mm, the other 8.0 mm. (a) What phase difference ϕ_1 between the two waves results in the smallest amplitude of the resultant wave? (b) What is that smallest amplitude? (c) What phase difference ϕ_2 results in the largest amplitude of the resultant wave? (d) What is that largest amplitude? (e) What is the resultant amplitude if the phase angle is $(\phi_1 - \phi_2)/2$?

••36   Four waves are to be sent along the same string, in the same direction:

$$y_1(x, t) = (4.00 \text{ mm}) \sin(2\pi x - 400\pi t)$$

$$y_2(x, t) = (4.00 \text{ mm}) \sin(2\pi x - 400\pi t + 0.7\pi)$$

$$y_3(x, t) = (4.00 \text{ mm}) \sin(2\pi x - 400\pi t + \pi)$$

$$y_4(x, t) = (4.00 \text{ mm}) \sin(2\pi x - 400\pi t + 1.7\pi).$$

What is the amplitude of the resultant wave?

••37   These two waves travel along the same string:

$$y_1(x, t) = (4.60 \text{ mm}) \sin(2\pi x - 400\pi t)$$

$$y_2(x, t) = (5.60 \text{ mm}) \sin(2\pi x - 400\pi t + 0.80\pi \text{ rad}).$$

What are (a) the amplitude and (b) the phase angle (relative to wave 1) of the resultant wave? (c) If a third wave of amplitude 5.00 mm is also to be sent along the string in the same direction as the first two waves, what should be its phase angle in order to maximize the amplitude of the new resultant wave?

sec. 16-13   Standing Waves and Resonance

•38   String A is stretched between two clamps separated by distance L. String B, with the same linear density and under the same tension as string A, is stretched between two clamps separated by distance $4L$. Consider the first eight harmonics of string B. For which of these eight harmonics of B (if any) does the frequency match the frequency of (a) A's first harmonic, (b) A's second harmonic, and (c) A's third harmonic?

•39   What are (a) the lowest frequency, (b) the second lowest frequency, and (c) the third lowest frequency for standing waves on a wire that is 10.0 m long, has a mass of 100 g, and is stretched under a tension of 250 N? SSM WWW

•40   A 125 cm length of string has a mass of 2.00 g. It is stretched with a tension of 7.00 N between fixed supports. (a) What is the wave speed for this string? (b) What is the lowest resonant frequency of this string?

•41   A string fixed at both ends is 8.40 m long and has a mass of 0.120 kg. It is subjected to a tension of 96.0 N and set oscillating. (a) What is the speed of the waves on the string? (b) What is the longest possible wavelength for a standing wave? (c) Give the frequency of that wave. SSM

•42   Two sinusoidal waves with identical wavelengths and amplitudes travel in opposite directions along a string with a speed of 10 cm/s. If the time interval between instants when the string is flat is 0.50 s, what is the wavelength of the waves?

•43   A nylon guitar string has a linear density of 7.20 g/m and is under a tension of 150 N. The fixed supports are distance $D = 90.0$ cm apart. The string is oscillating in the standing wave pattern shown in Fig. 16-39. Calculate the (a) speed, (b) wavelength, and (c) frequency of the traveling waves whose superposition gives this standing wave. ILW

Fig. 16-39   Problem 43.

•44   A string under tension τ_i oscillates in the third harmonic at frequency f_3, and the waves on the string have wavelength λ_3. If the tension is increased to $\tau_f = 4\tau_i$ and the string is again made to oscillate in the third harmonic, what then are (a) the frequency of oscillation in terms of f_3 and (b) the wavelength of the waves in terms of λ_3?

•45   A string that is stretched between fixed supports separated by 75.0 cm has resonant frequencies of 420 and 315 Hz, with no intermediate resonant frequencies. What are (a) the lowest resonant frequency and (b) the wave speed? SSM ILW

•46   One of the harmonic frequencies for a particular string under tension is 325 Hz. The next higher harmonic frequency is 390 Hz. What harmonic frequency is next higher after the harmonic frequency 195 Hz?

••47   A string oscillates according to the equation

$$y' = (0.50 \text{ cm}) \sin\left[\left(\frac{\pi}{3} \text{ cm}^{-1}\right)x\right] \cos[(40\pi \text{ s}^{-1})t].$$

What are the (a) amplitude and (b) speed of the two waves (identical except for direction of travel) whose superposition gives this oscillation? (c) What is the distance between nodes? (d) What is the transverse speed of a particle of the string at the position $x = 1.5$ cm when $t = \frac{9}{8}$ s? SSM

••48   A rope, under a tension of 200 N and fixed at both ends, oscillates in a second-harmonic standing wave pattern. The displacement of the rope is given by

$$y = (0.10 \text{ m})(\sin \pi x/2) \sin 12\pi t,$$

where $x = 0$ at one end of the rope, x is in meters, and t is in seconds. What are (a) the length of the rope, (b) the speed of the waves on the rope, and (c) the mass of the rope? (d) If the rope oscillates in a third-harmonic standing wave pattern, what will be the period of oscillation?

••49   Two waves are generated on a string of length 3.0 m to produce a three-loop standing wave with an amplitude of 1.0 cm. The wave speed is 100 m/s. Let the equation for one of the waves be of the form $y(x, t) = y_m \sin(kx + \omega t)$. In the equation for the other wave, what are (a) y_m, (b) k, (c) ω, and (d) the sign in front of ω? SSM WWW

••50   A standing wave pattern on a string is described by

$$y(x, t) = 0.040 (\sin 5\pi x)(\cos 40\pi t),$$

where x and y are in meters and t is in seconds. For $x \geq 0$, what is the location of the node with the (a) smallest, (b) second smallest, and (c) third smallest value of x? (d) What is the period of the oscillatory motion of any (nonnode) point? What are the (e) speed and (f) amplitude of the two traveling waves that interfere to produce this wave? For $t \geq 0$, what are the (g) first, (h) second, and (i) third time that all points on the string have zero transverse velocity?

••51   For a certain transverse standing wave on a long string, an antinode is at $x = 0$ and a node is at $x = 0.10$ m. The displacement $y(t)$ of the string particle at $x = 0$ is shown in Fig. 16-40. When $t = 0.50$ s, what is the displacement of the

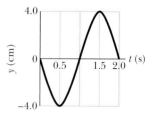

Fig. 16-40   Problem 51.

string particle at (a) $x = 0.20$ m and (b) $x = 0.30$ m? What is the transverse velocity of the string particle at $x = 0.20$ m at (c) $t = 0.50$ s and (d) $t = 1.0$ s? (e) Sketch the standing wave at $t = 0.50$ s for the range $x = 0$ to $x = 0.40$ m.

••52 The following two waves are sent in opposite directions on a horizontal string so as to create a standing wave in a vertical plane:

$$y_1(x, t) = (6.00 \text{ mm}) \sin(4.00\pi x - 400\pi t)$$

$$y_2(x, t) = (6.00 \text{ mm}) \sin(4.00\pi x + 400\pi t),$$

with x in meters and t in seconds. An antinode is located at point A. In the time interval that point takes to move from maximum upward displacement to maximum downward displacement, how far does each wave move along the string?

••53 A generator at one end of a very long string creates a wave given by

$$y = (6.0 \text{ cm}) \cos \frac{\pi}{2} [(2.00 \text{ m}^{-1})x + (8.00 \text{ s}^{-1})t],$$

and a generator at the other end creates the wave

$$y = (6.0 \text{ cm}) \cos \frac{\pi}{2} [(2.00 \text{ m}^{-1})x - (8.00 \text{ s}^{-1})t].$$

Calculate the (a) frequency, (b) wavelength, and (c) speed of each wave. For $x \geq 0$, what is the location of the node having the (d) smallest, (e) second smallest, and (f) third smallest value of x? For $x \geq 0$, what is the location of the antinode having the (g) smallest, (h) second smallest, and (i) third smallest value of x? **SSM**

••54 Two sinusoidal waves with the same amplitude and wavelength travel together along a string that is stretched along an x axis. Their resultant wave is shown twice in Fig. 16-41, as the antinode A travels from an extreme upward displacement to an extreme downward displacement in 6.0 ms. The tick marks along the axis are separated by 10 cm. Let the equation for one of the two waves be of the form $y(x, t) = y_m \sin (kx + \omega t)$. In the equation for the other wave, what are (a) y_m, (b) k, (c) ω, and (d) the sign in front of ω?

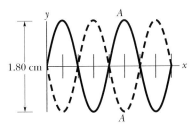

Fig. 16-41   Problem 54.

•••55 In Fig. 16-42, an aluminum wire, of length $L_1 = 60.0$ cm, cross-sectional area 1.00×10^{-2} cm^2, and density 2.60

Fig. 16-42   Problem 55.

g/cm^3, is joined to a steel wire, of density 7.80 g/cm^3 and the same cross-sectional area. The compound wire, loaded with a block of mass $m = 10.0$ kg, is arranged so that the distance L_2 from the joint to the supporting pulley is 86.6 cm. Transverse waves are set up on the wire by an external source of variable frequency; a node is located at the pulley. (a) Find the lowest frequency that generates a standing wave having the joint as one of the nodes. (b) How many nodes are observed at this frequency? **SSM**

•••56 Standing waves are set up with the equipment in Sample Problem 16-8 (Fig. 16-25) except that a different string is used. A standing wave appears when the mass of the hanging block is 286.1 g or 447.0 g, but not for any intermediate mass. What is the linear density of the string?

Additional Problems

57 Figure 16-43 shows the displacement y versus time t of the point on a string at $x = 0$, as a wave passes through that point. The wave has the form $y(x, t) = y_m \sin(kx - \omega t + \phi)$. What is ϕ? (*Caution:* A calculator does not always give the proper inverse trig function, so check your answer by substituting it and an assumed value of ω into $y(x, t)$ and then plotting the function.)

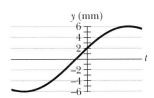

Fig. 16-43   Problem 57.

58 Figure 16-44 shows transverse acceleration a_y versus time t of the point on a string at $x = 0$, as a wave written in the general form $y(x, t) = y_m \sin(kx - \omega t + \phi)$ passes through that point. What is ϕ? (*Caution:* A calculator does not always give the proper inverse trig function, so check your answer by substituting it and an assumed value of ω into $y(x, t)$ and then plotting the function.)

Fig. 16-44   Problem 58.

59 A single pulse, whose wave shape is given by the function $h(x - 5.0t)$, is shown in Fig. 16-45 for $t = 0$. Here x is in centimeters and t is in seconds. What are the (a) speed and (b) direction of travel of the pulse? (c) Plot $h(x - 5t)$ as a function of x for $t = 2$ s. (d) Plot $h(x - 5t)$ as a function of t for $x = 10$ cm.

Fig. 16-45   Problem 59.

60 Two sinusoidal waves, identical except for phase, travel in the same direction along a string, producing a net wave $y'(x, t) = (3.0 \text{ mm}) \sin(20x - 4.0t + 0.820 \text{ rad})$, with x in meters and t in seconds. What are (a) the wavelength λ of the two waves, (b) the phase difference between them, and (c) their amplitude y_m?

61 Three sinusoidal waves of the same frequency travel along a string in the positive direction of an x axis. Their amplitudes are y_1, $y_1/2$, and $y_1/3$, and their phase constants are 0, $\pi/2$, and π, respectively. What are the (a) amplitude and (b) phase constant of the resultant wave? (c) Plot the wave form of the resultant wave at $t = 0$, and discuss its behavior as t increases.

62 Two sinusoidal 120 Hz waves, of the same frequency and amplitude, are to be sent in the positive direction of an *x* axis that is directed along a cord under tension. The waves can be sent in phase, or they can be phase-shifted. Figure 16-46 shows the amplitude y'

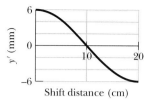

Fig. 16-46 Problem 62.

of the resulting wave versus the distance of the shift (how far one wave is shifted from the other wave). If the equations for the two waves are of the form $y(x, t) = y_m \sin(kx \pm \omega t)$, what are (a) y_m, (b) k, (c) ω, and (d) the correct choice of sign in front of ω?

63 At time $t = 0$ and at position $x = 0$ m along a string, a traveling sinusoidal wave with an angular frequency of 440 rad/s has displacement $y = +4.5$ mm and transverse velocity $u = -0.75$ m/s. If the wave has general form $y(x, t) = y_m \sin(kx - \omega t + \phi)$, what is phase constant ϕ?

64 Four sinusoidal waves travel in the positive direction of an *x* axis aligned along a string. Their frequencies are in the ratio 1:2:3:4, and their amplitudes are in the ratio $1:\frac{1}{2}:\frac{1}{3}:\frac{1}{4}$. When $t = 0$, at $x = 0$, the first and third waves are 180° out of phase with the second and fourth. Plot the resultant wave form when $t = 0$, and discuss its behavior as *t* increases.

65 In a demonstration, a 1.2 kg horizontal rope is fixed in place at its two ends ($x = 0$ and $x = 2.0$ m) and made to oscillate up and down in the fundamental mode, at frequency 5.0 Hz. At $t = 0$, the point at $x = 1.0$ m has zero displacement and is moving upward in the positive direction of a *y* axis with a transverse velocity of 5.0 m/s. What are (a) the amplitude of the motion of that point and (b) the tension in the rope? (c) Write the standing wave equation for the fundamental mode.

66 A transverse sinusoidal wave is generated at one end of a long, horizontal string by a bar that moves up and down through a distance of 1.00 cm. The motion is continuous and is repeated regularly 120 times per second. The string has linear density 120 g/m and is kept under a tension of 90.0 N. Find the maximum value of (a) the transverse speed *u* and (b) the transverse component of the tension τ.

(c) Show that the two maximum values calculated above occur at the same phase values for the wave. What is the transverse displacement *y* of the string at these phases? (d) What is the maximum rate of energy transfer along the string? (e) What is the transverse displacement *y* when this maximum transfer occurs? (f) What is the minimum rate of energy transfer along the string? (g) What is the transverse displacement *y* when this minimum transfer occurs?

67 A sinusoidal transverse wave of amplitude y_m and wavelength λ travels on a stretched cord. (a) Find the ratio of the maximum particle speed (the speed with which a single particle in the cord moves transverse to the wave) to the wave speed. (b) Does this ratio depend on the material of which the cord is made?

68 What is the fastest transverse wave that can be sent along a steel wire? For safety reasons, the maximum tensile stress to which steel wires should be subjected is 7.00×10^8 N/m². The density of steel is 7800 kg/m³. Show that your answer does not depend on the diameter of the wire.

69 A sinusoidal transverse wave traveling in the negative direction of an *x* axis has an amplitude of 1.00 cm, a frequency of 550 Hz, and a speed of 330 m/s. If the wave equation is of the form $y(x, t) = y_m \sin(kx \pm \omega t)$, what are (a) y_m, (b) ω, (c) k, and (d) the correct choice of sign in front of ω?

70 A sinusoidal transverse wave traveling in the positive direction of an *x* axis has an amplitude of 2.0 cm, a wavelength of 10 cm, and a frequency of 400 Hz. If the wave equation is of the form $y(x, t) = y_m \sin(kx \pm \omega t)$, what are (a) y_m, (b) k, (c) ω, and (d) the correct choice of sign in front of ω? What are (e) the maximum transverse speed of a point on the cord and (f) the speed of the wave?

71 Two sinusoidal waves of the same wavelength travel in the same direction along a stretched string. For wave 1, $y_m = 3.0$ mm and $\phi = 0$; for wave 2, $y_m = 5.0$ mm and $\phi = 70°$. What are the (a) amplitude and (b) phase constant of the resultant wave?

72 A wave on a string is described by

$$y(x, t) = 15.0 \sin(\pi x/8 - 4\pi t),$$

where *x* and *y* are in centimeters and *t* is in seconds. (a) What is the transverse speed for a point on the string at $x = 6.00$ cm when $t = 0.250$ s? (b) What is the maximum transverse speed of any point on the string? (c) What is the magnitude of the transverse acceleration for a point on the string at $x = 6.00$ cm when $t = 0.250$ s? (d) What is the magnitude of the maximum transverse acceleration for any point on the string?

73 A wave has a speed of 240 m/s and a wavelength of 3.2 m. What are the (a) frequency and (b) period of the wave?

74 When played in a certain manner, the lowest resonant frequency of a certain violin string is concert *A* (440 Hz). What is the frequency of the (a) second and (b) third harmonic of the string?

75 A 120 cm length of string is stretched between fixed supports. What are the (a) longest, (b) second longest, and (c) third longest wavelength for waves traveling on the string if standing waves are to be set up? (d) Sketch those standing waves.

76 The equation of a transverse wave traveling along a string is

$$y = 0.15 \sin(0.79x - 13t),$$

in which *x* and *y* are in meters and *t* is in seconds. (a) What is the displacement *y* at $x = 2.3$ m, $t = 0.16$ s? A second wave is to be added to the first wave to produce standing waves on the string. If the wave equation for the second wave is of the form $y(x, t) = y_m \sin(kx \pm \omega t)$, what are (b) y_m, (c) k, (d) ω, and (e) the correct choice of sign in front of ω for this second wave? (f) What is the displacement of the resultant standing wave at $x = 2.3$ m, $t = 0.16$ s?

77 A 1.50 m wire has a mass of 8.70 g and is under a tension of 120 N. The wire is held rigidly at both ends and set into oscillation. (a) What is the speed of waves on the wire? What is the wavelength of the waves that produce (b) one-loop and (c) two-loop standing waves? What is the frequency of the waves that produce (d) one-loop and (e) two-loop standing waves?

78 Energy is transmitted at rate P_1 by a wave of frequency f_1 on a string under tension τ_1. What is the new energy transmission rate P_2 in terms of P_1 (a) if the tension is increased to $\tau_2 = 4\tau_1$ and (b) if, instead, the frequency is decreased to $f_2 = f_1/2$?

79 The equation of a transverse wave traveling along a string is

$$y = (2.0 \text{ mm}) \sin[(20 \text{ m}^{-1})x - (600 \text{ s}^{-1})t].$$

Find the (a) amplitude, (b) frequency, (c) velocity (including sign), and (d) wavelength of the wave. (e) Find the maximum transverse speed of a particle in the string.

80 A certain transverse si-nusoidal wave of wavelength 20 cm is moving in the positive direction of an x axis. The transverse velocity of the particle at $x = 0$ as a function of time is shown in Fig. 16-47. What are the (a) wave speed, (b) amplitude, and (c) frequency? (d) Sketch the wave between $x = 0$ and $x = 20$ cm at $t = 2.0$ s.

Fig. 16-47   Problem 80.

81 Two waves are described by

$$y_1 = 0.30 \sin[\pi(5x - 200)t]$$

and $$y_2 = 0.30 \sin[\pi(5x - 200t) + \pi/3],$$

where y_1, y_2, and x are in meters and t is in seconds. When these two waves are combined, a traveling wave is produced. What are the (a) amplitude, (b) wave speed, and (c) wavelength of that traveling wave?

82 Oscillation of a 600 Hz tuning fork sets up standing waves in a string clamped at both ends. The wave speed for the string is 400 m/s. The standing wave has four loops and an amplitude of 2.0 mm. (a) What is the length of the string? (b) Write an equation for the displacement of the string as a function of position and time.

83 In an experiment on standing waves, a string 90 cm long is attached to the prong of an electrically driven tuning fork that oscillates perpendicular to the length of the string at a frequency of 60 Hz. The mass of the string is 0.044 kg. What tension must the string be under (weights are attached to the other end) if it is to oscillate in four loops?

84 A standing wave results from the sum of two transverse traveling waves given by

$$y_1 = 0.050 \cos(\pi x - 4\pi t)$$

and $$y_2 = 0.050 \cos(\pi x + 4\pi t),$$

where x, y_1, and y_2 are in meters and t is in seconds. (a) What is the smallest positive value of x that corresponds to a node? Beginning at $t = 0$, what is the value of the (b) first, (c) second, and (d) third time the particle at $x = 0$ has zero velocity?

85 *Body armor.* When a high-speed projectile such as a bullet or bomb fragment strikes modern body armor, the fabric of the armor stops the projectile and prevents penetration by quickly spreading the projectile's energy over a large area. This spreading is done by longitudinal and transverse pulses that move *radially* from the impact point, where the projectile pushes a cone-shaped dent into the fabric. The longitudinal pulse, racing along the fibers of the fabric at speed v_l ahead of the denting, causes the fibers to thin and stretch, with material flowing radially inward into the dent. One such radial

fiber is shown in Fig. 16-48a. Part of the projectile's energy goes into this motion and stretching. The transverse pulse, moving at a slower speed v_t, is due to the denting. As the projectile increases the dent's depth, the dent increases in radius, causing the material in the fibers to move in the same direction as the projectile (perpendicular to the transverse pulse's direction of travel). The rest of the projectile's energy goes into this motion. All the energy that does not eventually go into permanently deforming the fibers ends up as thermal energy.

Figure 16-48b is a graph of speed v versus time t for a bullet of mass 10.2 g fired from a .38 Special revolver directly into body armor. Take $v_l = 2000$ m/s, and assume that the half-angle θ of the conical dent is 60°. At the end of the collision, what are the radii of (a) the thinned region and (b) the dent (assuming that the person wearing the armor remains stationary)?

(a)

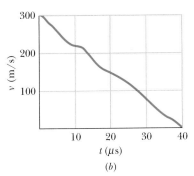

(b)

Fig. 16-48   Problem 85.

86 (a) Write an equation describing a sinusoidal transverse wave traveling on a cord in the positive direction of a y axis with an angular wave number of 60 cm^{-1}, a period of 0.20 s, and an amplitude of 3.0 mm. Take the transverse direction to be the z direction. (b) What is the maximum transverse speed of a point on the cord?

87 The type of rubber band used inside some baseballs and golf balls obeys Hooke's law over a wide range of elongation of the band. A segment of this material has an unstretched length ℓ and a mass m. When a force F is applied, the band stretches an additional length $\Delta\ell$. (a) What is the speed (in terms of m, $\Delta\ell$, and the spring constant k) of transverse waves on this stretched rubber band? (b) Using your answer to (a), show that the time required for a transverse pulse to travel the length of the rubber band is proportional to $1/\sqrt{\Delta\ell}$ if $\Delta\ell \ll \ell$ and is constant if $\Delta\ell \gg \ell$.

88 Two waves,

$$y_1 = (2.50 \text{ mm}) \sin[(25.1 \text{ rad/m})x - (440 \text{ rad/s})t]$$

and $y_2 = (1.50 \text{ mm}) \sin[(25.1 \text{ rad/m})x + (440 \text{ rad/s})t]$,

travel along a stretched string. (a) Plot the resultant wave as a function of t for $x = 0$, $\lambda/8$, $\lambda/4$, $3\lambda/8$, and $\lambda/2$, where λ is the wavelength. The graphs should extend from $t = 0$ to a little over one period. (b) The resultant wave is the superposition of a standing wave and a traveling wave. In which direction does the traveling wave move? (c) How can you change the original waves so the resultant wave is the superposition of standing and traveling waves with the same amplitudes as before but with the traveling wave moving in the opposite direction? Next, use your graphs to find the place at which the oscillation amplitude is (d) maximum and (e) minimum. (f) How is the maximum amplitude related to the amplitudes of the original two waves? (g) How is the minimum amplitude related to the amplitudes of the original two waves?

89 The speed of electromagnetic waves (which include visible light, radio, and x rays) in vacuum is 3.0×10^8 m/s. (a) Wavelengths of visible light waves range from about 400 nm in the violet to about 700 nm in the red. What is the range of frequencies of these waves? (b) The range of frequencies for shortwave radio (for example, FM radio and VHF television) is 1.5 to 300 MHz. What is the corresponding wavelength range? (c) X ray wavelengths range from about 5.0 nm to about 1.0×10^{-2} nm. What is the frequency range for x rays?

90 Two waves on a string are described by these equations:

$$y_1 = (0.10 \text{ m}) \sin 2\pi [(0.50 \text{ m}^{-1})x + (20 \text{ s}^{-1})t],$$

$$y_2 = (0.20 \text{ m}) \sin 2\pi [(0.50 \text{ m}^{-1})x - (20 \text{ s}^{-1})t].$$

Graph y versus t for the point on the string at $x = 3.0$ m.

91 A traveling wave on a string is described by

$$y = 2.0 \sin \left[2\pi \left(\frac{t}{0.40} + \frac{x}{80} \right) \right],$$

where x and y are in centimeters and t is in seconds. (a) For $t = 0$, plot y as a function of x for $0 \le x \le 160$ cm. (b) Repeat (a) for $t = 0.05$ s and $t = 0.10$ s. From your graphs, determine (c) the wave speed and (d) the direction in which the wave is traveling.

92 Show that the maximum kinetic energy in each loop of a standing wave produced by two traveling waves of identical amplitudes is $2\pi^2 \mu y_m^2 f v$.

93 In Fig. 16-49, two pulses travel along a string in opposite directions. The wave speed v is 2.0 m/s, and the pulses are distance $D = 6.0$ cm apart at $t = 0$. (a) Sketch the wave patterns when t is equal to 5.0, 10, 15, 20, and 25 ms. (b) In what form (or type) is the energy of the pulses at $t = 15$ ms?

Fig. 16-49 Problem 93.

94 The Le Pont de Normandie bridge across the Seine River in France has the longest cable-stayed span in the world (Fig. 16-50a). Soon after construction, standing waves were produced on some of the cables by even mild winds. Such oscillations are of concern because they can fatigue the metal strands in the cables, eventually leading to premature failure. To fix the problem, mountain climbers scaled the bridge to attach ropes between adjacent cables.

Consider the simple arrangement of two vertical cables in Fig. 16-50b. Cable A of length L and cable B of length $2L$ have the same linear density, are under the same tension, and are connected by a horizontal rope at the midpoint of the shorter cable. Initially the rope is neither slack nor under tension. Suppose that at time $t = 0$ a wind gust moves the cables rightward and thereby starts them oscillating in their fundamental modes. Let T_A be the oscillation period of cable A, and assume that the cables would oscillate with the same amplitude if not connected. For which of the following times is the rope (a) under tension, (b) back in its initial state, and (c) slack: $0.25T_A$, $0.5T_A$, $0.75T_A$, T_A, $1.25T_A$, $1.5T_A$, $1.75T_A$, $2T_A$? Figure 16-50c shows a second choice of rope placement. (d) Do the two designs damp out the cable oscillations if only the fundamental oscillations can occur?

(a)

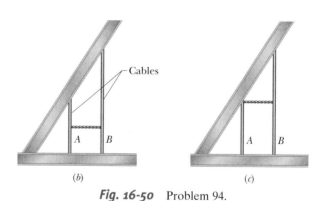

(b)   (c)

Fig. 16-50 Problem 94.

On-Line Simulation Problems

The website http://www.wiley.com/college/halliday has simulation problems about this chapter.

Waves — II

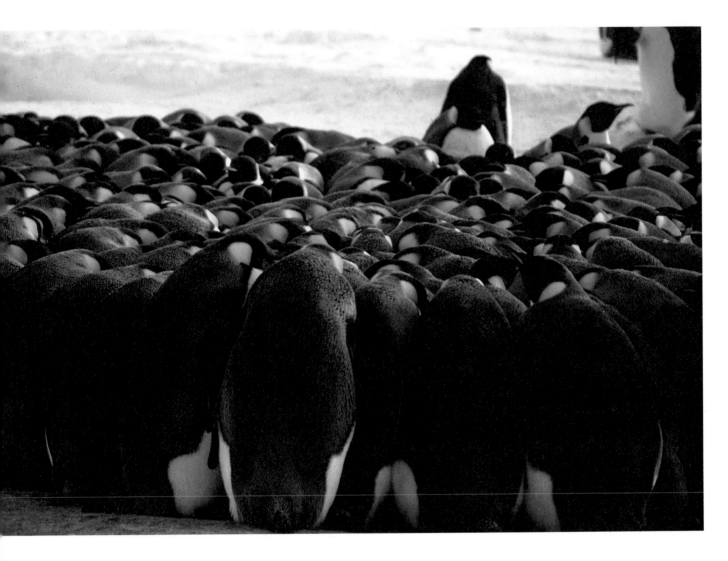

After diving into the water and eating, an emperor penguin must crawl back onto its home ice flow and return to its mate. In the winter, however, that mate can be anywhere among thousands of penguins huddled tightly together to avoid freezing in the harsh weather of the Antarctic, where temperatures can be −40°C and wind speeds 300 km/h. Besides, penguins all look about the same, even to other penguins, and so a penguin cannot visually recognize its mate.

How, then, does a penguin find its mate among the huddled thousands?

The answer is in this chapter.

Fig. 17-1 This image of a 6-month-old fetus is made with ultrasound (which has a frequency above your hearing range).

17-1   *What Is Physics?*

The physics of sound waves is the basis of countless studies in the research journals of many fields. Here are just a few examples. Some physiologists are concerned with how speech is produced, how speech impairment might be corrected, how hearing loss can be alleviated, and even how snoring is produced. Some acoustic engineers are concerned with improving the acoustics of cathedrals and concert halls, with reducing noise near freeways and road construction, and with reproducing music by speaker systems. Some aviation engineers are concerned with the shock waves produced by supersonic aircraft and the aircraft noise produced in communities near an airport. Some medical researchers are concerned with how noises produced by the heart and lungs can signal a medical problem in a patient. Some paleontologists are concerned with how a dinosaur's fossil might reveal the dinosaur's vocalizations. Some military engineers are concerned with how the sounds of sniper fire might allow a soldier to pinpoint the sniper's location, and, on the gentler side, some biologists are concerned with how a cat purrs.

To begin our discussion of the physics of sound, we must first answer the question "What *are* sound waves?"

17-2   *Sound Waves*

As we saw in Chapter 16, mechanical waves are waves that require a material medium to exist. There are two types of mechanical waves: *Transverse waves* involve oscillations perpendicular to the direction in which the wave travels; *longitudinal waves* involve oscillations parallel to the direction of wave travel.

In this book, a **sound wave** is defined roughly as any longitudinal wave. Seismic prospecting teams use such waves to probe Earth's crust for oil. Ships carry sound-ranging gear (sonar) to detect underwater obstacles. Submarines use sound waves to stalk other submarines, largely by listening for the characteristic noises produced by the propulsion system. Figure 17-1, a computer-processed image of a fetal head and arm, shows how sound waves can be used to explore the soft tissues of the human body. In this chapter we shall focus on sound waves that travel through the air and that are audible to people.

Figure 17-2 illustrates several ideas that we shall use in our discussions. Point *S* represents a tiny sound source, called a *point source*, that emits sound waves in all directions. The *wavefronts* and *rays* indicate the direction of travel and the spread of the sound waves. **Wavefronts** are surfaces over which the oscillations due to the sound wave have the same value; such surfaces are represented by whole or partial circles in a two-dimensional drawing for a point source. **Rays** are directed lines perpendicular to the wavefronts that indicate the direction of travel of the wavefronts. The short double arrows superimposed on the rays of Fig. 17-2 indicate that the longitudinal oscillations of the air are parallel to the rays.

Near a point source like that of Fig. 17-2, the wavefronts are spherical and spread out in three dimensions, and there the waves are said to be *spherical*. As the wavefronts move outward and their radii become larger, their curvature decreases. Far from the source, we approximate the wavefronts as planes (or lines on two-dimensional drawings), and the waves are said to be *planar*.

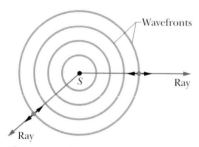

Fig. 17-2 A sound wave travels from a point source *S* through a three-dimensional medium. The wavefronts form spheres centered on *S*; the rays are radial to *S*. The short, double-headed arrows indicate that elements of the medium oscillate parallel to the rays.

17-3   *The Speed of Sound*

The speed of any mechanical wave, transverse or longitudinal, depends on both an inertial property of the medium (to store kinetic energy) and an elastic property of the medium (to store potential energy). Thus, we can generalize Eq. 16-26, which gives the speed of a transverse wave along a stretched string, by writing

$$v = \sqrt{\frac{\tau}{\mu}} = \sqrt{\frac{\text{elastic property}}{\text{inertial property}}}, \qquad (17\text{-}1)$$

where (for transverse waves) τ is the tension in the string and μ is the string's linear density. If the medium is air and the wave is longitudinal, we can guess that the inertial property, corresponding to μ, is the volume density ρ of air. What shall we put for the elastic property?

In a stretched string, potential energy is associated with the periodic stretching of the string elements as the wave passes through them. As a sound wave passes through air, potential energy is associated with periodic compressions and expansions of small volume elements of the air. The property that determines the extent to which an element of a medium changes in volume when the pressure (force per unit area) on it changes is the **bulk modulus** B, defined (from Eq. 12-25) as

$$B = -\frac{\Delta p}{\Delta V/V} \qquad \text{(definition of bulk modulus)}. \qquad (17\text{-}2)$$

Here $\Delta V/V$ is the fractional change in volume produced by a change in pressure Δp. As explained in Section 14-3, the SI unit for pressure is the newton per square meter, which is given a special name, the *pascal* (Pa). From Eq. 17-2 we see that the unit for B is also the pascal. The signs of Δp and ΔV are always opposite: When we increase the pressure on an element (Δp is positive), its volume decreases (ΔV is negative). We include a minus sign in Eq. 17-2 so that B is always a positive quantity. Now substituting B for τ and ρ for μ in Eq. 17-1 yields

$$v = \sqrt{\frac{B}{\rho}} \qquad \text{(speed of sound)} \qquad (17\text{-}3)$$

as the speed of sound in a medium with bulk modulus B and density ρ. Table 17-1 lists the speed of sound in various media.

The density of water is almost 1000 times greater than the density of air. If this were the only relevant factor, we would expect from Eq. 17-3 that the speed of sound in water would be considerably less than the speed of sound in air. However, Table 17-1 shows us that the reverse is true. We conclude (again from Eq. 17-3) that the bulk modulus of water must be more than 1000 times greater than that of air. This is indeed the case. Water is much more incompressible than air, which (see Eq. 17-2) is another way of saying that its bulk modulus is much greater.

Formal Derivation of Eq. 17-3

We now derive Eq. 17-3 by direct application of Newton's laws. Let a single pulse in which air is compressed travel (from right to left) with speed v through the air in a long tube, like that in Fig. 16-2. Let us run along with the pulse at that speed, so that the pulse appears to stand still in our reference frame. Figure 17-3a shows the situation as it is viewed from that frame. The pulse is standing still, and air is moving at speed v through it from left to right.

Let the pressure of the undisturbed air be p and the pressure inside the pulse be $p + \Delta p$, where Δp is positive due to the compression. Consider an element of air of thickness Δx and face area A, moving toward the pulse at speed v. As this element enters the pulse, the leading face of the element encounters a region of higher pressure, which slows the element to speed $v + \Delta v$, in which Δv is

TABLE 17-1
The Speed of Sound[a]

Medium	Speed (m/s)
Gases	
Air (0°C)	331
Air (20°C)	343
Helium	965
Hydrogen	1284
Liquids	
Water (0°C)	1402
Water (20°C)	1482
Seawater[b]	1522
Solids	
Aluminum	6420
Steel	5941
Granite	6000

[a]At 0°C and 1 atm pressure, except where noted.

[b]At 20°C and 3.5% salinity.

Fig. 17-3 A compression pulse is sent from right to left down a long air-filled tube. The reference frame of the figure is chosen so that the pulse is at rest and the air moves from left to right. (*a*) An element of air of width Δx moves toward the pulse with speed v. (*b*) The leading face of the element enters the pulse. The forces acting on the leading and trailing faces (due to air pressure) are shown.

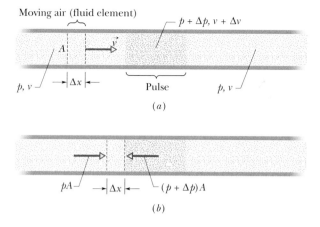

negative. This slowing is complete when the rear face of the element reaches the pulse, which requires time interval

$$\Delta t = \frac{\Delta x}{v}. \tag{17-4}$$

Let us apply Newton's second law to the element. During Δt, the average force on the element's trailing face is pA toward the right, and the average force on the leading face is $(p + \Delta p)A$ toward the left (Fig. 17-3*b*). Therefore, the average net force on the element during Δt is

$$
\begin{aligned}
F &= pA - (p + \Delta p)A \\
&= -\Delta p\, A \quad \text{(net force)}.
\end{aligned}
\tag{17-5}
$$

The minus sign indicates that the net force on the air element is directed to the left in Fig. 17-3*b*. The volume of the element is $A\,\Delta x$, so with the aid of Eq. 17-4, we can write its mass as

$$\Delta m = \rho\,\Delta V = \rho A\,\Delta x = \rho A v\,\Delta t \quad \text{(mass)}. \tag{17-6}$$

The average acceleration of the element during Δt is

$$a = \frac{\Delta v}{\Delta t} \quad \text{(acceleration)}. \tag{17-7}$$

Thus, from Newton's second law ($F = ma$), we have, from Eqs. 17-5, 17-6, and 17-7,

$$-\Delta p\, A = (\rho A v\,\Delta t)\frac{\Delta v}{\Delta t},$$

which we can write as

$$\rho v^2 = -\frac{\Delta p}{\Delta v/v}. \tag{17-8}$$

The air that occupies a volume $V\ (= Av\,\Delta t)$ outside the pulse is compressed by an amount $\Delta V\ (= A\,\Delta v\,\Delta t)$ as it enters the pulse. Thus,

$$\frac{\Delta V}{V} = \frac{A\,\Delta v\,\Delta t}{Av\,\Delta t} = \frac{\Delta v}{v}. \tag{17-9}$$

Substituting Eq. 17-9 and then Eq. 17-2 into Eq. 17-8 leads to

$$\rho v^2 = -\frac{\Delta p}{\Delta v/v} = -\frac{\Delta p}{\Delta V/V} = B.$$

Solving for v yields Eq. 17-3 for the speed of the air toward the right in Fig. 17-3, and thus for the actual speed of the pulse toward the left.

Sample Problem 17-1

One clue used by your brain to determine the direction of a source of sound is the time delay Δt between the arrival of the sound at the ear closer to the source and the arrival at the farther ear. Assume that the source is distant so that a wavefront from it is approximately planar when it reaches you, and let D represent the separation between your ears.

(a) Find an expression that gives Δt in terms of D and the angle θ between the direction of the source and the forward direction.

Solution: The situation is shown (from an overhead view) in Fig. 17-4, where wavefronts approach you from a source that is located in front of you and to your right. The **Key Idea** here is that the time delay Δt is due to the distance d that each wavefront must travel to reach your left ear (L) after it reaches your right ear (R). From Fig. 17-4, we find

$$\Delta t = \frac{d}{v} = \frac{D \sin \theta}{v}, \quad \text{(Answer)} \quad (17\text{-}10)$$

where v is the speed of sound in air. Based on a lifetime of experience, your brain correlates each detected value of Δt (from zero to the maximum value) with a value of θ (from zero to 90°) for the direction of the sound source.

(b) Suppose that you are submerged in water at 20°C when a wavefront arrives from directly to your right. Based on the time-delay clue, at what angle θ from the forward direction does the source seem to be?

Solution: The **Key Idea** here is that the speed of the sound is now the speed v_w in water, so in Eq. 17-10 we substitute

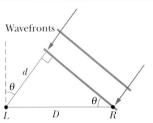

Fig. 17-4 A wavefront travels a distance d ($= D \sin \theta$) farther to reach the left ear (L) than to reach the right ear (R).

v_w for v and 90° for θ, finding that

$$\Delta t_w = \frac{D \sin 90°}{v_w} = \frac{D}{v_w}. \quad (17\text{-}11)$$

Since v_w is about four times v, delay Δt_w is about one-fourth the maximum time delay in air. Based on experience, your brain will process the water time delay as if it occurred in air. Thus, the sound source appears to be at an angle θ smaller than 90°. To find that apparent angle, we substitute the time delay D/v_w from Eq. 17-11 for Δt in Eq. 17-10, obtaining

$$\frac{D}{v_w} = \frac{D \sin \theta}{v}. \quad (17\text{-}12)$$

Then, to solve for θ we substitute $v = 343$ m/s and $v_w = 1482$ m/s (from Table 17-1) into Eq. 17-12, finding

$$\sin \theta = \frac{v}{v_w} = \frac{343 \text{ m/s}}{1482 \text{ m/s}} = 0.231$$

and thus

$$\theta = 13°. \quad \text{(Answer)}$$

17-4 Traveling Sound Waves

Here we examine the displacements and pressure variations associated with a sinusoidal sound wave traveling through air. Figure 17-5a displays such a wave traveling rightward through a long air-filled tube. Recall from Chapter 16 that we can produce such a wave by sinusoidally moving a piston at the left end of the tube (as in Fig. 16-2). The piston's rightward motion moves the element of

Fig. 17-5 (a) A sound wave, traveling through a long air-filled tube with speed v, consists of a moving, periodic pattern of expansions and compressions of the air. The wave is shown at an arbitrary instant. (b) A horizontally expanded view of a short piece of the tube. As the wave passes, an air element of thickness Δx oscillates left and right in simple harmonic motion about its equilibrium position. At the instant shown in (b), the element happens to be displaced a distance s to the right of its equilibrium position. Its maximum displacement, either right or left, is s_m.

(a) $s(x, t) = s_m \cos(kx - \omega t)$

Displacement amplitude / Oscillating term

(b) $\Delta p(x, t) = \Delta p_m \sin(kx - \omega t)$

Pressure amplitude / Pressure variation

Fig. 17-6 (*a*) The displacement function and (*b*) the pressure-variation function of a traveling sound wave consist of an amplitude and an oscillating term.

air next to the piston face and compresses that air; the piston's leftward motion allows the element of air to move back to the left and the pressure to decrease. As each element of air pushes on the next element in turn, the right–left motion of the air and the change in its pressure travel along the tube as a sound wave.

Consider the thin element of air of thickness Δx shown in Fig. 17-5*b*. As the wave travels through this portion of the tube, the element of air oscillates left and right in simple harmonic motion about its equilibrium position. Thus, the oscillations of each air element due to the traveling sound wave are like those of a string element due to a transverse wave, except that the air element oscillates *longitudinally* rather than *transversely*. Because string elements oscillate parallel to the *y* axis, we write their displacements in the form $y(x, t)$. Similarly, because air elements oscillate parallel to the *x* axis, we could write their displacements in the confusing form $x(x, t)$, but we shall use $s(x, t)$ instead.

To show that the displacements $s(x, t)$ are sinusoidal functions of *x* and *t*, we can use either a sine function or a cosine function. In this chapter we use a cosine function, writing

$$s(x, t) = s_m \cos(kx - \omega t). \tag{17-13}$$

Figure 17-6*a* labels the various parts of this equation. In it, s_m is the **displacement amplitude**—that is, the maximum displacement of the air element to either side of its equilibrium position (see Fig. 17-5*b*). The angular wave number *k*, angular frequency ω, frequency *f*, wavelength λ, speed *v*, and period *T* for a sound (longitudinal) wave are defined and interrelated exactly as for a transverse wave, except that λ is now the distance (again along the direction of travel) in which the pattern of compression and expansion due to the wave begins to repeat itself (see Fig. 17-5*a*). (We assume s_m is much less than λ.)

As the wave moves, the air pressure at any position *x* in Fig. 17-5*a* varies sinusoidally, as we prove next. To describe this variation we write

$$\Delta p(x, t) = \Delta p_m \sin(kx - \omega t). \tag{17-14}$$

Figure 17-6*b* labels the various parts of this equation. A negative value of Δp in Eq. 17-14 corresponds to an expansion of the air, and a positive value to a compression. Here Δp_m is the **pressure amplitude,** which is the maximum increase or decrease in pressure due to the wave; Δp_m is normally very much less than the pressure *p* present when there is no wave. As we shall prove, the pressure amplitude Δp_m is related to the displacement amplitude s_m in Eq. 17-13 by

$$\Delta p_m = (v \rho \omega) s_m. \tag{17-15}$$

Figure 17-7 shows plots of Eqs. 17-13 and 17-14 at $t = 0$; with time, the two curves would move rightward along the horizontal axes. Note that the displacement and pressure variation are $\pi/2$ rad (or 90°) out of phase. Thus, for example, the pressure variation Δp at any point along the wave is zero when the displacement there is a maximum.

(a)

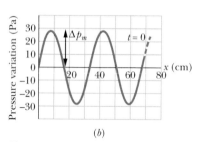

(b)

Fig. 17-7 (*a*) A plot of the displacement function (Eq. 17-13) for $t = 0$. (*b*) A similar plot of the pressure-variation function (Eq. 17-14). Both plots are for a 1000 Hz sound wave whose pressure amplitude is at the threshold of pain; see Sample Problem 17-2.

✔ **CHECKPOINT 1**   When the oscillating air element in Fig. 17-5*b* is moving rightward through the point of zero displacement, is the pressure in the element at its equilibrium value, just beginning to increase, or just beginning to decrease?

Derivation of Eqs. 17-14 and 17-15

Figure 17-5*b* shows an oscillating element of air of cross-sectional area *A* and thickness Δx, with its center displaced from its equilibrium position by distance *s*. From Eq. 17-2 we can write, for the pressure variation in the displaced element,

$$\Delta p = -B \frac{\Delta V}{V}. \tag{17-16}$$

The quantity V in Eq. 17-16 is the volume of the element, given by

$$V = A \, \Delta x. \qquad (17\text{-}17)$$

The quantity ΔV in Eq. 17-16 is the change in volume that occurs when the element is displaced. This volume change comes about because the displacements of the two faces of the element are not quite the same, differing by some amount Δs. Thus, we can write the change in volume as

$$\Delta V = A \, \Delta s. \qquad (17\text{-}18)$$

Substituting Eqs. 17-17 and 17-18 into Eq. 17-16 and passing to the differential limit yield

$$\Delta p = -B \frac{\Delta s}{\Delta x} = -B \frac{\partial s}{\partial x}. \qquad (17\text{-}19)$$

The symbols ∂ indicate that the derivative in Eq. 17-19 is a *partial derivative*, which tells us how s changes with x when the time t is fixed. From Eq. 17-13 we then have, treating t as a constant,

$$\frac{\partial s}{\partial x} = \frac{\partial}{\partial x} [s_m \cos(kx - \omega t)] = -k s_m \sin(kx - \omega t).$$

Substituting this quantity for the partial derivative in Eq. 17-19 yields

$$\Delta p = B k s_m \sin(kx - \omega t).$$

Setting $\Delta p_m = B k s_m$, this yields Eq. 17-14, which we set out to prove.
  Using Eq. 17-3, we can now write

$$\Delta p_m = (Bk)s_m = (v^2 \rho k)s_m.$$

Equation 17-15, which we also wanted to prove, follows at once if we substitute ω/v for k from Eq. 16-13.

Sample Problem 17-2

The maximum pressure amplitude Δp_m that the human ear can tolerate in loud sounds is about 28 Pa (which is very much less than the normal air pressure of about 10^5 Pa). What is the displacement amplitude s_m for such a sound in air of density $\rho = 1.21$ kg/m³, at a frequency of 1000 Hz and a speed of 343 m/s?

Solution: The **Key Idea** is that the displacement amplitude s_m of a sound wave is related to the pressure amplitude Δp_m of the wave according to Eq. 17-15. Solving that equation for s_m yields

$$s_m = \frac{\Delta p_m}{v \rho \omega} = \frac{\Delta p_m}{v \rho (2\pi f)}.$$

Substituting known data then gives us

$$s_m = \frac{28 \text{ Pa}}{(343 \text{ m/s})(1.21 \text{ kg/m}^3)(2\pi)(1000 \text{ Hz})}$$
$$= 1.1 \times 10^{-5} \text{ m} = 11 \ \mu\text{m}. \qquad \text{(Answer)}$$

That is only about one-seventh the thickness of this page. Obviously, the displacement amplitude of even the loudest sound that the ear can tolerate is very small.
  The pressure amplitude Δp_m for the *faintest* detectable sound at 1000 Hz is 2.8×10^{-5} Pa. Proceeding as above leads to $s_m = 1.1 \times 10^{-11}$ m or 11 pm, which is about one-tenth the radius of a typical atom. The ear is indeed a sensitive detector of sound waves.

17-5 Interference

Like transverse waves, sound waves can undergo interference. Let us consider, in particular, the interference between two identical sound waves traveling in the same direction. Figure 17-8 shows how we can set up such a situation: Two point sources S_1 and S_2 emit sound waves that are in phase and of identical wavelength λ. Thus, the sources themselves are said to be in phase; that is, as the waves emerge from the sources, their displacements are always identical. We are interested in the waves that then travel through point P in Fig. 17-8. We assume that the distance to P is much greater than the distance between the sources so that we can approximate the waves as traveling in the same direction at P.
  If the waves traveled along paths with identical lengths to reach point P, they would be in phase there. As with transverse waves, this means that they would

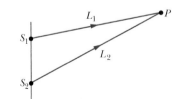

Fig. 17-8 Two point sources S_1 and S_2 emit spherical sound waves in phase. The rays indicate that the waves pass through a common point P.

undergo fully constructive interference there. However, in Fig. 17-8, path L_2 traveled by the wave from S_2 is longer than path L_1 traveled by the wave from S_1. The difference in path lengths means that the waves may not be in phase at point P. In other words, their phase difference ϕ at P depends on their **path length difference** $\Delta L = |L_2 - L_1|$.

To relate phase difference ϕ to path length difference ΔL, we recall (from Section 16-4) that a phase difference of 2π rad corresponds to one wavelength. Thus, we can write the proportion

$$\frac{\phi}{2\pi} = \frac{\Delta L}{\lambda}, \tag{17-20}$$

from which

$$\phi = \frac{\Delta L}{\lambda} 2\pi. \tag{17-21}$$

Fully constructive interference occurs when ϕ is zero, 2π, or any integer multiple of 2π. We can write this condition as

$$\phi = m(2\pi), \quad \text{for } m = 0, 1, 2, \ldots \quad \text{(fully constructive interference).} \tag{17-22}$$

From Eq. 17-21, this occurs when the ratio $\Delta L/\lambda$ is

$$\frac{\Delta L}{\lambda} = 0, 1, 2, \ldots \quad \text{(fully constructive interference).} \tag{17-23}$$

For example, if the path length difference $\Delta L = |L_2 - L_1|$ in Fig. 17-8 is equal to 2λ, then $\Delta L/\lambda = 2$ and the waves undergo fully constructive interference at point P. The interference is fully constructive because the wave from S_2 is phase-shifted relative to the wave from S_1 by 2λ, putting the two waves *exactly in phase* at P.

Fully destructive interference occurs when ϕ is an odd multiple of π, a condition we can write as

$$\phi = (2m + 1)\pi, \quad \text{for } m = 0, 1, 2, \ldots \quad \text{(fully destructive interference).} \tag{17-24}$$

From Eq. 17-21, this occurs when the ratio $\Delta L/\lambda$ is

$$\frac{\Delta L}{\lambda} = 0.5, 1.5, 2.5, \ldots \quad \text{(fully destructive interference).} \tag{17-25}$$

For example, if the path length difference $\Delta L = |L_2 - L_1|$ in Fig. 17-8 is equal to 2.5λ, then $\Delta L/\lambda = 2.5$ and the waves undergo fully destructive interference at point P. The interference is fully destructive because the wave from S_2 is phase-shifted relative to the wave from S_1 by 2.5 wavelengths, which puts the two waves *exactly out of phase* at P.

Of course, two waves could produce intermediate interference as, say, when $\Delta L/\lambda = 1.2$. This would be closer to fully constructive interference ($\Delta L/\lambda = 1.0$) than to fully destructive interference ($\Delta L/\lambda = 1.5$).

Sample Problem 17-3

In Fig. 17-9a, two point sources S_1 and S_2, which are in phase and separated by distance $D = 1.5\lambda$, emit identical sound waves of wavelength λ.

(a) What is the path length difference of the waves from S_1 and S_2 at point P_1, which lies on the perpendicular bisector of distance D, at a distance greater than D from the sources? What type of interference occurs at P_1?

Solution: The **Key Idea** here is that, because the waves travel identical distances to reach P_1, their path length difference is

$$\Delta L = 0. \quad \text{(Answer)}$$

From Eq. 17-23, this means that the waves undergo fully constructive interference at P_1.

(b) What are the path length difference and type of interference at point P_2 in Fig. 17-9a?

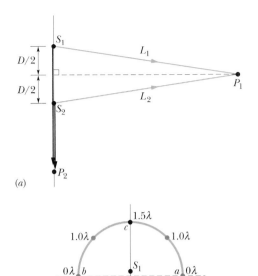

(a)

(b)

Fig. 17-9 (a) Two point sources S_1 and S_2, separated by distance D, emit spherical sound waves in phase. The waves travel equal distances to reach point P_1. Point P_2 is on the line extending through S_1 and S_2. (b) The path length difference (in terms of wavelength) between the waves from S_1 and S_2, at eight points on a large circle around the sources.

Solution: Now the **Key Idea** is that the wave from S_1 travels the extra distance D (= 1.5λ) to reach P_2. Thus, the path

length difference is

$$\Delta L = 1.5\lambda. \qquad \text{(Answer)}$$

From Eq. 17-25, this means that the waves are exactly out of phase at P_2 and undergo fully destructive interference there.

(c) Figure 17-9b shows a circle with a radius much greater than D, centered on the midpoint between sources S_1 and S_2. What is the number of points N around this circle at which the interference is fully constructive?

Solution: Imagine that, starting at point a, we move clockwise along the circle to point d. One **Key Idea** here is that as we move to point d, the path length difference ΔL increases and so the type of interference changes. From (a), we know that the path length difference is $\Delta L = 0\lambda$ at point a. From (b), we know that $\Delta L = 1.5\lambda$ at point d. Thus, there must be one point along the circle between a and d at which $\Delta L = \lambda$, as indicated in Fig. 17-9b. From Eq. 17-23, fully constructive interference occurs at that point. Also, there can be no other point along the way from point a to point d at which fully constructive interference occurs, because there is no other integer than 1 between 0 and 1.5.

Another **Key Idea** here is to use symmetry to locate the other points of fully constructive interference along the rest of the circle. Symmetry about line cd gives us point b, at which $\Delta L = 0\lambda$. Also, there are three more points at which $\Delta L = \lambda$. In all we have

$$N = 6. \qquad \text{(Answer)}$$

✓**CHECKPOINT 2** In this sample problem, if the distance D between sources S_1 and S_2 were, instead, equal to 4λ, what would be the path length difference and what type of interference would occur at (a) point P_1 and (b) point P_2?

17-6 *Intensity and Sound Level*

If you have ever tried to sleep while someone played loud music nearby, you are well aware that there is more to sound than frequency, wavelength, and speed. There is also intensity. The **intensity** I of a sound wave at a surface is the average rate per unit area at which energy is transferred by the wave through or onto the surface. We can write this as

$$I = \frac{P}{A}, \qquad (17\text{-}26)$$

where P is the time rate of energy transfer (the power) of the sound wave and A is the area of the surface intercepting the sound. As we shall derive shortly, the intensity I is related to the displacement amplitude s_m of the sound wave by

$$I = \tfrac{1}{2}\rho v \omega^2 s_m^2. \qquad (17\text{-}27)$$

Variation of Intensity with Distance

How intensity varies with distance from a real sound source is often complex. Some real sources (like loudspeakers) may transmit sound only in particular directions, and the environment usually produces echoes (reflected sound waves) that overlap the direct sound waves. In some situations, however, we can ignore echoes and assume that the sound source is a point source that emits the sound *isotropically*—that is, with equal intensity in all directions. The wavefronts

Sound can cause the wall of a drinking glass to oscillate. If the sound produces a standing wave of oscillations and if the intensity of the sound is large enough, the glass will shatter.

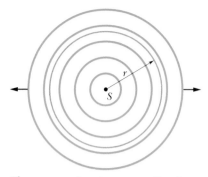

Fig. 17-10 A point source S emits sound waves uniformly in all directions. The waves pass through an imaginary sphere of radius r that is centered on S.

spreading from such an isotropic point source S at a particular instant are shown in Fig. 17-10.

Let us assume that the mechanical energy of the sound waves is conserved as they spread from this source. Let us also center an imaginary sphere of radius r on the source, as shown in Fig. 17-10. All the energy emitted by the source must pass through the surface of the sphere. Thus, the time rate at which energy is transferred through the surface by the sound waves must equal the time rate at which energy is emitted by the source (that is, the power P_s of the source). From Eq. 17-26, the intensity I at the sphere must then be

$$I = \frac{P_s}{4\pi r^2},\qquad(17\text{-}28)$$

where $4\pi r^2$ is the area of the sphere. Equation 17-28 tells us that the intensity of sound from an isotropic point source decreases with the square of the distance r from the source.

CHECKPOINT 3 The figure indicates three small patches 1, 2, and 3 that lie on the surfaces of two imaginary spheres; the spheres are centered on an isotropic point source S of sound. The rates at which energy is transmitted through the three patches by the sound waves are equal. Rank the patches according to (a) the intensity of the sound on them and (b) their area, greatest first.

The Decibel Scale

You saw in Sample Problem 17-2 that the displacement amplitude at the human ear ranges from about 10^{-5} m for the loudest tolerable sound to about 10^{-11} m for the faintest detectable sound, a ratio of 10^6. From Eq. 17-27 we see that the intensity of a sound varies as the *square* of its amplitude, so the ratio of intensities at these two limits of the human auditory system is 10^{12}. Humans can hear over an enormous range of intensities.

We deal with such an enormous range of values by using logarithms. Consider the relation

$$y = \log x,$$

in which x and y are variables. It is a property of this equation that if we *multiply* x by 10, then y increases by 1. To see this, we write

$$y' = \log(10x) = \log 10 + \log x = 1 + y.$$

Similarly, if we multiply x by 10^{12}, y increases by only 12.

Thus, instead of speaking of the intensity I of a sound wave, it is much more convenient to speak of its **sound level** β, defined as

$$\beta = (10\text{ dB}) \log \frac{I}{I_0}.\qquad(17\text{-}29)$$

TABLE 17-2
Some Sound Levels (dB)

Hearing threshold	0
Rustle of leaves	10
Conversation	60
Rock concert	110
Pain threshold	120
Jet engine	130

Here dB is the abbreviation for **decibel,** the unit of sound level, a name that was chosen to recognize the work of Alexander Graham Bell. I_0 in Eq. 17-29 is a standard reference intensity ($= 10^{-12}$ W/m²), chosen because it is near the lower limit of the human range of hearing. For $I = I_0$, Eq. 17-29 gives $\beta = 10 \log 1 = 0$, so our standard reference level corresponds to zero decibels. Then β increases by 10 dB every time the sound intensity increases by an order of magnitude (a factor of 10). Thus, $\beta = 40$ corresponds to an intensity that is 10^4 times the standard reference level. Table 17-2 lists the sound levels for a variety of environments.

Derivation of Eq. 17-27

Consider, in Fig. 17-5a, a thin slice of air of thickness dx, area A, and mass dm, oscillating back and forth as the sound wave of Eq. 17-13 passes through it. The kinetic energy dK of the slice of air is

$$dK = \tfrac{1}{2}dm \, v_s^2. \qquad (17\text{-}30)$$

Here v_s is not the speed of the wave but the speed of the oscillating element of air, obtained from Eq. 17-13 as

$$v_s = \frac{\partial s}{\partial t} = -\omega s_m \sin(kx - \omega t).$$

Using this relation and putting $dm = \rho A \, dx$ allow us to rewrite Eq. 17-30 as

$$dK = \tfrac{1}{2}(\rho A \, dx)(-\omega s_m)^2 \sin^2(kx - \omega t). \qquad (17\text{-}31)$$

Dividing Eq. 17-31 by dt gives the rate at which kinetic energy moves along with the wave. As we saw in Chapter 16 for transverse waves, dx/dt is the wave speed v, so we have

$$\frac{dK}{dt} = \tfrac{1}{2}\rho A v \omega^2 s_m^2 \sin^2(kx - \omega t). \qquad (17\text{-}32)$$

The *average* rate at which kinetic energy is transported is

$$\left(\frac{dK}{dt}\right)_{\text{avg}} = \tfrac{1}{2}\rho A v \omega^2 s_m^2 [\sin^2(kx - \omega t)]_{\text{avg}}$$

$$= \tfrac{1}{4}\rho A v \omega^2 s_m^2. \qquad (17\text{-}33)$$

To obtain this equation, we have used the fact that the average value of the square of a sine (or a cosine) function over one full oscillation is $\tfrac{1}{2}$.

We assume that *potential* energy is carried along with the wave at this same average rate. The wave intensity I, which is the average rate per unit area at which energy of both kinds is transmitted by the wave, is then, from Eq. 17-33,

$$I = \frac{2(dK/dt)_{\text{avg}}}{A} = \tfrac{1}{2}\rho v \omega^2 s_m^2,$$

which is Eq. 17-27, the equation we set out to derive.

Sample Problem 17-4

An electric spark jumps along a straight line of length $L = 10$ m, emitting a pulse of sound that travels radially outward from the spark. (The spark is said to be a *line source* of sound.) The power of the emission is $P_s = 1.6 \times 10^4$ W.

(a) What is the intensity I of the sound when it reaches a distance $r = 12$ m from the spark?

Solution: Let us center an imaginary cylinder of radius $r = 12$ m and length $L = 10$ m (open at both ends) on the spark, as shown in Fig. 17-11. One **Key Idea** here is that the intensity I at the cylindrical surface is the ratio P/A, where P is the time rate at which sound energy passes through the surface and A is the surface area. Another **Key Idea** is to assume that the principle of conservation of energy applies to the sound energy. This means that the rate P at which energy is transferred through the cylinder must equal the rate P_s at which energy is emitted by the source. Putting these ideas together and noting that the area of the cylindrical surface is $A = 2\pi rL$, we have

$$I = \frac{P}{A} = \frac{P_s}{2\pi rL}. \qquad (17\text{-}34)$$

This tells us that the intensity of the sound from a line source decreases with distance r (and not with the square of distance r as for a point source). Substituting the given data, we find

$$I = \frac{1.6 \times 10^4 \text{ W}}{2\pi(12 \text{ m})(10 \text{ m})} = 21.2 \text{ W/m}^2 \approx 21 \text{ W/m}^2. \qquad \text{(Answer)}$$

(b) At what time rate P_d is sound energy intercepted by an acoustic detector of area $A_d = 2.0$ cm^2, aimed at the spark and located a distance $r = 12$ m from the spark?

Solution: Applying the first **Key Idea** of part (a), we know that the intensity of sound at the detector is the ratio of the

Fig. 17-11 A spark along a straight line of length L emits sound waves radially outward. The waves pass through an imaginary cylinder of radius r and length L that is centered on the spark.

energy transfer rate P_d there to the detector's area A_d:

$$I = \frac{P_d}{A_d}. \qquad (17\text{-}35)$$

We can imagine that the detector lies on the cylindrical surface of (a). Then the sound intensity at the detector is the intensity I ($= 21.2$ W/m^2) at the cylindrical surface. Solving Eq. 17-35 for P_d gives us

$$P_d = (21.2 \text{ W/m}^2)(2.0 \times 10^{-4} \text{ m}^2) = 4.2 \text{ mW}. \qquad \text{(Answer)}$$

Sample Problem 17-5

Many veteran rockers suffer from acute hearing damage because of the high sound levels they endured for years while playing music near loudspeakers or listening to music on headphones. Some, like Ted Nugent, can no longer hear in a damaged ear. Others, like Peter Townshend of the Who, have a continuous ringing sensation (tinnitus). Recently, many rockers, such as Lars Ulrich of Metallica (Fig. 17-12), began wearing special earplugs to protect their hearing during performances. If an earplug decreases the sound level of the sound waves by 20 dB, what is the ratio of the final intensity I_f of the waves to their initial intensity I_i?

Solution: The **Key Idea** here is that for both the final and initial waves, the sound level β is related to the intensity by the definition of sound level in Eq. 17-29. For the final waves we have

$$\beta_f = (10 \text{ dB}) \log \frac{I_f}{I_0},$$

and for the initial waves we have

$$\beta_i = (10 \text{ dB}) \log \frac{I_i}{I_0}.$$

The difference in the sound levels is

$$\beta_f - \beta_i = (10 \text{ dB})\left(\log \frac{I_f}{I_0} - \log \frac{I_i}{I_0}\right). \qquad (17\text{-}36)$$

Using the identity

$$\log \frac{a}{b} - \log \frac{c}{d} = \log \frac{ad}{bc},$$

we can rewrite Eq. 17-36 as

$$\beta_f - \beta_i = (10 \text{ dB}) \log \frac{I_f}{I_i}. \qquad (17\text{-}37)$$

Rearranging and then substituting the given decrease in sound

Fig. 17-12 Lars Ulrich of Metallica is an advocate for the organization HEAR (Hearing Education and Awareness for Rockers), which warns about the damage high sound levels can have on hearing.

level as $\beta_f - \beta_i = -20$ dB, we find

$$\log \frac{I_f}{I_i} = \frac{\beta_f - \beta_i}{10 \text{ dB}} = \frac{-20 \text{ dB}}{10 \text{ dB}} = -2.0.$$

We next take the antilog of the far left and far right sides of this equation. (Although the antilog $10^{-2.0}$ can be evaluated mentally, you could use a calculator by keying in 10^-2.0 or using the 10^x key.) We find

$$\frac{I_f}{I_i} = \log^{-1}(-2.0) = 0.010. \qquad \text{(Answer)}$$

Thus, the earplug reduces the intensity of the sound waves to 0.010 of their initial intensity, which is a decrease of two orders of magnitude.

17-7 Sources of Musical Sound

Musical sounds can be set up by oscillating strings (guitar, piano, violin), membranes (kettledrum, snare drum), air columns (flute, oboe, pipe organ, and the horns of Fig. 17-13), wooden blocks or steel bars (marimba, xylophone), and many other oscillating bodies. Most instruments involve more than a single oscillating part. In the violin, for example, both the strings and the body of the instrument participate in producing the music.

Recall from Chapter 16 that standing waves can be set up on a stretched string that is fixed at both ends. They arise because waves traveling along the string are reflected back onto the string at each end. If the wavelength of the waves is suitably matched to the length of the string, the superposition of waves traveling in opposite directions produces a standing wave pattern (or oscillation

Fig. 17-13 The air column within a horn oscillates when the instrument is played.

mode). The wavelength required of the waves for such a match is one that cor-responds to a *resonant frequency* of the string. The advantage of setting up stand-ing waves is that the string then oscillates with a large, sustained amplitude, pushing back and forth against the surrounding air and thus generating a notice-able sound wave with the same frequency as the oscillations of the string. This production of sound is of obvious importance to, say, a guitarist.

We can set up standing waves of sound in an air-filled pipe in a similar way. As sound waves travel through the air in the pipe, they are reflected at each end and travel back through the pipe. (The reflection occurs even if an end is open, but the reflection is not as complete as when the end is closed.) If the wavelength of the sound waves is suitably matched to the length of the pipe, the superposition of waves traveling in opposite directions through the pipe sets up a standing wave pattern. The wavelength required of the sound waves for such a match is one that corresponds to a resonant frequency of the pipe. The advantage of such a standing wave is that the air in the pipe oscillates with a large, sustained ampli-tude, emitting at any open end a sound wave that has the same frequency as the oscillations in the pipe. This emission of sound is of obvious importance to, say, an organist.

Many other aspects of standing sound wave patterns are similar to those of string waves: The closed end of a pipe is like the fixed end of a string in that there must be a node (zero displacement) there, and the open end of a pipe is like the end of a string attached to a freely moving ring, as in Fig. 16-21*b*, in that there must be an antinode there. (Actually, the antinode for the open end of a pipe is located slightly beyond the end, but we shall not dwell on that detail.)

The simplest standing wave pattern that can be set up in a pipe with two open ends is shown in Fig. 17-14*a*. There is an antinode across each open end, as required. There is also a node across the middle of the pipe. An easier way of representing this standing longitudinal sound wave is shown in Fig. 17-14*b*—by drawing it as a standing transverse string wave.

The standing wave pattern of Fig. 17-14*a* is called the *fundamental mode* or *first harmonic*. For it to be set up, the sound waves in a pipe of length L must have a wavelength given by $L = \lambda/2$, so that $\lambda = 2L$. Several more standing sound wave patterns for a pipe with two open ends are shown in Fig. 17-15*a* using string wave representations. The *second harmonic* requires sound waves of wave-length $\lambda = L$, the *third harmonic* requires wavelength $\lambda = 2L/3$, and so on.

More generally, the resonant frequencies for a pipe of length L with two open ends correspond to the wavelengths

$$\lambda = \frac{2L}{n}, \qquad \text{for } n = 1, 2, 3, \ldots, \tag{17-38}$$

where n is called the *harmonic number*. The resonant frequencies for a pipe with two open ends are then given by

$$f = \frac{v}{\lambda} = \frac{nv}{2L}, \qquad \text{for } n = 1, 2, 3, \ldots \quad \text{(pipe, two open ends)}, \tag{17-39}$$

where v is the speed of sound.

Figure 17-15*b* shows (using string wave representations) some of the standing sound wave patterns that can be set up in a pipe with only one open end. As required, across the open end there is an antinode and across the closed end there is a node. The simplest pattern requires sound waves having a wavelength given by $L = \lambda/4$, so that $\lambda = 4L$. The next simplest pattern requires a wavelength given by $L = 3\lambda/4$, so that $\lambda = 4L/3$, and so on.

More generally, the resonant frequencies for a pipe of length L with only one open end correspond to the wavelengths

$$\lambda = \frac{4L}{n}, \qquad \text{for } n = 1, 3, 5, \ldots, \tag{17-40}$$

Fig. 17-14 (*a*) The simplest stand-ing wave pattern of displacement for (longitudinal) sound waves in a pipe with both ends open has an an-tinode (A) across each end and a node (N) across the middle. (The longitudinal displacements repre-sented by the double arrows are greatly exaggerated.) (*b*) The corre-sponding standing wave pattern for (transverse) string waves.

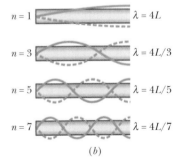

Fig. 17-15 Standing wave patterns for string waves superimposed on pipes to represent standing sound wave patterns in the pipes. (*a*) With *both* ends of the pipe open, any har-monic can be set up in the pipe. (*b*) With only *one* end open, only odd harmonics can be set up.

Fig. 17-16 The saxophone and violin families, showing the relations between instrument length and frequency range. The frequency range of each instrument is indicated by a horizontal bar along a frequency scale suggested by the keyboard at the bottom; the frequency increases toward the right.

in which the harmonic number *n must be an odd number.* The resonant frequencies are then given by

$$f = \frac{v}{\lambda} = \frac{nv}{4L}, \quad \text{for } n = 1, 3, 5, \ldots \quad \text{(pipe, one open end).} \quad (17\text{-}41)$$

Note again that only odd harmonics can exist in a pipe with one open end. For example, the second harmonic, with $n = 2$, cannot be set up in such a pipe. Note also that for such a pipe the adjective in a phrase such as "the third harmonic" still refers to the harmonic number *n* (and not to, say, the third possible harmonic).

The length of a musical instrument reflects the range of frequencies over which the instrument is designed to function, and smaller length implies higher frequencies. Figure 17-16, for example, shows the saxophone and violin families, with their frequency ranges suggested by the piano keyboard. Note that, for every instrument, there is overlap with its higher- and lower-frequency neighbors.

In any oscillating system that gives rise to a musical sound, whether it is a violin string or the air in an organ pipe, the fundamental and one or more of the higher harmonics are usually generated simultaneously. Thus, you hear them together—that is, superimposed as a net wave. When different instruments are played at the same note, they produce the same fundamental frequency but different intensities for the higher harmonics. For example, the fourth harmonic of middle C might be relatively loud on one instrument and relatively quiet or even missing on another. Thus, because different instruments produce different net waves, they sound different to you even when they are played at the same note. That would be the case for the two net waves shown in Fig. 17-17, which were produced at the same note by different instruments.

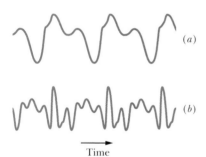

Fig. 17-17 The wave forms produced by (*a*) a flute and (*b*) an oboe when played at the same note, with the same first harmonic frequency.

✔ **CHECKPOINT 4** Pipe *A*, with length *L*, and pipe *B*, with length 2*L*, both have two open ends. Which harmonic of pipe *B* has the same frequency as the fundamental of pipe *A*?

Sample Problem 17-6

Weak background noises from a room set up the fundamental standing wave in a cardboard tube of length *L* = 67.0 cm with two open ends. Assume that the speed of sound in the air within the tube is 343 m/s.

(a) What frequency do you hear from the tube?

Solution: The **Key Idea** here is that, with both pipe ends open, we have a symmetric situation in which the standing wave has an antinode at each end of the tube. The standing wave pattern (in string wave style) is that of Fig. 17-14*b*. The frequency is given by Eq. 17-39 with $n = 1$ for the fundamental mode:

$$f = \frac{nv}{2L} = \frac{(1)(343 \text{ m/s})}{2(0.670 \text{ m})} = 256 \text{ Hz}. \quad \text{(Answer)}$$

If the background noises set up any higher harmonics, such as the second harmonic, you also hear frequencies that are *integer* multiples of 256 Hz.

(b) If you jam your ear against one end of the tube, what fundamental frequency do you hear from the tube?

Solution: The **Key Idea** now is that, with your ear effectively closing one end of the tube, we have an asymmetric situation—an antinode still exists at the open end, but a node is now at the other (closed) end. The standing wave pattern is the top one in Fig. 17-15*b*. The frequency is given by Eq. 17-41 with $n = 1$ for the fundamental mode:

$$f = \frac{nv}{4L} = \frac{(1)(343 \text{ m/s})}{4(0.670 \text{ m})} = 128 \text{ Hz}. \quad \text{(Answer)}$$

If the background noises set up any higher harmonics, they will be *odd* multiples of 128 Hz. That means that the frequency of 256 Hz (which is an even multiple) cannot now occur.

17-8 Beats

If we listen, a few minutes apart, to two sounds whose frequencies are, say, 552 and 564 Hz, most of us cannot tell one from the other. However, if the sounds reach our ears simultaneously, what we hear is a sound whose frequency is 558 Hz, the *average* of the two combining frequencies. We also hear a striking variation in the intensity of this sound—it increases and decreases in slow, wavering **beats** that repeat at a frequency of 12 Hz, the *difference* between the two combining frequencies. Figure 17-18 shows this beat phenomenon.

Let the time-dependent variations of the displacements due to two sound waves at a particular location be

$$s_1 = s_m \cos \omega_1 t \quad \text{and} \quad s_2 = s_m \cos \omega_2 t, \quad (17\text{-}42)$$

where $\omega_1 > \omega_2$. We have assumed, for simplicity, that the waves have the same amplitude. According to the superposition principle, the resultant displacement is

$$s = s_1 + s_2 = s_m(\cos \omega_1 t + \cos \omega_2 t).$$

Using the trigonometric identity (see Appendix E)

$$\cos \alpha + \cos \beta = 2 \cos \left[\tfrac{1}{2}(\alpha - \beta)\right] \cos \left[\tfrac{1}{2}(\alpha + \beta)\right]$$

allows us to write the resultant displacement as

$$s = 2s_m \cos \left[\tfrac{1}{2}(\omega_1 - \omega_2)t\right] \cos \left[\tfrac{1}{2}(\omega_1 + \omega_2)t\right]. \quad (17\text{-}43)$$

If we write

$$\omega' = \tfrac{1}{2}(\omega_1 - \omega_2) \quad \text{and} \quad \omega = \tfrac{1}{2}(\omega_1 + \omega_2), \quad (17\text{-}44)$$

we can then write Eq. 17-43 as

$$s(t) = [2s_m \cos \omega' t] \cos \omega t. \quad (17\text{-}45)$$

We now assume that the angular frequencies ω_1 and ω_2 of the combining waves are almost equal, which means that $\omega \gg \omega'$ in Eq. 17-44. We can then regard Eq. 17-45 as a cosine function whose angular frequency is ω and whose amplitude (which is not constant but varies with angular frequency ω') is the absolute value of the quantity in the brackets.

(*a*)   (*b*)   (*c*)

Time

Fig. 17-18 (*a*, *b*) The pressure variations Δp of two sound waves as they would be detected separately. The frequencies of the waves are nearly equal. (*c*) The resultant pressure variation if the two waves are detected simultaneously.

A maximum amplitude will occur whenever cos $\omega't$ in Eq. 17-45 has the value $+1$ or -1, which happens twice in each repetition of the cosine function. Because cos $\omega't$ has angular frequency ω', the angular frequency ω_{beat} at which beats occur is $\omega_{beat} = 2\omega'$. Then, with the aid of Eq. 17-44, we can write

$$\omega_{beat} = 2\omega' = (2)(\tfrac{1}{2})(\omega_1 - \omega_2) = \omega_1 - \omega_2.$$

Because $\omega = 2\pi f$, we can recast this as

$$f_{beat} = f_1 - f_2 \quad \text{(beat frequency)}. \qquad (17\text{-}46)$$

Musicians use the beat phenomenon in tuning instruments. If an instrument is sounded against a standard frequency (for example, the note called "concert A" played on an orchestra's first oboe) and tuned until the beat disappears, the instrument is in tune with that standard. In musical Vienna, concert A (440 Hz) is available as a telephone service for the city's many musicians.

Sample Problem 17-7

Most birds vocalize by using only one side of their two-sided vocal organ, called the *syrinx*. Emperor penguins, however, vocalize by using both sides simultaneously. Each side sets up acoustic standing waves in the bird's throat and mouth, much like in a pipe with two open ends. Suppose that the frequency of the first harmonic produced by side A is $f_{A1} = 432$ Hz and the frequency of the first harmonic produced by side B is $f_{B1} = 371$ Hz. What is the beat frequency between those two first-harmonic frequencies and between the two second-harmonic frequencies?

Solution: The **Key Idea** here is that the beat frequency between two frequencies is their difference, as given by Eq. 17-46 ($f_{beat} = f_1 - f_2$). For the two first-harmonic frequencies f_{A1} and f_{B1}, the beat frequency is

$$f_{beat,1} = f_{A1} - f_{B1} = 432 \text{ Hz} - 371 \text{ Hz}$$
$$= 61 \text{ Hz}. \qquad \text{(Answer)}$$

Because the standing waves in the penguin are effectively in a pipe with two open ends, the resonant frequencies are given by Eq. 17-39 ($f = nv/2L$), in which L is the (unknown) length of the effective pipe. The first-harmonic frequency is $f_1 = v/2L$, and the second-harmonic frequency is $f_2 = 2v/2L$. Comparing these two frequencies, we see that, in general,

$$f_2 = 2f_1.$$

For the penguin, the second harmonic of side A has frequency $f_{A2} = 2f_{A1}$ and the second harmonic of side B has frequency $f_{B2} = 2f_{B1}$. Using Eq. 17-46 with frequencies f_{A2} and f_{B2}, we find that the corresponding beat frequency is

$$f_{beat,2} = f_{A2} - f_{B2} = 2f_{A1} - 2f_{B1}$$
$$= 2(432 \text{ Hz}) - 2(371 \text{ Hz})$$
$$= 122 \text{ Hz}. \qquad \text{(Answer)}$$

Experiments indicate that penguins can perceive such large beat frequencies (humans cannot). Thus, a penguin's cry can be rich with different harmonics and different beat frequencies, allowing the voice to be recognized even among the voices of thousands of other, closely huddled penguins.

17-9 The Doppler Effect

A police car is parked by the side of the highway, sounding its 1000 Hz siren. If you are also parked by the highway, you will hear that same frequency. However, if there is relative motion between you and the police car, either toward or away from each other, you will hear a different frequency. For example, if you are driving *toward* the police car at 120 km/h (about 75 mi/h), you will hear a *higher* frequency (1096 Hz, an *increase* of 96 Hz). If you are driving *away from* the police car at that same speed, you will hear a *lower* frequency (904 Hz, a *decrease* of 96 Hz).

These motion-related frequency changes are examples of the **Doppler effect.** The effect was proposed (although not fully worked out) in 1842 by Austrian physicist Johann Christian Doppler. It was tested experimentally in 1845 by Buys Ballot in Holland, "using a locomotive drawing an open car with several trumpeters."

The Doppler effect holds not only for sound waves but also for electromagnetic waves, including microwaves, radio waves, and visible light. Here, however, we shall consider only sound waves, and we shall take as a reference frame the body of air through which these waves travel. This means that we shall measure the speeds of a source S of sound waves and a detector D of those waves *relative to that body of air.* (Unless otherwise stated, the body of air is stationary relative to the ground, so the speeds can also be measured relative to the ground.) We shall assume that S and D move either directly toward or directly away from each other, at speeds less than the speed of sound.

If either the detector or the source is moving, or both are moving, the emitted frequency f and the detected frequency f' are related by

$$f' = f\frac{v \pm v_D}{v \pm v_S}\qquad\text{(general Doppler effect),}\qquad(17\text{-}47)$$

where v is the speed of sound through the air, v_D is the detector's speed relative to the air, and v_S is the source's speed relative to the air. The choice of plus or minus signs is set by this rule:

> When the motion of detector or source is toward the other, the sign on its speed must give an upward shift in frequency. When the motion of detector or source is away from the other, the sign on its speed must give a downward shift in frequency.

In short, *toward* means *shift up,* and *away* means *shift down.*

Here are some examples of the rule. If the detector moves toward the source, use the plus sign in the numerator of Eq. 17-47 to get a shift up in the frequency. If it moves away, use the minus sign in the numerator to get a shift down. If it is stationary, substitute 0 for v_D. If the source moves toward the detector, use the minus sign in the denominator of Eq. 17-47 to get a shift up in the frequency. If it moves away, use the plus sign in the denominator to get a shift down. If the source is stationary, substitute 0 for v_S.

Next, we derive equations for the Doppler effect for the following two specific situations and then derive Eq. 17-47 for the general situation.

1. When the detector moves relative to the air and the source is stationary relative to the air, the motion changes the frequency at which the detector intercepts wavefronts and thus changes the detected frequency of the sound wave.

2. When the source moves relative to the air and the detector is stationary relative to the air, the motion changes the wavelength of the sound wave and thus changes the detected frequency (recall that frequency is related to wavelength).

Detector Moving, Source Stationary

In Fig. 17-19, a detector D (represented by an ear) is moving at speed v_D toward a stationary source S that emits spherical wavefronts, of wavelength λ and frequency f, moving at the speed v of sound in air. The wavefronts are drawn one wavelength apart. The frequency detected by detector D is the rate at which D intercepts wavefronts (or individual wavelengths). If D were stationary, that rate

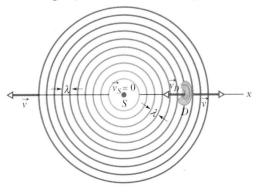

Fig. 17-19 A stationary source of sound S emits spherical wavefronts, shown one wavelength apart, that expand outward at speed v. A sound detector D, represented by an ear, moves with velocity $\vec{v}_D$ toward the source. The detector senses a higher frequency because of its motion.

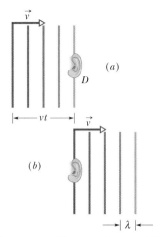

Fig. 17-20 The wavefronts of Fig. 17-19, assumed planar, (a) reach and (b) pass a stationary detector D; they move a distance vt to the right in time t.

would be f, but since D is moving into the wavefronts, the rate of interception is greater, and thus the detected frequency f' is greater than f.

Let us for the moment consider the situation in which D is stationary (Fig. 17-20). In time t, the wavefronts move to the right a distance vt. The number of wavelengths in that distance vt is the number of wavelengths intercepted by D in time t, and that number is vt/λ. The rate at which D intercepts wavelengths, which is the frequency f detected by D, is

$$f = \frac{vt/\lambda}{t} = \frac{v}{\lambda}. \qquad (17\text{-}48)$$

In this situation, with D stationary, there is no Doppler effect—the frequency detected by D is the frequency emitted by S.

Now let us again consider the situation in which D moves in the direction opposite the wavefront velocity (Fig. 17-21). In time t, the wavefronts move to the right a distance vt as previously, but now D moves to the left a distance $v_D t$. Thus, in this time t, the distance moved by the wavefronts relative to D is $vt + v_D t$. The number of wavelengths in this relative distance $vt + v_D t$ is the number of wavelengths intercepted by D in time t and is $(vt + v_D t)/\lambda$. The *rate* at which D intercepts wavelengths in this situation is the frequency f', given by

$$f' = \frac{(vt + v_D t)/\lambda}{t} = \frac{v + v_D}{\lambda}. \qquad (17\text{-}49)$$

From Eq. 17-48, we have $\lambda = v/f$. Then Eq. 17-49 becomes

$$f' = \frac{v + v_D}{v/f} = f\frac{v + v_D}{v}. \qquad (17\text{-}50)$$

Note that in Eq. 17-50, f' must be greater than f unless $v_D = 0$ (the detector is stationary).

Similarly, we can find the frequency detected by D if D moves away from the source. In this situation, the wavefronts move a distance $vt - v_D t$ relative to D in time t, and f' is given by

$$f' = f\frac{v - v_D}{v}. \qquad (17\text{-}51)$$

In Eq. 17-51, f' must be less than f unless $v_D = 0$.

We can summarize Eqs. 17-50 and 17-51 with

$$f' = f\frac{v \pm v_D}{v} \qquad \text{(detector moving, source stationary).} \qquad (17\text{-}52)$$

Source Moving, Detector Stationary

Let detector D be stationary with respect to the body of air, and let source S move toward D at speed v_S (Fig. 17-22). The motion of S changes the wavelength of the sound waves it emits and thus the frequency detected by D.

To see this change, let T $(= 1/f)$ be the time between the emission of any pair of successive wavefronts W_1 and W_2. During T, wavefront W_1 moves a distance vT and the source moves a distance $v_S T$. At the end of T, wavefront W_2 is emitted. In the direction in which S moves, the distance between W_1 and W_2, which is the wavelength λ' of the waves moving in that direction, is $vT - v_S T$. If D detects those waves, it detects frequency f' given by

$$f' = \frac{v}{\lambda'} = \frac{v}{vT - v_S T} = \frac{v}{v/f - v_S/f}$$
$$= f\frac{v}{v - v_S}. \qquad (17\text{-}53)$$

Note that f' must be greater than f unless $v_S = 0$.

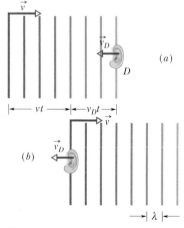

Fig. 17-21 Wavefronts traveling to the right (a) reach and (b) pass detector D, which moves in the opposite direction. In time t, the wavefronts move a distance vt to the right and D moves a distance $v_D t$ to the left.

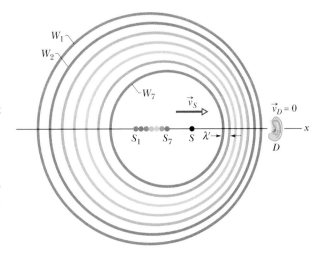

Fig. 17-22 A detector D is stationary, and a source S is moving toward it at speed v_S. Wavefront W_1 was emitted when the source was at S_1, wavefront W_7 when it was at S_7. At the moment depicted, the source is at S. The detector senses a higher frequency because the moving source, chasing its own wavefronts, emits a reduced wavelength λ' in the direction of its motion.

In the direction opposite that taken by S, the wavelength λ' of the waves is $vT + v_S T$. If D detects those waves, it detects frequency f' given by

$$f' = f \frac{v}{v + v_S}. \tag{17-54}$$

Now f' must be less than f unless $v_S = 0$.

We can summarize Eqs. 17-53 and 17-54 with

$$f' = f \frac{v}{v \pm v_S} \qquad \text{(source moving, detector stationary).} \tag{17-55}$$

General Doppler Effect Equation

We can now derive the general Doppler effect equation by replacing f in Eq. 17-55 (the source frequency) with f' of Eq. 17-52 (the frequency associated with motion of the detector). The result is Eq. 17-47 for the general Doppler effect.

That general equation holds not only when both detector and source are moving but also in the two specific situations we just discussed. For the situation in which the detector is moving and the source is stationary, substitution of $v_S = 0$ into Eq. 17-47 gives us Eq. 17-52, which we previously found. For the situation in which the source is moving and the detector is stationary, substitution of $v_D = 0$ into Eq. 17-47 gives us Eq. 17-55, which we previously found. Thus, Eq. 17-47 is the equation to remember.

✔CHECKPOINT 5   The figure indicates the directions of motion of a sound source and a detector for six situations in stationary air. For each situation, is the detected frequency greater than or less than the emitted frequency, or can't we tell without more information about the actual speeds?

	Source	Detector		Source	Detector
(a)	→	• 0 speed	(d)	←	←
(b)	←	• 0 speed	(e)	→	←
(c)	→	→	(f)	←	←

Sample Problem 17-8

A rocket moves at a speed of 242 m/s directly toward a stationary pole (through stationary air) while emitting sound waves at frequency $f = 1250$ Hz.

(a) What frequency f' is measured by a detector that is attached to the pole?

Solution: We can find f' with Eq. 17-47 for the general Doppler effect. The **Key Idea** here is that, because the sound source (the rocket) moves through the air *toward* the station-

ary detector on the pole, we need to choose the sign on v_S that gives a *shift up* in the frequency of the sound. Thus, in Eq. 17-47 we use the minus sign in the denominator. We then substitute 0 for the detector speed v_D, 242 m/s for the source speed v_S, 343 m/s for the speed of sound v (from Table 17-1), and 1250 Hz for the emitted frequency f. We find

$$f' = f \frac{v \pm v_D}{v \pm v_S} = (1250 \text{ Hz}) \frac{343 \text{ m/s} \pm 0}{343 \text{ m/s} - 242 \text{ m/s}}$$

$$= 4245 \text{ Hz} \approx 4250 \text{ Hz}, \qquad \text{(Answer)}$$

which, indeed, is a greater frequency than the emitted frequency.

(b) Some of the sound reaching the pole reflects back to the rocket as an echo. What frequency f'' does a detector on the rocket detect for the echo?

Solution: Two **Key Ideas** here are the following:

1. The pole is now the source of sound (because it is the source of the echo), and the rocket's detector is now the detector (because it detects the echo).
2. The frequency of the sound emitted by the source (the pole) is equal to f' from part (a), the frequency of the sound the pole intercepts and reflects.

We can rewrite Eq. 17-47 in terms of the source frequency f' and the detected frequency f'' as

$$f'' = f' \frac{v \pm v_D}{v \pm v_S}. \tag{17-56}$$

A third **Key Idea** here is that, because the detector (on the rocket) moves through the air *toward* the stationary source, we need to use the sign on v_D that gives a *shift up* in the frequency of the sound. Thus, we use the plus sign in the numerator of Eq. 17-56. Also, we substitute $v_D = 242$ m/s, $v_S = 0$, $v = 343$ m/s, and $f' = 4245$ Hz. We find

$$f'' = (4245 \text{ Hz}) \frac{343 \text{ m/s} + 242 \text{ m/s}}{343 \text{ m/s} \pm 0}$$
$$= 7240 \text{ Hz}, \qquad \text{(Answer)}$$

which, indeed, is greater than the frequency of the sound reflected by the pole.

✓**CHECKPOINT 6** If the air in this sample problem is moving toward the pole at 20 m/s, (a) what value for v_S should be used in the solution of part (a), and (b) what value for v_D should be used in the solution of part (b)?

17-10 Supersonic Speeds, Shock Waves

If a source is moving toward a stationary detector at a speed equal to the speed of sound — that is, if $v_S = v$ — Eqs. 17-47 and 17-55 predict that the detected frequency f' will be infinitely great. This means that the source is moving so fast that it keeps pace with its own spherical wavefronts, as Fig. 17-23a suggests. What happens when the speed of the source *exceeds* the speed of sound?

For such *supersonic* speeds, Eqs. 17-47 and 17-55 no longer apply. Figure 17-23b depicts the spherical wavefronts that originated at various positions of the source. The radius of any wavefront in this figure is vt, where v is the speed of sound and t is the time that has elapsed since the source emitted that wavefront. Note that all the wavefronts bunch along a V-shaped envelope in the two-dimensional drawing of Fig. 17-23b. The wavefronts actually extend in three dimensions, and the bunching actually forms a cone called the *Mach cone*. A *shock wave* is said to exist along the surface of this cone, because the bunching of wavefronts causes an abrupt rise and fall of air pressure as the surface passes through any point. From Fig. 17-23b, we see that the half-angle θ of the cone, called the *Mach cone angle*, is given by

$$\sin \theta = \frac{vt}{v_S t} = \frac{v}{v_S} \qquad \text{(Mach cone angle).} \tag{17-57}$$

The ratio v_S/v is called the *Mach number*. When you hear that a particular plane has flown at Mach 2.3, it means that its speed was 2.3 times the speed of

Fig. 17-23 (a) A source of sound S moves at speed v_S equal to the speed of sound and thus as fast as the wavefronts it generates. (b) A source S moves at speed v_S faster than the speed of sound and thus faster than the wavefronts. When the source was at position S_1 it generated wavefront W_1, and at position S_6 it generated W_6. All the spherical wavefronts expand at the speed of sound v and bunch along the surface of a cone called the Mach cone, forming a shock wave. The surface of the cone has half-angle θ and is tangent to all the wavefronts.

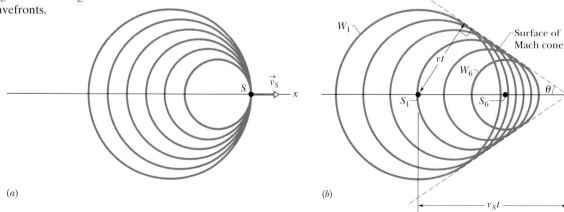

(a)

(b)

sound in the air through which the plane was flying. The shock wave generated by a supersonic aircraft (Fig. 17-24) or projectile produces a burst of sound, called a *sonic boom*, in which the air pressure first suddenly increases and then suddenly decreases below normal before returning to normal. Part of the sound that is heard when a rifle is fired is the sonic boom produced by the bullet. A sonic boom can also be heard from a long bullwhip when it is snapped quickly: Near the end of the whip's motion, its tip is moving faster than sound and produces a small sonic boom—the *crack* of the whip.

The photograph that opens Chapter 11 shows the jet-powered car *Thrust SSC* as it was traveling faster than the speed of sound in the Black Rock Desert of Nevada. The shock wave generated by the car sent a sonic boom across the desert as a sure signal that the car was supersonic. You can see evidence for the shock wave just above the car in the photograph. Our view of the background in that direction is distorted because light from the background encountered significant changes in air pressure and density as it passed into and out of the shock wave to reach the camera. These changes caused slight deflections in the light, distorting the background in the photograph taken by the camera.

Fig. 17-24 Shock waves produced by the wings of a Navy FA 18 jet. The shock waves are visible because the sudden decrease in air pressure in them caused water molecules in the air to condense, forming a fog.

Review & Summary

Sound Waves Sound waves are longitudinal mechanical waves that can travel through solids, liquids, or gases. The speed v of a sound wave in a medium having **bulk modulus** B and density ρ is

$$v = \sqrt{\frac{B}{\rho}} \quad \text{(speed of sound)}. \qquad (17\text{-}3)$$

In air at 20°C, the speed of sound is 343 m/s.

A sound wave causes a longitudinal displacement s of a mass element in a medium as given by

$$s = s_m \cos(kx - \omega t), \qquad (17\text{-}13)$$

where s_m is the **displacement amplitude** (maximum displacement) from equilibrium, $k = 2\pi/\lambda$, and $\omega = 2\pi f$, λ and f being the wavelength and frequency, respectively, of the sound wave. The sound wave also causes a pressure change Δp of the medium from the equilibrium pressure:

$$\Delta p = \Delta p_m \sin(kx - \omega t), \qquad (17\text{-}14)$$

where the **pressure amplitude** is

$$\Delta p_m = (v\rho\omega)s_m. \qquad (17\text{-}15)$$

Interference The interference of two sound waves with identical wavelengths passing through a common point depends on their phase difference ϕ there. If the sound waves were emitted in phase and are traveling in approximately the same direction, ϕ is given by

$$\phi = \frac{\Delta L}{\lambda} 2\pi, \qquad (17\text{-}21)$$

where ΔL is their **path length difference** (the difference in the distances traveled by the waves to reach the common point). Fully constructive interference occurs when ϕ is an integer multiple of 2π,

$$\phi = m(2\pi), \quad \text{for } m = 0, 1, 2, \ldots, \qquad (17\text{-}22)$$

and, equivalently, when ΔL is related to wavelength λ by

$$\frac{\Delta L}{\lambda} = 0, 1, 2, \ldots . \qquad (17\text{-}23)$$

Fully destructive interference occurs when ϕ is an odd multiple of π,

$$\phi = (2m + 1)\pi, \quad \text{for } m = 0, 1, 2, \ldots, \qquad (17\text{-}24)$$

and, equivalently, when ΔL is related to λ by

$$\frac{\Delta L}{\lambda} = 0.5, 1.5, 2.5, \ldots . \qquad (17\text{-}25)$$

Sound Intensity The **intensity** I of a sound wave at a surface is the average rate per unit area at which energy is transferred by the wave through or onto the surface:

$$I = \frac{P}{A}, \qquad (17\text{-}26)$$

where P is the time rate of energy transfer (power) of the sound wave and A is the area of the surface intercepting the sound. The intensity I is related to the displacement amplitude s_m of the sound wave by

$$I = \tfrac{1}{2}\rho v \omega^2 s_m^2. \qquad (17\text{-}27)$$

The intensity at a distance r from a point source that emits

sound waves of power P_s is

$$I = \frac{P_s}{4\pi r^2}. \tag{17-28}$$

Sound Level in Decibels   The *sound level* β in *decibels* (dB) is defined as

$$\beta = (10 \text{ dB}) \log \frac{I}{I_0}, \tag{17-29}$$

where $I_0 (= 10^{-12} \text{ W/m}^2)$ is a reference intensity level to which all intensities are compared. For every factor-of-10 increase in intensity, 10 dB is added to the sound level.

Standing Wave Patterns in Pipes   Standing sound wave patterns can be set up in pipes. A pipe open at both ends will resonate at frequencies

$$f = \frac{v}{\lambda} = \frac{nv}{2L}, \qquad n = 1, 2, 3, \ldots, \tag{17-39}$$

where v is the speed of sound in the air in the pipe. For a pipe closed at one end and open at the other, the resonant frequencies are

$$f = \frac{v}{\lambda} = \frac{nv}{4L}, \qquad n = 1, 3, 5, \ldots. \tag{17-41}$$

Beats   *Beats* arise when two waves having slightly different frequencies, f_1 and f_2, are detected together. The beat frequency is

$$f_{\text{beat}} = f_1 - f_2. \tag{17-46}$$

The Doppler Effect   The *Doppler effect* is a change in the observed frequency of a wave when the source or the detector moves relative to the transmitting medium (such as air). For sound the observed frequency f' is given in terms of the source frequency f by

$$f' = f \frac{v \pm v_D}{v \pm v_S} \qquad \text{(general Doppler effect)}, \tag{17-47}$$

where v_D is the speed of the detector relative to the medium, v_S is that of the source, and v is the speed of sound in the medium. The signs are chosen such that f' tends to be *greater* for motion toward and *less* for motion away.

Shock Wave   If the speed of a source relative to the medium exceeds the speed of sound in the medium, the Doppler equation no longer applies. In such a case, shock waves result. The half angle θ of the Mach cone is given by

$$\sin \theta = \frac{v}{v_S} \qquad \text{(Mach cone angle).} \tag{17-57}$$

Questions

1 In Fig. 17-25, two point sources S_1 and S_2, which are in phase, emit identical sound waves of wavelength 2.0 m. In terms of wavelengths, what is the phase difference between the waves arriving at point P if (a) $L_1 = 38$ m and $L_2 = 34$ m, and (b) $L_1 = 39$ m and $L_2 = 36$ m? (c) Assuming that the source separation is much smaller than L_1 and L_2, what type of interference occurs at P in situations (a) and (b)?

Fig. 17-25   Question 1.

2 In Fig. 17-26, three long tubes (A, B, and C) are filled with different gases under different pressures. The ratio of the bulk modulus to the density is indicated for each gas in terms of a basic value B_0/ρ_0. Each tube has a piston at its left end that can send a sound pulse through the tube (as in Fig. 16-2). The three pulses are sent simultaneously. Rank the tubes according to the time of arrival of the pulses at the open right ends of the tubes, earliest first.

Fig. 17-26   Question 2.

3 In a first experiment, a sinusoidal sound wave is sent through a long tube of air, transporting energy at the average rate of $P_{\text{avg},1}$. In a second experiment, two other sound waves, identical to the first one, are to be sent simultaneously through the tube with a phase difference ϕ of either 0, 0.2 wavelength, or 0.5 wavelength between the waves. (a) With only mental calculation, rank those choices of ϕ according to the average

rate at which the waves will transport energy, greatest first. (b) For the first choice of ϕ, what is the average rate in terms of $P_{\text{avg},1}$?

4 The sixth harmonic is set up in a pipe. (a) How many open ends does the pipe have (it has at least one)? (b) Is there a node, antinode, or some intermediate state at the midpoint?

5 For a particular tube, here are four of the six harmonic frequencies below 1000 Hz: 300, 600, 750, and 900 Hz. What two frequencies are missing from the list?

6 Pipe A has length L and one open end. Pipe B has length $2L$ and two open ends. Which harmonics of pipe B have a frequency that matches a resonant frequency of pipe A?

7 Figure 17-27 shows a stretched string of length L and pipes a, b, c, and d of lengths L, $2L$, $L/2$, and $L/2$, respectively. The string's tension is adjusted until the speed of waves on the string equals the speed of sound waves in the air. The fundamental mode of oscillation is then set up on the string. In which pipe will the sound produced by the string cause resonance, and what oscillation mode will that sound set up?

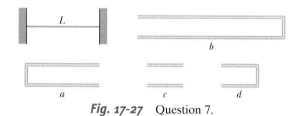

Fig. 17-27   Question 7.

8 In Fig. 17-28, pipe A is made to oscillate in its third harmonic by a small internal sound source. Sound emitted at the right end happens to resonate four nearby pipes, each with only one open end (they are *not* drawn to scale). Pipe *B* oscillates in its lowest harmonic, pipe *C* in its second lowest harmonic, pipe *D* in its third lowest harmonic, and pipe *E* in its fourth lowest harmonic. Without computation, rank all five pipes according to their length, greatest first. (*Hint:* Draw the standing waves to scale and then draw the pipes to scale.)

Fig. 17-28 Question 8.

9 Figure 17-29 shows a moving sound source *S* that emits at a certain frequency, and four stationary sound detectors. Rank the detectors according to the frequency of the sound they detect from the source, greatest first.

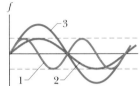

Fig. 17-29 Question 9.

10 A friend rides, in turn, the rims of three fast merry-go-rounds while holding a sound source that emits isotropically at a certain frequency. You stand far from each merry-go-round. The frequency you hear for each of your friend's three rides varies as the merry-go-round rotates. The variations in frequency for the three rides are given by the three curves in Fig. 17-30. Rank the curves according to (a) the linear speed *v* of the sound source, (b) the angular speeds *ω* of the merry-go-rounds, and (c) the radii *r* of the merry-go-rounds, greatest first.

Fig. 17-30 Question 10.

Problems

Where needed in the problems, use

$$\text{speed of sound in air} = 343 \text{ m/s}$$

and $\qquad$ density of air = 1.21 kg/m³

unless otherwise specified.

sec. 17-3 The Speed of Sound

•1 A column of soldiers, marching at 120 paces per minute, keep in step with the beat of a drummer at the head of the column. It is observed that the soldiers in the rear end of the column are striding forward with the left foot when the drummer is advancing with the right foot. What is the approximate length of the column?

•2 When the door of the Chapel of the Mausoleum in Hamilton, Scotland, is slammed shut, the last echo heard by someone standing just inside the door reportedly comes 15 s later. (a) If that echo were due to a single reflection off a wall opposite the door, how far from the door would that wall be? (b) If, instead, the wall is 25.7 m away, how many reflections (back and forth) correspond to the last echo?

•3 Two spectators at a soccer game in Montjuic Stadium see, and a moment later hear, the ball being kicked on the playing field. The time delay for spectator *A* is 0.23 s, and for spectator *B* it is 0.12 s. Sight lines from the two spectators to the player kicking the ball meet at an angle of 90°. How far are (a) spectator *A* and (b) spectator *B* from the player? (c) How far are the spectators from each other? SSM

•4 What is the bulk modulus of oxygen if 32.0 g of oxygen occupies 22.4 L and the speed of sound in the oxygen is 317 m/s?

••5 A stone is dropped into a well. The splash is heard 3.00 s later. What is the depth of the well? SSM WWW

••6 A man strikes one end of a thin rod with a hammer. The speed of sound in the rod is 15 times the speed of sound in air. A woman, at the other end with her ear close to the rod, hears the sound of the blow twice with a 0.12 s interval between; one sound comes through the rod and the other comes through the air alongside the rod. If the speed of sound in air is 343 m/s, what is the length of the rod?

••7 Earthquakes generate sound waves inside Earth. Unlike a gas, Earth can experience both transverse (S) and longitudinal (P) sound waves. Typically, the speed of S waves is about 4.5 km/s, and that of P waves 8.0 km/s. A seismograph records P and S waves from an earthquake. The first P waves arrive 3.0 min before the first S waves. If the waves travel in a straight line, how far away does the earthquake occur? SSM ILW

sec. 17-4 Traveling Sound Waves

•8 The pressure in a traveling sound wave is given by the equation

$$\Delta p = (1.50 \text{ Pa}) \sin \pi[(0.900 \text{ m}^{-1})x - (315 \text{ s}^{-1})t].$$

Find the (a) pressure amplitude, (b) frequency, (c) wavelength, and (d) speed of the wave.

•9 Diagnostic ultrasound of frequency 4.50 MHz is used to examine tumors in soft tissue. (a) What is the wavelength in air of such a sound wave? (b) If the speed of sound in tissue is 1500 m/s, what is the wavelength of this wave in tissue? SSM

•10 If the form of a sound wave traveling through air is

$$s(x, t) = (6.0 \text{ nm}) \cos(kx + (3000 \text{ rad/s})t + \phi),$$

how much time does any given air molecule along the path take to move between displacements *s* = +2.0 nm and *s* = −2.0 nm?

••11 A hand-clap on stage in an amphitheater sends out sound waves that scatter from terraces of width *w* = 0.75 m

(Fig. 17-31). The sound returns to the stage as a periodic series of pulses, one from each terrace; the parade of pulses sounds like a played note. (a) Assuming that all the rays in Fig. 17-31 are horizontal, find the frequency at which the pulses return (that is, the frequency of the perceived note). (b) If the width w of the terraces were smaller, would the frequency be higher or lower? **SSM**

Fig. 17-31   Problem 11.

••12   A sound wave of the form $s = s_m \cos(kx - \omega t + \phi)$ travels at 343 m/s through air in a long horizontal tube. At one instant, air molecule A at $x = 2.000$ m is at its maximum positive displacement of 6.00 nm and air molecule B at $x = 2.070$ m is at a positive displacement of 2.00 nm. All the molecules between A and B are at intermediate displacements. What is the frequency of the wave?

••13   Figure 17-32 shows the output from a pressure monitor mounted at a point along the path taken by a sound wave of a single frequency traveling at 343 m/s through air with a uniform density of 1.21 kg/m³. If the *displacement* function of the wave is written as $s(x, t) = s_m \cos(kx - \omega t)$, what are (a) s_m, (b) k, and (c) ω? The air is then cooled so that its density is 1.35 kg/m³ and the speed of a sound wave through it is 320 m/s. The sound source again emits the sound wave at the same frequency and same pressure amplitude. What now are (d) s_m, (e) k, and (f) ω?

Fig. 17-32   Problem 13.

sec. 17-5   Interference

•14   Two sound waves, from two different sources with the same frequency, 540 Hz, travel in the same direction at 330 m/s. The sources are in phase. What is the phase difference of the waves at a point that is 4.40 m from one source and 4.00 m from the other?

••15   Figure 17-33 shows two isotropic point sources of sound, S_1 and S_2. The sources emit waves in phase at wavelength 0.50 m; they are separated by $D = 1.75$ m. If we move a sound detector along a large circle centered at the midpoint between the sources, at how many points do waves arrive at the detector (a) exactly in phase and (b) exactly out of phase?

Fig. 17-33   Problems 15 and 100.

••16   In Fig. 17-34, sound with a 40.0 cm wavelength travels rightward from a source and through a tube that consists of a straight portion and a half-circle. Part of the sound wave travels through the half-circle and then rejoins the rest of the wave, which goes directly through the straight portion. This rejoining results in interference. What is the smallest radius r that results in an intensity minimum at the detector?

Fig. 17-34   Problem 16.

••17   In Fig. 17-35, two speakers separated by distance $d_1 = 2.00$ m are in phase. Assume the amplitudes of the sound waves from the speakers are approximately the same at the listener's ear at distance $d_2 = 3.75$ m directly in front of one speaker. Consider the full audible range for normal hearing, 20 Hz to 20 kHz. (a) What is the lowest frequency $f_{min,1}$ that gives minimum signal (destructive interference) at the listener's ear? By what number must $f_{min,1}$ be multiplied to get (b) the second lowest frequency $f_{min,2}$ that gives minimum signal and (c) the third lowest frequency $f_{min,3}$ that gives minimum signal? (d) What is the lowest frequency $f_{max,1}$ that gives maximum signal (constructive interference) at the listener's ear? By what number must $f_{max,1}$ be multiplied to get (e) the second lowest frequency $f_{max,2}$ that gives maximum signal and (f) the third lowest frequency $f_{max,3}$ that gives maximum signal? **SSM**

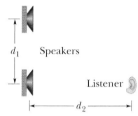

Fig. 17-35   Problem 17.

••18   Figure 17-36 shows four isotropic point sources of sound that are uniformly spaced on an x axis. The sources emit sound at the same wavelength λ and same amplitude s_m, and they emit in phase. A point P is shown on the x axis. Assume that as the sound waves travel to P, the decrease in their amplitude is negligible. What multiple of s_m is the amplitude of the net wave at P if distance d in the figure is (a) $\lambda/4$, (b) $\lambda/2$, and (c) λ?

Fig. 17-36   Problem 18.

••19   Two loudspeakers are located 3.35 m apart on an outdoor stage. A listener is 18.3 m from one and 19.5 m from the other. During the sound check, a signal generator drives the two speakers in phase with the same amplitude and frequency. The transmitted frequency is swept through the audible range (20 Hz to 20 kHz). (a) What is the lowest frequency $f_{min,1}$ that gives minimum signal (destructive interference) at the listener's location? By what number must $f_{min,1}$ be multiplied to get (b) the second lowest frequency $f_{min,2}$ that gives minimum signal and (c) the third lowest frequency $f_{min,3}$ that gives minimum signal? (d) What is the lowest frequency $f_{max,1}$ that gives maximum signal (constructive interference) at the listener's location? By what number must $f_{max,1}$ be multiplied to get (e) the second lowest frequency $f_{max,2}$ that gives maximum signal and (f) the third lowest frequency $f_{max,3}$ that gives maximum signal? **ILW**

•••20   Figure 17-37 shows two point sources S_1 and S_2 that emit sound of wavelength $\lambda = 2.00$ m. The emissions are isotropic and in phase, and the separation between the sources is $d = 16.0$ m. At any point P on the x axis, the wave from S_1 and the wave from S_2 interfere. When P is very far away ($x \approx \infty$), what are (a) the phase difference between the arriving waves from S_1 and S_2 and (b) the type of interference they produce? Now move point P along the x axis toward

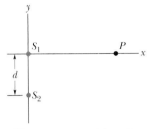

Fig. 17-37 Problem 20.

S_1. (c) Does the phase difference between the waves increase or decrease? At what distance x do the waves have a phase difference of (d) 0.50λ, (e) 1.00λ, and (f) 1.50λ?

sec. 17-6 Intensity and Sound Level

•21 A source emits sound waves isotropically. The intensity of the waves 2.50 m from the source is 1.91×10^{-4} W/m². Assuming that the energy of the waves is conserved, find the power of the source. **SSM**

•22 A 1.0 W point source emits sound waves isotropically. Assuming that the energy of the waves is conserved, find the intensity (a) 1.0 m from the source and (b) 2.5 m from the source.

•23 A sound wave of frequency 300 Hz has an intensity of 1.00 μW/m². What is the amplitude of the air oscillations caused by this wave?

•24 Two sounds differ in sound level by 1.00 dB. What is the ratio of the greater intensity to the smaller intensity?

•25 A certain sound source is increased in sound level by 30.0 dB. By what multiple is (a) its intensity increased and (b) its pressure amplitude increased? **SSM WWW**

•26 The source of a sound wave has a power of 1.00 μW. If it is a point source, (a) what is the intensity 3.00 m away and (b) what is the sound level in decibels at that distance?

••27 Suppose that the sound level of a conversation is initially at an angry 70 dB and then drops to a soothing 50 dB (Fig. 17-38). Assuming that the frequency of the sound is 500 Hz, determine the (a) initial and (b) final sound intensities and the (c) initial and (d) final sound wave amplitudes.

Fig. 17-38 Problem 27.

••28 Approximately a third of people with normal hearing have ears that continuously emit a low-intensity sound outward through the ear canal. A person with such *spontaneous otoacoustic emission* is rarely aware of the sound, except perhaps in a noise-free environment, but occasionally the emission is loud enough to be heard by someone else nearby. In one observation, the sound wave had a frequency of 1665 Hz and a pressure amplitude of 1.13×10^{-3} Pa. What were (a) the displacement amplitude and (b) the intensity of the wave emitted by the ear?

••29 Two atmospheric sound sources A and B emit isotropically at constant power. The sound levels β of their emissions are plotted in Fig. 17-39 versus the radial distance r from the sources. What are (a) the difference in their sound levels at

$r = 10$ m and (b) the ratio of the larger power to the smaller power?

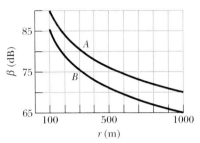

Fig. 17-39 Problem 29.

••30 A point source emits 30.0 W of sound isotropically. A small microphone intercepts the sound in an area of 0.750 cm², 200 m from the source. Calculate (a) the sound intensity there and (b) the power intercepted by the microphone.

•••31 A sound source sends a sinusoidal sound wave of angular frequency 3000 rad/s and amplitude 12.0 nm through a tube of air. The internal radius of the tube is 2.00 cm. (a) What is the average rate at which energy (the sum of the kinetic and potential energies) is transported to the opposite end of the tube? (b) If, simultaneously, an identical wave travels along an adjacent, identical tube, what is the total average rate at which energy is transported to the opposite ends of the two tubes by the waves? If, instead, those two waves are sent along the *same* tube simultaneously, what is the total average rate at which they transport energy when their phase difference is (c) 0, (d) 0.40π rad, and (e) π rad?

sec. 17-7 Sources of Musical Sound

•32 The crest of a *Parasaurolophus* dinosaur skull contains a nasal passage in the shape of a long, bent tube open at both ends. The dinosaur may have used the passage to produce sound by setting up the fundamental mode in it. (a) If the nasal passage in a certain *Parasaurolophus* fossil is 2.0 m long, what frequency would have been produced? (b) If that dinosaur could be recreated (as in *Jurassic Park*), would a person with a hearing range of 60 Hz to 20 kHz be able to hear that fundamental mode and, if so, would the sound be high or low frequency? Fossil skulls that contain shorter nasal passages are thought to be those of the female *Parasaurolophus*. (c) Would that make the female's fundamental frequency higher or lower than the male's?

•33 In pipe A, the ratio of a particular harmonic frequency to the next lower harmonic frequency is 1.2. In pipe B, the ratio of a particular harmonic frequency to the next lower harmonic frequency is 1.4. How many open ends are in (a) pipe A and (b) pipe B?

•34 A sound wave in a fluid medium is reflected at a barrier so that a standing wave is formed. The distance between nodes is 3.8 cm, and the speed of propagation is 1500 m/s. Find the frequency of the sound wave.

•35 A violin string 15.0 cm long and fixed at both ends oscillates in its $n = 1$ mode. The speed of waves on the string is 250 m/s, and the speed of sound in air is 348 m/s. What are the (a) frequency and (b) wavelength of the emitted sound wave?

•36 Organ pipe A, with both ends open, has a fundamental frequency of 300 Hz. The third harmonic of organ pipe B, with one end open, has the same frequency as the second harmonic of pipe A. How long are (a) pipe A and (b) pipe B?

•37 (a) Find the speed of waves on a violin string of mass 800 mg and length 22.0 cm if the fundamental frequency is 920 Hz. (b) What is the tension in the string? For the fundamental, what is the wavelength of (c) the waves on the string and (d) the sound waves emitted by the string? **SSM ILW**

•38 The water level in a vertical glass tube 1.00 m long can be adjusted to any position in the tube. A tuning fork vibrating at 686 Hz is held just over the open top end of the tube, to set up a standing wave of sound in the air-filled top portion of the tube. (That air-filled top portion acts as a tube with one end closed and the other end open.) (a) For how many different positions of the water level will sound from the fork set up resonance in the tube's air-filled portion, which acts as a pipe with one end closed (by the water) and the other end open? What are the (b) least and (c) second least water heights in the tube for resonance to occur?

•39 In Fig. 17-40, S is a small loudspeaker driven by an audio oscillator with a frequency that is varied from 1000 Hz to 2000 Hz, and D is a cylindrical pipe with two open ends and a length of 45.7 cm. The speed of sound in the air-filled pipe is 344 m/s. (a) At how many frequencies does the sound from the loudspeaker set up resonance in the pipe? What are the (b) lowest and (c) second lowest frequencies at which resonance occurs? **SSM**

Fig. 17-40 Problem 39.

••40 One of the harmonic frequencies of tube A with two open ends is 325 Hz. The next-highest harmonic frequency is 390 Hz. (a) What harmonic frequency is next highest after the harmonic frequency 195 Hz? (b) What is the number of this next-highest harmonic?

One of the harmonic frequencies of tube B with only one open end is 1080 Hz. The next-highest harmonic frequency is 1320 Hz. (c) What harmonic frequency is next highest after the harmonic frequency 600 Hz? (d) What is the number of this next-highest harmonic?

••41 A well with vertical sides and water at the bottom resonates at 7.00 Hz and at no lower frequency. (The air-filled portion of the well acts as a tube with one closed end and one open end.) The air in the well has a density of 1.10 kg/m^3 and a bulk modulus of 1.33×10^5 Pa. How far down in the well is the water surface?

••42 A tube 1.20 m long is closed at one end. A stretched wire is placed near the open end. The wire is 0.330 m long and has a mass of 9.60 g. It is fixed at both ends and oscillates in its fundamental mode. By resonance, it sets the air column in the tube into oscillation at that column's fundamental frequency. Find (a) that frequency and (b) the tension in the wire.

••43 A violin string 30.0 cm long with linear density 0.650 g/m is placed near a loudspeaker that is fed by an audio oscillator of variable frequency. It is found that the string is set into oscillation only at the frequencies 880 and 1320 Hz as the frequency of the oscillator is varied over the range 500–1500 Hz. What is the tension in the string? **SSM**

••44 Pipe A, which is 1.20 m long and open at both ends, oscillates at its third lowest harmonic frequency. It is filled with air for which the speed of sound is 343 m/s. Pipe B, which is closed at one end, oscillates at its second lowest harmonic frequency. This frequency of B happens to match the frequency of A. An x axis extends along the interior of A, with x = 0 at one end. (a) How many nodes are along that axis? What are the (b) smallest and (c) second smallest value of x locating those nodes?

sec. 17-8 Beats

•45 A tuning fork of unknown frequency makes 3.00 beats per second with a standard fork of frequency 384 Hz. The beat frequency decreases when a small piece of wax is put on a prong of the first fork. What is the frequency of this fork?

•46 The A string of a violin is a little too tightly stretched. Beats at 4.00 per second are heard when the string is sounded together with a tuning fork that is oscillating accurately at concert A (440 Hz). What is the period of the violin string oscillation?

••47 Two identical piano wires have a fundamental frequency of 600 Hz when kept under the same tension. What fractional increase in the tension of one wire will lead to the occurrence of 6.0 beats/s when both wires oscillate simultaneously? **SSM**

••48 You have five tuning forks that oscillate at close but different frequencies. What are the (a) maximum and (b) minimum number of different beat frequencies you can produce by sounding the forks two at a time, depending on how the frequencies differ?

sec. 17-9 The Doppler Effect

•49 A whistle of frequency 540 Hz moves in a circle of radius 60.0 cm at an angular speed of 15.0 rad/s. What are the (a) lowest and (b) highest frequencies heard by a listener a long distance away, at rest with respect to the center of the circle? **ILW**

•50 A state trooper chases a speeder along a straight stretch of road; both vehicles move at 160 km/h. The siren on the trooper's vehicle produces sound at a frequency of 500 Hz. What is the Doppler shift in the frequency heard by the speeder?

•51 An ambulance with a siren emitting a whine at 1600 Hz overtakes and passes a cyclist pedaling a bike at 2.44 m/s. After being passed, the cyclist hears a frequency of 1590 Hz. How fast is the ambulance moving?

••52 A stationary motion detector sends sound waves of frequency 0.150 MHz toward a truck approaching at a speed of 45.0 m/s. What is the frequency of the waves reflected back to the detector?

••53 An acoustic burglar alarm consists of a source emitting waves of frequency 28.0 kHz. What is the beat frequency between the source waves and the waves reflected from an intruder walking at an average speed of 0.950 m/s directly away from the alarm? **ILW**

••54 In Fig. 17-41, a French submarine and a U.S. submarine move toward each other during maneuvers in motionless water in the North Atlantic. The French sub moves at speed v_F = 50.0 km/h, and the U.S. sub at v_{US} = 70.0 km/h. The

French sub sends out a sonar signal (sound wave in water) at 1000 Hz. Sonar waves travel at 5470 km/h. (a) What is the signal's frequency as detected by the U.S. sub? (b) What frequency is detected by the French sub in the signal reflected back to it by the U.S. sub?

Fig. 17-41   Problem 54.

••**55** A sound source A and a reflecting surface B move directly toward each other. Relative to the air, the speed of source A is 29.9 m/s, the speed of surface B is 65.8 m/s, and the speed of sound is 329 m/s. The source emits waves at frequency 1200 Hz as measured in the source frame. In the reflector frame, what are the (a) frequency and (b) wavelength of the arriving sound waves? In the source frame, what are the (c) frequency and (d) wavelength of the sound waves reflected back to the source?

••**56** A stationary detector measures the frequency of a sound source that first moves at constant velocity directly toward the detector and then (after passing the detector) directly away from it. The emitted frequency is f. During the approach the detected frequency is f'_{app} and during the recession it is f'_{rec}. If $(f'_{app} - f'_{rec})/f = 0.500$, what is the ratio v_s/v of the speed of the source to the speed of sound?

••**57** A bat is flitting about in a cave, navigating via ultrasonic bleeps. Assume that the sound emission frequency of the bat is 39 000 Hz. During one fast swoop directly toward a flat wall surface, the bat is moving at 0.025 times the speed of sound in air. What frequency does the bat hear reflected off the wall?

••**58** Figure 17-42 shows four tubes with lengths 1.0 m or 2.0 m, with one or two open ends as drawn. The third harmonic is set up in each tube, and some of the sound that escapes from them is detected by detector D, which moves directly away from the tubes. In terms of the speed of sound v, what speed must the detector have such that the detected frequency of the sound from (a) tube 1, (b) tube 2, (c) tube 3, and (d) tube 4 is equal to the tube's fundamental frequency?

Fig. 17-42   Problem 58.

•••**59** A girl is sitting near the open window of a train that is moving at a velocity of 10.00 m/s to the east. The girl's uncle stands near the tracks and watches the train move away. The locomotive whistle emits sound at frequency 500.0 Hz. The air is still. (a) What frequency does the uncle hear? (b) What frequency does the girl hear? A wind begins to blow from the east at 10.00 m/s. (c) What frequency does the uncle now hear? (d) What frequency does the girl now hear? **SSM WWW**

•••**60** A 2000 Hz siren and a civil defense official are both at rest with respect to the ground. What frequency does the official hear if the wind is blowing at 12 m/s (a) from source to official and (b) from official to source?

•••**61** Two trains are traveling toward each other at 30.5 m/s relative to the ground. One train is blowing a whistle at 500 Hz. (a) What frequency is heard on the other train in still air? (b) What frequency is heard on the other train if the wind is blowing at 30.5 m/s toward the whistle and away from the listener? (c) What frequency is heard if the wind direction is reversed?

sec. 17-10   **Supersonic Speeds, Shock Waves**

•**62** The shock wave off the cockpit of the FA 18 in Fig. 17-24 has an angle of about 60°. The airplane was traveling at about 1350 km/h when the photograph was taken. Approximately what was the speed of sound at the airplane's altitude?

••**63** A jet plane passes over you at a height of 5000 m and a speed of Mach 1.5. (a) Find the Mach cone angle. (b) How long after the jet passes directly overhead does the shock wave reach you? Use 331 m/s for the speed of sound. **SSM**

••**64** A plane flies at 1.25 times the speed of sound. Its sonic boom reaches a man on the ground 1.00 min after the plane passes directly overhead. What is the altitude of the plane? Assume the speed of sound to be 330 m/s.

Additional Problems

65 A detector initially moves at constant velocity directly toward a stationary sound source and then (after passing it) directly from it. The emitted frequency is f. During the approach the detected frequency is f'_{app} and during the recession it is f'_{rec}. If the frequencies are related by $(f'_{app} - f'_{rec})/f = 0.500$, what is the ratio v_D/v of the speed of the detector to the speed of sound?

66 A sinusoidal sound wave moves at 343 m/s through air in the positive direction of an x axis. At one instant, air molecule A is at its maximum displacement in the negative direction of the axis while air molecule B is at its equilibrium position. The separation between those molecules is 15.0 cm, and the molecules between A and B have intermediate displacements in the negative direction of the axis. (a) What is the frequency of the sound wave?

In a similar arrangement, for a different sinusoidal sound wave, air molecule C is at its maximum displacement in the positive direction while molecule D is at its maximum displacement in the negative direction. The separation between the molecules is again 15.0 cm, and the molecules between C and D have intermediate displacements. (b) What is the frequency of the sound wave?

67 Straight line AB connects two point sources that are 5.00 m apart, emit 300 Hz sound waves of the same amplitude, and emit exactly out of phase. (a) What is the shortest distance between the midpoint of AB and a point on AB where the interfering waves cause maximum oscillation of the air molecules? What are the (b) second and (c) third shortest distances?

68 Two sound waves with an amplitude of 12 nm and a wavelength of 35 cm travel in the same direction through a long tube, with a phase difference of $\pi/3$ rad. What are the (a) amplitude and (b) wavelength of the net sound wave produced by their interference? If, instead, the sound waves travel through the tube in opposite directions, what are the (c) amplitude and (d) wavelength of the net wave?

69 Four sound waves are to be sent through the same tube of air, in the same direction:

$$s_1(x, t) = (9.00 \text{ nm}) \cos(2\pi x - 700\pi t)$$

$$s_2(x, t) = (9.00 \text{ nm}) \cos(2\pi x - 700\pi t + 0.7\pi)$$

$$s_3(x, t) = (9.00 \text{ nm}) \cos(2\pi x - 700\pi t + \pi)$$

$$s_4(x, t) = (9.00 \text{ nm}) \cos(2\pi x - 700\pi t + 1.7\pi).$$

What is the amplitude of the resultant wave? (*Hint:* Use a phasor diagram to simplify the problem.)

70 A sound source moves along an x axis, between detectors A and B. The wavelength of the sound detected at A is 0.500 that of the sound detected at B. What is the ratio v_s/v of the speed of the source to the speed of sound?

71 Pipe A has only one open end; pipe B is four times as long and has two open ends. Of the lowest 10 harmonic numbers n_B of pipe B, what are the (a) smallest, (b) second smallest, and (c) third smallest values at which a harmonic frequency of B matches one of the harmonic frequencies of A?

72 In Fig. 17-43, sound of wavelength 0.850 m is emitted isotropically by point source S. Sound ray 1 extends directly to detector D, at distance $L = 10.0$ m. Sound ray 2 extends to D via a reflection (effectively, a "bouncing") of the sound at a flat surface. That reflection occurs on a perpendicular bisector to the SD line, at distance d from the line. Assume that the reflection shifts the sound wave by 0.500λ. For what least value of d (other than zero) do the direct sound and the reflected sound arrive at D (a) exactly out of phase and (b) exactly in phase?

Fig. 17-43   Problem 72.

73 In Fig. 17-44, sound waves A and B, both of wavelength λ, are initially in phase and traveling rightward, as indicated by the two rays. Wave A is reflected from four surfaces but ends up traveling in its original direction. What multiple of wavelength λ is the smallest value of distance L in the figure that puts A and B exactly out of phase with each other after the reflections?

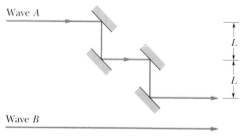

Fig. 17-44   Problem 73.

74 In Fig. 17-45, sound waves A and B, both of wavelength λ, are initially in phase and traveling rightward, as indicated by the two rays. Wave A is reflected from four surfaces but ends up traveling in its original direction. Wave B ends in that direction after reflecting from two surfaces. Let distance L in the figure be expressed as a multiple q of λ: $L = q\lambda$. What are the (a) smallest and (b) second smallest values of q that put A and B exactly out of phase with each other after the reflections?

Fig. 17-45   Problem 74.

75 The speed of sound in a certain metal is v_m. One end of a long pipe of that metal of length L is struck a hard blow. A listener at the other end hears two sounds, one from the wave that travels along the pipe's metal wall and the other from the wave that travels through the air inside the pipe. (a) If v is the speed of sound in air, what is the time interval t between the arrivals of the two sounds at the listener's ear? (b) If $t = 1.00$ s and the metal is steel, what is the length L?

76 A pipe 0.60 m long and closed at one end is filled with an unknown gas. The third lowest harmonic frequency for the pipe is 750 Hz. (a) What is the speed of sound in the unknown gas? (b) What is the fundamental frequency for this pipe when it is filled with the unknown gas?

77 A siren emitting a sound of frequency 1000 Hz moves away from you toward the face of a cliff at a speed of 10 m/s. Take the speed of sound in air as 330 m/s. (a) What is the frequency of the sound you hear coming directly from the siren? (b) What is the frequency of the sound you hear reflected off the cliff? (c) What is the beat frequency between the two sounds? Is it perceptible (it must be less than 20 Hz to be perceived)?

78 A trumpet player on a moving railroad flatcar and a second trumpet player standing alongside the track both play a 440 Hz note. The sound waves heard by a stationary observer between the two players have a beat frequency of 4.0 beats/s. What is the flatcar's speed?

79 *Waterfall acoustics.* The turbulent impact of the water in a waterfall causes the surrounding ground to oscillate in a wide range of low frequencies. If the water falls freely (instead of hitting rock on the way down), the oscillations are greatest in amplitude at a particular frequency f_m. This fact suggests that acoustic resonance is involved and f_m is the fundamental frequency. The following table gives, for nine U.S. and Canadian waterfalls, measured values for f_m and for the length L of the water's free fall. Determine how to plot the data to get the speed of sound in the water of a waterfall. From the plot, find the speed of sound if waterfall resonance is effectively like that in a tube with (a) two open ends and (b) only one open end. The speed of sound in turbulent water filled with air bubbles can be about 25% less than the speed of 1400 m/s in still water. (c) From the answers to (a) and (b), determine how many open ends are effectively involved in waterfall resonance?

WATERFALL	1	2	3	4	5	6	7	8	9
f_m (Hz)	5.6	3.8	8.0	6.1	8.8	6.0	19	21	40
L (m)	97	71	53	49	35	24	13	11	8

80 A certain loudspeaker system emits sound isotropically with a frequency of 2000 Hz and an intensity of 0.960 mW/m² at a distance of 6.10 m. Assume that there are no reflections. (a) What is the intensity at 30.0 m? At 6.10 m, what are (b) the displacement amplitude and (c) the pressure amplitude of the sound?

81 At a certain point, two waves produce pressure variations given by $\Delta p_1 = \Delta p_m \sin \omega t$ and $\Delta p_2 = \Delta p_m \sin(\omega t - \phi)$. At this point, what is the ratio $\Delta p_r/\Delta p_m$, where Δp_r is the pressure amplitude of the resultant wave, if ϕ is (a) 0, (b) $\pi/2$, (c) $\pi/3$, and (d) $\pi/4$?

82 The average density of Earth's crust 10 km beneath the continents is 2.7 g/cm³. The speed of longitudinal seismic waves at that depth, found by timing their arrival from distant earthquakes, is 5.4 km/s. Use this information to find the bulk modulus of Earth's crust at that depth. For comparison, the bulk modulus of steel is about 16×10^{10} Pa.

83 The sound intensity is 0.0080 W/m² at a distance of 10 m from an isotropic point source of sound. (a) What is the power of the source? (b) What is the sound intensity 5.0 m from the source? (c) What is the sound level 10 m from the source?

84 A person on a railroad car blows a trumpet note at 440 Hz. The car is moving toward a wall at 20.0 m/s. Calculate the frequency of (a) the sound as received at the wall and (b) the reflected sound arriving back at the trumpeter.

85 Two identical tuning forks can oscillate at 440 Hz. A person is located somewhere on the line between them. Calculate the beat frequency as measured by this individual if (a) she is standing still and the tuning forks move in the same direction along the line at 3.00 m/s, and (b) the tuning forks are stationary and the listener moves along the line at 3.00 m/s.

86 An avalanche of sand along some rare desert sand dunes can produce a booming that is loud enough to be heard 10 km away. The booming apparently results from a periodic oscillation of the sliding layer of sand—the layer's thickness expands and contracts. If the emitted frequency is 90 Hz, what are (a) the period of the thickness oscillation and (b) the wavelength of the sound?

87 Male *Rana catesbeiana* bullfrogs are known for their loud mating call. The call is emitted not by the frog's mouth but by its eardrums, which lie on the surface of the head. If the emitted sound has a frequency of 260 Hz and a sound level of 85 dB (near the eardrum), what is the amplitude of the eardrum's oscillation? The air density is 1.21 kg/m³.

88 When you "crack" a knuckle, you suddenly widen the knuckle cavity, allowing more volume for the synovial fluid inside it and causing a gas bubble suddenly to appear in the fluid. The sudden production of the bubble, called "cavitation," produces a sound pulse—the cracking sound. Assume that the sound is transmitted uniformly in all directions and that it fully passes from the knuckle interior to the outside. If the pulse has a sound level of 62 dB at your ear, estimate the rate at which energy is produced by the cavitation.

89 A police car is chasing a speeding Porsche 911. Assume that the Porsche's maximum speed is 80.0 m/s and the police car's is 54.0 m/s. At the moment both cars reach their maximum speed, what frequency will the Porsche driver hear if the frequency of the police car's siren is 440 Hz? Take the speed of sound in air to be 340 m/s.

90 Figure 17-46a shows two point sound sources S_1 and S_2, located on a line. The sources emit sound isotropically and at the same wavelength λ and same amplitude. Detection point P_1 is on a perpendicular bisector to the line between the sources; waves arrive there from the two sources with zero phase difference. Detection point P_2 is on the line through the sources; waves arrive there with a phase difference of 5.0 wavelengths. (a) In terms of λ, what is the distance between the sources? (b) What type of interference occurs at P_1?

The sources are now both moved directly away from P_1 by distance $\lambda/2$, at an angle $\theta = 30°$ (Fig. 17-46b). (c) In terms of λ, what is the phase difference between the waves arriving at point P_3, which is on the new line through the sources? (d) What type of interference occurs at P_3?

Fig. 17-46 Problem 90.

91 You can estimate your distance from a lightning stroke by counting the seconds between the flash you see and the thunder you later hear. By what integer should you divide the number of seconds to get the distance in kilometers?

92 A continuous sinusoidal longitudinal wave is sent along a very long coiled spring from an attached oscillating source. The wave travels in the negative direction of an x axis; the source frequency is 25 Hz; at any instant the distance between successive points of maximum expansion in the spring is 24 cm; the maximum longitudinal displacement of a spring particle is 0.30 cm; and the particle at $x = 0$ has zero displacement at time $t = 0$. If the wave is written in the form $s(x, t) = s_m \cos (kx \pm \omega t)$, what are (a) s_m, (b) k, (c) ω, (d) the wave speed, and (e) the correct choice of sign in front of ω?

93 (a) If two sound waves, one in air and one in (fresh) water, are equal in intensity, what is the ratio of the pressure amplitude of the wave in water to that of the wave in air? Assume the water and the air are at 20°C. (See Table 14-1.) (b) If the pressure amplitudes are equal instead, what is the ratio of the intensities of the waves?

94 A sound wave travels out uniformly in all directions from a point source. (a) Justify the following expression for the displacement s of the transmitting medium at any distance r from the source:

$$s = \frac{b}{r} \sin k(r - vt),$$

where b is a constant. Consider the speed, direction of propagation, periodicity, and intensity of the wave. (b) What is the dimension of the constant b?

95 Figure 17-47 shows an air-filled, acoustic interferometer, used to demonstrate the interference of sound waves. Sound source S is an oscillating diaphragm; D is a sound detector, such as the ear or a microphone. Path SBD can be varied in length, but path SAD is fixed. At D, the sound wave coming along path *SBD* interferes with that coming along path *SAD*. In one demonstration, the sound intensity at *D* has a minimum value of 100 units at one position of the movable arm and continuously climbs to a maximum value of 900 units when that arm is shifted by 1.65 cm. Find (a) the frequency of the sound emitted by the source and (b) the ratio of the amplitude at *D* of the *SAD* wave to that of the *SBD* wave. (c) How can it happen that these waves have different amplitudes, considering that they originate at the same source?

Fig. 17-47   Problem 95.

96   Find the ratios (greater to smaller) of the (a) intensities, (b) pressure amplitudes, and (c) particle displacement amplitudes for two sounds whose sound levels differ by 37 dB.

97   A bullet is fired with a speed of 685 m/s. Find the angle made by the shock cone with the line of motion of the bullet.

98   In Fig. 17-48, a point source *S* of sound waves lies near a reflecting wall *AB*. A sound detector *D* intercepts sound ray R_1 traveling directly from *S*. It also intercepts sound ray R_2 that reflects from the wall such that the *angle of incidence* θ_i is equal to the *angle of reflection* θ_r. Assume that the reflection of sound by the wall causes a phase shift of 0.500λ. If the distances in Fig. 17-48 are $d_1 = 2.50$ m, $d_2 = 20.0$ m, and $d_3 = 12.5$ m, what are the (a) lowest and (b) second lowest frequency at which R_1 and R_2 are in phase at *D*?

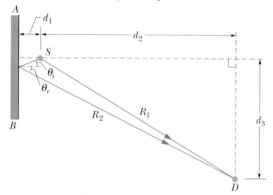

Fig. 17-48   Problem 98.

99   A point source that is stationary on an *x* axis emits a sinusoidal sound wave at a frequency of 686 Hz and speed 343 m/s. The wave travels radially outward from the source, causing air molecules to oscillate radially inward and outward. Let us define a wavefront as a line that connects points where the air molecules have the maximum, radially outward displacement. At any given instant, the wavefronts are concentric circles that are centered on the source. (a) Along *x*, what is the adjacent wavefront separation? Next, the source moves along *x* at a speed of 110 m/s. Along *x*, what are the wavefront separations (b) in front of and (c) behind the source?

100   In Fig. 17-33, S_1 and S_2 are two isotropic point sources of sound. They emit waves in phase at wavelength 0.50 m; they are separated by $D = 1.60$ m. If we move a sound detector along a large circle centered at the midpoint between the sources, at how many points do waves arrive at the detector (a) exactly in phase and (b) exactly out of phase?

101   Suppose a spherical loudspeaker emits sound isotropically at 10 W into a room with completely absorbent walls, floor, and ceiling (an *anechoic chamber*). (a) What is the intensity of the sound at distance $d = 3.0$ m from the center of the source? (b) What is the ratio of the wave amplitude at $d = 4.0$ m to that at $d = 3.0$ m?

102   On July 10, 1996, a granite block broke away from a wall in Yosemite Valley and, as it began to slide down the wall, was launched into projectile motion. Seismic waves produced by its impact with the ground triggered seismographs as far away as 200 km. Later measurements indicated that the block had a mass between 7.3×10^7 kg and 1.7×10^8 kg and that it landed 500 m vertically below the launch point and 30 m horizontally from it. (The launch angle is not known.) (a) Estimate the block's kinetic energy just before it landed.

Consider two types of seismic waves that spread from the impact point—a hemispherical *body wave* traveled through the ground in an expanding hemisphere and a cylindrical *surface wave* traveled along the ground in an expanding shallow vertical cylinder (Fig. 17-49). Assume that the impact lasted 0.50 s, the vertical cylinder had a depth *d* of 5.0 m, and each wave type received 20% of the energy the block had just before impact. Neglecting any mechanical energy loss the waves experienced as they traveled, determine the intensities of (b) the body wave and (c) the surface wave when they reached a seismograph 200 km away. (d) On the basis of these results, which wave is more easily detected on a distant seismograph?

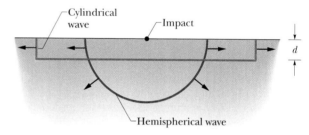

Fig. 17-49   Problem 102.

103   A sperm whale (Fig. 17-50*a*) vocalizes by producing a series of clicks. Actually, the whale makes only a single sound

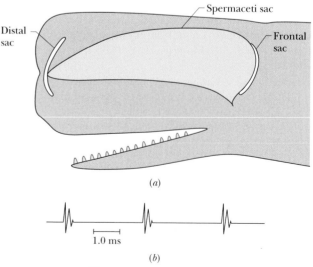

Fig. 17-50   Problem 103.

near the front of its head to start the series. Part of that sound then emerges from the head into the water to become the first click of the series. The rest of the sound travels backward through the spermaceti sac (a body of fat), reflects from the frontal sac (an air layer), and then travels forward through the spermaceti sac. When it reaches the distal sac (another air layer) at the front of the head, some of the sound escapes into the water to form the second click, and the rest is sent back through the spermaceti sac (and ends up forming later clicks).

Figure 17-50*b* shows a strip-chart recording of a series of clicks detected by an underwater microphone. A unit time interval of 1.0 ms is indicated on the chart. Assuming that the speed of sound in the spermaceti sac is 1372 m/s, find the length of the spermaceti sac. From such a calculation, marine scientists estimate the length of a whale from its click series.

104 Passengers in an auto traveling at 16.0 m/s toward the east hear a siren frequency of 950 Hz from an emergency vehicle approaching them from behind at a speed (relative to the air and ground) of 40.0 m/s. The speed of sound in air is 340 m/s. (a) What siren frequency does a passenger riding in the emergency vehicle hear? (b) What frequency do the passengers in the auto hear after the emergency vehicle passes them?

105 You are standing at a distance D from an isotropic point source of sound. You walk 50.0 m toward the source and observe that the intensity of the sound has doubled. Calculate the distance D.

106 To search for a fossilized dinosaur embedded in rock, paleontologists can use sound waves to produce a computer image of the dinosaur. The image then guides the paleontologists as they dig the dinosaur out of the rock. (The technique is shown in the opening scenes of the movie *Jurassic Park*.) The basic idea of the detection technique is that a strong pulse

Fig. 17-51 Problem 106.

of sound is emitted by a source (a seismic gun) at ground level and then detected by hydrophones that lie at evenly spaced depths in a bore hole drilled into the ground. The source and one hydrophone are shown in Fig. 17-51. If the sound wave travels from the source to the hydrophone through only rock as in Fig. 17-51, it travels at a known speed V and takes a certain time T. If, instead, it travels through a fossilized bone along the way, it takes slightly more time because it travels more slowly in the bone than in the rock. By measuring the difference Δt between the expected and measured travel times, the distance d traveled in the bone can be determined. After this procedure is repeated for many locations of the source and hydrophones, a computer can transform the many computed distances d into an image of the fossil.

(a) Let the speed of sound through fossilized bone be $V - \Delta V$, where ΔV is small relative to V. Show that the distance d is given by

$$d \approx \frac{V^2 \, \Delta t}{\Delta V}.$$

(b) For $V = 5000$ m/s and $\Delta V = 200$ m/s, what typical value of Δt can be expected if the sound passes along the diameter

of a leg bone of an adult *T. rex*? (Estimate the bone's diameter.)

107 Ultrasound, which consists of sound waves with frequencies above the human audible range, can be used to produce an image of the interior of a human body. Moreover, ultrasound can be used to measure the speed of the blood

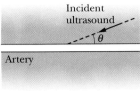

Fig. 17-52 Problem 107.

in the body; it does so by comparing the frequency of the ultrasound sent into the body with the frequency of the ultrasound reflected back to the body's surface by the blood. As the blood pulses, this detected frequency varies.

Suppose that an ultrasound image of the arm of a patient shows an artery that is angled at $\theta = 20°$ to the ultrasound's line of travel (Fig. 17-52). Suppose also that the frequency of the ultrasound reflected by the blood in the artery is increased by a maximum of 5495 Hz from the original ultrasound frequency of 5.000 000 MHz. (a) In Fig. 17-52, is the direction of the blood flow rightward or leftward? (b) The speed of sound in the human arm is 1540 m/s. What is the maximum speed of the blood? (*Hint:* The Doppler effect is caused by the component of the blood's velocity along the ultrasound's direction of travel.) (c) If angle θ were greater, would the reflected frequency be greater or less?

108 The period of a pulsating variable star may be estimated by considering the star to be executing *radial* longitudinal pulsations in the fundamental standing wave mode; that is, the star's radius varies periodically with time, with a displacement antinode at the star's surface. (a) Would you expect the center of the star to be a displacement node or antinode? (b) By analogy with a pipe with one open end, show that the period of pulsation T is given by $T = 4R/v$, where R is the equilibrium radius of the star and v is the average sound speed in the material of the star. (c) Typical white dwarf stars are composed of material with a bulk modulus of 1.33×10^{22} Pa and a density of 10^{10} kg/m³. They have radii equal to 9.0×10^{-3} solar radius. What is the approximate pulsation period of a white dwarf?

109 At a distance of 10 km, a 100 Hz horn, assumed to be an isotropic point source, is barely audible. At what distance would it begin to cause pain?

110 A listener at rest (with respect to the air and the ground) hears a signal of frequency f_1 from a source moving toward him with a velocity of 15 m/s, due east. If the listener then moves toward the approaching source with a velocity of 25 m/s, due west, he hears a frequency f_2 that differs from f_1 by 37 Hz. What is the frequency of the source? (Take the speed of sound in air to be 340 m/s.)

111 A guitar player tunes the fundamental frequency of a guitar string to 440 Hz. (a) What will be the fundamental frequency if she then increases the tension in the string by 20%? (b) What will it be if, instead, she decreases the length along which the string oscillates by sliding her finger from the tuning key one-third of the way down the string toward the bridge at the lower end?

On-Line Simulation Problems

The website http://www.wiley.com/college/halliday has simulation problems about this chapter.

18 Temperature, Heat, and the First Law of Thermodynamics

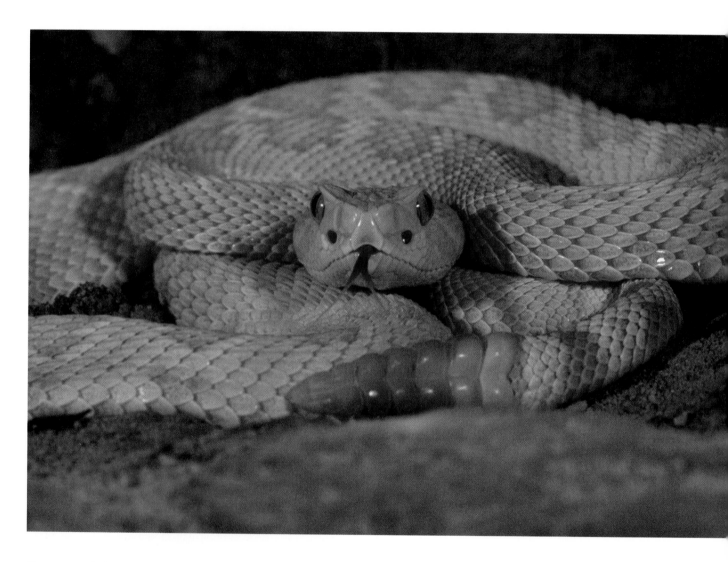

Because of their highly poisonous venom, rattlesnakes pose a danger to humans. When the snakes are discovered in residential areas, they are usually killed. However, the death of a rattlesnake does not immediately decrease its danger. Indeed, numerous people have made the error of reaching for a dead snake to remove it. Although the snake may have been dead for as long as 30 minutes, it can still strike a person, burying its fangs into a reaching hand and delivering its venom.

How can a dead rattlesnake strike a reaching hand?

The answer is in this chapter.

18-1 What Is Physics?

One of the principal branches of physics and engineering is **thermodynamics**, which is the study and application of the *thermal energy* (often called the *internal energy*) of systems. One of the central concepts of thermodynamics is *temperature*, which we begin to explore in the next section. Since childhood, you have been developing a working knowledge of thermal energy and temperature. For example, you know to be cautious with hot foods and hot stoves and to store perishable foods in cool or cold compartments. You also know how to control the temperature inside home and car, and how to protect yourself from wind chill and sunburn.

Examples of how thermodynamics figures into everyday engineering and science are countless. Automobile engineers are concerned with the heating of a car engine, such as during a NASCAR race. Food engineers are concerned both with the proper heating of foods, such as pizzas being microwaved, and with the proper cooling of foods, such as TV dinners being quickly frozen at a processing plant. Geologists are concerned with the transfer of thermal energy in an El Nino event and in the gradual warming of ice expanses in the Arctic and Antarctic. Agricultural engineers are concerned with the weather conditions that determine whether the agriculture of a country thrives or vanishes. Medical engineers are concerned with how a patient's temperature might distinguish between a benign viral infection and a cancerous growth.

The starting point in our discussion of thermodynamics is the concept of temperature and how it is measured.

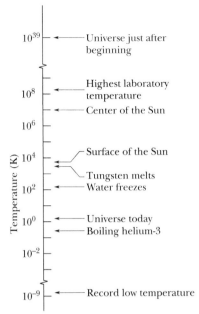

Fig. 18-1 Some temperatures on the Kelvin scale. Temperature $T = 0$ corresponds to $10^{-\infty}$ and cannot be plotted on this logarithmic scale.

18-2 Temperature

Temperature is one of the seven SI base quantities. Physicists measure temperature on the **Kelvin scale,** which is marked in units called *kelvins*. Although the temperature of a body apparently has no upper limit, it does have a lower limit; this limiting low temperature is taken as the zero of the Kelvin temperature scale. Room temperature is about 290 kelvins, or 290 K as we write it, above this *absolute zero*. Figure 18-1 shows a wide range of temperatures.

When the universe began 13.7 billion years ago, its temperature was about 10^{39} K. As the universe expanded it cooled, and it has now reached an average temperature of about 3 K. We on Earth are a little warmer than that because we happen to live near a star. Without our Sun, we too would be at 3 K (or, rather, we could not exist).

18-3 The Zeroth Law of Thermodynamics

The properties of many bodies change as we alter their temperature, perhaps by moving them from a refrigerator to a warm oven. To give a few examples: As their temperature increases, the volume of a liquid increases, a metal rod grows a little longer, and the electrical resistance of a wire increases, as does the pressure exerted by a confined gas. We can use any one of these properties as the basis of an instrument that will help us pin down the concept of temperature.

Figure 18-2 shows such an instrument. Any resourceful engineer could design and construct it, using any one of the properties listed above. The instrument is fitted with a digital readout display and has the following properties: If you heat it (say, with a Bunsen burner), the displayed number starts to increase; if you then put it into a refrigerator, the displayed number starts to decrease. The instrument is not calibrated in any way, and the numbers have (as yet) no physical meaning. The device is a *thermoscope* but not (as yet) a *thermometer*.

Fig. 18-2 A thermoscope. The numbers increase when the device is heated and decrease when it is cooled. The thermally sensitive element could be—among many possibilities—a coil of wire whose electrical resistance is measured and displayed.

Fig. 18-3   (a) Body T (a thermoscope) and body A are in thermal equilibrium. (Body S is a thermally insulating screen.) (b) Body T and body B are also in thermal equilibrium, at the same reading of the thermoscope. (c) If (a) and (b) are true, the zeroth law of thermodynamics states that body A and body B are also in thermal equilibrium.

Suppose that, as in Fig. 18-3a, we put the thermoscope (which we shall call body T) into intimate contact with another body (body A). The entire system is confined within a thick-walled insulating box. The numbers displayed by the thermoscope roll by until, eventually, they come to rest (let us say the reading is "137.04") and no further change takes place. In fact, we suppose that every measurable property of body T and of body A has assumed a stable, unchanging value. Then we say that the two bodies are in *thermal equilibrium* with each other. Even though the displayed readings for body T have not been calibrated, we conclude that bodies T and A must be at the same (unknown) temperature.

Suppose that we next put body T into intimate contact with body B (Fig. 18-3b) and find that the two bodies come to thermal equilibrium *at the same reading of the thermoscope.* Then bodies T and B must be at the same (still unknown) temperature. If we now put bodies A and B into intimate contact (Fig. 18-3c), are they immediately in thermal equilibrium with each other? Experimentally, we find that they are.

The experimental fact shown in Fig. 18-3 is summed up in the **zeroth law of thermodynamics:**

> If bodies A and B are each in thermal equilibrium with a third body T, then A and B are in thermal equilibrium with each other.

In less formal language, the message of the zeroth law is: "Every body has a property called **temperature.** When two bodies are in thermal equilibrium, their temperatures are equal. And vice versa." We can now make our thermoscope (the third body T) into a thermometer, confident that its readings will have physical meaning. All we have to do is calibrate it.

We use the zeroth law constantly in the laboratory. If we want to know whether the liquids in two beakers are at the same temperature, we measure the temperature of each with a thermometer. We do not need to bring the two liquids into intimate contact and observe whether they are or are not in thermal equilibrium.

The zeroth law, which has been called a logical afterthought, came to light only in the 1930s, long after the first and second laws of thermodynamics had been discovered and numbered. Because the concept of temperature is fundamental to those two laws, the law that establishes temperature as a valid concept should have the lowest number—hence the zero.

18-4   Measuring Temperature

Here we first define and measure temperatures on the Kelvin scale. Then we calibrate a thermoscope so as to make it a thermometer.

The Triple Point of Water

To set up a temperature scale, we pick some reproducible thermal phenomenon and, quite arbitrarily, assign a certain Kelvin temperature to its environment; that is, we select a *standard fixed point* and give it a standard fixed-point *temperature.* We could, for example, select the freezing point or the boiling point of water but, for technical reasons, we select instead the **triple point of water.**

Liquid water, solid ice, and water vapor (gaseous water) can coexist, in thermal equilibrium, at only one set of values of pressure and temperature. Figure 18-4 shows a triple-point cell, in which this so-called triple point of water can be achieved in the laboratory. By international agreement, the triple point of water has been assigned a value of 273.16 K as the standard fixed-point temperature for the calibration of thermometers; that is,

$$T_3 = 273.16 \text{ K} \quad \text{(triple-point temperature),} \quad (18\text{-}1)$$

in which the subscript 3 means "triple point." This agreement also sets the size of the kelvin as 1/273.16 of the difference between absolute zero and the triple-point temperature of water.

Note that we do not use a degree mark in reporting Kelvin temperatures. It is 300 K (not 300°K), and it is read "300 kelvins" (not "300 degrees Kelvin"). The usual SI prefixes apply. Thus, 0.0035 K is 3.5 mK. No distinction in nomenclature is made between Kelvin temperatures and temperature differences, so we can write, "the boiling point of sulfur is 717.8 K" and "the temperature of this water bath was raised by 8.5 K."

The Constant-Volume Gas Thermometer

The standard thermometer, against which all other thermometers are calibrated, is based on the pressure of a gas in a fixed volume. Figure 18-5 shows such a **constant-volume gas thermometer;** it consists of a gas-filled bulb connected by a tube to a mercury manometer. By raising and lowering reservoir R, the mercury level in the left arm of the U-tube can always be brought to the zero of the scale to keep the gas volume constant (variations in the gas volume can affect temperature measurements).

The temperature of any body in thermal contact with the bulb (such as the liquid surrounding the bulb in Fig. 18-5) is then defined to be

$$T = Cp, \tag{18-2}$$

in which p is the pressure exerted by the gas and C is a constant. From Eq. 14-10, the pressure p is

$$p = p_0 - \rho g h, \tag{18-3}$$

in which p_0 is the atmospheric pressure, ρ is the density of the mercury in the manometer, and h is the measured difference between the mercury levels in the two arms of the tube.* (The minus sign is used in Eq. 18-3 because pressure p is measured *above* the level at which the pressure is p_0.)

If we next put the bulb in a triple-point cell (Fig. 18-4), the temperature now being measured is

$$T_3 = Cp_3, \tag{18-4}$$

in which p_3 is the gas pressure now. Eliminating C between Eqs. 18-2 and 18-4 gives us the temperature as

$$T = T_3 \left(\frac{p}{p_3} \right) = (273.16 \text{ K}) \left(\frac{p}{p_3} \right) \qquad \text{(provisional).} \tag{18-5}$$

We still have a problem with this thermometer. If we use it to measure, say, the boiling point of water, we find that different gases in the bulb give slightly different results. However, as we use smaller and smaller amounts of gas to fill the bulb, the readings converge nicely to a single temperature, no matter what gas we use. Figure 18-6 shows this convergence for three gases.

Thus the recipe for measuring a temperature with a gas thermometer is

$$T = (273.16 \text{ K}) \left(\lim_{\text{gas} \to 0} \frac{p}{p_3} \right). \tag{18-6}$$

The recipe instructs us to measure an unknown temperature T as follows: Fill the thermometer bulb with an arbitrary amount of *any* gas (for example, nitrogen) and measure p_3 (using a triple-point cell) and p, the gas pressure at the temperature being measured. (Keep the gas volume the same.) Calculate the

---

*For pressure units, we shall use units introduced in Section 14-3. The SI unit for pressure is the newton per square meter, which is called the pascal (Pa). The pascal is related to other common pressure units by

$$1 \text{ atm} = 1.01 \times 10^5 \text{ Pa} = 760 \text{ torr} = 14.7 \text{ lb/in.}^2.$$

Fig. 18-4 A triple-point cell, in which solid ice, liquid water, and water vapor coexist in thermal equilibrium. By international agreement, the temperature of this mixture has been defined to be 273.16 K. The bulb of a constant-volume gas thermometer is shown inserted into the well of the cell.

Fig. 18-5 A constant-volume gas thermometer, its bulb immersed in a liquid whose temperature T is to be measured.

Fig. 18-6 Temperatures measured by a constant-volume gas thermometer, with its bulb immersed in boiling water. For temperature calculations using Eq. 18-5, pressure p_3 was measured at the triple point of water. Three different gases in the thermometer bulb gave generally different results at different gas pressures, but as the amount of gas was decreased (decreasing p_3), all three curves converged to 373.125 K.

ratio p/p_3. Then repeat both measurements with a smaller amount of gas in the bulb, and again calculate this ratio. Continue this way, using smaller and smaller amounts of gas, until you can extrapolate to the ratio p/p_3 that you would find if there were approximately no gas in the bulb. Calculate the temperature T by substituting that extrapolated ratio into Eq. 18-6. (The temperature is called the *ideal gas temperature*.)

18-5 The Celsius and Fahrenheit Scales

So far, we have discussed only the Kelvin scale, used in basic scientific work. In nearly all countries of the world, the Celsius scale (formerly called the centigrade scale) is the scale of choice for popular and commercial use and much scientific use. Celsius temperatures are measured in degrees, and the Celsius degree has the same size as the kelvin. However, the zero of the Celsius scale is shifted to a more convenient value than absolute zero. If T_C represents a Celsius temperature and T a Kelvin temperature, then

$$T_C = T - 273.15°. \tag{18-7}$$

In expressing temperatures on the Celsius scale, the degree symbol is commonly used. Thus, we write 20.00°C for a Celsius reading but 293.15 K for a Kelvin reading.

The Fahrenheit scale, used in the United States, employs a smaller degree than the Celsius scale and a different zero of temperature. You can easily verify both these differences by examining an ordinary room thermometer on which both scales are marked. The relation between the Celsius and Fahrenheit scales is

$$T_F = \tfrac{9}{5}T_C + 32°, \tag{18-8}$$

where T_F is Fahrenheit temperature. Converting between these two scales can be done easily by remembering a few corresponding points, such as the freezing and boiling points of water (Table 18-1). Figure 18-7 compares the Kelvin, Celsius, and Fahrenheit scales.

TABLE 18-1
Some Corresponding Temperatures

Temperature	°C	°F
Boiling point of water[a]	100	212
Normal body temperature	37.0	98.6
Accepted comfort level	20	68
Freezing point of water[a]	0	32
Zero of Fahrenheit scale	≈ −18	0
Scales coincide	−40	−40

[a]Strictly, the boiling point of water on the Celsius scale is 99.975°C, and the freezing point is 0.00°C. Thus, there is slightly less than 100 C° between those two points.

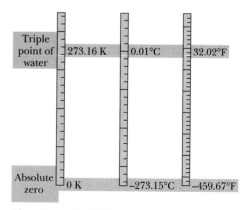

Fig. 18-7 The Kelvin, Celsius, and Fahrenheit temperature scales compared.

We use the letters C and F to distinguish measurements and degrees on the two scales. Thus,

$$0°C = 32°F$$

means that 0° on the Celsius scale measures the same temperature as 32° on the Fahrenheit scale, whereas

$$5 C° = 9 F°$$

means that a temperature difference of 5 Celsius degrees (note the degree symbol appears *after* C) is equivalent to a temperature difference of 9 Fahrenheit degrees.

Sample Problem 18-1

Suppose you come across old scientific notes that describe a temperature scale called Z on which the boiling point of water is 65.0°Z and the freezing point is −14.0°Z. To what temperature on the Fahrenheit scale would a temperature of $T = -98.0°Z$ correspond? Assume that the Z scale is linear; that is, the size of a Z degree is the same everywhere on the Z scale.

Solution: One **Key Idea** here is to relate the given temperature T to *either* known temperature on the Z scale. Since $T = -98.0°Z$ is closer to the freezing point (−14.0°Z) than to the boiling point (65.0°Z), we use the freezing point. Then we note that T is *below this point* by $-14.0°Z - (-98.0°Z) = 84.0 Z°$ (Fig. 18-8). (Read this difference as "84.0 Z degrees.")

Another **Key Idea** is to set up a conversion factor between the Z and Fahrenheit scales to convert this difference. To do so, we use *both* known temperatures on the Z scale and the corresponding temperatures on the Fahrenheit scale. On the Z scale, the difference between the boiling and freezing points is $65.0°Z - (-14.0°Z) = 79.0 Z°$. On the Fahrenheit scale, it is $212°F - 32.0°F = 180 F°$. Thus, a temperature difference of 79.0 Z° is equivalent to a temperature difference of 180 F° (Fig. 18-8), and we can use the ratio (180 F°)/(79.0 Z°) as our conversion factor.

Now, since T is below the freezing point by 84.0 Z°, it must also be below the freezing point by

$$(84.0 \, Z°) \frac{180 \, F°}{79.0 \, Z°} = 191 \, F°.$$

Fig. 18-8 An unknown temperature scale compared with the Fahrenheit temperature scale.

Because the freezing point is at 32.0°F, this means that

$$T = 32.0°F - 191 \, F° = -159°F. \quad \text{(Answer)}$$

CHECKPOINT 1 The figure here shows three linear temperature scales with the freezing and boiling points of water indicated. (a) Rank the degrees on these scales by size, greatest first. (b) Rank the following temperatures, highest first: 50°X, 50°W, and 50°Y.

PROBLEM-SOLVING TACTICS

TACTIC 1: *Temperature Changes*

Between the boiling and freezing points of water, there are (approximately) 100 kelvins and 100 Celsius degrees. Thus, a kelvin is the same size as a Celsius degree. From this or from Eq. 18-7, we then know that any temperature change is the same number whether expressed in kelvins or Celsius degrees. For example, a temperature change of 10 K is exactly equivalent to a temperature change of 10 C°.

Between the boiling and freezing points of water, there are 180 Fahrenheit degrees. Thus, 180 F° = 100 K, and a Fahrenheit degree must be 100/180, or 5/9, the size of a kelvin or Celsius degree. From this or from Eq. 18-8, we then know that any temperature change expressed in Fahrenheit degrees must

be $\frac{9}{5}$ times that same temperature change expressed in either kelvins or Celsius degrees. For example, in Fahrenheit degrees, a temperature change of 10 K is (9/5)(10 K), or 18 F°.

Take care not to confuse a *temperature* with a temperature *change* or *difference*. A temperature of 10 K is certainly not the same as one of 10°C or 18°F but, as noted above, a temperature *change* of 10 K is the same as one of 10 C° or 18 F°. This distinction is very important in an equation containing a temperature T instead of a temperature change or difference such as $T_2 - T_1$: A temperature T by itself should generally be in kelvins and not degrees Celsius or Fahrenheit. In short, beware the "bare T."

18-6 Thermal Expansion

You can often loosen a tight metal jar lid by holding it under a stream of hot water. Both the metal of the lid and the glass of the jar expand as the hot water adds energy to their atoms. (With the added energy, the atoms can move a bit farther from one another than usual, against the spring-like interatomic forces that hold every solid together.) However, because the atoms in the metal move farther apart than those in the glass, the lid expands more than the jar and thus is loosened.

Such **thermal expansion** of materials with an increase in temperature must be anticipated in many common situations. When a bridge is subject to large seasonal changes in temperature, for example, sections of the bridge are separated by *expansion slots* so that the sections have room to expand on hot days without the bridge buckling. When a dental cavity is filled, the filling material must have the same thermal expansion properties as the surrounding tooth; otherwise, consuming cold ice cream and then hot coffee would be very painful. When the Concorde aircraft (Fig. 18-9) was built, the design had to allow for the thermal expansion of the fuselage during supersonic flight because of frictional heating by the passing air.

The thermal expansion properties of some materials can be put to common use. Thermometers and thermostats may be based on the differences in expansion between the components of a *bimetal strip* (Fig. 18-10). Also, the familiar liquid-in-glass thermometers are based on the fact that liquids such as mercury and alcohol expand to a different (greater) extent than their glass containers.

Fig. 18-9 When a Concorde flew faster than the speed of sound, thermal expansion due to the rubbing by passing air increased the aircraft's length by about 12.5 cm. (The temperature increased to about 128°C at the aircraft nose and about 90°C at the tail, and cabin windows were noticeably warm to the touch.)

Linear Expansion

If the temperature of a metal rod of length L is raised by an amount ΔT, its length is found to increase by an amount

$$\Delta L = L\alpha \, \Delta T, \tag{18-9}$$

in which α is a constant called the **coefficient of linear expansion.** The coefficient α has the unit "per degree" or "per kelvin" and depends on the material. Although α varies somewhat with temperature, for most practical purposes it can be taken as constant for a particular material. Table 18-2 shows some coefficients of linear expansion. Note that the unit C° there could be replaced with the unit K.

The thermal expansion of a solid is like photographic enlargement except it is in three dimensions. Figure 18-11*b* shows the (exaggerated) thermal expansion of a steel ruler. Equation 18-9 applies to every linear dimension of the ruler, including its edge, thickness, diagonals, and the diameters of the circle etched on it and the circular hole cut in it. If the disk cut from that hole originally fits snugly in the hole, it will continue to fit snugly if it undergoes the same temperature increase as the ruler.

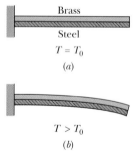

Fig. 18-10 (*a*) A bimetal strip, consisting of a strip of brass and a strip of steel welded together, at temperature T_0. (*b*) The strip bends as shown at temperatures above this reference temperature. Below the reference temperature the strip bends the other way. Many thermostats operate on this principle, making and breaking an electrical contact as the temperature rises and falls.

Volume Expansion

If all dimensions of a solid expand with temperature, the volume of that solid must also expand. For liquids, volume expansion is the only meaningful expansion parameter. If the temperature of a solid or liquid whose volume is V is increased by an amount ΔT, the increase in volume is found to be

$$\Delta V = V\beta \, \Delta T, \tag{18-10}$$

where β is the **coefficient of volume expansion** of the solid or liquid. The coef-

TABLE 18-2
Some Coefficients of Linear Expansion[a]

Substance	$\alpha \ (10^{-6}/C°)$	Substance	$\alpha \ (10^{-6}/C°)$
Ice (at 0°C)	51	Steel	11
Lead	29	Glass (ordinary)	9
Aluminum	23	Glass (Pyrex)	3.2
Brass	19	Diamond	1.2
Copper	17	Invar[b]	0.7
Concrete	12	Fused quartz	0.5

[a]Room temperature values except for the listing for ice.

[b]This alloy was designed to have a low coefficient of expansion. The word is a shortened form of "invariable."

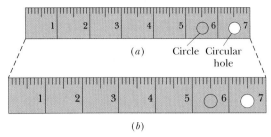

Fig. 18-11 The same steel ruler at two different temperatures. When it expands, the scale, the numbers, the thickness, and the diameters of the circle and circular hole are all increased by the same factor. (The expansion has been exaggerated for clarity.)

ficients of volume expansion and linear expansion for a solid are related by

$$\beta = 3\alpha. \tag{18-11}$$

The most common liquid, water, does not behave like other liquids. Above about 4°C, water expands as the temperature rises, as we would expect. Between 0 and about 4°C, however, water *contracts* with increasing temperature. Thus, at about 4°C, the density of water passes through a maximum. At all other temperatures, the density of water is less than this maximum value.

This behavior of water is the reason lakes freeze from the top down rather than from the bottom up. As water on the surface is cooled from, say, 10°C toward the freezing point, it becomes denser ("heavier") than lower water and sinks to the bottom. Below 4°C, however, further cooling makes the water then on the surface *less* dense ("lighter") than the lower water, so it stays on the surface until it freezes. Thus the surface freezes while the lower water is still liquid. If lakes froze from the bottom up, the ice so formed would tend not to melt completely during the summer, because it would be insulated by the water above. After a few years, many bodies of open water in the temperate zones of Earth would be frozen solid all year round—and aquatic life could not exist.

✓ CHECKPOINT 2 The figure here shows four rectangular metal plates, with sides of L, $2L$, or $3L$. They are all made of the same material, and their temperature is to be increased by the same amount. Rank the plates according to the expected increase in (a) their vertical heights and (b) their areas, greatest first.

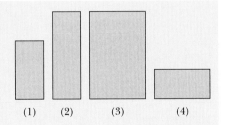

(1)   (2)   (3)   (4)

Sample Problem 18-2

On a hot day in Las Vegas, an oil trucker loaded 37 000 L of diesel fuel. He encountered cold weather on the way to Payson, Utah, where the temperature was 23.0 K lower than in Las Vegas, and where he delivered his entire load. How many liters did he deliver? The coefficient of volume expansion for diesel fuel is $9.50 \times 10^{-4}/C°$, and the coefficient of linear expansion for his steel truck tank is $11 \times 10^{-6}/C°$.

Solution: The **Key Idea** here is that the volume of the diesel fuel depends directly on the temperature. Thus, because the temperature decreased, the volume of the fuel did also. From

Eq. 18-10, the volume change is

$$\Delta V = V\beta \, \Delta T$$
$$= (37\,000 \text{ L})(9.50 \times 10^{-4}/C°)(-23.0 \text{ K}) = -808 \text{ L}.$$

Thus, the amount delivered was

$$V_{\text{del}} = V + \Delta V = 37\,000 \text{ L} - 808 \text{ L}$$
$$= 36\,190 \text{ L}. \quad \text{(Answer)}$$

Note that the thermal expansion of the steel tank has nothing to do with the problem. Question: Who paid for the "missing" diesel fuel?

Fig. 18-12 If the temperature of a system exceeds that of its environment as in (*a*), heat Q is lost by the system to the environment until thermal equilibrium (*b*) is established. (*c*) If the temperature of the system is below that of the environment, heat is absorbed by the system until thermal equilibrium is established.

18-7 Temperature and Heat

If you take a can of cola from the refrigerator and leave it on the kitchen table, its temperature will rise—rapidly at first but then more slowly—until the temperature of the cola equals that of the room (the two are then in thermal equilibrium). In the same way, the temperature of a cup of hot coffee, left sitting on the table, will fall until it also reaches room temperature.

In generalizing this situation, we describe the cola or the coffee as a *system* (with temperature T_S) and the relevant parts of the kitchen as the *environment* (with temperature T_E) of that system. Our observation is that if T_S is not equal to T_E, then T_S will change (T_E can also change some) until the two temperatures are equal and thus thermal equilibrium is reached.

Such a change in temperature is due to a change in the thermal energy of the system because of a transfer of energy between the system and the system's environment. (Recall that *thermal energy* is an internal energy that consists of the kinetic and potential energies associated with the random motions of the atoms, molecules, and other microscopic bodies within an object.) The transferred energy is called **heat** and is symbolized Q. Heat is *positive* when energy is transferred to a system's thermal energy from its environment (we say that heat is absorbed by the system). Heat is *negative* when energy is transferred from a system's thermal energy to its environment (we say that heat is released or lost by the system).

This transfer of energy is shown in Fig. 18-12. In the situation of Fig. 18-12*a*, in which $T_S > T_E$, energy is transferred from the system to the environment, so Q is negative. In Fig. 18-12*b*, in which $T_S = T_E$, there is no such transfer, Q is zero, and heat is neither released nor absorbed. In Fig. 18-12*c*, in which $T_S < T_E$, the transfer is to the system from the environment; so Q is positive.

We are led then to this definition of heat:

> Heat is the energy transferred between a system and its environment because of a temperature difference that exists between them.

Recall that energy can also be transferred between a system and its environment as *work W* via a force acting on a system. Heat and work, unlike temperature, pressure, and volume, are not intrinsic properties of a system. They have meaning only as they describe the transfer of energy into or out of a system. Similarly, the phrase "a \$600 transfer" has meaning if it describes the transfer to or from an account, not what is in the account, because the account holds money, not a transfer. Here, it is proper to say: "During the last 3 min, 15 J of heat was transferred to the system from its environment" or "During the last minute, 12 J of work was done on the system by its environment." It is meaningless to say: "This system contains 450 J of heat" or "This system contains 385 J of work."

Before scientists realized that heat is transferred energy, heat was measured in terms of its ability to raise the temperature of water. Thus, the **calorie** (cal) was defined as the amount of heat that would raise the temperature of 1 g of water from 14.5°C to 15.5°C. In the British system, the corresponding unit of heat was the **British thermal unit** (Btu), defined as the amount of heat that would raise the temperature of 1 lb of water from 63°F to 64°F.

In 1948, the scientific community decided that since heat (like work) is transferred energy, the SI unit for heat should be the one we use for energy—namely, the **joule.** The calorie is now defined to be 4.1868 J (exactly), with no reference to the heating of water. (The "calorie" used in nutrition, sometimes called the Calorie (Cal), is really a kilocalorie.) The relations among the various heat units are

$$1 \text{ cal} = 3.969 \times 10^{-3} \text{ Btu} = 4.1868 \text{ J}. \qquad (18\text{-}12)$$

18-8 The Absorption of Heat by Solids and Liquids

Heat Capacity

The **heat capacity** C of an object is the proportionality constant between the heat Q that the object absorbs or loses and the resulting temperature change ΔT of the object; that is,

$$Q = C \, \Delta T = C(T_f - T_i), \qquad (18\text{-}13)$$

in which T_i and T_f are the initial and final temperatures of the object. Heat capacity C has the unit of energy per degree or energy per kelvin. The heat capacity C of, say, a marble slab used in a bun warmer might be 179 cal/C°, which we can also write as 179 cal/K or as 749 J/K.

The word "capacity" in this context is really misleading in that it suggests analogy with the capacity of a bucket to hold water. *That analogy is false,* and you should not think of the object as "containing" heat or being limited in its ability to absorb heat. Heat transfer can proceed without limit as long as the necessary temperature difference is maintained. The object may, of course, melt or vaporize during the process.

Specific Heat

Two objects made of the same material—say, marble—will have heat capacities proportional to their masses. It is therefore convenient to define a "heat capacity per unit mass" or **specific heat** c that refers not to an object but to a unit mass of the material of which the object is made. Equation 18-13 then becomes

$$Q = cm \, \Delta T = cm(T_f - T_i). \qquad (18\text{-}14)$$

Through experiment we would find that although the heat capacity of a particular marble slab might be 179 cal/C° (or 749 J/K), the specific heat of marble itself (in that slab or in any other marble object) is 0.21 cal/g · C° (or 880 J/kg · K).

From the way the calorie and the British thermal unit were initially defined, the specific heat of water is

$$c = 1 \text{ cal/g} \cdot \text{C}° = 1 \text{ Btu/lb} \cdot \text{F}° = 4190 \text{ J/kg} \cdot \text{K}. \qquad (18\text{-}15)$$

Table 18-3 shows the specific heats of some substances at room temperature. Note that the value for water is relatively high. The specific heat of any substance actually depends somewhat on temperature, but the values in Table 18-3 apply reasonably well in a range of temperatures near room temperature.

> ✓ **CHECKPOINT 3**   A certain amount of heat Q will warm 1 g of material A by 3 C° and 1 g of material B by 4 C°. Which material has the greater specific heat?

Molar Specific Heat

In many instances the most convenient unit for specifying the amount of a substance is the mole (mol), where

$$1 \text{ mol} = 6.02 \times 10^{23} \text{ elementary units}$$

of *any* substance. Thus 1 mol of aluminum means 6.02×10^{23} atoms (the atom is the elementary unit), and 1 mol of aluminum oxide means 6.02×10^{23} molecules (the molecule is the elementary unit of the compound).

When quantities are expressed in moles, specific heats must also involve moles (rather than a mass unit); they are then called **molar specific heats.** Table 18-3 shows the values for some elemental solids (each consisting of a single element) at room temperature.

TABLE 18-3

Some Specific Heats and Molar Specific Heats at Room Temperature

Substance	Specific Heat		Molar Specific Heat
	cal / g · K	J / kg · K	J / mol · K
Elemental Solids			
Lead	0.0305	128	26.5
Tungsten	0.0321	134	24.8
Silver	0.0564	236	25.5
Copper	0.0923	386	24.5
Aluminum	0.215	900	24.4
Other Solids			
Brass	0.092	380	
Granite	0.19	790	
Glass	0.20	840	
Ice (−10°C)	0.530	2220	
Liquids			
Mercury	0.033	140	
Ethyl alcohol	0.58	2430	
Seawater	0.93	3900	
Water	1.00	4190	

An Important Point

In determining and then using the specific heat of any substance, we need to know the conditions under which energy is transferred as heat. For solids and liquids, we usually assume that the sample is under constant pressure (usually atmospheric) during the transfer. It is also conceivable that the sample is held at constant volume while the heat is absorbed. This means that thermal expansion of the sample is prevented by applying external pressure. For solids and liquids, this is very hard to arrange experimentally, but the effect can be calculated, and it turns out that the specific heats under constant pressure and constant volume for any solid or liquid differ usually by no more than a few percent. Gases, as you will see, have quite different values for their specific heats under constant-pressure conditions and under constant-volume conditions.

Heats of Transformation

When energy is absorbed as heat by a solid or liquid, the temperature of the sample does not necessarily rise. Instead, the sample may change from one *phase*, or *state*, to another. Matter can exist in three common states: In the *solid state*, the molecules of a sample are locked into a fairly rigid structure by their mutual attraction. In the *liquid state*, the molecules have more energy and move about more. They may form brief clusters, but the sample does not have a rigid structure and can flow or settle into a container. In the *gas*, or *vapor*, *state*, the molecules have even more energy, are free of one another, and can fill up the full volume of a container.

To *melt* a solid means to change it from the solid state to the liquid state. The process requires energy because the molecules of the solid must be freed from their rigid structure. Melting an ice cube to form liquid water is a common example. To *freeze* a liquid to form a solid is the reverse of melting and requires that energy be removed from the liquid, so that the molecules can settle into a rigid structure.

To *vaporize* a liquid means to change it from the liquid state to the vapor (gas) state. This process, like melting, requires energy because the molecules must be freed from their clusters. Boiling liquid water to transfer it to water vapor (or steam—a gas of individual water molecules) is a common example. *Condensing* a gas to form a liquid is the reverse of vaporizing; it requires that energy be removed from the gas, so that the molecules can cluster instead of flying away from one another.

The amount of energy per unit mass that must be transferred as heat when a sample completely undergoes a phase change is called the **heat of transformation** L. Thus, when a sample of mass m completely undergoes a phase change, the total energy transferred is

$$Q = Lm. \tag{18-16}$$

When the phase change is from liquid to gas (then the sample must absorb heat) or from gas to liquid (then the sample must release heat), the heat of transformation is called the **heat of vaporization** L_V. For water at its normal boiling or condensation temperature,

$$L_V = 539 \text{ cal/g} = 40.7 \text{ kJ/mol} = 2256 \text{ kJ/kg}. \tag{18-17}$$

When the phase change is from solid to liquid (then the sample must absorb heat) or from liquid to solid (then the sample must release heat), the heat of transformation is called the **heat of fusion** L_F. For water at its normal freezing or melting temperature,

$$L_F = 79.5 \text{ cal/g} = 6.01 \text{ kJ/mol} = 333 \text{ kJ/kg}. \tag{18-18}$$

Table 18-4 shows the heats of transformation for some substances.

TABLE 18-4
Some Heats of Transformation

	Melting		Boiling	
Substance	Melting Point (K)	Heat of Fusion L_F (kJ/kg)	Boiling Point (K)	Heat of Vaporization L_V (kJ/kg)
Hydrogen	14.0	58.0	20.3	455
Oxygen	54.8	13.9	90.2	213
Mercury	234	11.4	630	296
Water	273	333	373	2256
Lead	601	23.2	2017	858
Silver	1235	105	2323	2336
Copper	1356	207	2868	4730

Sample Problem 18-3

(a) How much heat must be absorbed by ice of mass $m = 720$ g at $-10°C$ to take it to liquid state at $15°C$?

Solution: The first **Key Idea** is that the heating process is accomplished in three steps.

Step 1   The **Key Idea** here is that the ice cannot melt at a temperature below the freezing point—so initially, any energy transferred to the ice as heat can only increase the temperature of the ice. The heat Q_1 needed to increase that temperature from the initial value $T_i = -10°C$ to a final value $T_f = 0°C$ (so that the ice can then melt) is given by Eq. 18-14 ($Q = cm\,\Delta T$). Using the specific heat of ice c_{ice} in Table 18-3 gives us

$$Q_1 = c_{ice}m(T_f - T_i)$$
$$= (2220 \text{ J/kg} \cdot \text{K})(0.720 \text{ kg})[0°C - (-10°C)]$$
$$= 15\,984 \text{ J} \approx 15.98 \text{ kJ}.$$

Step 2   The next **Key Idea** is that the temperature cannot increase from $0°C$ until all the ice melts—so any energy transferred to the ice as heat now can only change ice to liquid water. The heat Q_2 needed to melt all the ice is given by Eq. 18-16 ($Q = Lm$). Here L is the heat of fusion L_F, with the value given in Eq. 18-18 and Table 18-4. We find

$$Q_2 = L_Fm = (333 \text{ kJ/kg})(0.720 \text{ kg}) \approx 239.8 \text{ kJ}.$$

Step 3   Now we have liquid water at $0°C$. The next **Key Idea** is that the energy transferred to the liquid water as heat now can only increase the temperature of the liquid water. The heat Q_3 needed to increase the temperature of the water from the initial value $T_i = 0°C$ to the final value $T_f = 15°C$ is given by Eq. 18-14 (with the specific heat of liquid water c_{liq}):

$$Q_3 = c_{liq}m(T_f - T_i)$$
$$= (4190 \text{ J/kg} \cdot \text{K})(0.720 \text{ kg})(15°C - 0°C)$$
$$= 45\,252 \text{ J} \approx 45.25 \text{ kJ}.$$

The total required heat Q_{tot} is the sum of the amounts required in the three steps:

$$Q_{tot} = Q_1 + Q_2 + Q_3$$
$$= 15.98 \text{ kJ} + 239.8 \text{ kJ} + 45.25 \text{ kJ}$$
$$\approx 300 \text{ kJ}. \qquad \text{(Answer)}$$

Note that the heat required to melt the ice is much greater than the heat required to raise the temperature of either the ice or the liquid water.

(b) If we supply the ice with a total energy of only 210 kJ (as heat), what then are the final state and temperature of the water?

Solution: From step 1, we know that 15.98 kJ is needed to raise the temperature of the ice to the melting point. The remaining heat Q_{rem} is then 210 kJ $-$ 15.98 kJ, or about 194 kJ. From step 2, we can see that this amount of heat is insufficient to melt all the ice. Then this **Key Idea** becomes important: Because the melting of the ice is incomplete, we must end up with a mixture of ice and liquid; the temperature of the mixture must be the freezing point, $0°C$.

We can find the mass m of ice that is melted by the available energy Q_{rem} by using Eq. 18-16 with L_F:

$$m = \frac{Q_{rem}}{L_F} = \frac{194 \text{ kJ}}{333 \text{ kJ/kg}} = 0.583 \text{ kg} \approx 580 \text{ g}.$$

Thus, the mass of the ice that remains is 720 g $-$ 580 g, or 140 g, and we have

580 g water   and   140 g ice,   at $0°C$.   (Answer)

Sample Problem 18-4

A copper slug whose mass m_c is 75 g is heated in a laboratory oven to a temperature T of $312°C$. The slug is then dropped into a glass beaker containing a mass $m_w = 220$ g of water. The heat capacity C_b of the beaker is 45 cal/K. The initial temperature T_i of the water and the beaker is 12°C. Assuming that the slug, beaker, and water are an isolated system and the water does not vaporize, find the final temperature T_f of the system at thermal equilibrium.

Solution: One **Key Idea** here is that, with the system isolated, only internal transfers of energy can occur. There are three such transfers, all as heat. The slug loses energy, the water gains energy, and the beaker gains energy. Another **Key Idea** is that, because these transfers do not involve a phase change, the energy transfers can only change the temperatures. To relate the transfers to the temperature changes, we can use Eqs. 18-13 and 18-14 to write

$$\text{for the water:} \quad Q_w = c_w m_w (T_f - T_i); \quad (18\text{-}19)$$

$$\text{for the beaker:} \quad Q_b = C_b (T_f - T_i); \quad (18\text{-}20)$$

$$\text{for the copper:} \quad Q_c = c_c m_c (T_f - T). \quad (18\text{-}21)$$

A third **Key Idea** is that, with the system isolated, the total energy of the system cannot change. This means that the sum of these three energy transfers is zero:

$$Q_w + Q_b + Q_c = 0. \quad (18\text{-}22)$$

Substituting Eqs. 18-19 through 18-21 into Eq. 18-22 yields

$$c_w m_w (T_f - T_i) + C_b (T_f - T_i) + c_c m_c (T_f - T) = 0. \quad (18\text{-}23)$$

Temperatures are contained in Eq. 18-23 only as differences. Thus, because the differences on the Celsius and Kelvin scales are identical, we can use either of those scales in this equation. Solving it for T_f, we obtain

$$T_f = \frac{c_c m_c T + C_b T_i + c_w m_w T_i}{c_w m_w + C_b + c_c m_c}.$$

Using Celsius temperatures and taking values for c_c and c_w from Table 18-3, we find the numerator to be

$$(0.0923 \text{ cal/g} \cdot \text{K})(75 \text{ g})(312°C) + (45 \text{ cal/K})(12°C)$$
$$+ (1.00 \text{ cal/g} \cdot \text{K})(220 \text{ g})(12°C) = 5339.8 \text{ cal},$$

and the denominator to be

$$(1.00 \text{ cal/g} \cdot \text{K})(220 \text{ g}) + 45 \text{ cal/K}$$
$$+ (0.0923 \text{ cal/g} \cdot \text{K})(75 \text{ g}) = 271.9 \text{ cal/C°}.$$

We then have

$$T_f = \frac{5339.8 \text{ cal}}{271.9 \text{ cal/C°}} = 19.6°C \approx 20°C. \quad \text{(Answer)}$$

From the given data you can show that

$$Q_w \approx 1670 \text{ cal}, \qquad Q_b \approx 342 \text{ cal}, \qquad Q_c \approx -2020 \text{ cal}.$$

Apart from rounding errors, the algebraic sum of these three heat transfers is indeed zero, as Eq. 18-22 requires.

18-9 A Closer Look at Heat and Work

Here we look in some detail at how energy can be transferred as heat and work between a system and its environment. Let us take as our system a gas confined to a cylinder with a movable piston, as in Fig. 18-13. The upward force on the piston due to the pressure of the confined gas is equal to the weight of lead shot loaded onto the top of the piston. The walls of the cylinder are made of insulating material that does not allow any transfer of energy as heat. The bottom of the cylinder rests on a reservoir for thermal energy, a *thermal reservoir* (perhaps a hot plate) whose temperature T you can control by turning a knob.

The system (the gas) starts from an *initial state i*, described by a pressure p_i, a volume V_i, and a temperature T_i. You want to change the system to a *final state f*, described by a pressure p_f, a volume V_f, and a temperature T_f. The procedure by which you change the system from its initial state to its final state is called a *thermodynamic process*. During such a process, energy may be transferred into the system from the thermal reservoir (positive heat) or vice versa (negative heat). Also, work can be done by the system to raise the loaded piston (positive work) or lower it (negative work). We assume that all such changes occur slowly, with the result that the system is always in (approximate) thermal equilibrium (that is, every part of the system is always in thermal equilibrium with every other part).

Suppose that you remove a few lead shot from the piston of Fig. 18-13, allowing the gas to push the piston and remaining shot upward through a differential displacement $d\vec{s}$ with an upward force $\vec{F}$. Since the displacement is tiny, we can assume that $\vec{F}$ is constant during the displacement. Then $\vec{F}$ has a magnitude that is equal to pA, where p is the pressure of the gas and A is the face area of the piston. The differential work dW done by the gas during the displacement is

$$dW = \vec{F} \cdot d\vec{s} = (pA)(ds) = p(A \, ds)$$
$$= p \, dV, \quad (18\text{-}24)$$

Fig. 18-13 A gas is confined to a cylinder with a movable piston. Heat Q can be added to or withdrawn from the gas by regulating the temperature T of the adjustable thermal reservoir. Work W can be done by the gas by raising or lowering the piston.

in which dV is the differential change in the volume of the gas due to the movement of the piston. When you have removed enough shot to allow the gas to change its volume from V_i to V_f, the total work done by the gas is

$$W = \int dW = \int_{V_i}^{V_f} p \, dV. \qquad (18\text{-}25)$$

During the change in volume, the pressure and temperature of the gas may also change. To evaluate the integral in Eq. 18-25 directly, we would need to know how pressure varies with volume for the actual process by which the system changes from state i to state f.

There are actually many ways to take the gas from state i to state f. One way is shown in Fig. 18-14a, which is a plot of the pressure of the gas versus its volume and which is called a p-V diagram. In Fig. 18-14a, the curve indicates that the pressure decreases as the volume increases. The integral in Eq. 18-25 (and thus the work W done by the gas) is represented by the shaded area under the curve between points i and f. Regardless of what exactly we do to take the gas along the curve, that work is positive, due to the fact that the gas increases its volume by forcing the piston upward.

Another way to get from state i to state f is shown in Fig. 18-14b. There the change takes place in two steps—the first from state i to state a, and the second from state a to state f.

Step ia of this process is carried out at constant pressure, which means that you leave undisturbed the lead shot that ride on top of the piston in Fig. 18-13. You cause the volume to increase (from V_i to V_f) by slowly turning up the temperature control knob, raising the temperature of the gas to some higher value T_a. (Increasing the temperature increases the force from the gas on the piston, moving it upward.) During this step, positive work is done by the expanding gas (to lift the loaded piston) and heat is absorbed by the system from the thermal reservoir (in response to the arbitrarily small temperature differences that you create as you turn up the temperature). This heat is positive because it is added to the system.

Step af of the process of Fig. 18-14b is carried out at constant volume, so you must wedge the piston, preventing it from moving. Then as you use the control knob to decrease the temperature, you find that the pressure drops from p_a to its final value p_f. During this step, heat is lost by the system to the thermal reservoir.

For the overall process iaf, the work W, which is positive and is carried out only during step ia, is represented by the shaded area under the curve. Energy is transferred as heat during both steps ia and af, with a net energy transfer Q.

Figure 18-14c shows a process in which the previous two steps are carried out in reverse order. The work W in this case is smaller than for Fig. 18-14b, as is the net heat absorbed. Figure 18-14d suggests that you can make the work done by the gas as small as you want (by following a path like $icdf$) or as large as you want (by following a path like $ighf$).

To sum up: A system can be taken from a given initial state to a given final state by an infinite number of processes. Heat may or may not be involved, and in general, the work W and the heat Q will have different values for different processes. We say that heat and work are *path-dependent* quantities.

Fig. 18-14 (a) The shaded area represents the work W done by a system as it goes from an initial state i to a final state f. Work W is positive because the system's volume increases. (b) W is still positive, but now greater. (c) W is still positive, but now smaller. (d) W can be even smaller (path $icdf$) or larger (path $ighf$). (e) Here the system goes from state f to state i as the gas is compressed to less volume by an external force. The work W done *by* the system is now negative. (f) The net work W_{net} done by the system during a complete cycle is represented by the shaded area.

Figure 18-14*e* shows an example in which negative work is done by a system as some external force compresses the system, reducing its volume. The absolute value of the work done is still equal to the area beneath the curve, but because the gas is *compressed,* the work done by the gas is negative.

Figure 18-14*f* shows a *thermodynamic cycle* in which the system is taken from some initial state *i* to some other state *f* and then back to *i*. The net work done by the system during the cycle is the sum of the *positive* work done during the expansion and the *negative* work done during the compression. In Fig. 18-14*f*, the net work is positive because the area under the expansion curve (*i* to *f*) is greater than the area under the compression curve (*f* to *i*).

CHECKPOINT 4   The *p-V* diagram here shows six curved paths (connected by vertical paths) that can be followed by a gas. Which two of the curved paths should be part of a closed cycle (those curved paths plus connecting vertical paths) if the net work done by the gas during the cycle is to be at its maximum positive value?

18-10   *The First Law of Thermodynamics*

You have just seen that when a system changes from a given initial state to a given final state, both the work W and the heat Q depend on the nature of the process. Experimentally, however, we find a surprising thing. *The quantity $Q - W$ is the same for all processes.* It depends only on the initial and final states and does not depend at all on how the system gets from one to the other. All other combinations of Q and W, including Q alone, W alone, $Q + W$, and $Q - 2W$, are *path dependent;* only the quantity $Q - W$ is not.

The quantity $Q - W$ must represent a change in some intrinsic property of the system. We call this property the *internal energy* E_{int} and we write

$$\Delta E_{int} = E_{int,f} - E_{int,i} = Q - W \quad \text{(first law).} \quad (18\text{-}26)$$

Equation 18-26 is the **first law of thermodynamics.** If the thermodynamic system undergoes only a differential change, we can write the first law as*

$$dE_{int} = dQ - dW \quad \text{(first law).} \quad (18\text{-}27)$$

The internal energy E_{int} of a system tends to increase if energy is added as heat Q and tends to decrease if energy is lost as work W done by the system.

In Chapter 8, we discussed the principle of energy conservation as it applies to isolated systems—that is, to systems in which no energy enters or leaves the system. The first law of thermodynamics is an extension of that principle to systems that are *not* isolated. In such cases, energy may be transferred into or out of the system as either work W or heat Q. In our statement of the first law of thermodynamics above, we assume that there are no changes in the kinetic energy or the potential energy of the system as a whole; that is, $\Delta K = \Delta U = 0$.

Before this chapter, the term *work* and the symbol W always meant the work done *on* a system. However, starting with Eq. 18-24 and continuing through the

---

*Here dQ and dW, unlike dE_{int}, are not true differentials; that is, there are no such functions as $Q(p, V)$ and $W(p, V)$ that depend only on the state of the system. The quantities dQ and dW are called *inexact differentials* and are usually represented by the symbols σQ and σW. For our purposes, we can treat them simply as infinitesimally small energy transfers.

next two chapters about thermodynamics, we focus on the work done *by* a system, such as the gas in Fig. 18-13.

The work done *on* a system is always the negative of the work done *by* the system, so if we rewrite Eq. 18-26 in terms of the work W_{on} done *on* the system, we have $\Delta E_{int} = Q + W_{on}$. This tells us the following: The internal energy of a system tends to increase if heat is absorbed by the system or if positive work is done *on* the system. Conversely, the internal energy tends to decrease if heat is lost by the system or if negative work is done *on* the system.

> **CHECKPOINT 5** The figure here shows four paths on a *p-V* diagram along which a gas can be taken from state *i* to state *f*. Rank the paths according to (a) the change ΔE_{int} in the internal energy of the gas, (b) the work *W* done by the gas, and (c) the magnitude of the energy transferred as heat *Q* between the gas and its environment, greatest first.

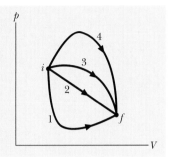

18-11 *Some Special Cases of the First Law of Thermodynamics*

Here we look at four different thermodynamic processes, in each of which a certain restriction is imposed on the system. We then see what consequences follow when we apply the first law of thermodynamics to the process. The results are summarized in Table 18-5.

1. *Adiabatic processes.* An adiabatic process is one that occurs so rapidly or occurs in a system that is so well insulated that *no transfer of energy as heat* occurs between the system and its environment. Putting $Q = 0$ in the first law (Eq. 18-26) yields

$$\Delta E_{int} = -W \qquad \text{(adiabatic process).} \qquad (18\text{-}28)$$

This tells us that if work is done *by* the system (that is, if *W* is positive), the internal energy of the system decreases by the amount of work. Conversely, if work is done *on* the system (that is, if *W* is negative), the internal energy of the system increases by that amount.

Figure 18-15 shows an idealized adiabatic process. Heat cannot enter or leave the system because of the insulation. Thus, the only way energy can be transferred between the system and its environment is by work. If we remove shot from the piston and allow the gas to expand, the work done by the system

Fig. 18-15 An adiabatic expansion can be carried out by slowly removing lead shot from the top of the piston. Adding lead shot reverses the process at any stage.

TABLE 18-5
The First Law of Thermodynamics: Four Special Cases

The Law: $\Delta E_{int} = Q - W$ (Eq. 18-26)		
Process	Restriction	Consequence
Adiabatic	$Q = 0$	$\Delta E_{int} = -W$
Constant volume	$W = 0$	$\Delta E_{int} = Q$
Closed cycle	$\Delta E_{int} = 0$	$Q = W$
Free expansion	$Q = W = 0$	$\Delta E_{int} = 0$

Stopcock

Vacuum

Insulation

Fig. 18-16 The initial stage of a free-expansion process. After the stopcock is opened, the gas fills both chambers and eventually reaches an equilibrium state.

(the gas) is positive and the internal energy of the gas decreases. If, instead, we add shot and compress the gas, the work done by the system is negative and the internal energy of the gas increases.

2. **Constant-volume processes.** If the volume of a system (such as a gas) is held constant, that system can do no work. Putting $W = 0$ in the first law (Eq. 18-26) yields

$$\Delta E_{int} = Q \qquad \text{(constant-volume process)}. \tag{18-29}$$

Thus, if heat is absorbed by a system (that is, if Q is positive), the internal energy of the system increases. Conversely, if heat is lost during the process (that is, if Q is negative), the internal energy of the system must decrease.

3. **Cyclical processes.** There are processes in which, after certain interchanges of heat and work, the system is restored to its initial state. In that case, no intrinsic property of the system—including its internal energy—can possibly change. Putting $\Delta E_{int} = 0$ in the first law (Eq. 18-26) yields

$$Q = W \qquad \text{(cyclical process)}. \tag{18-30}$$

Thus, the net work done during the process must exactly equal the net amount of energy transferred as heat; the store of internal energy of the system remains unchanged. Cyclical processes form a closed loop on a p-V plot, as shown in Fig. 18-14f. We discuss such processes in detail in Chapter 20.

4. **Free expansions.** These are adiabatic processes in which no transfer of heat occurs between the system and its environment and no work is done on or by the system. Thus, $Q = W = 0$, and the first law requires that

$$\Delta E_{int} = 0 \qquad \text{(free expansion)}. \tag{18-31}$$

Figure 18-16 shows how such an expansion can be carried out. A gas, which is in thermal equilibrium within itself, is initially confined by a closed stopcock to one half of an insulated double chamber; the other half is evacuated. The stopcock is opened, and the gas expands freely to fill both halves of the chamber. No heat is transferred to or from the gas because of the insulation. No work is done by the gas because it rushes into a vacuum and thus does not meet any pressure.

A free expansion differs from all other processes we have considered because it cannot be done slowly and in a controlled way. As a result, at any given instant during the sudden expansion, the gas is not in thermal equilibrium and its pressure is not uniform. Thus, although we can plot the initial and final states on a p-V diagram, we cannot plot the expansion itself.

CHECKPOINT 6 For one complete cycle as shown in the p-V diagram here, are (a) ΔE_{int} for the gas and (b) the net energy transferred as heat Q positive, negative, or zero?

Sample Problem 18-5

Let 1.00 kg of liquid water at 100°C be converted to steam at 100°C by boiling at standard atmospheric pressure (which is 1.00 atm or 1.01×10^5 Pa) in the arrangement of Fig. 18-17. The volume of that water changes from an initial value of 1.00×10^{-3} m^3 as a liquid to 1.671 m^3 as steam.

(a) How much work is done by the system during this process?

Solution: The **Key Idea** here is that the system must do positive work because the volume increases. In the general case

we would calculate the work W done by integrating the pressure with respect to the volume (Eq. 18-25). However, here the pressure is constant at 1.01×10^5 Pa, so we can take p outside the integral. We then have

$$W = \int_{V_i}^{V_f} p \, dV = p \int_{V_i}^{V_f} dV = p(V_f - V_i)$$
$$= (1.01 \times 10^5 \text{ Pa})(1.671 \text{ m}^3 - 1.00 \times 10^{-3} \text{ m}^3)$$
$$= 1.69 \times 10^5 \text{ J} = 169 \text{ kJ}. \qquad \text{(Answer)}$$

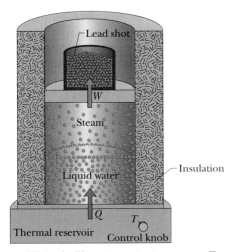

Fig. 18-17 Water boiling at constant pressure. Energy is transferred from the thermal reservoir as heat until the liquid water has changed completely into steam. Work is done by the expanding gas as it lifts the loaded piston.

(b) How much energy is transferred as heat during the process?

Solution: The Key Idea here is that the heat causes only a phase change and not a change in temperature, so it is given

fully by Eq. 18-16 ($Q = Lm$). Because the change is from liquid to gaseous phase, L is the heat of vaporization L_V, with the value given in Eq. 18-17 and Table 18-4. We find

$$Q = L_V m = (2256 \text{ kJ/kg})(1.00 \text{ kg})$$
$$= 2256 \text{ kJ} \approx 2260 \text{ kJ}. \qquad \text{(Answer)}$$

(c) What is the change in the system's internal energy during the process?

Solution: The Key Idea here is that the change in the system's internal energy is related to the heat (here, this is energy transferred into the system) and the work (here, this is energy transferred out of the system) by the first law of thermodynamics (Eq. 18-26). Thus, we can write

$$\Delta E_{\text{int}} = Q - W = 2256 \text{ kJ} - 169 \text{ kJ}$$
$$\approx 2090 \text{ kJ} = 2.09 \text{ MJ}. \qquad \text{(Answer)}$$

This quantity is positive, indicating that the internal energy of the system has increased during the boiling process. This energy goes into separating the H_2O molecules, which strongly attract one another in the liquid state. We see that, when water is boiled, about 7.5% ($= 169 \text{ kJ}/2260 \text{ kJ}$) of the heat goes into the work of pushing back the atmosphere. The rest of the heat goes into the system's internal energy.

18-12 Heat Transfer Mechanisms

We have discussed the transfer of energy as heat between a system and its environment, but we have not yet described how that transfer takes place. There are three transfer mechanisms: conduction, convection, and radiation.

Conduction

If you leave the end of a metal poker in a fire for enough time, its handle will get hot. Energy is transferred from the fire to the handle by (thermal) **conduction** along the length of the poker. The vibration amplitudes of the atoms and electrons of the metal at the fire end of the poker become relatively large because of the high temperature of their environment. These increased vibrational amplitudes, and thus the associated energy, are passed along the poker, from atom to atom, during collisions between adjacent atoms. In this way, a region of rising temperature extends itself along the poker to the handle.

Consider a slab of face area A and thickness L, whose faces are maintained at temperatures T_H and T_C by a hot reservoir and a cold reservoir, as in Fig. 18-18. Let Q be the energy that is transferred as heat through the slab, from its hot face to its cold face, in time t. Experiment shows that the *conduction rate* P_{cond} (the amount of energy transferred per unit time) is

$$P_{\text{cond}} = \frac{Q}{t} = kA \frac{T_H - T_C}{L}, \qquad (18\text{-}32)$$

Fig. 18-18 Thermal conduction. Energy is transferred as heat from a reservoir at temperature T_H to a cooler reservoir at temperature T_C through a conducting slab of thickness L and thermal conductivity k.

TABLE 18-6
Some Thermal Conductivities

Substance	k (W/m·K)
Metals	
Stainless steel	14
Lead	35
Iron	67
Brass	109
Aluminum	235
Copper	401
Silver	428
Gases	
Air (dry)	0.026
Helium	0.15
Hydrogen	0.18
Building Materials	
Polyurethane form	0.024
Rock wool	0.043
Fiberglass	0.048
White pine	0.11
Window glass	1.0

in which k, called the *thermal conductivity,* is a constant that depends on the material of which the slab is made. A material that readily transfers energy by conduction is a *good thermal conductor* and has a high value of k. Table 18-6 gives the thermal conductivities of some common metals, gases, and building materials.

Thermal Resistance to Conduction (R-Value)

If you are interested in insulating your house or in keeping cola cans cold on a picnic, you are more concerned with poor heat conductors than with good ones. For this reason, the concept of *thermal resistance R* has been introduced into engineering practice. The R-value of a slab of thickness L is defined as

$$R = \frac{L}{k}. \tag{18-33}$$

The lower the thermal conductivity of the material of which a slab is made, the higher the R-value of the slab; so something that has a high R-value is a *poor thermal conductor* and thus a *good thermal insulator.*

Note that R is a property attributed to a slab of a specified thickness, not to a material. The commonly used unit for R (which, in the United States at least, is almost never stated) is the square foot–Fahrenheit degree–hour per British thermal unit (ft² · F° · h/Btu). (Now you know why the unit is rarely stated.)

Conduction Through a Composite Slab

Figure 18-19 shows a composite slab, consisting of two materials having different thicknesses L_1 and L_2 and different thermal conductivities k_1 and k_2. The temperatures of the outer surfaces of the slab are T_H and T_C. Each face of the slab has area A. Let us derive an expression for the conduction rate through the slab under the assumption that the transfer is a *steady-state* process; that is, the temperatures everywhere in the slab and the rate of energy transfer do not change with time.

In the steady state, the conduction rates through the two materials must be equal. This is the same as saying that the energy transferred through one material in a certain time must be equal to that transferred through the other material in the same time. If this were not true, temperatures in the slab would be changing and we would not have a steady-state situation. Letting T_X be the temperature of the interface between the two materials, we can now use Eq. 18-32 to write

$$P_{\text{cond}} = \frac{k_2 A(T_H - T_X)}{L_2} = \frac{k_1 A(T_X - T_C)}{L_1}. \tag{18-34}$$

Solving Eq. 18-34 for T_X yields, after a little algebra,

$$T_X = \frac{k_1 L_2 T_C + k_2 L_1 T_H}{k_1 L_2 + k_2 L_1}. \tag{18-35}$$

Substituting this expression for T_X into either equality of Eq. 18-34 yields

$$P_{\text{cond}} = \frac{A(T_H - T_C)}{L_1/k_1 + L_2/k_2}. \tag{18-36}$$

We can extend Eq. 18-36 to apply to any number n of materials making up a slab:

$$P_{\text{cond}} = \frac{A(T_H - T_C)}{\Sigma \,(L/k)}. \tag{18-37}$$

The summation sign in the denominator tells us to add the values of L/k for all the materials.

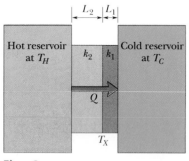

Fig. 18-19 Heat is transferred at a steady rate through a composite slab made up of two different materials with different thicknesses and different thermal conductivities. The steady-state temperature at the interface of the two materials is T_X.

CHECKPOINT 7 The figure shows the face and interface temperatures of a composite slab consisting of four materials, of identical thicknesses, through which the heat transfer is steady. Rank the materials according to their thermal conductivities, greatest first.

$$25°C \quad 15°C \quad 10°C \quad -5.0°C \quad -10°C$$
$$\quad\quad\quad a \quad\quad b \quad\quad c \quad\quad d$$

Convection

When you look at the flame of a candle or a match, you are watching thermal energy being transported upward by **convection.** Such energy transfer occurs when a fluid, such as air or water, comes in contact with an object whose temperature is higher than that of the fluid. The temperature of the part of the fluid that is in contact with the hot object increases, and (in most cases) that fluid expands and thus becomes less dense. Because this expanded fluid is now lighter than the surrounding cooler fluid, buoyant forces cause it to rise. Some of the surrounding cooler fluid then flows so as to take the place of the rising warmer fluid, and the process can then continue.

Convection is part of many natural processes. Atmospheric convection plays a fundamental role in determining global climate patterns and daily weather variations. Glider pilots and birds alike seek rising thermals (convection currents of warm air) that keep them aloft. Huge energy transfers take place within the oceans by the same process. Finally, energy is transported to the surface of the Sun from the nuclear furnace at its core by enormous cells of convection, in which hot gas rises to the surface along the cell core and cooler gas around the core descends below the surface.

Radiation

The third method by which an object and its environment can exchange energy as heat is via electromagnetic waves (visible light is one kind of electromagnetic wave). Energy transferred in this way is often called **thermal radiation** to distinguish it from electromagnetic *signals* (as in, say, television broadcasts) and from nuclear radiation (energy and particles emitted by nuclei). (To "radiate" generally means to emit.) When you stand in front of a big fire, you are warmed by absorbing thermal radiation from the fire; that is, your thermal energy increases as the fire's thermal energy decreases. No medium is required for heat transfer via radiation—the radiation can travel through vacuum from, say, the Sun to you.

The rate P_{rad} at which an object emits energy via electromagnetic radiation depends on the object's surface area A and the temperature T of that area in kelvins and is given by

$$P_{rad} = \sigma \varepsilon A T^4. \tag{18-38}$$

Here $\sigma = 5.6703 \times 10^{-8} \text{ W/m}^2 \cdot \text{K}^4$ is called the *Stefan–Boltzmann constant* after Josef Stefan (who discovered Eq. 18-38 experimentally in 1879) and Ludwig Boltzmann (who derived it theoretically soon after). The symbol ε represents the *emissivity* of the object's surface, which has a value between 0 and 1, depending on the composition of the surface. A surface with the maximum emissivity of 1.0 is said to be a *blackbody radiator,* but such a surface is an ideal limit and does not occur in nature. Note again that the temperature in Eq. 18-38 must be in kelvins so that a temperature of absolute zero corresponds to no radiation. Note also that every object whose temperature is above 0 K—including you—emits thermal radiation. (See Fig. 18-20.)

Fig. 18-20 A false-color thermogram reveals the rate at which energy is radiated by a house. The rates, from largest to smallest, are color-coded as white, yellow, red, purple, green, and black. You can tell that the roof and windows are poorly insulated.

The rate P_{abs} at which an object absorbs energy via thermal radiation from its environment, which we take to be at uniform temperature T_{env} (in kelvins), is

$$P_{abs} = \sigma \varepsilon A T_{env}^4. \tag{18-39}$$

The emissivity ε in Eq. 18-39 is the same as that in Eq. 18-38. An idealized blackbody radiator, with $\varepsilon = 1$, will absorb all the radiated energy it intercepts (rather than sending a portion back away from itself through reflection or scattering).

Because an object will radiate energy to the environment while it absorbs energy from the environment, the object's net rate P_{net} of energy exchange due to thermal radiation is

$$P_{net} = P_{abs} - P_{rad} = \sigma \varepsilon A(T_{env}^4 - T^4). \tag{18-40}$$

P_{net} is positive if net energy is being absorbed via radiation and negative if it is being lost via radiation.

Thermal radiation is involved in the numerous cases of a dead rattlesnake striking a hand reaching toward it, as discussed in the opening story of this chapter. Pits between each eye and nostril of a rattlesnake serve as sensors of thermal radiation (note them in this chapter's opening photograph). When, say, a mouse moves close to a rattlesnake's head, the thermal radiation from the mouse triggers these sensors, causing a reflex action in which the snake strikes the mouse with its fangs and injects its venom. The thermal radiation from a reaching hand can cause the same reflex action even if the snake has been dead for a while because the snake's nervous system continues to function. As one snake expert advised, if you must remove a recently killed rattlesnake, use a long stick rather than your hand.

Sample Problem 18-6

Figure 18-21 shows the cross section of a wall made of white pine of thickness L_a and brick of thickness L_d ($= 2.0L_a$), sandwiching two layers of unknown material with identical thicknesses and thermal conductivities. The thermal conductivity of the pine is k_a and that of the brick is k_d ($= 5.0k_a$). The face area A of the wall is unknown. Thermal conduction through the wall has reached the steady state; the only known interface temperatures are $T_1 = 25°C$, $T_2 = 20°C$, and $T_5 = -10°C$. What is interface temperature T_4?

Solution: One **Key Idea** here is that temperature T_4 helps determine the rate P_d at which energy is conducted through the brick, as given by Eq. 18-32. However, we lack enough data to solve Eq. 18-32 for T_4. A second **Key Idea** is that because the conduction is steady, the conduction rate P_d through the brick must equal the conduction rate P_a through the pine. From Eq. 18-32 and Fig. 18-21, we can write

$$P_a = k_a A \frac{T_1 - T_2}{L_a} \quad \text{and} \quad P_d = k_d A \frac{T_4 - T_5}{L_d}.$$

Setting $P_a = P_d$ and solving for T_4 yield

$$T_4 = \frac{k_a L_d}{k_d L_a}(T_1 - T_2) + T_5.$$

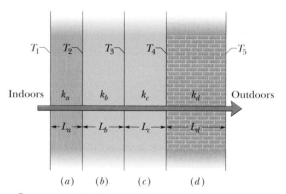

Fig. 18-21 A wall of four layers through which there is steady-state heat transfer.

Letting $L_d = 2.0L_a$ and $k_d = 5.0k_a$, and inserting the known temperatures, we find

$$T_4 = \frac{k_a(2.0L_a)}{(5.0k_a)L_a}(25°C - 20°C) + (-10°C)$$

$$= -8.0°C. \tag{Answer}$$

Sample Problem 18-7

During an extended wilderness hike, you have a terrific craving for ice. Unfortunately, the air temperature drops to only 6.0°C each night—too high to freeze water. However, because a clear, moonless night sky acts like a blackbody radiator at a temperature of $T_s = -23$°C, perhaps you can make ice by letting a shallow layer of water radiate energy to such a sky. To start, you thermally insulate a container from the ground by placing a poorly conducting layer of, say, foam rubber or straw beneath it. Then you pour water into the container, forming a thin, uniform layer with mass $m = 4.5$ g, top surface area $A = 9.0$ cm^2, depth $d = 5.0$ mm, emissivity $\varepsilon = 0.90$, and initial temperature 6.0°C. Find the time required for the water to freeze via radiation. Can the freezing be accomplished during one night?

Solution: One **Key Idea** is that the water cannot freeze at a temperature above the freezing point. Therefore, the radiation must first reduce the water temperature from 6.0°C to the freezing point of 0°C. Using Eq. 18-14 and Table 18-3, we find the associated energy loss to be

$$Q_1 = cm(T_f - T_i)$$
$$= (4190 \text{ J/kg} \cdot \text{K})(4.5 \times 10^{-3} \text{ kg})(0°C - 6.0°C)$$
$$= -113 \text{ J}.$$

Thus, 113 J must be radiated away by the water to drop its temperature to the freezing point.

A second **Key Idea** is that the water must next have an additional energy loss Q_2 for all of it to freeze. Using Eq. 18-16 ($Q = mL$) with the value of L being L_F from Eq. 18-18 or Table 18-4, and inserting a minus sign to indicate an energy loss, we find

$$Q_2 = -mL_F = -(4.5 \times 10^{-3} \text{ kg})(3.33 \times 10^5 \text{ J/kg})$$
$$= -1499 \text{ J}.$$

The total required energy loss is thus

$$Q_{tot} = Q_1 + Q_2 = -113 \text{ J} - 1499 \text{ J} = -1612 \text{ J}.$$

A third **Key Idea** is that, while the water loses energy by radiating to the sky, it also absorbs energy radiated to it from the sky. In a total time t, we want the net energy of this exchange to be the energy loss Q_{tot}; so we want the power of this exchange to be

$$\text{power} = \frac{\text{net energy}}{\text{time}} = \frac{Q_{tot}}{t}. \quad (18\text{-}41)$$

The power of such an energy exchange is also the net rate P_{net} of thermal radiation, as given by Eq. 18-40; so the time t required for the energy loss to be Q_{tot} is

$$t = \frac{Q}{P_{net}} = \frac{Q}{\sigma \varepsilon A(T_s^4 - T^4)}. \quad (18\text{-}42)$$

Although the temperature T of the water decreases slightly while the water is cooling, we can approximate T as being the freezing point, 273 K. With $T_s = 250$ K, the denominator of Eq. 18-42 is

$$(5.67 \times 10^{-8} \text{ W/m}^2 \cdot \text{K}^4)(0.90)(9.0 \times 10^{-4} \text{ m}^2)$$
$$\times [(250 \text{ K})^4 - (273 \text{ K})^4] = -7.57 \times 10^{-2} \text{ J/s},$$

and Eq. 18-42 gives us

$$t = \frac{-1612 \text{ J}}{-7.57 \times 10^{-2} \text{ J/s}} = 2.13 \times 10^4 \text{ s} = 5.9 \text{ h}. \quad (\text{Answer})$$

Because t is less than a night, freezing water by having it radiate to the dark sky is feasible. In fact, in some parts of the world people used this technique long before the introduction of electric freezers.

Review & Summary

Temperature; Thermometers Temperature is an SI base quantity related to our sense of hot and cold. It is measured with a thermometer, which contains a working substance with a measurable property, such as length or pressure, that changes in a regular way as the substance becomes hotter or colder.

Zeroth Law of Thermodynamics When a thermometer and some other object are placed in contact with each other, they eventually reach thermal equilibrium. The reading of the thermometer is then taken to be the temperature of the other object. The process provides consistent and useful temperature measurements because of the **zeroth law of thermodynamics:** If bodies A and B are each in thermal equilibrium with a third body C (the thermometer), then A and B are in thermal equilibrium with each other.

The Kelvin Temperature Scale In the SI system, temperature is measured on the **Kelvin scale,** which is based on the

triple point of water (273.16 K). Other temperatures are then defined by use of a *constant-volume gas thermometer*, in which a sample of gas is maintained at constant volume so its pressure is proportional to its temperature. We define the *temperature T* as measured with a gas thermometer to be

$$T = (273.16 \text{ K})\left(\lim_{\text{gas} \to 0} \frac{p}{p_3} \right). \quad (18\text{-}6)$$

Here T is in kelvins, and p_3 and p are the pressures of the gas at 273.16 K and the measured temperature, respectively.

Celsius and Fahrenheit Scales The Celsius temperature scale is defined by

$$T_C = T - 273.15°, \quad (18\text{-}7)$$

with T in kelvins. The Fahrenheit temperature scale is defined by

$$T_F = \tfrac{9}{5}T_C + 32°. \quad (18\text{-}8)$$

Thermal Expansion All objects change size with changes in temperature. For a temperature change ΔT, a change ΔL in any linear dimension L is given by

$$\Delta L = L\alpha \, \Delta T, \tag{18-9}$$

in which α is the **coefficient of linear expansion.** The change ΔV in the volume V of a solid or liquid is

$$\Delta V = V\beta \, \Delta T. \tag{18-10}$$

Here $\beta = 3\alpha$ is the material's **coefficient of volume expansion.**

Heat Heat Q is energy that is transferred between a system and its environment because of a temperature difference between them. It can be measured in **joules** (J), **calories** (cal), **kilocalories** (Cal or kcal), or **British thermal units** (Btu), with

$$1 \text{ cal} = 3.969 \times 10^{-3} \text{ Btu} = 4.1868 \text{ J}. \tag{18-12}$$

Heat Capacity and Specific Heat If heat Q is absorbed by an object, the object's temperature change $T_f - T_i$ is related to Q by

$$Q = C(T_f - T_i), \tag{18-13}$$

in which C is the **heat capacity** of the object. If the object has mass m, then

$$Q = cm(T_f - T_i), \tag{18-14}$$

where c is the **specific heat** of the material making up the object. The **molar specific heat** of a material is the heat capacity per mole, which means per 6.02×10^{23} elementary units of the material.

Heat of Transformation Heat absorbed by a material may change the material's physical state—for example, from solid to liquid or from liquid to gas. The amount of energy required per unit mass to change the state (but not the temperature) of a particular material is its **heat of transformation** L. Thus,

$$Q = Lm. \tag{18-16}$$

The **heat of vaporization** L_V is the amount of energy per unit mass that must be added to vaporize a liquid or that must be removed to condense a gas. The **heat of fusion** L_F is the amount of energy per unit mass that must be added to melt a solid or that must be removed to freeze a liquid.

Work Associated with Volume Change A gas may exchange energy with its surroundings through work. The amount of work W done *by* a gas as it expands or contracts from an initial volume V_i to a final volume V_f is given by

$$W = \int dW = \int_{V_i}^{V_f} p \, dV. \tag{18-25}$$

The integration is necessary because the pressure p may vary during the volume change.

First Law of Thermodynamics The principle of conservation of energy for a thermodynamic process is expressed in the **first law of thermodynamics,** which may assume either of the forms

$$\Delta E_{\text{int}} = E_{\text{int},f} - E_{\text{int},i} = Q - W \quad \text{(first law)} \tag{18-26}$$

or

$$dE_{\text{int}} = dQ - dW \quad \text{(first law).} \tag{18-27}$$

E_{int} represents the internal energy of the material, which depends only on the material's state (temperature, pressure, and volume). Q represents the energy exchanged as heat between the system and its surroundings; Q is positive if the system absorbs heat and negative if the system loses heat. W is the work done *by* the system; W is positive if the system expands against an external force from the surroundings and negative if the system contracts because of an external force. Q *and* W *are path dependent;* ΔE_{int} *is path independent.*

Applications of the First Law The first law of thermodynamics finds application in several special cases:

$$\begin{aligned} \textit{adiabatic processes:} \quad & Q = 0, \quad \Delta E_{\text{int}} = -W \\ \textit{constant-volume processes:} \quad & W = 0, \quad \Delta E_{\text{int}} = Q \\ \textit{cyclical processes:} \quad & \Delta E_{\text{int}} = 0, \quad Q = W \\ \textit{free expansions:} \quad & Q = W = \Delta E_{\text{int}} = 0 \end{aligned}$$

Conduction, Convection, and Radiation The rate P_{cond} at which energy is *conducted* through a slab whose faces are maintained at temperatures T_H and T_C is

$$P_{\text{cond}} = \frac{Q}{t} = kA \frac{T_H - T_C}{L}, \tag{18-32}$$

in which A and L are the face area and length of the slab, and k is the thermal conductivity of the material.

Convection occurs when temperature differences cause an energy transfer by motion within a fluid. *Radiation* is an energy transfer via the emission of electromagnetic energy. The rate P_{rad} at which an object emits energy via thermal radiation is

$$P_{\text{rad}} = \sigma \varepsilon A T^4, \tag{18-38}$$

where σ ($= 5.6703 \times 10^{-8}$ W/m$^2 \cdot$K^4) is the Stefan–Boltzmann constant, ε is the emissivity of the object's surface, A is its surface area, and T is its surface temperature (in kelvins). The rate P_{abs} at which an object absorbs energy via thermal radiation from its environment, which is at the uniform temperature T_{env} (in kelvins), is

$$P_{\text{abs}} = \sigma \varepsilon A T_{\text{env}}^4. \tag{18-39}$$

Questions

1 Figure 18-22 shows three linear temperature scales, with the freezing and boiling points of water indicated. Rank the three scales according to the size of one degree on them, greatest first.

2 The initial length L, change in temperature ΔT, and change

Fig. 18-22 Question 1.

in length ΔL of four rods are given in the following table. Rank the rods according to their coefficients of thermal expansion, greatest first.

Rod	L (m)	ΔT (C°)	ΔL (m)
a	2	10	4×10^{-4}
b	1	20	4×10^{-4}
c	2	10	8×10^{-4}
d	4	5	4×10^{-4}

3 Materials A, B, and C are solids that are at their melting temperatures. Material A requires 200 J to melt 4 kg, material B requires 300 J to melt 5 kg, and material C requires 300 J to melt 6 kg. Rank the materials according to their heats of fusion, greatest first.

4 A hot object is dropped into a thermally insulated container of water, and the object and water are then allowed to come to thermal equilibrium. The experiment is repeated twice, with different hot objects. All three objects have the same mass and initial temperature, and the mass and initial temperature of the water are the same in the three experiments. For each of the experiments, Fig. 18-23 gives graphs of the temperatures T of the object and the water versus time t. Rank the graphs according to the specific heats of the objects, greatest first.

Fig. 18-23   Question 4.

5 Figure 18-24 shows two closed cycles on p-V diagrams for a gas. The three parts of cycle 1 are of the same length and shape as those of cycle 2. For each cycle, should the cycle be traversed clockwise or counterclockwise if (a) the net work W done by the gas is to be positive and (b) the net energy transferred by the gas as heat Q is to be positive?

Fig. 18-24   Questions 5 and 6.

6 For which cycle in Fig. 18-24, traversed clockwise, is (a) W greater and (b) Q greater?

7 Figure 18-25 shows three different arrangements of mate-

Fig. 18-25   Question 7.

rials 1, 2, and 3 to form a wall. The thermal conductivities are $k_1 > k_2 > k_3$. The left side of the wall is 20 C° higher than the right side. Rank the arrangements according to (a) the (steady state) rate of energy conduction through the wall and (b) the temperature difference across material 1, greatest first.

8 A solid cube of edge length r, a solid sphere of radius r, and a solid hemisphere of radius r, all made of the same material, are maintained at temperature 300 K in an environment at temperature 350 K. Rank the objects according to the net rate at which thermal radiation is exchanged with the environment, greatest first.

9 Three different materials of identical mass are placed one at a time in a special freezer that can extract energy from a material at a certain constant rate. During the cooling process, each material begins in the liquid state and ends

Fig. 18-26   Question 9.

in the solid state; Fig. 18-26 shows the temperature T versus time t. (a) For material 1, is the specific heat for the liquid state greater than or less than that for the solid state? Rank the materials according to (b) freezing-point temperature, (c) specific heat in the liquid state, (d) specific heat in the solid state, and (e) heat of fusion, all greatest first.

10 A sample A of liquid water and a sample B of ice, of identical mass, are placed in a thermally insulated container and allowed to come to thermal equilibrium. Figure 18-27a is a sketch of the temperature T of the samples versus time t. (a) Is the equilibrium temperature above, below, or at the freezing point of water? (b) In reaching equilibrium, does the liquid partly freeze, fully freeze, or undergo no freezing? (c) Does the ice partly melt, fully melt, or undergo no melting?

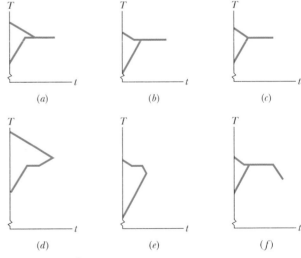

Fig. 18-27   Questions 10 and 11.

11 Question 10 continued: Graphs b through f of Fig. 18-27 are additional sketches of T versus t, of which one or more are impossible to produce. (a) Which is impossible and why? (b) In the possible ones, is the equilibrium temperature above, below, or at the freezing point of water? (c) As the possible situations reach equilibrium, does the liquid partly freeze, fully freeze, or undergo no freezing? Does the ice partly melt, fully melt, or undergo no melting?

Problems

sec. 18-4 Measuring Temperature

•1 Two constant-volume gas thermometers are assembled, one with nitrogen and the other with hydrogen. Both contain enough gas so that $p_3 = 80$ kPa. (a) What is the difference between the pressures in the two thermometers if both bulbs are in boiling water? (*Hint:* See Fig. 18-6.) (b) Which gas is at higher pressure?

•2 Suppose the temperature of a gas is 373.15 K when it is at the boiling point of water. What then is the limiting value of the ratio of the pressure of the gas at that boiling point to its pressure at the triple point of water? (Assume the volume of the gas is the same at both temperatures.)

•3 A gas thermometer is constructed of two gas-containing bulbs, each in a water bath, as shown in Fig. 18-28. The pressure difference between the two bulbs is measured by a mercury manometer as shown. Appropriate reservoirs, not shown in the diagram, maintain constant gas volume in the two bulbs. There is no difference in pressure when both baths are at the triple point of water. The pressure difference is 120 torr when one bath is at the triple point and the other is at the boiling point of water. It is 90.0 torr when one bath is at the triple point and the other is at an unknown temperature to be measured. What is the unknown temperature?

Fig. 18-28 Problem 3.

sec. 18-5 The Celsius and Fahrenheit Scales

•4 At what temperature is the Fahrenheit scale reading equal to (a) twice that of the Celsius scale and (b) half that of the Celsius scale?

•5 (a) In 1964, the temperature in the Siberian village of Oymyakon reached −71°C. What temperature is this on the Fahrenheit scale? (b) The highest officially recorded temperature in the continental United States was 134°F in Death Valley, California. What is this temperature on the Celsius scale?

••6 On a linear X temperature scale, water freezes at −125.0°X and boils at 375.0°X. On a linear Y temperature scale, water freezes at −70.00°Y and boils at −30.00°Y. A temperature of 50.00°Y corresponds to what temperature on the X scale?

••7 Suppose that on a linear temperature scale X, water boils at −53.5°X and freezes at −170°X. What is a temperature of 340 K on the X scale? (Approximate water's boiling point as 373 K.) ILW

sec. 18-6 Thermal Expansion

•8 An aluminum-alloy rod has a length of 10.000 cm at 20.000°C and a length of 10.015 cm at the boiling point of water. (a) What is the length of the rod at the freezing point of water? (b) What is the temperature if the length of the rod is 10.009 cm?

•9 A circular hole in an aluminum plate is 2.725 cm in diameter at 0.000°C. What is its diameter when the temperature of the plate is raised to 100.0°C? ILW

•10 An aluminum flagpole is 33 m high. By how much does its length increase as the temperature increases by 15 C°?

•11 Find the change in volume of an aluminum sphere with an initial radius of 10 cm when the sphere is heated from 0.0°C to 100°C. SSM

•12 What is the volume of a lead ball at 30.00°C if the ball's volume at 60.00°C is 50.00 cm³?

•13 At 20°C, a brass cube has an edge length of 30 cm. What is the increase in the cube's surface area when it is heated from 20°C to 75°C?

••14 At 20°C, a rod is exactly 20.05 cm long on a steel ruler. Both the rod and the ruler are placed in an oven at 270°C, where the rod now measures 20.11 cm on the same ruler. What is the coefficient of linear expansion for the material of which the rod is made?

••15 An aluminum cup of 100 cm³ capacity is completely filled with glycerin at 22°C. How much glycerin, if any, will spill out of the cup if the temperature of both the cup and the glycerin is increased to 28°C? (The coefficient of volume expansion of glycerin is 5.1×10^{-4}/C°.) SSM WWW

••16 When the temperature of a metal cylinder is raised from 0.0°C to 100°C, its length increases by 0.23%. (a) Find the percent change in density. (b) What is the metal? Use Table 18-2.

••17 A steel rod is 3.000 cm in diameter at 25.00°C. A brass ring has an interior diameter of 2.992 cm at 25.00°C. At what common temperature will the ring just slide onto the rod? ILW

••18 When the temperature of a copper coin is raised by 100 C°, its diameter increases by 0.18%. To two significant figures, give the percent increase in (a) the area of a face, (b) the thickness, (c) the volume, and (d) the mass of the coin. (e) Calculate the coefficient of linear expansion of the coin.

••19 A 1.28-m-long vertical glass tube is half filled with a liquid at 20°C. How much will the height of the liquid column change when the tube is heated to 30°C? Take $\alpha_{glass} = 1.0 \times 10^{-5}$/K and $\beta_{liquid} = 4.0 \times 10^{-5}$/K.

••20 In a certain experiment, a small radioactive source must move at selected, extremely slow speeds. This motion is accomplished by fastening the source to one end of an aluminum rod and heating the central section of the rod in a controlled way. If the effective heated section of the rod in Fig. 18-29 has length $d = 2.00$ cm, at what constant rate must the temperature of the rod be changed if the source is to move at a constant speed of 100 nm/s?

Fig. 18-29 Problem 20.

•••21 As a result of a temperature rise of 32 C°, a bar with a crack at its center buckles upward (Fig. 18-30). If the fixed distance L_0 is 3.77 m and the coefficient of linear expansion of the bar is 25 × 10⁻⁶/C°, find the rise x of the center. **SSM ILW**

Fig. 18-30 Problem 21.

sec. 18-8 **The Absorption of Heat by Solids and Liquids**

•22 How much water remains unfrozen after 50.2 kJ is transferred as heat from 260 g of liquid water initially at its freezing point?

•23 A certain substance has a mass per mole of 50.0 g/mol. When 314 J is added as heat to a 30.0 g sample, the sample's temperature rises from 25.0°C to 45.0°C. What are the (a) specific heat and (b) molar specific heat of this substance? (c) How many moles are present?

•24 A certain diet doctor encourages people to diet by drinking ice water. His theory is that the body must burn off enough fat to raise the temperature of the water from 0.00°C to the body temperature of 37.0°C. How many liters of ice water would have to be consumed to burn off 454 g (about 1 lb) of fat, assuming that burning this much fat requires 3500 Cal be transferred to the ice water? Why is it not advisable to follow this diet? (One liter = 10³ cm³. The density of water is 1.00 g/cm³.)

•25 Calculate the minimum amount of energy, in joules, required to completely melt 130 g of silver initially at 15.0°C. **SSM**

•26 What mass of butter, which has a usable energy content of 6.0 Cal/g (= 6000 cal/g), would be equivalent to the change in gravitational potential energy of a 73.0 kg man who ascends from sea level to the top of Mt. Everest, at elevation 8.84 km? Assume that the average g for the ascent is 9.80 m/s².

•27 A small electric immersion heater is used to heat 100 g of water for a cup of instant coffee. The heater is labeled "200 watts" (it converts electrical energy to thermal energy at this rate). Calculate the time required to bring all this water from 23.0°C to 100°C, ignoring any heat losses. **SSM**

•28 One way to keep the contents of a garage from becoming too cold on a night when a severe subfreezing temperature is forecast is to put a tub of water in the garage. If the mass of the water is 125 kg and its initial temperature is 20°C, (a) how much energy must the water transfer to its surroundings in order to freeze completely and (b) what is the lowest possible temperature of the water and its surroundings until that happens?

••29 *Nonmetric version:* How long does a 2.0 × 10⁵ Btu/h water heater take to raise the temperature of 40 gal of water from 70°F to 100°F? *Metric version:* How long does a 59 kW water heater take to raise the temperature of 150 L of water from 21°C to 38°C?

••30 A 150 g copper bowl contains 220 g of water, both at 20.0°C. A very hot 300 g copper cylinder is dropped into the water, causing the water to boil, with 5.00 g being converted to steam. The final temperature of the system is 100°C. Neglect energy transfers with the environment. (a) How much energy (in calories) is transferred to the water as heat? (b) How much

to the bowl? (c) What is the original temperature of the cylinder?

••31 What mass of steam at 100°C must be mixed with 150 g of ice at its melting point, in a thermally insulated container, to produce liquid water at 50°C? **ILW**

••32 A 0.400 kg sample is placed in a cooling apparatus that removes energy as heat at a constant rate. Figure 18-31 gives the temperature T of the sample versus time t; the sample freezes during the energy removal. The specific heat of the sample in its initial liquid phase is 3000 J/kg·K. What are (a) the sample's heat of fusion and (b) its specific heat in the frozen phase?

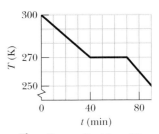

Fig. 18-31 Problem 32.

••33 In a solar water heater, energy from the Sun is gathered by water that circulates through tubes in a rooftop collector. The solar radiation enters the collector through a transparent cover and warms the water in the tubes; this water is pumped into a holding tank. Assume that the efficiency of the overall system is 20% (that is, 80% of the incident solar energy is lost from the system). What collector area is necessary to raise the temperature of 200 L of water in the tank from 20°C to 40°C in 1.0 h when the intensity of incident sunlight is 700 W/m²?

••34 Samples A and B are at different initial temperatures when they are placed in a thermally insulated container and allowed to come to thermal equilibrium. Figure 18-32a gives their temperatures T versus time t. Sample A has a mass of 5.0 kg; sample B has a mass of 1.5 kg. Figure 18-32b is a general plot for the material of sample B. It shows the temperature change ΔT that the material undergoes when energy is transferred to it as heat Q. The change ΔT is plotted versus the energy Q per unit mass of the material. What is the specific heat of sample A?

(a)   (b)

Fig. 18-32 Problem 34.

••35 Ethyl alcohol has a boiling point of 78.0°C, a freezing point of −114°C, a heat of vaporization of 879 kJ/kg, a heat of fusion of 109 kJ/kg, and a specific heat of 2.43 kJ/kg·K. How much energy must be removed from 0.510 kg of ethyl alcohol that is initially a gas at 78.0°C so that it becomes a solid at −114°C?

••36 A 0.530 kg sample of liquid water and a sample of ice are placed in a thermally insulated container. The container also contains a device that transfers energy as heat from the liquid water to the ice at a constant rate P, until thermal equilibrium is reached. The temperatures T of the liquid water and the ice are given in Fig. 18-33 as functions of time t. (a) What is rate P? (b) What is the initial mass of the ice in the container? (c) When thermal equilibrium is reached, what is the mass of the ice produced in this process?

Fig. 18-33
Problem 36.

••37 A person makes a quantity of iced tea by mixing 500 g of hot tea (essentially water) with an equal mass of ice at its melting point. Assume the mixture has negligible energy exchanges with its environment. If the tea's initial temperature is $T_i = 90°C$, when thermal equilibrium is reached what are (a) the mixture's temperature T_f and (b) the remaining mass m_f of ice? If $T_i = 70°C$, when thermal equilibrium is reached what are (c) T_f and (d) m_f?

••38 The specific heat of a substance varies with temperature according to $c = 0.20 + 0.14T + 0.023T^2$, with T in °C and c in cal/g·K. Find the energy required to raise the temperature of 2.0 g of this substance from 5.0°C to 15°C.

••39 (a) Two 50 g ice cubes are dropped into 200 g of water in a thermally insulated container. If the water is initially at 25°C, and the ice comes directly from a freezer at −15°C, what is the final temperature at thermal equilibrium? (b) What is the final temperature if only one ice cube is used? **SSM WWW**

••40 An insulated Thermos contains 130 cm³ of hot coffee at 80.0°C. You put in a 12.0 g ice cube at its melting point to cool the coffee. By how many degrees has your coffee cooled once the ice has melted? Treat the coffee as though it were pure water and neglect energy exchanges with the environment.

•••41 A 20.0 g copper ring at 0.000°C has an inner diameter of $D = 2.54000$ cm. An aluminum sphere at 100.0°C has a diameter of $d = 2.54508$ cm. The sphere is placed on top of the ring (Fig. 18-34), and the two are allowed to come to thermal equilibrium, with no heat lost to the surroundings. The sphere just passes through the ring at the equilibrium temperature. What is the mass of the sphere? **SSM**

Fig. 18-34 Problem 41.

sec. 18-11 Some Special Cases of the First Law of Thermodynamics

•42 A thermodynamic system is taken from state A to state B to state C, and then back to A, as shown in the p-V diagram of Fig. 18-35a. (a)–(g) Complete the table in Fig. 18-35b by inserting a plus sign, a minus sign, or a zero in each indicated cell. (h) What is the net work done by the system as it moves once through the cycle $ABCA$?

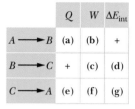

	Q	W	ΔE_{int}
$A \longrightarrow B$	(a)	(b)	+
$B \longrightarrow C$	+	(c)	(d)
$C \longrightarrow A$	(e)	(f)	(g)

(a)   (b)

Fig. 18-35 Problem 42.

•43 A sample of gas expands from 1.0 m³ to 4.0 m³ while its pressure decreases from 40 Pa to 10 Pa. How much work is done by the gas if its pressure changes with volume via (a) path A, (b) path B, and (c) path C in Fig. 18-36?

Fig. 18-36 Problem 43.

•44 Suppose 200 J of work is done on a system and 70.0 cal is extracted from the system as heat. In the sense of the first law of thermodynamics, what are the values (including algebraic signs) of (a) W, (b) Q, and (c) ΔE_{int}?

•45 A gas within a closed chamber undergoes the cycle shown in the p-V diagram of Fig. 18-37. Calculate the net energy added to the system as heat during one complete cycle. **SSM ILW**

Fig. 18-37 Problem 45.

•46 A sample of gas is taken through cycle $abca$ shown in the p-V diagram of Fig. 18-38. The net work done is +1.2 J. Along path ab, the change in the internal energy is +3.0 J and the magnitude of the work done is 5.0 J. Along path ca, the energy transferred to the gas as heat is +2.5 J. How much energy is transferred as heat along (a) path ab and (b) path bc?

Fig. 18-38 Problem 46.

•47 Figure 18-39 displays a closed cycle for a gas (the figure is not drawn to scale). The change in the internal energy of the gas as it moves from a to c along the path abc is −200 J. As it moves from c to d, 180 J must be transferred to it as

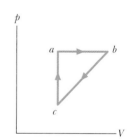

Fig. 18-39 Problem 47.

heat. An additional transfer of 80 J as heat is needed as it moves from *d* to *a*. How much work is done on the gas as it moves from *c* to *d*?

•48 Gas within a chamber passes through the cycle shown in Fig. 18-40. Determine the energy transferred by the system as heat during process *CA* if the energy added as heat Q_{AB} during process *AB* is 20.0 J, no energy is transferred as heat during process *BC*, and the net work done during the cycle is 15.0 J.

Fig. 18-40   Problem 48.

••49 When a system is taken from state *i* to state *f* along path *iaf* in Fig. 18-41, $Q = 50$ cal and $W = 20$ cal. Along path *ibf*, $Q = 36$ cal. (a) What is W along path *ibf*? (b) If $W = -13$ cal for the return path *fi*, what is Q for this path? (c) If $E_{int,i} = 10$ cal, what is $E_{int,f}$? If $E_{int,b} = 22$ cal, what is Q for (d) path *ib* and (e) path *bf*?

SSM WWW

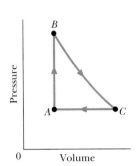

Fig. 18-41   Problem 49.

sec. 18-12   Heat Transfer Mechanisms

•50 The ceiling of a single-family dwelling in a cold climate should have an *R*-value of 30. To give such insulation, how thick would a layer of (a) polyurethane foam and (b) silver have to be?

•51 Consider the slab shown in Fig. 18-18. Suppose that $L = 25.0$ cm, $A = 90.0$ cm^2, and the material is copper. If $T_H = 125°C$, $T_C = 10.0°C$, and a steady state is reached, find the conduction rate through the slab. **SSM**

•52 If you were to walk briefly in space without a spacesuit while far from the Sun (as an astronaut does in the movie *2001*), you would feel the cold of space—while you radiated energy, you would absorb almost none from your environment. (a) At what rate would you lose energy? (b) How much energy would you lose in 30 s? Assume that your emissivity is 0.90, and estimate other data needed in the calculations.

•53 A cylindrical copper rod of length 1.2 m and cross-sectional area 4.8 cm^2 is insulated to prevent heat loss through its surface. The ends are maintained at a temperature difference of 100 C° by having one end in a water–ice mixture and the other in a mixture of boiling water and steam. (a) At what rate is energy conducted along the rod? (b) At what rate does ice melt at the cold end? **ILW**

•54 A sphere of radius 0.500 m, temperature 27.0°C, and emissivity 0.850 is located in an environment of temperature 77.0°C. At what rate does the sphere (a) emit and (b) absorb thermal radiation? (c) What is the sphere's net rate of energy exchange?

••55 In Fig. 18-42*a*, two iden-

Fig. 18-42   Problem 55.

tical rectangular rods of metal are welded end to end, with a temperature of $T_1 = 0°C$ on the left side and a temperature of $T_2 = 100°C$ on the right side. In 2.0 min, 10 J is conducted at a constant rate from the right side to the left side. How much time would be required to conduct 10 J if the rods were welded side to side as in Fig. 18-42*b*?

••56 A solid cylinder of radius $r_1 = 2.5$ cm, length $h_1 = 5.0$ cm, emissivity 0.85, and temperature 30°C is suspended in an environment of temperature 50°C. (a) What is the cylinder's net thermal radiation transfer rate P_1? (b) If the cylinder is stretched until its radius is $r_2 = 0.50$ cm, its net thermal radiation transfer rate becomes P_2. What is the ratio P_2/P_1?

••57 (a) What is the rate of energy loss in watts per square meter through a glass window 3.0 mm thick if the outside temperature is $-20°F$ and the inside temperature is $+72°F$? (b) A storm window having the same thickness of glass is installed parallel to the first window, with an air gap of 7.5 cm between the two windows. What now is the rate of energy loss if conduction is the only important energy-loss mechanism? **ILW**

••58 Figure 18-43 shows the cross section of a wall made of three layers. The thicknesses of the layers are L_1, $L_2 = 0.700L_1$, and $L_3 = 0.350L_1$. The thermal conductivities are k_1, $k_2 = 0.900k_1$, and $k_3 = 0.800k_1$. The temperatures at the left and right sides of the

Fig. 18-43   Problem 58.

wall are 30.0°C and −15.0°C, respectively. Thermal conduction through the wall has reached the steady state. (a) What is the temperature difference ΔT_2 across layer 2 (between the left and right sides of the layer)? If k_2 were, instead, equal to $1.1k_1$, (b) would the rate at which energy is conducted through the wall be greater than, less than, or the same as previously, and (c) what would be the value of ΔT_2?

••59 Figure 18-44 shows (in cross section) a wall consisting of four layers, with thermal conductivities $k_1 = 0.060$ W/m · K, $k_3 = 0.040$ W/m · K, and $k_4 = 0.12$ W/m · K (k_2 is not known). The layer thicknesses are $L_1 = 1.5$ cm, $L_3 = 2.8$ cm, and $L_4 = 3.5$ cm (L_2 is not known). Energy transfer through the wall is steady. What is the temperature of the interface between layers 3 and 4?

Fig. 18-44   Problem 59.

••60 Ice has formed on a shallow pond, and a steady state has been reached, with the air above the ice at −5.0°C and the bottom of the pond at 4.0°C. If the total depth of *ice + water* is 1.4 m, how thick is the ice? (Assume that the thermal conductivities of ice and water are 0.40 and 0.12 cal/m · C° · s, respectively.)

••61 A tank of water has been outdoors in cold weather, and a slab of ice 5.0 cm thick has formed on its surface (Fig.

18-45). The air above the ice is at $-10°C$. Calculate the rate of ice formation (in centimeters per hour) on the ice slab. Take the thermal conductivity of ice to be 0.0040 cal/s · cm · C° and its density to be 0.92 g/cm³. Assume no energy transfer through the tank walls or bottom. **SSM**

Fig. 18-45 Problem 61.

Additional Problems

62 In the extrusion of cold chocolate from a tube, work is done on the chocolate by the pressure applied by a ram forcing the chocolate through the tube. The work per unit mass of extruded chocolate is equal to p/ρ, where p is the difference between the applied pressure and the pressure where the chocolate emerges from the tube, and ρ is the density of the chocolate. Rather than increasing the temperature of the chocolate, this work melts cocoa fats in the chocolate. These fats have a heat of fusion of 150 kJ/kg. Assume that all of the work goes into that melting and that these fats make up 30% of the chocolate's mass. What percentage of the fats melt during the extrusion if $p = 5.5$ MPa and $\rho = 1200$ kg/m³?

63 The p-V diagram in Fig. 18-46 shows two paths along which a sample of gas can be taken from state a to state b. Path 1 requires that energy equal to $5.0p_1V_1$ be transferred to the gas as heat. Path 2 requires that energy equal to $5.5p_1V_1$ be transferred to the gas as heat. What is the ratio p_2/p_1?

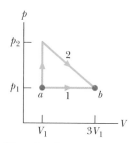

Fig. 18-46 Problem 63.

64 A solid aluminum cube 20 cm on an edge floats face-down on mercury. How much higher on the cube's sides will the mercury level be when the temperature rises from 270 to 320 K? (The coefficient of volume expansion for mercury is 1.8×10^{-4}/K.)

65 Equation 18-9 ($\Delta L = \alpha L \, \Delta T$, with L interpreted as the initial length in any given process) only approximates a change in length due to a change in temperature. Let's check the approximation. Suppose that 20.000 00 kJ is transferred as heat to a 0.400 000 kg copper rod with an initial length of $L = 3.000\,000$ m. Assume that $\alpha = 17 \times 10^{-6}$/C° is the exact coefficient of linear expansion for copper and 386 J/kg · K is the exact specific heat for copper. According to Eq. 18-9, what are (a) the increase in the rod's length and (b) the new length? Next, the same amount of energy is transferred as heat *from* the rod (so that the rod returns to its initial state). According to Eq. 18-9, what are (c) the decrease in the rod's length and (d) the new length? (e) What is the difference between the initial length and the answer to (d)?

66 In a series of experiments, block B is to be placed in a thermally insulated container with block A, which has the same mass as block B. In each experiment, block B is initially at a certain temperature T_B, but temperature T_A of block A

is changed from experiment to experiment. Let T_f represent the final temperature of the two blocks when they reach thermal equilibrium in any of the experiments. Figure 18-47 gives temperature T_f versus the initial temperature T_A for a range of possible values of T_A. What are (a) temperature T_B and (b) the ratio c_B/c_A of the specific heats of the blocks?

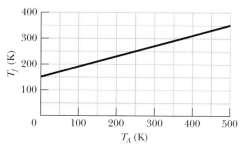

Fig. 18-47 Problem 66.

67 A recruit can join the semi-secret "300 F" club at the Amundsen–Scott South Pole Station only when the outside temperature is below $-70°C$. On such a day, the recruit first basks in a hot sauna and then runs outside wearing only shoes. (This is, of course, extremely dangerous, but the rite is effectively a protest against the constant danger of the winter cold at the south pole.)

Assume that upon stepping out of the sauna, the recruit's skin temperature is 102°F and the walls, ceiling, and floor of the sauna room have a temperature of 30°C. Estimate the recruit's surface area, and take the skin emissivity to be 0.80. (a) What is the approximate net rate P_{net} at which the recruit loses energy via thermal radiation exchanges with the room? Next, assume that when outdoors, half the recruit's surface area exchanges thermal radiation with the sky at a temperature of $-25°C$ and the other half exchanges thermal radiation with the snow and ground at a temperature of $-80°C$. What is the approximate net rate at which the recruit loses energy via thermal radiation exchanges with (b) the sky and (c) the snow and ground?

68 A 0.300 kg sample is placed in a cooling apparatus that removes energy as heat at a constant rate of 2.81 W. Figure 18-48 gives the temperature T of the sample versus time t. What is the specific heat of the sample?

Fig. 18-48 Problem 68.

69 (a) The temperature of the surface of the Sun is about 6000 K. Express this on the Fahrenheit scale. (b) Express normal human body temperature, 98.6°F, on the Celsius scale. (c) In the continental United States, the lowest officially recorded temperature is $-70°F$ at Rogers Pass, Montana. Express this on the Celsius scale. (d) Express the normal boiling point of oxygen, $-183°C$, on the Fahrenheit scale.

70 By how much does the volume of an aluminum cube 5.00 cm on an edge increase when the cube is heated from 10.0°C to 60.0°C?

71 Calculate the specific heat of a metal from the following data. A container made of the metal has a mass of 3.6 kg and contains 14 kg of water. A 1.8 kg piece of the metal initially at a temperature of 180°C is dropped into the water. The container and water initially have a temperature of 16.0°C, and the final temperature of the entire system is 18.0°C.

72 A sample of gas expands from 1.0 m³ to 4.0 m³ along path *B* in the *p-V* diagram in Fig. 18-49. It is then compressed back to 1.0 m³ along either path *A* or path *C*. Compute the net work done by the gas for the complete cycle along (a) path *BA* and (b) path *BC*.

Fig. 18-49   Problem 72.

73 An athlete needs to lose weight and decides to do it by "pumping iron." (a) How many times must an 80.0 kg weight be lifted a distance of 1.00 m in order to burn off 1.00 lb of fat, assuming that that much fat is equivalent to 3500 Cal? (b) If the weight is lifted once every 2.00 s, how long does the task take?

74 A copper rod, an aluminum rod, and a brass rod, each of 6.00 m length and 1.00 cm diameter, are placed end to end with the aluminum rod between the other two. The free end of the copper rod is maintained at water's boiling point, and the free end of the brass rod is maintained at water's freezing point. What is the steady-state temperature of (a) the copper–aluminum junction and (b) the aluminum–brass junction?

75 Soon after Earth was formed, heat released by the decay of radioactive elements raised the average internal temperature from 300 to 3000 K, at about which value it remains today. Assuming an average coefficient of volume expansion of 3.0×10^{-5} K⁻¹, by how much has the radius of Earth increased since the planet was formed?

76 Icebergs in the North Atlantic present hazards to shipping, causing the lengths of shipping routes to be increased by about 30% during the iceberg season. Attempts to destroy icebergs include planting explosives, bombing, torpedoing, shelling, ramming, and coating with black soot. Suppose that direct melting of the iceberg, by placing heat sources in the ice, is tried. How much energy as heat is required to melt 10% of an iceberg that has a mass of 200 000 metric tons? (Use 1 metric ton = 1000 kg.)

77 A sample of gas undergoes a transition from an initial state *a* to a final state *b* by three different paths (processes), as shown in the *p-V* diagram in Fig. 18-50. The energy transferred to the gas as heat in process 1 is $10p_iV_i$. In terms of p_iV_i, what are (a) the energy transferred to the gas as heat in process 2 and (b) the change in internal energy that the gas undergoes in process 3?

Fig. 18-50   Problem 77.

78 The average rate at which energy is conducted outward through the ground surface in North America is 54.0 mW/m², and the average thermal conductivity of the near-surface rocks is 2.50 W/m · K. Assuming a surface temperature of 10.0°C, find the temperature at a depth of 35.0 km (near the base of the crust). Ignore the heat generated by the presence of radioactive elements.

79 The temperature of a Pyrex disk is changed from 10.0°C to 60.0°C. Its initial radius is 8.00 cm; its initial thickness is 0.500 cm. Take these data as being exact. What is the change in the volume of the disk? (See Table 18-2.)

80 In a certain solar house, energy from the Sun is stored in barrels filled with water. In a particular winter stretch of five cloudy days, 1.00×10^6 kcal is needed to maintain the inside of the house at 22.0°C. Assuming that the water in the barrels is at 50.0°C and that the water has a density of 1.00×10^3 kg/m³, what volume of water is required?

81 A sample of gas expands from an initial pressure and volume of 10 Pa and 1.0 m³ to a final volume of 2.0 m³. During the expansion, the pressure and volume are related by the equation $p = aV^2$, where $a = 10$ N/m⁸. Determine the work done by the gas during this expansion.

82 Figure 18-51 displays a closed cycle for a gas. From *c* to *b*, 40 J is transferred from the gas as heat. From *b* to *a*, 130 J is transferred from the gas as heat, and the magnitude of the work done by the gas is 80 J. From *a* to *c*, 400 J is transferred to the gas as heat. What is the work done by the gas from *a* to *c*? (*Hint:* You need to supply the plus and minus signs for the given data.)

Fig. 18-51   Problem 82.

83 Figure 18-52 displays a closed cycle for a gas. The change in internal energy along path *ca* is −160 J. The energy transferred to the gas as heat is 200 J along path *ab*, and 40 J along path *bc*. How much work is done by the gas along (a) path *abc* and (b) path *ab*?

Fig. 18-52   Problem 83.

84 A steel rod has a length of exactly 20 cm at 30°C. How much longer is it at 50°C?

85 An object of mass 6.00 kg falls through a height of 50.0 m and, by means of a mechanical linkage, rotates a paddle wheel that stirs 0.600 kg of water. Assume that the initial gravitational potential energy of the object is fully transferred to thermal energy of the water, which is initially at 15.0°C. What is the temperature rise of the water?

86 (a) Calculate the rate at which body heat is conducted through the clothing of a skier in a steady-state process, given the following data: the body surface area is 1.8 m², and the clothing is 1.0 cm thick; the skin surface temperature is 33°C and the outer surface of the clothing is at 1.0°C; the thermal conductivity of the clothing is 0.040 W/m · K. (b) If, after a fall, the skier's clothes became soaked with water of thermal conductivity 0.60 W/m · K, by how much is the rate of conduction multiplied?

87 A cube of edge length 6.0×10^{-6} m, emissivity 0.75, and temperature −100°C floats in an environment at −150°C. What is the cube's net thermal radiation transfer rate?

88 A glass window pane is exactly 20 cm by 30 cm at 10°C. By how much has its area increased when its temperature is 40°C, assuming that it can expand freely?

89 A 2.50 kg lump of aluminum is heated to 92.0°C and then dropped into 8.00 kg of water at 5.00°C. Assuming that the lump–water system is thermally isolated, what is the system's equilibrium temperature?

90 Figure 18-53a shows a cylinder containing gas and closed by a movable piston. The cylinder is kept submerged in an ice–water mixture. The piston is *quickly* pushed down from position 1 to position 2 and then held at position 2 until the gas is again at the temperature of the ice–water mixture; it then is *slowly* raised back to position 1. Figure 18-53b is a *p-V* diagram for the process. If 100 g of ice is melted during the cycle, how much work has been done *on* the gas?

(a)
(b)

Fig. 18-53 Problem 90.

91 A large cylindrical water tank with a bottom 1.7 m in diameter is made of iron boilerplate 5.2 mm thick. Water in the tank is heated from below by a gas burner that is able to maintain a temperature difference of 2.3 C° between the top and bottom surfaces of the bottom plate. How much energy is conducted through that plate in 5.0 min?

92 A steel rod at 25.0°C is bolted at both ends and then cooled. At what temperature will it rupture? Use Table 12-1.

93 The temperature of a 0.700 kg cube of ice is decreased to −150°C. Then energy is gradually transferred to the cube as heat while it is otherwise thermally isolated from its environment. The total transfer is 0.6993 MJ. Assume the value of c_{ice} given in Table 18-3 is valid for temperatures from −150°C to 0°C. What is the final temperature of the water?

94 Suppose that you intercept 5.0×10^{-3} of the energy radiated by a hot sphere that has a radius of 0.020 m, an emissivity of 0.80, and a surface temperature of 500 K. How much energy do you intercept in 2.0 min?

95 (a) What is the coefficient of linear expansion of aluminum per Fahrenheit degree? (b) Use your answer in (a) to calculate the change in length of a 6.0 m aluminum rod if the rod is heated from 40°F to 95°F.

96 The thermal conductivity of Pyrex glass at temperature 0°C is 2.9×10^{-3} cal/cm · C° · s. Express this quantity in (a) W/m · K and (b) Btu/ft · F° · h. (c) Using your result in (a), find the *R*-value for a Pyrex sheet of thickness 6.4 mm.

97 A rectangular plate of glass initially has the dimensions 0.200 m by 0.300 m. The coefficient of linear expansion for the glass is 9.00×10^{-6}/K. What is the change in the plate's area if its temperature is increased by 20.0 K?

98 A thermometer of mass 0.0550 kg and of specific heat 0.837 kJ/kg · K reads 15.0°C. It is then completely immersed in 0.300 kg of water, and it comes to the same final temperature as the water. If the thermometer then reads 44.4°C, what was the temperature of the water before insertion of the thermometer?

99 An idealized representation of the air temperature as a function of distance from a single-pane window on a calm winter day is shown in Fig. 18-54. The window dimensions are 60 cm × 60 cm × 0.50 cm. Assume that energy is conducted along a path that is perpendicular to the window, from points 8.0 cm from the window on one side to points 8.0 cm from it on the other side. (a) At what rate is energy conducted through the window? (*Hint:* The temperature drop across the window glass is very small.) (b) Estimate the difference in temperature between the inner and outer glass surfaces.

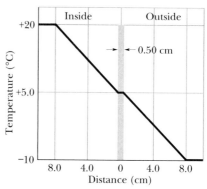

Fig. 18-54 Problem 99.

100 Three equal-length straight rods, of aluminum, Invar, and steel, all at 20.0°C, form an equilateral triangle with hinge pins at the vertices. At what temperature will the angle opposite the Invar rod be 59.95°? See Appendix E for needed trigonometric formulas and Table 18-2 for needed data.

101 It is possible to melt ice by rubbing one block of it against another. How much work, in joules, would you have to do to get 1.00 g of ice to melt?

102 A *flow calorimeter* is a device used to measure the specific heat of a liquid. Energy is added as heat at a known rate to a stream of the liquid as it passes through the calorimeter at a known rate. Measurement of the resulting temperature difference between the inflow and the outflow points of the liquid stream enables us to compute the specific heat of the liquid. Suppose a liquid of density 0.85 g/cm³ flows through a calorimeter at the rate of 8.0 cm³/s. When energy is added at the rate of 250 W by means of an electric heating coil, a temperature difference of 15 C° is established in steady-state conditions between the inflow and the outflow points. What is the specific heat of the liquid?

On-Line Simulation Problems

The website http://www.wiley.com/college/halliday has simulation problems for this chapter.

The Kinetic Theory of Gases

The interior of this railroad tank car was being cleaned with steam by a crew late one afternoon. Because the job was unfinished at the end of their work shift, they sealed the car and left for the night. When they returned the next morning, they discovered that something had crushed the car in spite of its extremely strong steel walls, as if some giant creature from a grade B science fiction movie had stepped on it during a rampage that night.

What crushed the tank car during the night?

The answer is in this chapter.

19-1   What Is Physics?

One of the main subjects in thermodynamics is the physics of gases. A gas consists of atoms (either individually or bound together as molecules) that fill their container's volume and exert pressure on the container's walls. We can usually assign a temperature to such a contained gas. These three variables associated with a gas—volume, pressure, and temperature—are all a consequence of the motion of the atoms. The volume is a result of the freedom the atoms have to spread throughout the container, the pressure is a result of the collisions of the atoms with the container's walls, and the temperature has to do with the kinetic energy of the atoms. The **kinetic theory of gases**, the focus of this chapter, relates the motion of the atoms to the volume, pressure, and temperature of the gas.

Applications of the kinetic theory of gases are countless. Automobile engineers are concerned with the combustion of vaporized fuel (a gas) in the automobile engines. Food engineers are concerned with the production rate of the fermentation gas that causes bread to rise as it bakes. Beverage engineers are concerned with how gas can produce the head in a glass of beer or shoot a cork from a champagne bottle. Medical engineers and physiologists are concerned with calculating how long a scuba driver must pause during ascent to eliminate nitrogen gas from the bloodstream (to avoid the *bends*). Environmental scientists are concerned with how heat exchanges between the oceans and the atmosphere can affect weather conditions.

The first step in our discussion of the kinetic theory of gases deals with measuring the amount of a gas present in a sample, for which we use Avogadro's number.

19-2   Avogadro's Number

When our thinking is slanted toward atoms and molecules, it makes sense to measure the sizes of our samples in moles. If we do so, we can be certain that we are comparing samples that contain the same number of atoms or molecules. The *mole* is one of the seven SI base units and is defined as follows:

 One mole is the number of atoms in a 12 g sample of carbon-12.

The obvious question now is: "How many atoms or molecules are there in a mole?" The answer is determined experimentally and, as you saw in Chapter 18, is

$$N_A = 6.02 \times 10^{23} \text{ mol}^{-1} \qquad \text{(Avogadro's number)}, \qquad (19\text{-}1)$$

where mol^{-1} represents the inverse mole or "per mole," and mol is the abbreviation for mole. The number N_A is called **Avogadro's number** after Italian scientist Amadeo Avogadro (1776–1856), who suggested that all gases occupying the same volume under the same conditions of temperature and pressure contain the same number of atoms or molecules.

The number of moles n contained in a sample of any substance is equal to the ratio of the number of molecules N in the sample to the number of molecules N_A in 1 mol:

$$n = \frac{N}{N_A}. \qquad (19\text{-}2)$$

(*Caution:* The three symbols in this equation can easily be confused with one another, so you should sort them with their meanings now, before you end in "N-confusion.") We can find the number of moles n in a sample from the mass M_{sam} of the sample and either the *molar mass M* (the mass of 1 mol) or the molecular mass m (the mass of one molecule):

$$n = \frac{M_{sam}}{M} = \frac{M_{sam}}{mN_A}. \qquad (19\text{-}3)$$

In Eq. 19-3, we used the fact that the mass M of 1 mol is the product of the mass m of one molecule and the number of molecules N_A in 1 mol:

$$M = mN_A. \qquad (19\text{-}4)$$

PROBLEM-SOLVING TACTICS

TACTIC 1: *Avogadro's Number of What?*

In Eq. 19-1, Avogadro's number is expressed in terms of mol^{-1}, which is the inverse mole, or 1/mol. We could instead explicitly state the elementary unit involved in a given situation. For example, if the elementary unit is an atom, we might write $N_A = 6.02 \times 10^{23}$ atoms/mol. If, instead, the elementary unit is a molecule, then we might write $N_A = 6.02 \times 10^{23}$ molecules/mol.

19-3   *Ideal Gases*

Our goal in this chapter is to explain the macroscopic properties of a gas—such as its pressure and its temperature—in terms of the behavior of the molecules that make it up. However, there is an immediate problem: which gas? Should it be hydrogen, oxygen, or methane, or perhaps uranium hexafluoride? They are all different. Experimenters have found, though, that if we confine 1 mol samples of various gases in boxes of identical volume and hold the gases at the same temperature, then their measured pressures are nearly—though not exactly—the same. If we repeat the measurements at lower gas densities, then these small differences in the measured pressures tend to disappear. Further experiments show that, at low enough densities, all real gases tend to obey the relation

$$pV = nRT \qquad \text{(ideal gas law)}, \qquad (19\text{-}5)$$

in which p is the absolute (not gauge) pressure, n is the number of moles of gas present, and T is the temperature in kelvins. The symbol R is a constant called the **gas constant** that has the same value for all gases—namely,

$$R = 8.31 \text{ J/mol} \cdot \text{K}. \qquad (19\text{-}6)$$

Equation 19-5 is called the **ideal gas law.** Provided the gas density is low, this law holds for any single gas or for any mixture of different gases. (For a mixture, n is the total number of moles in the mixture.)

We can rewrite Eq. 19-5 in an alternative form, in terms of a constant called the **Boltzmann constant** k, which is defined as

$$k = \frac{R}{N_A} = \frac{8.31 \text{ J/mol} \cdot \text{K}}{6.02 \times 10^{23} \text{ mol}^{-1}} = 1.38 \times 10^{-23} \text{ J/K}. \qquad (19\text{-}7)$$

This allows us to write $R = kN_A$. Then, with Eq. 19-2 ($n = N/N_A$), we see that

$$nR = Nk. \qquad (19\text{-}8)$$

Substituting this into Eq. 19-5 gives a second expression for the ideal gas law:

$$pV = NkT \qquad \text{(ideal gas law)}. \qquad (19\text{-}9)$$

(*Caution:* Note the difference between the two expressions for the ideal gas law—Eq. 19-5 involves the number of moles n, and Eq. 19-9 involves the number of molecules N.)

You may well ask, "What is an *ideal gas*, and what is so 'ideal' about it?" The answer lies in the simplicity of the law (Eqs. 19-5 and 19-9) that governs its macroscopic properties. Using this law—as you will see—we can deduce many properties of the ideal gas in a simple way. Although there is no such thing in nature as a truly ideal gas, *all real* gases approach the ideal state at low enough densities—that is, under conditions in which their molecules are far enough apart that they do not interact with one another. Thus, the ideal gas concept allows us to gain useful insights into the limiting behavior of real gases.

With Eq. 19-9, we can explain what crushed the railroad tank car in this chapter's opening photograph. When the car was being cleaned, its interior was filled with very hot steam, which is a gas of water molecules. The cleaning crew left the steam inside the car when they closed all the valves on the car at the end of their work shift. At that point the pressure of the gas in the car was equal to atmospheric pressure because the valves had been opened to the atmosphere during the cleaning. As the car cooled during the night, the steam cooled and much of it condensed, which means that the number N of gas molecules and the temperature T of the gas both decreased. Thus, the right side of Eq. 19-9 decreased, and because volume V was constant, the gas pressure p on the left side also decreased. At some point during the night, the gas pressure inside the car reached such a low value that the external atmospheric pressure was able to crush the car's steel walls. The cleaning crew could have prevented this accident by leaving the valves open, so that air could enter the car to keep the internal pressure equal to the external atmospheric pressure.

Work Done by an Ideal Gas at Constant Temperature

Suppose we put an ideal gas in a piston–cylinder arrangement like those in Chapter 18. Suppose also that we allow the gas to expand from an initial volume V_i to a final volume V_f while we keep the temperature T of the gas constant. Such a process, at *constant temperature*, is called an **isothermal expansion** (and the reverse is called an **isothermal compression**).

On a p-V diagram, an *isotherm* is a curve that connects points that have the same temperature. Thus, it is a graph of pressure versus volume for a gas whose temperature T is held constant. For n moles of an ideal gas, it is a graph of the equation

$$p = nRT \frac{1}{V} = (\text{a constant})\frac{1}{V}. \tag{19-10}$$

Figure 19-1 shows three isotherms, each corresponding to a different (constant) value of T. (Note that the values of T for the isotherms increase upward to the right.) Superimposed on the middle isotherm is the path followed by a gas during an isothermal expansion from state i to state f at a constant temperature of 310 K.

To find the work done by an ideal gas during an isothermal expansion, we start with Eq. 18-25,

$$W = \int_{V_i}^{V_f} p \, dV. \tag{19-11}$$

This is a general expression for the work done during any change in volume of any gas. For an ideal gas, we can use Eq. 19-5 ($pV = nRT$) to substitute for p, obtaining

$$W = \int_{V_i}^{V_f} \frac{nRT}{V} \, dV. \tag{19-12}$$

Because we are considering an isothermal expansion, T is constant, so we can move it in front of the integral sign to write

$$W = nRT \int_{V_i}^{V_f} \frac{dV}{V} = nRT \left[\ln V \right]_{V_i}^{V_f}. \tag{19-13}$$

By evaluating the expression in brackets at the limits and then using the relationship $\ln a - \ln b = \ln (a/b)$, we find that

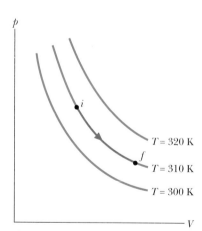

Fig. 19-1 Three isotherms on a p-V diagram. The path shown along the middle isotherm represents an isothermal expansion of a gas from an initial state i to a final state f. The path from f to i along the isotherm would represent the reverse process—that is, an isothermal compression.

$$W = nRT \ln \frac{V_f}{V_i} \quad \text{(ideal gas, isothermal process).} \quad (19\text{-}14)$$

Recall that the symbol ln specifies a *natural* logarithm, which has base *e*.

For an expansion, V_f is greater than V_i, so the ratio V_f/V_i in Eq. 19-14 is greater than unity. The natural logarithm of a quantity greater than unity is positive, and so the work W done by an ideal gas during an isothermal expansion is positive, as we expect. For a compression, V_f is less than V_i, so the ratio of volumes in Eq. 19-14 is less than unity. The natural logarithm in that equation—hence the work W—is negative, again as we expect.

Work Done at Constant Volume and at Constant Pressure

Equation 19-14 does not give the work W done by an ideal gas during *every* thermodynamic process. Instead, it gives the work only for a process in which the temperature is held constant. If the temperature varies, then the symbol T in Eq. 19-12 cannot be moved in front of the integral symbol as in Eq. 19-13, and thus we do not end up with Eq. 19-14.

However, we can always go back to Eq. 19-11 to find the work W done by an ideal gas (or any other gas) during any process, such as a constant-volume process and a constant-pressure process. If the volume of the gas is constant, then Eq. 19-11 yields

$$W = 0 \quad \text{(constant-volume process).} \quad (19\text{-}15)$$

If, instead, the volume changes while the pressure p of the gas is held constant, then Eq. 19-11 becomes

$$W = p(V_f - V_i) = p\,\Delta V \quad \text{(constant-pressure process).} \quad (19\text{-}16)$$

CHECKPOINT 1 An ideal gas has an initial pressure of 3 pressure units and an initial volume of 4 volume units. The table gives the final pressure and volume of the gas (in those same units) in five processes. Which processes start and end on the same isotherm?

	a	b	c	d	e
p	12	6	5	4	1
V	1	2	7	3	12

Sample Problem 19-1

A cylinder contains 12 L of oxygen at 20°C and 15 atm. The temperature is raised to 35°C, and the volume is reduced to 8.5 L. What is the final pressure of the gas in atmospheres? Assume that the gas is ideal.

Solution: The **Key Idea** here is that, because the gas is ideal, its pressure, volume, temperature, and number of moles are related by the ideal gas law, both in the initial state i and in the final state f (after the changes). Thus, from Eq. 19-5 we can write

$$p_i V_i = nRT_i \quad \text{and} \quad p_f V_f = nRT_f.$$

Dividing the second equation by the first equation and solving for p_f yields

$$p_f = \frac{p_i T_f V_i}{T_i V_f}. \quad (19\text{-}17)$$

Note here that if we converted the given initial and final volumes from liters to the proper units of cubic meters, the multiplying conversion factors would cancel out of Eq. 19-17. The same would be true for conversion factors that convert the pressures from atmospheres to the proper pascals. However, to convert the given temperatures to kelvins requires the addition of an amount that would not cancel and thus must be included. Hence, we must write

$$T_i = (273 + 20)\ \text{K} = 293\ \text{K}$$

and

$$T_f = (273 + 35)\ \text{K} = 308\ \text{K}.$$

Inserting the given data into Eq. 19-17 then yields

$$p_f = \frac{(15\ \text{atm})(308\ \text{K})(12\ \text{L})}{(293\ \text{K})(8.5\ \text{L})} = 22\ \text{atm}. \quad \text{(Answer)}$$

Sample Problem 19-2

One mole of oxygen (assume it to be an ideal gas) expands at a constant temperature T of 310 K from an initial volume V_i of 12 L to a final volume V_f of 19 L. How much work is done by the gas during the expansion?

Solution: The **Key Idea** is this: Generally we find the work by integrating the gas pressure with respect to the gas volume, using Eq. 19-11. However, because the gas here is ideal and the expansion is isothermal, that integration leads to Eq.

19-14. Therefore, we can write

$$W = nRT \ln \frac{V_f}{V_i}$$

$$= (1 \text{ mol})(8.31 \text{ J/mol} \cdot \text{K})(310 \text{ K}) \ln \frac{19 \text{ L}}{12 \text{ L}}$$

$$= 1180 \text{ J}. \hspace{2cm} \text{(Answer)}$$

The expansion is graphed in the *p*-*V* diagram of Fig. 19-2. The work done by the gas during the expansion is represented by the area beneath the curve *if*.

You can show that if the expansion is now reversed, with the gas undergoing an isothermal compression from 19 L to 12 L, the work done by the gas will be −1180 J. Thus, an external force would have to do 1180 J of work on the gas to compress it.

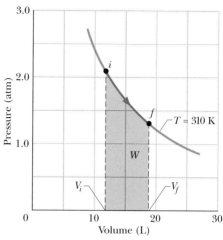

Fig. 19-2   The shaded area represents the work done by 1 mol of oxygen in expanding from V_i to V_f at a constant temperature *T* of 310 K.

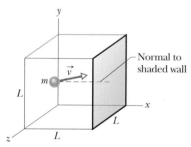

Fig. 19-3   A cubical box of edge length *L*, containing *n* moles of an ideal gas. A molecule of mass *m* and velocity $\vec{v}$ is about to collide with the shaded wall of area L^2. A normal to that wall is shown.

19-4   Pressure, Temperature, and RMS Speed

Here is our first kinetic theory problem. Let *n* moles of an ideal gas be confined in a cubical box of volume *V*, as in Fig. 19-3. The walls of the box are held at temperature *T*. What is the connection between the pressure *p* exerted by the gas on the walls and the speeds of the molecules?

The molecules of gas in the box are moving in all directions and with various speeds, bumping into one another and bouncing from the walls of the box like balls in a racquetball court. We ignore (for the time being) collisions of the molecules with one another and consider only elastic collisions with the walls.

Figure 19-3 shows a typical gas molecule, of mass *m* and velocity $\vec{v}$, that is about to collide with the shaded wall. Because we assume that any collision of a molecule with a wall is elastic, when this molecule collides with the shaded wall, the only component of its velocity that is changed is the *x* component, and that component is reversed. This means that the only change in the particle's momentum is along the *x* axis, and that change is

$$\Delta p_x = (-mv_x) - (mv_x) = -2mv_x.$$

Hence, the momentum Δp_x delivered to the wall by the molecule during the collision is $+2mv_x$. (Because in this book the symbol *p* represents both momentum and pressure, we must be careful to note that here *p* represents momentum and is a vector quantity.)

The molecule of Fig. 19-3 will hit the shaded wall repeatedly. The time Δt between collisions is the time the molecule takes to travel to the opposite wall and back again (a distance 2L) at speed v_x. Thus, Δt is equal to $2L/v_x$. (Note that this result holds even if the molecule bounces off any of the other walls along the way, because those walls are parallel to *x* and so cannot change v_x.) Therefore, the average rate at which momentum is delivered to the shaded wall by this single molecule is

$$\frac{\Delta p_x}{\Delta t} = \frac{2mv_x}{2L/v_x} = \frac{mv_x^2}{L}.$$

From Newton's second law ($\vec{F} = d\vec{p}/dt$), the rate at which momentum is delivered to the wall is the force acting on that wall. To find the total force, we must add up the contributions of all the molecules that strike the wall, allowing for the possibility that they all have different speeds. Dividing the magnitude of the total force F_x by the area of the wall ($= L^2$) then gives the pressure *p* on that wall, where now and in the rest of this discussion, *p* represents pressure. Thus, using the expression for $\Delta p_x/\Delta t$, we can write this pressure as

$$p = \frac{F_x}{L^2} = \frac{mv_{x1}^2/L + mv_{x2}^2/L + \cdots + mv_{xN}^2/L}{L^2}$$

$$= \left(\frac{m}{L^3}\right)(v_{x1}^2 + v_{x2}^2 + \cdots + v_{xN}^2), \qquad (19\text{-}18)$$

where N is the number of molecules in the box.

Since $N = nN_A$, there are nN_A terms in the second set of parentheses of Eq. 19-18. We can replace that quantity by $nN_A(v_x^2)_{avg}$, where $(v_x^2)_{avg}$ is the average value of the square of the x components of all the molecular speeds. Equation 19-18 then becomes

$$p = \frac{nmN_A}{L^3}(v_x^2)_{avg}.$$

However, mN_A is the molar mass M of the gas (that is, the mass of 1 mol of the gas). Also, L^3 is the volume of the box, so

$$p = \frac{nM(v_x^2)_{avg}}{V}. \qquad (19\text{-}19)$$

For any molecule, $v^2 = v_x^2 + v_y^2 + v_z^2$. Because there are many molecules and because they are all moving in random directions, the average values of the squares of their velocity components are equal, so that $v_x^2 = \frac{1}{3}v^2$. Thus, Eq. 19-19 becomes

$$p = \frac{nM(v^2)_{avg}}{3V}. \qquad (19\text{-}20)$$

The square root of $(v^2)_{avg}$ is a kind of average speed, called the **root-mean-square speed** of the molecules and symbolized by v_{rms}. Its name describes it rather well: You *square* each speed, you find the *mean* (that is, the average) of all these squared speeds, and then you take the square *root* of that mean. With $\sqrt{(v^2)_{avg}} = v_{rms}$, we can then write Eq. 19-20 as

$$p = \frac{nMv_{rms}^2}{3V}. \qquad (19\text{-}21)$$

Equation 19-21 is very much in the spirit of kinetic theory. It tells us how the pressure of the gas (a purely macroscopic quantity) depends on the speed of the molecules (a purely microscopic quantity).

We can turn Eq. 19-21 around and use it to calculate v_{rms}. Combining Eq. 19-21 with the ideal gas law ($pV = nRT$) leads to

$$v_{rms} = \sqrt{\frac{3RT}{M}}. \qquad (19\text{-}22)$$

Table 19-1 shows some rms speeds calculated from Eq. 19-22. The speeds are surprisingly high. For hydrogen molecules at room temperature (300 K), the rms speed is 1920 m/s, or 4300 mi/h—faster than a speeding bullet! On the surface of the Sun, where the temperature is 2×10^6 K, the rms speed of hydrogen molecules would be 82 times greater than at room temperature were it not for the fact that at such high speeds, the molecules cannot survive collisions among themselves. Remember too that the rms speed is only a kind of average speed; many molecules move much faster than this, and some much slower.

The speed of sound in a gas is closely related to the rms speed of the molecules of that gas. In a sound wave, the disturbance is passed on from molecule to molecule by means of collisions. The wave cannot move any faster than the "average" speed of the molecules. In fact, the speed of sound must be somewhat

TABLE 19-1

Some RMS Speeds at Room Temperature ($T = $ 300 K)[a]

Gas	Molar Mass (10^{-3} kg/mol)	v_{rms} (m/s)
Hydrogen (H_2)	2.02	1920
Helium (He)	4.0	1370
Water vapor (H_2O)	18.0	645
Nitrogen (N_2)	28.0	517
Oxygen (O_2)	32.0	483
Carbon dioxide (CO_2)	44.0	412
Sulfur dioxide (SO_2)	64.1	342

[a]For convenience, we often set room temperature equal to 300 K even though (at 27°C or 81°F) that represents a fairly warm room.

less than this "average" molecular speed because not all molecules are moving in exactly the same direction as the wave. As examples, at room temperature, the rms speeds of hydrogen and nitrogen molecules are 1920 m/s and 517 m/s, respectively. The speeds of sound in these two gases at this temperature are 1350 m/s and 350 m/s, respectively.

A question often arises: If molecules move so fast, why does it take as long as a minute or so before you can smell perfume when someone opens a bottle across a room? The answer is that, as we shall discuss in Section 19-6, each perfume molecule may have a high speed but it moves away from the bottle only very slowly because its repeated collisions with other molecules prevent it from moving directly across the room to you.

Sample Problem 19-3

Here are five numbers: 5, 11, 32, 67, and 89.

(a) What is the average value n_{avg} of these numbers?

Solution: We find this from

$$n_{avg} = \frac{5 + 11 + 32 + 67 + 89}{5} = 40.8.$$ (Answer)

(b) What is the rms value n_{rms} of these numbers?

Solution: We find this from

$$n_{rms} = \sqrt{\frac{5^2 + 11^2 + 32^2 + 67^2 + 89^2}{5}} = 52.1.$$ (Answer)

The rms value is greater than the average value because the larger numbers—being squared—are relatively more important in forming the rms value. To test this, let us replace 89 in our set of five numbers by 300. The average value of the new set of five numbers (as you should show) is 2.0 times the previous average value. The rms value, however, is 2.7 times the previous rms value.

19-5 Translational Kinetic Energy

We again consider a single molecule of an ideal gas as it moves around in the box of Fig. 19-3, but we now assume that its speed changes when it collides with other molecules. Its translational kinetic energy at any instant is $\frac{1}{2}mv^2$. Its *average* translational kinetic energy over the time that we watch it is

$$K_{avg} = \left(\tfrac{1}{2}mv^2\right)_{avg} = \tfrac{1}{2}m(v^2)_{avg} = \tfrac{1}{2}mv_{rms}^2,$$ (19-23)

in which we make the assumption that the average speed of the molecule during our observation is the same as the average speed of all the molecules at any given time. (Provided the total energy of the gas is not changing and provided we observe our molecule for long enough, this assumption is appropriate.) Substituting for v_{rms} from Eq. 19-22 leads to

$$K_{avg} = \left(\tfrac{1}{2}m\right)\frac{3RT}{M}.$$

However, M/m, the molar mass divided by the mass of a molecule, is simply Avogadro's number. Thus,

$$K_{avg} = \frac{3RT}{2N_A}.$$

Using Eq. 19-7 ($k = R/N_A$), we can then write

$$K_{avg} = \tfrac{3}{2}kT.$$ (19-24)

This equation tells us something unexpected:

▶ At a given temperature T, all ideal gas molecules—no matter what their mass—have the same average translational kinetic energy—namely, $\frac{3}{2}kT$. When we measure the temperature of a gas, we are also measuring the average translational kinetic energy of its molecules.

A gas mixture consists of molecules of types 1, 2, and 3, with molecular masses $m_1 > m_2 > m_3$. Rank the three types according to (a) average kinetic energy and (b) rms speed, greatest first.

Fig. 19-4 A molecule traveling through a gas, colliding with other gas molecules in its path. Although the other molecules are shown as stationary, they are also moving in a similar fashion.

19-6 Mean Free Path

We continue to examine the motion of molecules in an ideal gas. Figure 19-4 shows the path of a typical molecule as it moves through the gas, changing both speed and direction abruptly as it collides elastically with other molecules. Between collisions, the molecule moves in a straight line at constant speed. Although the figure shows the other molecules as stationary, they are also moving.

One useful parameter to describe this random motion is the **mean free path** λ of the molecules. As its name implies, λ is the average distance traversed by a molecule between collisions. We expect λ to vary inversely with N/V, the number of molecules per unit volume (or density of molecules). The larger N/V is, the more collisions there should be and the smaller the mean free path. We also expect λ to vary inversely with the size of the molecules—with their diameter d, say. (If the molecules were points, as we have assumed them to be, they would never collide and the mean free path would be infinite.) Thus, the larger the molecules are, the smaller the mean free path. We can even predict that λ should vary (inversely) as the *square* of the molecular diameter because the cross section of a molecule—not its diameter—determines its effective target area.

The expression for the mean free path does, in fact, turn out to be

$$\lambda = \frac{1}{\sqrt{2}\pi d^2\, N/V} \qquad \text{(mean free path).} \qquad (19\text{-}25)$$

To justify Eq. 19-25, we focus attention on a single molecule and assume—as Fig. 19-4 suggests—that our molecule is traveling with a constant speed v and that all the other molecules are at rest. Later, we shall relax this assumption.

We assume further that the molecules are spheres of diameter d. A collision will then take place if the centers of two molecules come within a distance d of each other, as in Fig. 19-5a. Another, more helpful way to look at the situation is to consider our single molecule to have a *radius* of d and all the other molecules to be *points,* as in Fig. 19-5b. This does not change our criterion for a collision.

As our single molecule zigzags through the gas, it sweeps out a short cylinder of cross-sectional area πd^2 between successive collisions. If we watch this molecule for a time interval Δt, it moves a distance $v\,\Delta t$, where v is its assumed speed. Thus, if we align all the short cylinders swept out in interval Δt, we form a composite cylinder (Fig. 19-6) of length $v\,\Delta t$ and volume $(\pi d^2)(v\,\Delta t)$. The number of collisions that occur in time Δt is then equal to the number of (point) molecules that lie within this cylinder.

Since N/V is the number of molecules per unit volume, the number of molecules in the cylinder is N/V times the volume of the cylinder, or $(N/V)(\pi d^2 v\,\Delta t)$. This is also the number of collisions in time Δt. The mean free path is the length of the path (and of the cylinder) divided by this number:

$$\lambda = \frac{\text{length of path during } \Delta t}{\text{number of collisions in } \Delta t} \approx \frac{v\,\Delta t}{\pi d^2 v\,\Delta t\, N/V}$$

$$= \frac{1}{\pi d^2\, N/V}. \qquad (19\text{-}26)$$

This equation is only approximate because it is based on the assumption that all the molecules except one are at rest. In fact, *all* the molecules are moving; when this is taken properly into account, Eq. 19-25 results. Note that it differs from the (approximate) Eq. 19-26 only by a factor of $1/\sqrt{2}$.

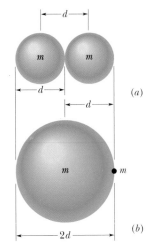

Fig. 19-5 (a) A collision occurs when the centers of two molecules come within a distance d of each other, d being the molecular diameter. (b) An equivalent but more convenient representation is to think of the moving molecule as having a *radius* d and all other molecules as being points. The condition for a collision is unchanged.

Fig. 19-6 In time Δt the moving molecule effectively sweeps out a cylinder of length $v\,\Delta t$ and radius d.

We can even get a glimpse of what is "approximate" about Eq. 19-26. The v in the numerator and that in the denominator are—strictly—not the same. The v in the numerator is v_{avg}, the mean speed of the molecules *relative to the container.* The v in the denominator is v_{rel}, the mean speed of our single molecule *relative to the other molecules,* which are moving. It is this latter average speed that determines the number of collisions. A detailed calculation, taking into account the actual speed distribution of the molecules, gives $v_{rel} = \sqrt{2} v_{avg}$ and thus the factor $\sqrt{2}$.

The mean free path of air molecules at sea level is about 0.1 μm. At an altitude of 100 km, the density of air has dropped to such an extent that the mean free path rises to about 16 cm. At 300 km, the mean free path is about 20 km. A problem faced by those who would study the physics and chemistry of the upper atmosphere in the laboratory is the unavailability of containers large enough to hold gas samples that simulate upper atmospheric conditions. Yet studies of the concentrations of Freon, carbon dioxide, and ozone in the upper atmosphere are of vital public concern.

Sample Problem 19-4

(a) What is the mean free path λ for oxygen molecules at temperature $T = 300$ K and pressure $p = 1.0$ atm? Assume that the molecular diameter is $d = 290$ pm and the gas is ideal.

Solution: The **Key Idea** here is that each oxygen molecule moves among other *moving* oxygen molecules in a zigzag path due to the resulting collisions. Thus, we use Eq. 19-25 for the mean free path, for which we need the number of molecules per unit volume, N/V. Because we assume the gas is ideal, we can use the ideal gas law of Eq. 19-9 ($pV = NkT$) to write $N/V = p/kT$. Substituting this into Eq. 19-25, we find

$$\lambda = \frac{1}{\sqrt{2}\pi d^2 \, N/V} = \frac{kT}{\sqrt{2}\pi d^2 p}$$

$$= \frac{(1.38 \times 10^{-23} \text{ J/K})(300 \text{ K})}{\sqrt{2}\pi (2.9 \times 10^{-10} \text{ m})^2 (1.01 \times 10^5 \text{ Pa})}$$

$$= 1.1 \times 10^{-7} \text{ m.} \qquad \text{(Answer)}$$

This is about 380 molecular diameters.

(b) Assume the average speed of the oxygen molecules is $v = 450$ m/s. What is the average time t between successive collisions for any given molecule? At what rate does the molecule collide; that is, what is the frequency f of its collisions?

Solution: To find the time t between collisions, we use this **Key Idea**: Between collisions, the molecule travels, on average,

the mean free path λ at speed v. Thus, the average time between collisions is

$$t = \frac{\text{distance}}{\text{speed}} = \frac{\lambda}{v} = \frac{1.1 \times 10^{-7} \text{ m}}{450 \text{ m/s}}$$

$$= 2.44 \times 10^{-10} \text{ s} \approx 0.24 \text{ ns.} \qquad \text{(Answer)}$$

This tells us that, on average, any given oxygen molecule has less than a nanosecond between collisions.

To find the frequency f of the collisions, we use this **Key Idea**: The average rate or frequency at which the collisions occur is the inverse of the time t between collisions. Thus,

$$f = \frac{1}{t} = \frac{1}{2.44 \times 10^{-10} \text{ s}} = 4.1 \times 10^9 \text{ s}^{-1}. \qquad \text{(Answer)}$$

This tells us that, on average, any given oxygen molecule makes about 4 billion collisions per second.

✔**CHECKPOINT 3** One mole of gas A, with molecular diameter $2d_0$ and average molecular speed v_0, is placed inside a certain container. One mole of gas B, with molecular diameter d_0 and average molecular speed $2v_0$ (the molecules of B are smaller but faster), is placed in an identical container. Which gas has the greater average collision rate within its container?

19-7 The Distribution of Molecular Speeds

The root-mean-square speed v_{rms} gives us a general idea of molecular speeds in a gas at a given temperature. We often want to know more. For example, what fraction of the molecules have speeds greater than the rms value? What fraction have speeds greater than twice the rms value? To answer such questions, we need to know how the possible values of speed are distributed among the molecules. Figure 19-7a shows this distribution for oxygen molecules at room temperature ($T = 300$ K); Fig. 19-7b compares it with the distribution at $T = 80$ K.

In 1852, Scottish physicist James Clerk Maxwell first solved the problem of finding the speed distribution of gas molecules. His result, known as **Maxwell's speed distribution law,** is

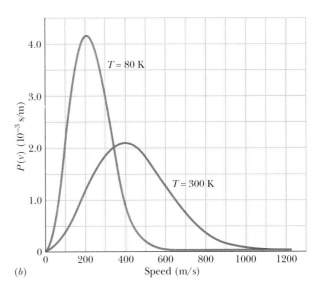

Fig. 19-7 (a) The Maxwell speed distribution for oxygen molecules at $T = 300$ K. The three characteristic speeds are marked. (b) The curves for 300 K and 80 K. Note that the molecules move more slowly at the lower temperature. Because these are probability distributions, the area under each curve has a numerical value of unity.

$$P(v) = 4\pi \left(\frac{M}{2\pi RT}\right)^{3/2} v^2 e^{-Mv^2/2RT}. \qquad (19\text{-}27)$$

Here M is the molar mass of the gas, R is the gas constant, T is the gas temperature, and v is the molecular speed. It is this equation that is plotted in Fig. 19-7a, b. The quantity $P(v)$ in Eq. 19-27 and Fig. 19-7 is a *probability distribution function*: For any speed v, the product $P(v)\, dv$ (a dimensionless quantity) is the fraction of molecules with speeds in the interval dv centered on speed v.

As Fig. 19-7a shows, this fraction is equal to the area of a strip with height $P(v)$ and width dv. The total area under the distribution curve corresponds to the fraction of the molecules whose speeds lie between zero and infinity. All molecules fall into this category, so the value of this total area is unity; that is,

$$\int_0^\infty P(v)\, dv = 1. \qquad (19\text{-}28)$$

The fraction (frac) of molecules with speeds in an interval of, say, v_1 to v_2 is then

$$\text{frac} = \int_{v_1}^{v_2} P(v)\, dv. \qquad (19\text{-}29)$$

Average, RMS, and Most Probable Speeds

In principle, we can find the **average speed** v_{avg} of the molecules in a gas with the following procedure: We *weight* each value of v in the distribution; that is, we multiply it by the fraction $P(v)\, dv$ of molecules with speeds in a differential interval dv centered on v. Then we add up all these values of $v\, P(v)\, dv$. The result is v_{avg}. In practice, we do all this by evaluating

$$v_{\text{avg}} = \int_0^\infty v\, P(v)\, dv. \qquad (19\text{-}30)$$

Substituting for $P(v)$ from Eq. 19-27 and using generic integral 20 from the list of integrals in Appendix E, we find

$$v_{\text{avg}} = \sqrt{\frac{8RT}{\pi M}} \qquad \text{(average speed).} \qquad (19\text{-}31)$$

Similarly, we can find the average of the square of the speeds $(v^2)_{\text{avg}}$ with

$$(v^2)_{\text{avg}} = \int_0^\infty v^2\, P(v)\, dv. \qquad (19\text{-}32)$$

Substituting for $P(v)$ from Eq. 19-27 and using generic integral 16 from the list of integrals in Appendix E, we find

$$(v^2)_{avg} = \frac{3RT}{M} .$$ (19-33)

The square root of $(v^2)_{avg}$ is the root-mean-square speed v_{rms}. Thus,

$$v_{rms} = \sqrt{\frac{3RT}{M}} \quad \text{(rms speed),}$$ (19-34)

which agrees with Eq. 19-22.

The **most probable speed** v_P is the speed at which $P(v)$ is maximum (see Fig. 19-7a). To calculate v_P, we set $dP/dv = 0$ (the slope of the curve in Fig. 19-7a is zero at the maximum of the curve) and then solve for v. Doing so, we find

$$v_P = \sqrt{\frac{2RT}{M}} \quad \text{(most probable speed).}$$ (19-35)

A molecule is more likely to have speed v_P than any other speed, but some molecules will have speeds that are many times v_P. These molecules lie in the *high-speed tail* of a distribution curve like that in Fig. 19-7a. We should be thankful for these few, higher speed molecules because they make possible both rain and sunshine (without which we could not exist). We next see why.

Rain   The speed distribution of water molecules in, say, a pond at summertime temperatures can be represented by a curve similar to that of Fig. 19-7a. Most of the molecules do not have nearly enough kinetic energy to escape from the water through its surface. However, small numbers of very fast molecules with speeds far out in the high-speed tail of the curve can do so. It is these water molecules that evaporate, making clouds and rain a possibility.

As the fast water molecules leave the surface, carrying energy with them, the temperature of the remaining water is maintained by heat transfer from the surroundings. Other fast molecules—produced in particularly favorable collisions— quickly take the place of those that have left, and the speed distribution is maintained.

Sunshine   Let the distribution curve of Fig. 19-7a now refer to protons in the core of the Sun. The Sun's energy is supplied by a nuclear fusion process that starts with the merging of two protons. However, protons repel each other because of their electrical charges, and protons of average speed do not have enough kinetic energy to overcome the repulsion and get close enough to merge. Very fast protons with speeds in the high-speed tail of the distribution curve can do so, however, and for that reason the Sun can shine.

Sample Problem 19-5

A container is filled with oxygen gas maintained at room temperature (300 K). What fraction of the molecules have speeds in the interval 599 to 601 m/s? The molar mass M of oxygen is 0.0320 kg/mol.

Solution: The Key Ideas here are the following:

1. The speeds of the molecules are distributed over a wide range of values, with the distribution $P(v)$ of Eq. 19-27.

2. The fraction of molecules with speeds in a differential interval dv is $P(v) \, dv$.

3. For a larger interval, the fraction is found by integrating $P(v)$ over the interval.

4. However, the interval $\Delta v = 2$ m/s here is small compared to the speed $v = 600$ m/s on which it is centered.

Thus, we can avoid the integration by approximating the fraction as

$$\text{frac} = P(v) \, \Delta v = 4\pi \left(\frac{M}{2\pi RT} \right)^{3/2} v^2 e^{-Mv^2/2RT} \, \Delta v.$$

The function $P(v)$ is plotted in Fig. 19-7a. The total area between the curve and the horizontal axis represents the total fraction of molecules (unity). The area of the thin gold strip represents the fraction we seek.

To evaluate frac in parts, we can write

$$\text{frac} = (4\pi)(A)(v^2)(e^B)(\Delta v), \qquad (19\text{-}36)$$

where

$$A = \left(\frac{M}{2\pi RT}\right)^{3/2} = \left(\frac{0.0320 \text{ kg/mol}}{(2\pi)(8.31 \text{ J/mol} \cdot \text{K})(300 \text{ K})}\right)^{3/2}$$
$$= 2.92 \times 10^{-9} \text{ s}^3/\text{m}^3$$

and $B = -\dfrac{Mv^2}{2RT} = -\dfrac{(0.0320 \text{ kg/mol})(600 \text{ m/s})^2}{(2)(8.31 \text{ J/mol} \cdot \text{K})(300 \text{ K})} = -2.31.$

Substituting A and B into Eq. 19-36 yields

$$\begin{aligned}\text{frac} &= (4\pi)(A)(v^2)(e^B)(\Delta v) \\ &= (4\pi)(2.92 \times 10^{-9} \text{ s}^3/\text{m}^3)(600 \text{ m/s})^2(e^{-2.31})(2 \text{ m/s}) \\ &= 2.62 \times 10^{-3}. \qquad \text{(Answer)}\end{aligned}$$

Thus, at room temperature, 0.262% of the oxygen molecules will have speeds that lie in the narrow range between 599 and 601 m/s. If the gold strip of Fig. 19-7a were drawn to the scale of this problem, it would be a very thin strip indeed.

Sample Problem 19-6

The molar mass M of oxygen is 0.0320 kg/mol.

(a) What is the average speed v_{avg} of oxygen gas molecules at $T = 300$ K?

Solution: The Key Idea here is that to find the average speed, we must weight speed v with the distribution function $P(v)$ of Eq. 19-27 and then integrate the resulting expression over the range of possible speeds (0 to ∞). That leads to Eq. 19-31, which gives us

$$\begin{aligned}v_{avg} &= \sqrt{\frac{8RT}{\pi M}} \\ &= \sqrt{\frac{8(8.31 \text{ J/mol} \cdot \text{K})(300 \text{ K})}{\pi(0.0320 \text{ kg/mol})}} = 445 \text{ m/s}. \quad \text{(Answer)}\end{aligned}$$

This result is plotted in Fig. 19-7a.

(b) What is the root-mean-square speed v_{rms} at 300 K?

Solution: The Key Idea here is that to find v_{rms}, we must first find $(v^2)_{avg}$ by weighting v^2 with the distribution function $P(v)$ of Eq. 19-27 and then integrating the expression over the range of possible speeds. Then we must take the square root

of the result. That leads to Eq. 19-34, which gives us

$$\begin{aligned}v_{rms} &= \sqrt{\frac{3RT}{M}} \\ &= \sqrt{\frac{3(8.31 \text{ J/mol} \cdot \text{K})(300 \text{ K})}{0.0320 \text{ kg/mol}}} = 483 \text{ m/s}. \quad \text{(Answer)}\end{aligned}$$

This result, plotted in Fig. 19-7a, is greater than v_{avg} because the greater speed values influence the calculation more when we integrate the v^2 values than when we integrate the v values.

(c) What is the most probable speed v_P at 300 K?

Solution: The Key Idea here is that v_P corresponds to the maximum of the distribution function $P(v)$, which we obtain by setting the derivative $dP/dv = 0$ and solving the result for v. That leads to Eq. 19-35, which gives us

$$\begin{aligned}v_P &= \sqrt{\frac{2RT}{M}} \\ &= \sqrt{\frac{2(8.31 \text{ J/mol} \cdot \text{K})(300 \text{ K})}{0.0320 \text{ kg/mol}}} = 395 \text{ m/s}. \quad \text{(Answer)}\end{aligned}$$

This result is also plotted in Fig. 19-7a.

19-8 The Molar Specific Heats of an Ideal Gas

In this section, we want to derive from molecular considerations an expression for the internal energy E_{int} of an ideal gas. In other words, we want an expression for the energy associated with the random motions of the atoms or molecules in the gas. We shall then use that expression to derive the molar specific heats of an ideal gas.

Internal Energy E_{int}

Let us first assume that our ideal gas is a *monatomic gas* (which has individual atoms rather than molecules), such as helium, neon, or argon. Let us also assume that the internal energy E_{int} of our ideal gas is simply the sum of the translational kinetic energies of its atoms. (As explained by quantum theory, individual atoms do not have rotational kinetic energy.)

The average translational kinetic energy of a single atom depends only on the gas temperature and is given by Eq. 19-24 as $K_{avg} = \frac{3}{2}kT$. A sample of n

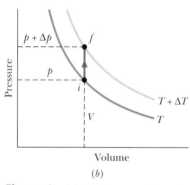

Fig. 19-8 (a) The temperature of an ideal gas is raised from T to $T + \Delta T$ in a constant-volume process. Heat is added, but no work is done. (b) The process on a p-V diagram.

moles of such a gas contains nN_A atoms. The internal energy E_{int} of the sample is then

$$E_{int} = (nN_A)K_{avg} = (nN_A)(\tfrac{3}{2}kT). \tag{19-37}$$

Using Eq. 19-7 ($k = R/N_A$), we can rewrite this as

$$E_{int} = \tfrac{3}{2}nRT \qquad \text{(monatomic ideal gas).} \tag{19-38}$$

Thus,

> The internal energy E_{int} of an ideal gas is a function of the gas temperature *only*; it does not depend on any other variable.

With Eq. 19-38 in hand, we are now able to derive an expression for the molar specific heat of an ideal gas. Actually, we shall derive two expressions. One is for the case in which the volume of the gas remains constant as energy is transferred to or from it as heat. The other is for the case in which the pressure of the gas remains constant as energy is transferred to or from it as heat. The symbols for these two molar specific heats are C_V and C_p, respectively. (By convention, the capital letter C is used in both cases, even though C_V and C_p represent types of specific heat and not heat capacities.)

Molar Specific Heat at Constant Volume

Figure 19-8a shows n moles of an ideal gas at pressure p and temperature T, confined to a cylinder of fixed volume V. This *initial state i* of the gas is marked on the p-V diagram of Fig. 19-8b. Suppose now that you add a small amount of energy to the gas as heat Q by slowly turning up the temperature of the thermal reservoir. The gas temperature rises a small amount to $T + \Delta T$, and its pressure rises to $p + \Delta p$, bringing the gas to *final state f*.

In such experiments, we would find that the heat Q is related to the temperature change ΔT by

$$Q = nC_V\,\Delta T \qquad \text{(constant volume),} \tag{19-39}$$

where C_V is a constant called the **molar specific heat at constant volume.** Substituting this expression for Q into the first law of thermodynamics as given by Eq. 18-26 ($\Delta E_{int} = Q - W$) yields

$$\Delta E_{int} = nC_V\,\Delta T - W. \tag{19-40}$$

With the volume held constant, the gas cannot expand and thus cannot do any work. Therefore, $W = 0$, and Eq. 19-40 gives us

$$C_V = \frac{\Delta E_{int}}{n\,\Delta T}. \tag{19-41}$$

From Eq. 19-38 we know that $E_{int} = \tfrac{3}{2}nRT$, so the change in internal energy must be

$$\Delta E_{int} = \tfrac{3}{2}nR\,\Delta T. \tag{19-42}$$

Substituting this result into Eq. 19-41 yields

$$C_V = \tfrac{3}{2}R = 12.5 \text{ J/mol} \cdot \text{K} \qquad \text{(monatomic gas).} \tag{19-43}$$

As Table 19-2 shows, this prediction of the kinetic theory (for ideal gases) agrees very well with experiment for real monatomic gases, the case that we have assumed. The (predicted and) experimental values of C_V for *diatomic gases* (which have molecules with two atoms) and *polyatomic gases* (which have molecules with more than two atoms) are greater than those for monatomic gases for reasons that will be suggested in Section 19-9.

TABLE 19-2
Molar Specific Heats at Constant Volume

Molecule	Example		C_V (J/mol · K)
Monatomic	Ideal		$\frac{3}{2}R = 12.5$
	Real	He	12.5
		Ar	12.6
Diatomic	Ideal		$\frac{5}{2}R = 20.8$
	Real	N_2	20.7
		O_2	20.8
Polyatomic	Ideal		$3R = 24.9$
	Real	NH_4	29.0
		CO_2	29.7

We can now generalize Eq. 19-38 for the internal energy of any ideal gas by substituting C_V for $\frac{3}{2}R$; we get

$$E_{int} = nC_V T \qquad \text{(any ideal gas).} \qquad (19\text{-}44)$$

This equation applies not only to an ideal monatomic gas but also to diatomic and polyatomic ideal gases, provided the appropriate value of C_V is used. Just as with Eq. 19-38, we see that the internal energy of a gas depends on the temperature of the gas but not on its pressure or density.

When an ideal gas that is confined to a container undergoes a temperature change ΔT, then from either Eq. 19-41 or Eq. 19-44 we can write the resulting change in its internal energy as

$$\Delta E_{int} = nC_V \Delta T \qquad \text{(ideal gas, any process).} \qquad (19\text{-}45)$$

This equation tells us:

> A change in the internal energy E_{int} of a confined ideal gas depends on the change in the gas temperature only; it does *not* depend on what type of process produces the change in the temperature.

As examples, consider the three paths between the two isotherms in the p-V diagram of Fig. 19-9. Path 1 represents a constant-volume process. Path 2 represents a constant-pressure process (that we are about to examine). Path 3 represents a process in which no heat is exchanged with the system's environment (we discuss this in Section 19-11). Although the values of heat Q and work W associated with these three paths differ, as do p_f and V_f, the values of ΔE_{int} associated with the three paths are identical and are all given by Eq. 19-45, because they all involve the same temperature change ΔT. Therefore, no matter what path is actually taken between T and $T + \Delta T$, we can *always* use path 1 and Eq. 19-45 to compute ΔE_{int} easily.

Molar Specific Heat at Constant Pressure

We now assume that the temperature of our ideal gas is increased by the same small amount ΔT as previously but now the necessary energy (heat Q) is added with the gas under constant pressure. An experiment for doing this is shown in Fig. 19-10a; the p-V diagram for the process is plotted in Fig. 19-10b. From such experiments we find that the heat Q is related to the temperature change ΔT by

$$Q = nC_p \Delta T \qquad \text{(constant pressure),} \qquad (19\text{-}46)$$

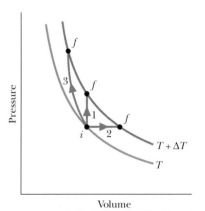

Fig. 19-9 Three paths representing three different processes that take an ideal gas from an initial state i at temperature T to some final state f at temperature $T + \Delta T$. The change ΔE_{int} in the internal energy of the gas is the same for these three processes and for any others that result in the same change of temperature.

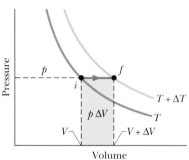

(a)

(b)

Fig. 19-10 (a) The temperature of an ideal gas is raised from T to $T + \Delta T$ in a constant-pressure process. Heat is added and work is done in lifting the loaded piston. (b) The process on a p-V diagram. The work $p\, \Delta V$ is given by the shaded area.

where C_p is a constant called the **molar specific heat at constant pressure.** This C_p is *greater* than the molar specific heat at constant volume C_V, because energy must now be supplied not only to raise the temperature of the gas but also for the gas to do work—that is, to lift the weighted piston of Fig. 19-10a.

To relate molar specific heats C_p and C_V, we start with the first law of thermodynamics (Eq. 19-26):

$$\Delta E_{\text{int}} = Q - W. \qquad (19\text{-}47)$$

We next replace each term in Eq. 19-47. For ΔE_{int}, we substitute from Eq. 19-45. For Q, we substitute from Eq. 19-46. To replace W, we first note that since the pressure remains constant, Eq. 19-16 tells us that $W = p\, \Delta V$. Then we note that, using the ideal gas equation ($pV = nRT$), we can write

$$W = p\, \Delta V = nR\, \Delta T. \qquad (19\text{-}48)$$

Making these substitutions in Eq. 19-47 and then dividing through by $n\, \Delta T$, we find

$$C_V = C_p - R$$

and then

$$C_p = C_V + R. \qquad (19\text{-}49)$$

This prediction of kinetic theory agrees well with experiment, not only for monatomic gases but for gases in general, as long as their density is low enough so that we may treat them as ideal.

✔**CHECKPOINT 4** The figure here shows five paths traversed by a gas on a p-V diagram. Rank the paths according to the change in internal energy of the gas, greatest first.

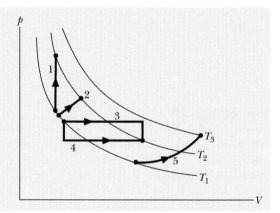

Sample Problem 19-7

A bubble of 5.00 mol of helium is submerged at a certain depth in liquid water when the water (and thus the helium) undergoes a temperature increase ΔT of 20.0 C° at constant pressure. As a result, the bubble expands. The helium is monatomic and ideal.

(a) How much energy is added to the helium as heat during the increase and expansion?

Solution: One Key Idea here is that the heat Q is related to the temperature change ΔT by a molar specific heat of the gas. Because the pressure p is held constant during the addition of energy, we use the molar specific heat at constant pressure C_p and Eq. 19-46,

$$Q = nC_p\, \Delta T, \qquad (19\text{-}50)$$

to find Q. To evaluate C_p we go to Eq. 19-49, which tells us that for any ideal gas, $C_p = C_V + R$. Then from Eq. 19-43, we know that for any *monatomic* gas (like the helium here),

$C_V = \frac{3}{2}R$. Thus, Eq. 19-50 gives us

$$Q = n(C_V + R)\, \Delta T = n(\tfrac{3}{2}R + R)\, \Delta T = n(\tfrac{5}{2}R)\, \Delta T$$
$$= (5.00\ \text{mol})(2.5)(8.31\ \text{J/mol} \cdot \text{K})(20.0\ \text{C}°)$$
$$= 2077.5\ \text{J} \approx 2080\ \text{J}. \qquad \text{(Answer)}$$

(b) What is the change ΔE_{int} in the internal energy of the helium during the temperature increase?

Solution: Because the bubble expands, this is not a constant-volume process. However, the helium is nonetheless confined (to the bubble). Thus, a Key Idea here is that the change ΔE_{int} is the same as *would occur* in a constant-volume process with the same temperature change ΔT. We can easily find the constant-volume change ΔE_{int} with Eq. 19-45:

$$\Delta E_{\text{int}} = nC_V\, \Delta T = n(\tfrac{3}{2}R)\, \Delta T$$
$$= (5.00\ \text{mol})(1.5)(8.31\ \text{J/mol} \cdot \text{K})(20.0\ \text{C}°)$$
$$= 1246.5\ \text{J} \approx 1250\ \text{J}. \qquad \text{(Answer)}$$

(c) How much work W is done by the helium as it expands against the pressure of the surrounding water during the temperature increase?

Solution: One Key Idea here is that the work done by *any* gas expanding against the pressure from its environment is given by Eq. 19-11, which tells us to integrate $p \, dV$. When the pressure is constant (as here), we can simplify that to $W = p \, \Delta V$. When the gas is *ideal* (as here), we can use the ideal gas law (Eq. 19-5) to write $p \, \Delta V = nR \, \Delta T$. We end up with

$$W = nR \, \Delta T$$
$$= (5.00 \text{ mol})(8.31 \text{ J/mol} \cdot \text{K})(20.0 \text{ C}°)$$
$$= 831 \text{ J}. \quad \text{(Answer)}$$

Because we happen to know Q and ΔE_{int}, we can work

this problem another way: The Key Idea now is that we can account for the energy changes of the gas with the first law of thermodynamics, writing

$$W = Q - \Delta E_{int} = 2077.5 \text{ J} - 1246.5 \text{ J}$$
$$= 831 \text{ J}. \quad \text{(Answer)}$$

Note that during the temperature increase, only a portion (1250 J) of the energy (2080 J) that is transferred to the helium as heat goes to increasing the internal energy of the helium and thus the temperature of the helium. The rest (831 J) is transferred out of the helium as work that the helium does during the expansion. If the water were frozen, it would not allow that expansion. Then the same temperature increase of 20.0 C° would require only 1250 J of heat, because no work would be done by the helium.

19-9 Degrees of Freedom and Molar Specific Heats

As Table 19-2 shows, the prediction that $C_V = \frac{3}{2}R$ agrees with experiment for monatomic gases but fails for diatomic and polyatomic gases. Let us try to explain the discrepancy by considering the possibility that molecules with more than one atom can store internal energy in forms other than translational kinetic energy.

Figure 19-11 shows common models of helium (a *monatomic* molecule, containing a single atom), oxygen (a *diatomic* molecule, containing two atoms), and methane (a *polyatomic* molecule). From such models, we would assume that all three types of molecules can have translational motions (say, moving left–right and up–down) and rotational motions (spinning about an axis like a top). In addition, we would assume that the diatomic and polyatomic molecules can have oscillatory motions, with the atoms oscillating slightly toward and away from one another, as if attached to opposite ends of a spring.

To keep account of the various ways in which energy can be stored in a gas, James Clerk Maxwell introduced the theorem of the **equipartition of energy**:

> Every kind of molecule has a certain number f of *degrees of freedom*, which are independent ways in which the molecule can store energy. Each such degree of freedom has associated with it—on average—an energy of $\frac{1}{2}kT$ per molecule (or $\frac{1}{2}RT$ per mole).

Let us apply the theorem to the translational and rotational motions of the molecules in Fig. 19-11. (We discuss oscillatory motion in the next section.) For the translational motion, superimpose an *xyz* coordinate system on any gas. The molecules will, in general, have velocity components along all three axes. Thus, gas molecules of all types have three degrees of translational freedom (three ways to move in translation) and, on average, an associated energy of $3(\frac{1}{2}kT)$ per molecule.

For the rotational motion, imagine the origin of our *xyz* coordinate system at the center of each molecule in Fig. 19-11. In a gas, each molecule should be able to rotate with an angular velocity component along each of the three axes, so each gas should have three degrees of rotational freedom and, on average, an additional energy of $3(\frac{1}{2}kT)$ per molecule. *However*, experiment shows this is true only for the polyatomic molecules. According to *quantum theory*, the physics dealing with the allowed motions and energies of molecules and atoms, a monatomic gas molecule does not rotate and so has no rotational energy (a single atom cannot rotate like a top). A diatomic molecule can rotate like a top only about axes perpendicular to the line connecting the atoms (the axes are shown in Fig. 19-11b) and not about that line itself. Therefore, a diatomic molecule can

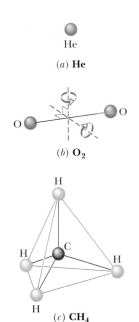

Fig. 19-11 Models of molecules as used in kinetic theory: (*a*) helium, a typical monatomic molecule; (*b*) oxygen, a typical diatomic molecule; and (*c*) methane, a typical polyatomic molecule. The spheres represent atoms, and the lines between them represent bonds. Two rotation axes are shown for the oxygen molecule.

TABLE 19-3 Degrees of Freedom for Various Molecules

| Molecule | Example | Degrees of Freedom | | | Predicted Molar Specific Heats | |
		Translational	Rotational	Total (f)	C_V (Eq. 19-51)	$C_p = C_V + R$
Monatomic	He	3	0	3	$\frac{3}{2}R$	$\frac{5}{2}R$
Diatomic	O_2	3	2	5	$\frac{5}{2}R$	$\frac{7}{2}R$
Polyatomic	CH_4	3	3	6	$3R$	$4R$

have only two degrees of rotational freedom and a rotational energy of only $2(\frac{1}{2}kT)$ per molecule.

To extend our analysis of molar specific heats (C_p and C_V, in Section 19-8) to ideal diatomic and polyatomic gases, it is necessary to retrace the derivations of that analysis in detail. First, we replace Eq. 19-38 ($E_{int} = \frac{3}{2}nRT$) with $E_{int} = (f/2)nRT$, where f is the number of degrees of freedom listed in Table 19-3. Doing so leads to the prediction

$$C_V = \left(\frac{f}{2}\right)R = 4.16f \text{ J/mol} \cdot \text{K},\qquad (19\text{-}51)$$

which agrees—as it must—with Eq. 19-43 for monatomic gases ($f = 3$). As Table 19-2 shows, this prediction also agrees with experiment for diatomic gases ($f = 5$), but it is too low for polyatomic gases ($f = 6$ for molecules comparable to CH_4).

Sample Problem 19-8

A cabin of volume V is filled with air (which we consider to be an ideal diatomic gas) at an initial low temperature T_1. After you light a wood stove, the air temperature increases to T_2. What is the resulting change ΔE_{int} in the internal energy of the air in the cabin?

Solution: As the air temperature increases, the air pressure p cannot change but must always be equal to the air pressure outside the room. The reason is that, because the room is not airtight, the air is not confined. As the temperature increases, air molecules leave through various openings and thus the number of moles n of air in the room decreases. Thus, one **Key Idea** here is that we *cannot* use Eq. 19-45 ($\Delta E_{int} = nC_V \Delta T$) to find ΔE_{int}, because it requires constant n.

A second **Key Idea** is that we *can* relate the internal energy E_{int} at any instant to n and the temperature T with Eq. 19-44 ($E_{int} = nC_V T$). From that equation we can then write

$$\Delta E_{int} = \Delta(nC_V T) = C_V \Delta(nT).$$

Next, using Eq. 19-5 ($pV = nRT$), we can replace nT with pV/R, obtaining

$$\Delta E_{int} = C_V \Delta\left(\frac{pV}{R}\right).\qquad (19\text{-}52)$$

Now, because p, V, and R are all constants, Eq. 19-52 yields

$$\Delta E_{int} = 0,\qquad \text{(Answer)}$$

even though the temperature changes.

Why does the cabin feel more comfortable at the higher temperature? There are at least two factors involved: (1) You exchange electromagnetic radiation (thermal radiation) with surfaces inside the room, and (2) you exchange energy with air molecules that collide with you. When the room temperature is increased, (1) the amount of thermal radiation emitted by the surfaces and absorbed by you is increased, and (2) the amount of energy you gain through the collisions of air molecules with you is increased.

19-10 A Hint of Quantum Theory

We can improve the agreement of kinetic theory with experiment by including the oscillations of the atoms in a gas of diatomic or polyatomic molecules. For example, the two atoms in the O_2 molecule of Fig. 19-11b can oscillate toward and away from each other, with the interconnecting bond acting like a spring. However, experiment shows that such oscillations occur only at relatively high temperatures of the gas—the motion is "turned on" only when the gas molecules have relatively large energies. Rotational motion is also subject to such "turning on," but at a lower temperature.

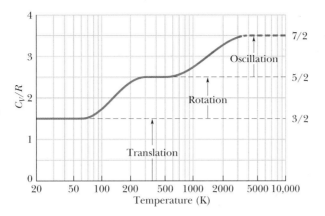

Fig. 19-12 C_V/R versus temperature for (diatomic) hydrogen gas. Because rotational and oscillatory motions begin at certain energies, only translation is possible at very low temperatures. As the temperature increases, rotational motion can begin. At still higher temperatures, oscillatory motion can begin.

Figure 19-12 is of help in seeing this turning on of rotational motion and oscillatory motion. The ratio C_V/R for diatomic hydrogen gas (H_2) is plotted there against temperature, with the temperature scale logarithmic to cover several orders of magnitude. Below about 80 K, we find that $C_V/R = 1.5$. This result implies that only the three translational degrees of freedom of hydrogen are involved in the specific heat.

As the temperature increases, the value of C_V/R gradually increases to 2.5, implying that two additional degrees of freedom have become involved. Quantum theory shows that these two degrees of freedom are associated with the rotational motion of the hydrogen molecules and that this motion requires a certain minimum amount of energy. At very low temperatures (below 80 K), the molecules do not have enough energy to rotate. As the temperature increases from 80 K, first a few molecules and then more and more obtain enough energy to rotate, and C_V/R increases, until all of them are rotating and $C_V/R = 2.5$.

Similarly, quantum theory shows that oscillatory motion of the molecules requires a certain (higher) minimum amount of energy. This minimum amount is not met until the molecules reach a temperature of about 1000 K, as shown in Fig. 19-12. As the temperature increases beyond 1000 K, more molecules have enough energy to oscillate and C_V/R increases, until all of them are oscillating and $C_V/R = 3.5$. (In Fig. 19-12, the plotted curve stops at 3200 K because there the atoms of a hydrogen molecule oscillate so much that they overwhelm their bond, and the molecule then *dissociates* into two separate atoms.)

19-11 *The Adiabatic Expansion of an Ideal Gas*

We saw in Section 17-4 that sound waves are propagated through air and other gases as a series of compressions and expansions; these variations in the transmission medium take place so rapidly that there is no time for energy to be transferred from one part of the medium to another as heat. As we saw in Section 18-11, a process for which $Q = 0$ is an *adiabatic process*. We can ensure that $Q = 0$ either by carrying out the process very quickly (as in sound waves) or by doing it (at any rate) in a well-insulated container. Let us see what the kinetic theory has to say about adiabatic processes.

Figure 19-13a shows our usual insulated cylinder, now containing an ideal gas and resting on an insulating stand. By removing mass from the piston, we can allow the gas to expand adiabatically. As the volume increases, both the pressure and the temperature drop. We shall prove next that the relation between the pressure and the volume during such an adiabatic process is

$$pV^\gamma = \text{a constant} \qquad \text{(adiabatic process)}, \qquad (19\text{-}53)$$

in which $\gamma = C_p/C_V$, the ratio of the molar specific heats for the gas. On a p-V diagram such as that in Fig. 19-13b, the process occurs along a line (called an

Fig. 19-13 (a) The volume of an ideal gas is increased by removing mass from the piston. The process is adiabatic ($Q = 0$). (b) The process proceeds from i to f along an adiabat on a p-V diagram.

adiabat) that has the equation $p = $ (a constant)$/V^\gamma$. Since the gas goes from an initial state i to a final state f, we can rewrite Eq. 19-53 as

$$p_i V_i^\gamma = p_f V_f^\gamma \qquad \text{(adiabatic process).} \qquad (19\text{-}54)$$

We can also write an equation for an adiabatic process in terms of T and V. To do so, we use the ideal gas equation ($pV = nRT$) to eliminate p from Eq. 19-53, finding

$$\left(\frac{nRT}{V}\right) V^\gamma = \text{a constant.}$$

Because n and R are constants, we can rewrite this in the alternative form

$$T V^{\gamma-1} = \text{a constant} \qquad \text{(adiabatic process),} \qquad (19\text{-}55)$$

in which the constant is different from that in Eq. 19-53. When the gas goes from an initial state i to a final state f, we can rewrite Eq. 19-55 as

$$T_i V_i^{\gamma-1} = T_f V_f^{\gamma-1} \qquad \text{(adiabatic process).} \qquad (19\text{-}56)$$

Understanding adiabatic processes allows you to understand why popping the cork on a cold bottle of champagne or the tab on a cold can of soda causes a slight fog to form at the opening of the container. At the top of any unopened carbonated drink sits a gas of carbon dioxide and water vapor. Because the gas pressure is greater than atmospheric pressure, the gas expands out into the atmosphere when the container is opened. Thus, the gas volume increases, but that means the gas must do work pushing against the atmosphere. Because the expansion is rapid, it is adiabatic, and the only source of energy for the work is the internal energy of the gas. Because the internal energy decreases, the temperature of the gas also decreases, which causes the water vapor in the gas to condense into tiny drops of fog. (Note that Eq. 19-56 also tells us that the temperature must decrease during an adiabatic expansion: V_f is greater than V_i, and so T_f must be less than T_i.)

Proof of Eq. 19-53

Suppose that you remove some shot from the piston of Fig. 19-13a, allowing the ideal gas to push the piston and the remaining shot upward and thus to increase the volume by a differential amount dV. Since the volume change is tiny, we may assume that the pressure p of the gas on the piston is constant during the change. This assumption allows us to say that the work dW done by the gas during the volume increase is equal to $p\,dV$. From Eq. 18-27, the first law of thermodynamics can then be written as

$$dE_{\text{int}} = Q - p\,dV. \qquad (19\text{-}57)$$

Since the gas is thermally insulated (and thus the expansion is adiabatic), we substitute 0 for Q. Then we use Eq. 19-45 to substitute $nC_V\,dT$ for dE_{int}. With these substitutions, and after some rearranging, we have

$$n\,dT = -\left(\frac{p}{C_V}\right)dV. \qquad (19\text{-}58)$$

Now from the ideal gas law ($pV = nRT$) we have

$$p\,dV + V\,dp = nR\,dT. \qquad (19\text{-}59)$$

Replacing R with its equal, $C_p - C_V$, in Eq. 19-59 yields

$$n\,dT = \frac{p\,dV + V\,dp}{C_p - C_V}. \qquad (19\text{-}60)$$

Equating Eqs. 19-58 and 19-60 and rearranging then give

$$\frac{dp}{p} + \left(\frac{C_p}{C_V}\right)\frac{dV}{V} = 0.$$

Replacing the ratio of the molar specific heats with γ and integrating (see integral 5 in Appendix E) yield

$$\ln p + \gamma \ln V = \text{a constant.}$$

Rewriting the left side as $\ln pV^\gamma$ and then taking the antilog of both sides, we find

$$pV^\gamma = \text{a constant,} \qquad (19\text{-}61)$$

which is what we set out to prove.

Free Expansions

Recall from Section 18-11 that a free expansion of a gas is an adiabatic process that involves no work done on or by the gas, and no change in the internal energy of the gas. A free expansion is thus quite different from the type of adiabatic process described by Eqs. 19-53 through 19-61, in which work is done and the internal energy changes. Those equations then do *not* apply to a free expansion, even though such an expansion is adiabatic.

Also recall that in a free expansion, a gas is in equilibrium only at its initial and final points; thus, we can plot only those points, but not the expansion itself, on a p-V diagram. In addition, because $\Delta E_{int} = 0$, the temperature of the final state must be that of the initial state. Thus, the initial and final points on a p-V diagram must be on the same isotherm, and instead of Eq. 19-56 we have

$$T_i = T_f \quad \text{(free expansion).} \qquad (19\text{-}62)$$

If we next assume that the gas is ideal (so that $pV = nRT$), then because there is no change in temperature, there can be no change in the product pV. Thus, instead of Eq. 19-53 a free expansion involves the relation

$$p_iV_i = p_fV_f \quad \text{(free expansion).} \qquad (19\text{-}63)$$

Sample Problem 19-9

In Sample Problem 19-2, 1 mol of oxygen (assumed to be an ideal gas) expands isothermally (at 310 K) from an initial volume of 12 L to a final volume of 19 L.

(a) What would be the final temperature if the gas had expanded adiabatically to this same final volume? Oxygen (O_2) is diatomic and here has rotation but not oscillation.

Solution: The Key Ideas here are as follows:

1. When a gas expands against the pressure of its environment, it must do work.

2. When the process is adiabatic (no energy is transferred as heat), then the energy required for the work can come only from the internal energy of the gas.

3. Because the internal energy decreases, the temperature T must also decrease.

We can relate the initial and final temperatures and volumes with Eq. 19-56:

$$T_i V_i^{\gamma-1} = T_f V_f^{\gamma-1}. \qquad (19\text{-}64)$$

Because the molecules are diatomic and have rotation but not oscillation, we can take the molar specific heats from Table 19-3. Thus,

$$\gamma = \frac{C_p}{C_V} = \frac{\tfrac{7}{2}R}{\tfrac{5}{2}R} = 1.40.$$

Solving Eq. 19-64 for T_f and inserting known data then yield

$$T_f = \frac{T_i V_i^{\gamma-1}}{V_f^{\gamma-1}} = \frac{(310\ \text{K})(12\ \text{L})^{1.40-1}}{(19\ \text{L})^{1.40-1}}$$

$$= (310\ \text{K})(\tfrac{12}{19})^{0.40} = 258\ \text{K}. \qquad \text{(Answer)}$$

(b) What would be the final temperature and pressure if, instead, the gas had expanded freely to the new volume, from an initial pressure of 2.0 Pa?

Solution: Here the **Key Idea** is that the temperature does not change in a free expansion:

$$T_f = T_i = 310\ \text{K}. \qquad \text{(Answer)}$$

We find the new pressure using Eq. 19-63, which gives us

$$p_f = p_i \frac{V_i}{V_f} = (2.0\ \text{Pa})\frac{12\ \text{L}}{19\ \text{L}} = 1.3\ \text{Pa}. \qquad \text{(Answer)}$$

PROBLEM-SOLVING TACTICS

TACTIC 2: *A Graphical Summary of Four Gas Processes*

In this chapter we have discussed four special processes that an ideal gas can undergo. An example of each (for a monatomic ideal gas) is shown in Fig. 19-14, and some associated characteristics are given in Table 19-4, including two process names (isobaric and isochoric) that we have not used but that you might see in other courses.

Fig. 19-14 A p-V diagram representing four special processes for an ideal monatomic gas.

TABLE 19-4
Four Special Processes

Path in Fig. 19-14	Constant Quantity	Process Type	Some Special Results ($\Delta E_{int} = Q - W$ and $\Delta E_{int} = nC_V \Delta T$ for all paths)
1	p	Isobaric	$Q = nC_p \Delta T;$ $W = p \Delta V$
2	T	Isothermal	$Q = W = nRT \ln(V_f/V_i);$ $\Delta E_{int} = 0$
3	$pV^{\gamma}, TV^{\gamma-1}$	Adiabatic	$Q = 0;$ $W = -\Delta E_{int}$
4	V	Isochoric	$Q = \Delta E_{int} = nC_V \Delta T;$ $W = 0$

CHECKPOINT 5 Rank paths 1, 2, and 3 in Fig. 19-14 according to the energy transfer to the gas as heat, greatest first.

Review & Summary

Kinetic Theory of Gases The *kinetic theory of gases* relates the *macroscopic* properties of gases (for example, pressure and temperature) to the *microscopic* properties of gas molecules (for example, speed and kinetic energy).

Avogadro's Number One mole of a substance contains N_A (*Avogadro's number*) elementary units (usually atoms or molecules), where N_A is found experimentally to be

$$N_A = 6.02 \times 10^{23}\ \text{mol}^{-1} \qquad \text{(Avogadro's number).} \quad (19\text{-}1)$$

One molar mass M of any substance is the mass of one mole of the substance. It is related to the mass m of the individual molecules of the substance by

$$M = mN_A. \qquad (19\text{-}4)$$

The number of moles n contained in a sample of mass M_{sam}, consisting of N molecules, is given by

$$n = \frac{N}{N_A} = \frac{M_{sam}}{M} = \frac{M_{sam}}{mN_A}. \qquad (19\text{-}2,\ 19\text{-}3)$$

Ideal Gas An *ideal gas* is one for which the pressure p, volume V, and temperature T are related by

$$pV = nRT \qquad \text{(ideal gas law).} \qquad (19\text{-}5)$$

Here n is the number of moles of the gas present and R is a constant ($8.31\ \text{J/mol} \cdot \text{K}$) called the **gas constant.** The ideal gas law can also be written as

$$pV = NkT, \qquad (19\text{-}9)$$

where the **Boltzmann constant** k is

$$k = \frac{R}{N_A} = 1.38 \times 10^{-23}\ \text{J/K}. \qquad (19\text{-}7)$$

Work in an Isothermal Volume Change The work done *by* an ideal gas during an **isothermal** (constant-temperature) change from volume V_i to volume V_f is

$$W = nRT \ln \frac{V_f}{V_i} \qquad \text{(ideal gas, isothermal process).} \qquad (19\text{-}14)$$

Pressure, Temperature, and Molecular Speed The pressure exerted by n moles of an ideal gas, in terms of the speed of its molecules, is

$$p = \frac{nM v_{rms}^2}{3V}, \quad (19\text{-}21)$$

where $v_{rms} = \sqrt{(v^2)_{avg}}$ is the **root-mean-square speed** of the molecules of the gas. With Eq. 19-5 this gives

$$v_{rms} = \sqrt{\frac{3RT}{M}}. \quad (19\text{-}22)$$

Temperature and Kinetic Energy The average translational kinetic energy K_{avg} per molecule of an ideal gas is

$$K_{avg} = \tfrac{3}{2}kT. \quad (19\text{-}24)$$

Mean Free Path The *mean free path* λ of a gas molecule is its average path length between collisions and is given by

$$\lambda = \frac{1}{\sqrt{2}\,\pi d^2\, N/V}, \quad (19\text{-}25)$$

where N/V is the number of molecules per unit volume and d is the molecular diameter.

Maxwell Speed Distribution The *Maxwell speed distribution* $P(v)$ is a function such that $P(v)\,dv$ gives the *fraction* of molecules with speeds in the interval dv at speed v:

$$P(v) = 4\pi \left(\frac{M}{2\pi RT}\right)^{3/2} v^2 e^{-Mv^2/2RT}. \quad (19\text{-}27)$$

Three measures of the distribution of speeds among the molecules of a gas are

$$v_{avg} = \sqrt{\frac{8RT}{\pi M}} \quad \text{(average speed)}, \quad (19\text{-}31)$$

$$v_P = \sqrt{\frac{2RT}{M}} \quad \text{(most probable speed)}, \quad (19\text{-}35)$$

and the rms speed defined above in Eq. 19-22.

Molar Specific Heats The molar specific heat C_V of a gas at constant volume is defined as

$$C_V = \frac{Q}{n\,\Delta T} = \frac{\Delta E_{int}}{n\,\Delta T}, \quad (19\text{-}39,\ 19\text{-}41)$$

in which Q is the energy transferred as heat to or from a sam-

ple of n moles of the gas, ΔT is the resulting temperature change of the gas, and ΔE_{int} is the resulting change in the internal energy of the gas. For an ideal monatomic gas,

$$C_V = \tfrac{3}{2}R = 12.5 \text{ J/mol} \cdot \text{K}. \quad (19\text{-}43)$$

The molar specific heat C_p of a gas at constant pressure is defined to be

$$C_p = \frac{Q}{n\,\Delta T}, \quad (19\text{-}46)$$

in which Q, n, and ΔT are defined as above. C_p is also given by

$$C_p = C_V + R. \quad (19\text{-}49)$$

For n moles of an ideal gas,

$$E_{int} = nC_V T \quad \text{(ideal gas).} \quad (19\text{-}44)$$

If n moles of a confined ideal gas undergo a temperature change ΔT due to *any* process, the change in the internal energy of the gas is

$$\Delta E_{int} = nC_V\,\Delta T \quad \text{(ideal gas, any process),} \quad (19\text{-}45)$$

in which the appropriate value of C_V must be substituted, according to the type of ideal gas.

Degrees of Freedom and C_V We find C_V by using the *equipartition of energy* theorem, which states that every *degree of freedom* of a molecule (that is, every independent way it can store energy) has associated with it—on average—an energy $\tfrac{1}{2}kT$ per molecule $(= \tfrac{1}{2}RT$ per mole). If f is the number of degrees of freedom, then $E_{int} = (f/2)nRT$ and

$$C_V = \left(\frac{f}{2}\right)R = 4.16f \text{ J/mol} \cdot \text{K}. \quad (19\text{-}51)$$

For monatomic gases $f = 3$ (three translational degrees); for diatomic gases $f = 5$ (three translational and two rotational degrees).

Adiabatic Process When an ideal gas undergoes a slow adiabatic volume change (a change for which $Q = 0$), its pressure and volume are related by

$$pV^\gamma = \text{a constant} \quad \text{(adiabatic process),} \quad (19\text{-}53)$$

in which $\gamma\ (= C_p/C_V)$ is the ratio of molar specific heats for the gas. For a free expansion, however, $pV = $ a constant.

Questions

1 For four situations for an ideal gas, the table gives the energy transferred to or from the gas as heat Q and either the work W done by the gas or the work W_{on} done on the gas, all in joules. Rank the four situations in terms of the temperature change of the gas, most positive first.

	a	b	c	d
Q	-50	$+35$	-15	$+20$
W	-50	$+35$		
W_{on}			-40	$+40$

2 The dot in Fig. 19-15a represents the initial state of a gas, and the vertical line through the dot divides the p-V diagram

into regions 1 and 2. For the following processes, determine whether the work W done by the gas is positive, negative, or

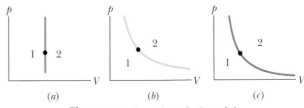

Fig. 19-15 Questions 2, 4, and 6.

zero: (a) the gas moves up along the vertical line, (b) it moves down along the vertical line, (c) it moves to anywhere in region 1, and (d) it moves to anywhere in region 2.

3 For a temperature increase of ΔT_1, a certain amount of an ideal gas requires 30 J when heated at constant volume and 50 J when heated at constant pressure. How much work is done by the gas in the second situation?

4 The dot in Fig. 19-15b represents the initial state of a gas, and the isotherm through the dot divides the p-V diagram into regions 1 and 2. For the following processes, determine whether the change ΔE_{int} in the internal energy of the gas is positive, negative, or zero: (a) the gas moves up along the isotherm, (b) it moves down along the isotherm, (c) it moves to anywhere in region 1, and (d) it moves to anywhere in region 2.

5 An ideal diatomic gas, with molecular rotation but not oscillation, loses energy as heat Q. Is the resulting decrease in the internal energy of the gas greater if the loss occurs in a constant-volume process or in a constant-pressure process?

6 The dot in Fig. 19-15c represents the initial state of a gas, and the adiabat through the dot divides the p-V diagram into regions 1 and 2. For the following processes, determine whether the corresponding heat Q is positive, negative, or zero: (a) the gas moves up along the adiabat, (b) it moves down along the adiabat, (c) it moves to anywhere in region 1, and (d) it moves to anywhere in region 2.

7 A certain amount of energy is to be transferred as heat to 1 mol of a monatomic gas (a) at constant pressure and (b) at constant volume, and to 1 mol of a diatomic gas (c) at constant pressure and (d) at constant volume. Figure 19-16 shows four paths from an initial point to four final points on a p-V diagram. Which path goes with which process? (e) Are the molecules of the diatomic gas rotating?

Fig. 19-16 Question 7.

8 Does the temperature of an ideal gas increase, decrease, or stay the same during (a) an isothermal expansion, (b) an expansion at constant pressure, (c) an adiabatic expansion, and (d) an increase in pressure at constant volume?

9 (a) Rank the four paths of Fig. 19-14 according to the work done by the gas, greatest first. (b) Rank paths 1, 2, and 3 according to the change in the internal energy of the gas, most positive first and most negative last.

10 In the p-V diagram of Fig. 19-17, the gas does 5 J of work when taken along isotherm ab and 4 J when taken along adiabat bc. What is the change in the internal energy of the gas when it is taken along the straight path from a to c?

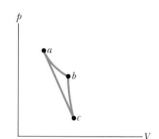

Fig. 19-17 Question 10.

Problems

sec. 19-2 Avogadro's Number

•**1** Find the mass in kilograms of 7.50×10^{24} atoms of arsenic, which has a molar mass of 74.9 g/mol.

•**2** Gold has a molar mass of 197 g/mol. (a) How many moles of gold are in a 2.50 g sample of pure gold? (b) How many atoms are in the sample?

sec. 19-3 Ideal Gases

•**3** Compute (a) the number of moles and (b) the number of molecules in 1.00 cm³ of an ideal gas at a pressure of 100 Pa and a temperature of 220 K.

•**4** The best laboratory vacuum has a pressure of about 1.00×10^{-18} atm, or 1.01×10^{-13} Pa. How many gas molecules are there per cubic centimeter in such a vacuum at 293 K?

•**5** Oxygen gas having a volume of 1000 cm³ at 40.0°C and 1.01×10^5 Pa expands until its volume is 1500 cm³ and its pressure is 1.06×10^5 Pa. Find (a) the number of moles of oxygen present and (b) the final temperature of the sample. **SSM**

•**6** An automobile tire has a volume of 1.64×10^{-2} m³ and contains air at a gauge pressure (pressure above atmospheric pressure) of 165 kPa when the temperature is 0.00°C. What is the gauge pressure of the air in the tires when its temperature rises to 27.0°C and its volume increases to 1.67×10^{-2} m³? Assume atmospheric pressure is 1.01×10^5 Pa.

•**7** A quantity of ideal gas at 10.0°C and 100 kPa occupies a volume of 2.50 m³. (a) How many moles of the gas are present? (b) If the pressure is now raised to 300 kPa and the temperature is raised to 30.0°C, how much volume does the gas occupy? Assume no leaks.

•**8** Suppose 1.80 mol of an ideal gas is taken from a volume of 3.00 m³ to a volume of 1.50 m³ via an isothermal compression at 30°C. (a) How much energy is transferred as heat during the compression, and (b) is the transfer *to* or *from* the gas?

•**9** A container encloses 2 mol of an ideal gas that has molar mass M_1 and 0.5 mol of a second ideal gas that has molar mass $M_2 = 3M_1$. What fraction of the total pressure on the container wall is attributable to the second gas? (The kinetic theory explanation of pressure leads to the experimentally discovered law of partial pressures for a mixture of gases that do not react chemically: *The total pressure exerted by the mixture is equal to the sum of the pressures that the several gases would exert separately if each were to occupy the vessel alone.*)

••**10** Suppose 0.825 mol of an ideal gas undergoes an isothermal expansion as energy is added to it as heat Q. If Fig.

19-18 shows the final volume V_f versus Q, what is the gas temperature?

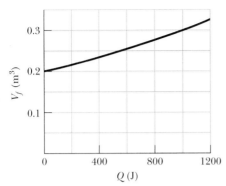

Fig. 19-18   Problem 10.

••**11**   In the temperature range 310 K to 330 K, the pressure p of a certain nonideal gas is related to volume V and temperature T by

$$p = (24.9 \text{ J/K}) \frac{T}{V} - (0.00662 \text{ J/K}^2) \frac{T^2}{V}.$$

How much work is done by the gas if its temperature is raised from 315 K to 325 K while the pressure is held constant?

••**12**   A sample of an ideal gas is taken through the cyclic process $abca$ shown in Fig. 19-19; at point a, $T = 200$ K. (a) How many moles of gas are in the sample? What are (b) the temperature of the gas at point b, (c) the temperature of the gas at point c, and (d) the net energy added to the gas as heat during the cycle?

Fig. 19-19   Problem 12.

••**13**   Air that initially occupies 0.140 m³ at a gauge pressure of 103.0 kPa is expanded isothermally to a pressure of 101.3 kPa and then cooled at constant pressure until it reaches its initial volume. Compute the work done by the air. (Gauge pressure is the difference between the actual pressure and atmospheric pressure.) **SSM ILW WWW**

•••**14**   An air bubble of volume 20 cm³ is at the bottom of a lake 40 m deep, where the temperature is 4.0°C. The bubble rises to the surface, which is at a temperature of 20°C. Take the temperature of the bubble's air to be the same as that of the surrounding water. Just as the bubble reaches the surface, what is its volume?

•••**15**   Container A in Fig. 19-20 holds an ideal gas at a pressure of 5.0×10^5 Pa and a temperature of 300 K. It is connected by a thin tube (and a closed valve) to container B, with four times the volume of A. Container B holds the same ideal gas at a pressure of 1.0×10^5 Pa and a temperature of 400 K. The valve is opened to allow the pressures to equalize,

Fig. 19-20   Problem 15.

but the temperature of each container is maintained. What then is the pressure in the two containers? **ILW**

sec. 19-4   Pressure, Temperature, and RMS Speed

•**16**   Find the rms speed of argon atoms at 313 K. See Appendix F for the molar mass of argon atoms.

•**17**   The lowest possible temperature in outer space is 2.7 K. What is the rms speed of hydrogen molecules at this temperature? (The molar mass of hydrogen molecules (H_2) is given in Table 19-1.) **SSM**

•**18**   Calculate the rms speed of helium atoms at 1000 K. See Appendix F for the molar mass of helium atoms.

•**19**   (a) Compute the rms of a nitrogen molecule at 20.0°C. The molar mass of nitrogen molecules (N_2) is given in Table 19-1. At what temperatures will the rms speed be (b) half that value and (c) twice that value?

•**20**   The temperature and pressure in the Sun's atmosphere are 2.00×10^6 K and 0.0300 Pa. Calculate the rms speed of free electrons (mass 9.11×10^{-31} kg) there, assuming they are an ideal gas.

••**21**   A beam of hydrogen molecules (H_2) is directed toward a wall, at an angle of 55° with the normal to the wall. Each molecule in the beam has a speed of 1.0 km/s and a mass of 3.3×10^{-24} g. The beam strikes the wall over an area of 2.0 cm², at the rate of 10^{23} molecules per second. What is the beam's pressure on the wall? **SSM**

••**22**   At 273 K and 1.00×10^{-2} atm, the density of a gas is 1.24×10^{-5} g/cm³. (a) Find v_{rms} for the gas molecules. (b) Find the molar mass of the gas and (c) identify the gas. (*Hint:* The gas is listed in Table 19-1.)

sec. 19-5   Translational Kinetic Energy

•**23**   What is the average translational kinetic energy of nitrogen molecules at 1600 K?

•**24**   Determine the average value of the translational kinetic energy of the molecules of an ideal gas at (a) 0.00°C and (b) 100°C. What is the translational kinetic energy per mole of an ideal gas at (c) 0.00°C and (d) 100°C?

••**25**   Water standing in the open at 32.0°C evaporates because of the escape of some of the surface molecules. The heat of vaporization (539 cal/g) is approximately equal to εn, where ε is the average energy of the escaping molecules and n is the number of molecules per gram. (a) Find ε. (b) What is the ratio of ε to the average kinetic energy of H_2O molecules, assuming the latter is related to temperature in the same way as it is for gases?

sec. 19-6   Mean Free Path

•**26**   At what frequency would the wavelength of sound in air be equal to the mean free path of oxygen molecules at 1.0 atm pressure and 0.00°C? Take the diameter of an oxygen molecule to be 3.0×10^{-8} cm.

•**27**   The atmospheric density at an altitude of 2500 km is about 1 molecule/cm³. (a) Assuming the molecular diameter of 2.0×10^{-8} cm, find the mean free path predicted by Eq. 19-25. (b) Explain whether the predicted value is meaningful. **SSM**

•**28**   The mean free path of nitrogen molecules at 0.0°C and 1.0 atm is 0.80×10^{-5} cm. At this temperature and pressure there are 2.7×10^{19} molecules/cm³. What is the molecular diameter?

••29 In a certain particle accelerator, protons travel around a circular path of diameter 23.0 m in an evacuated chamber, whose residual gas is at 295 K and 1.00×10^{-6} torr pressure. (a) Calculate the number of gas molecules per cubic centimeter at this pressure. (b) What is the mean free path of the gas molecules if the molecular diameter is 2.00×10^{-8} cm?

••30 At 20°C and 750 torr pressure, the mean free paths for argon gas (Ar) and nitrogen gas (N_2) are $\lambda_{Ar} = 9.9 \times 10^{-6}$ cm and $\lambda_{N_2} = 27.5 \times 10^{-6}$ cm. (a) Find the ratio of the diameter of an Ar atom to that of an N_2 molecule. What is the mean free path of argon at (b) 20°C and 150 torr, and (c) −40°C and 750 torr?

sec. 19-7   The Distribution of Molecular Speeds

•31 The speeds of 10 molecules are 2.0, 3.0, 4.0, . . . , 11 km/s. What are their (a) average speed and (b) rms speed? **SSM**

•32 The speeds of 22 particles are as follows (N_i represents the number of particles that have speed v_i):

N_i	2	4	6	8	2
v_i (cm/s)	1.0	2.0	3.0	4.0	5.0

What are (a) v_{avg}, (b) v_{rms}, and (c) v_P?

•33 Ten particles are moving with the following speeds: four at 200 m/s, two at 500 m/s, and four at 600 m/s. Calculate their (a) average and (b) rms speeds. (c) Is $v_{rms} > v_{avg}$?

••34 Figure 19-21 gives the probability distribution for nitrogen gas. What are the (a) gas temperature and (b) rms speed of the molecules?

Fig. 19-21   Problem 34.

••35 At what temperature does the rms speed of (a) H_2 (molecular hydrogen) and (b) O_2 (molecular oxygen) equal the escape speed from Earth (Table 13-2)? At what temperature does the rms speed of (c) H_2 and (d) O_2 equal the escape speed from the Moon (where the gravitational acceleration at the surface has magnitude 0.16g)? Considering the answers to parts (a) and (b), should there be much (e) hydrogen and (f) oxygen high in Earth's upper atmosphere, where the temperature is about 1000 K?

••36 It is found that the most probable speed of molecules in a gas when it has (uniform) temperature T_2 is the same as the rms speed of the molecules in this gas when it has (uniform) temperature T_1. Calculate T_2/T_1.

••37 A hydrogen molecule (diameter 1.0×10^{-8} cm), traveling at the rms speed, escapes from a 4000 K furnace into a chamber containing *cold* argon atoms (diameter 3.0×10^{-8} cm) at a density of 4.0×10^{19} atoms/cm³. (a) What is the speed of the hydrogen molecule? (b) If it collides with an argon atom, what is the closest their centers can be, considering each as spherical? (c) What is the initial number of collisions per second experienced by the hydrogen molecule? (*Hint:* Assume that the argon atoms are stationary. Then the mean free path of the hydrogen molecule is given by Eq. 19-26 and not Eq. 19-25.)

••38 Two containers are at the same temperature. The first contains gas with pressure p_1, molecular mass m_1, and rms speed v_{rms1}. The second contains gas with pressure $2.0p_1$, molecular mass m_2, and average speed $v_{avg2} = 2.0v_{rms1}$. Find the mass ratio m_1/m_2.

••39 Figure 19-22 shows a hypothetical speed distribution for a sample of N gas particles (note that $P(v) = 0$ for speed $v > 2v_0$). What are the values of (a) av_0, (b) v_{avg}/v_0, and (c) v_{rms}/v_0? (d) What fraction of the particles has a speed between $1.5v_0$ and $2.0v_0$? **SSM WWW**

Fig. 19-22   Problem 39.

sec. 19-8   The Molar Specific Heats of an Ideal Gas

•40 What is the internal energy of 1.0 mol of an ideal monatomic gas at 273 K?

••41 The temperature of 2.00 mol of an ideal monatomic gas is raised 15.0 K at constant volume. What are (a) the work W done by the gas, (b) the energy transferred as heat Q, (c) the change ΔE_{int} in the internal energy of the gas, and (d) the change ΔK in the average kinetic energy per atom?

••42 Under constant pressure, the temperature of 2.00 mol of an ideal monatomic gas is raised 15.0 K. What are (a) the work W done by the gas, (b) the energy transferred as heat Q, (c) the change ΔE_{int} in the internal energy of the gas, and (d) the change ΔK in the average kinetic energy per atom?

••43 A container holds a mixture of three nonreacting gases: 2.40 mol of gas 1 with $C_{v1} = 12.0$ J/mol · K, 1.50 mol of gas 2 with $C_{v2} = 12.8$ J/mol · K, and 3.20 mol of gas 3 with $C_{v3} = 20.0$ J/mol · K. What is C_v of the mixture? **SSM**

••44 When 20.9 J was added as heat to a particular ideal gas, the volume of the gas changed from 50.0 cm³ to 100 cm³ while the pressure remained at 1.00 atm. (a) By how much did the internal energy of the gas change? If the quantity of gas present was 2.00×10^{-3} mol, find (b) C_p and (c) C_V.

••45 The mass of a gas molecule can be computed from its specific heat at constant volume c_V. (Note that this is not C_V.) Take $c_V = 0.075$ cal/g · C° for argon and calculate (a) the mass of an argon atom and (b) the molar mass of argon. **ILW**

••46 One mole of an ideal diatomic gas goes from a to c along the diagonal path in Fig. 19-23. During the transition,

Fig. 19-23   Problem 46.

(a) what is the change in internal energy of the gas, and (b) how much energy is added to the gas as heat? (c) How much heat is required if the gas goes from *a* to *c* along the indirect path *abc*?

•••**47** In an industrial process the volume of 25.0 mol of a monatomic ideal gas is reduced at a uniform rate from 0.616 m^3 to 0.308 m^3 in 2.00 h while its temperature is increased at a uniform rate from 27.0°C to 450°C. Throughout the process, the gas passes through thermodynamic equilibrium states. What are (a) the cumulative work done on the gas, (b) the cumulative energy absorbed by the gas as heat, and (c) the molar specific heat for the process? (*Hint:* To evaluate the integral for the work, you might use

$$\int \frac{a + bx}{A + Bx}\,dx = \frac{bx}{B} + \frac{aB - bA}{B^2}\ln(A + Bx),$$

an indefinite integral.) Suppose the process is replaced with a two-step process that reaches the same final state. In step 1, the gas volume is reduced at constant temperature, and in step 2 the temperature is increased at constant volume. For this process, what are (d) the cumulative work done on the gas, (e) the cumulative energy absorbed by the gas as heat, and (f) the molar specific heat for the process?

sec. 19-9 Degrees of Freedom and Molar Specific Heats

•**48** We give 70 J as heat to a diatomic gas, which then expands at constant pressure. The gas molecules rotate but do not oscillate. By how much does the internal energy of the gas increase?

•**49** When 1.0 mol of oxygen (O_2) gas is heated at constant pressure starting at 0°C, how much energy must be added to the gas as heat to double its volume? (The molecules rotate but do not oscillate.) **ILW**

••**50** Suppose 12.0 g of oxygen (O_2) gas is heated at constant atmospheric pressure from 25.0°C to 125°C. (a) How many moles of oxygen are present? (See Table 19-1 for the molar mass.) (b) How much energy is transferred to the oxygen as heat? (The molecules rotate but do not oscillate.) (c) What fraction of the heat is used to raise the internal energy of the oxygen?

••**51** Suppose 4.00 mol of an ideal diatomic gas, with molecular rotation but not oscillation, experienced a temperature increase of 60.0 K under constant-pressure conditions. What are (a) the energy transferred as heat Q, (b) the change ΔE_{int} in internal energy of the gas, (c) the work W done by the gas, and (d) the change ΔK in the total translational kinetic energy of the gas? **SSM WWW**

sec. 19-11 The Adiabatic Expansion of an Ideal Gas

•**52** We know that for an adiabatic process pV^γ = a constant. Evaluate "a constant" for an adiabatic process involving exactly 2.0 mol of an ideal gas passing through the state having exactly p = 1.0 atm and T = 300 K. Assume a diatomic gas whose molecules rotate but do not oscillate.

•**53** A certain gas occupies a volume of 4.3 L at a pressure of 1.2 atm and a temperature of 310 K. It is compressed adiabatically to a volume of 0.76 L. Determine (a) the final pressure and (b) the final temperature, assuming the gas to be an ideal gas for which γ = 1.4.

•**54** Suppose 1.00 L of a gas with γ = 1.30, initially at 273 K and 1.00 atm, is suddenly compressed adiabatically to half its

initial volume. Find its final (a) pressure and (b) temperature. (c) If the gas is then cooled to 273 K at constant pressure, what is its final volume?

••**55** Figure 19-24 shows two paths that may be taken by a gas from an initial point *i* to a final point *f*. Path 1 consists of an isothermal expansion (work is 50 J in magnitude), an adiabatic expansion (work is 40 J in magnitude), an isothermal compression (work is 30 J in magnitude), and then an adiabatic compression (work is 25 J in magnitude). What is the change in the internal energy of the gas if the gas goes from point *i* to point *f* along path 2?

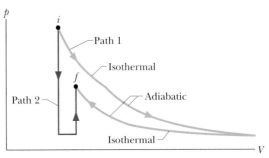

Fig. 19-24 Problem 55.

••**56** The volume of an ideal gas is adiabatically reduced from 200 L to 74.3 L. The initial pressure and temperature are 1.00 atm and 300 K. The final pressure is 4.00 atm. (a) Is the gas monatomic, diatomic, or polyatomic? (b) What is the final temperature? (c) How many moles are in the gas?

••**57** A gas is to be expanded from initial state *i* to final state *f* along either path 1 or path 2 on a *p-V* diagram. Path 1 consists of three steps: an isothermal expansion (work is 40 J in magnitude), an adiabatic expansion (work is 20 J in magnitude), and another isothermal expansion (work is 30 J in magnitude). Path 2 consists of two steps: a pressure reduction at constant volume and an expansion at constant pressure. What is the change in the internal energy of the gas along path 2?

••**58** (a) An ideal gas initially at pressure p_0 undergoes a free expansion until its volume is 3.00 times its initial volume. What then is the ratio of its pressure to p_0? (b) The gas is next slowly and adiabatically compressed back to its original volume. The pressure after compression is $(3.00)^{1/3}p_0$. Is the gas monatomic, diatomic, or polyatomic? (c) What is the ratio of the average kinetic energy per molecule in this final state to that in the initial state?

•••**59** Figure 19-25 shows a cycle undergone by 1.00 mol of an ideal monatomic gas. For 1 → 2, what are (a) heat Q, (b) the change in internal energy ΔE_{int}, and (c) the work done W? For 2 → 3, what are (d) Q, (e) ΔE_{int}, and (f) W? For 3 → 1, what are (g) Q, (h) ΔE_{int}, and (i) W? For the full cycle, what are (j) Q, (k) ΔE_{int}, and (l) W? The initial pressure at point 1 is 1.00 atm ($= 1.013 \times 10^5$ Pa). What are the (m) volume and (n) pressure at point 2 and the (o) volume and (p) pressure at point 3? **SSM**

Fig. 19-25 Problem 59.

Additional Problems

60 An ideal gas is taken through a complete cycle in three steps: adiabatic expansion with work equal to 125 J, isothermal contraction at 325 K, and increase in pressure at constant volume. (a) Draw a p-V diagram for the three steps. (b) How much energy is transferred as heat in step 3, and (c) is it transferred *to* or *from* the gas?

61 (a) What is the volume occupied by 1.00 mol of an ideal gas at standard conditions—that is, 1.00 atm (= 1.01×10^5 Pa) and 273 K? (b) Show that the number of molecules per cubic centimeter (the *Loschmidt number*) at standard conditions is 2.69×10^9.

62 In an interstellar gas cloud at 50.0 K, the pressure is 1.00×10^{-8} Pa. Assuming that the molecular diameters of the gases in the cloud are all 20.0 nm, what is their mean free path?

63 The envelope and basket of a hot-air balloon have a combined weight of 2.45 kN, and the envelope has a capacity (volume) of 2.18×10^3 m³. When it is fully inflated, what should be the temperature of the enclosed air to give the balloon a *lifting capacity* (force) of 2.67 kN (in addition to the balloon's weight)? Assume that the surrounding air, at 20.0°C, has a weight per unit volume of 11.9 N/m³ and a molecular mass of 0.028 kg/mole, and is at a pressure of 1.0 atm.

64 A container holds a gas of molecular hydrogen (H_2) at 250 K. What are (a) the most probable speed v_P of the molecules and (b) the maximum value P_{max} of the probability distribution function $P(v)$? (c) With a graphing calculator or computer math package, determine what percentage of the molecules have speeds between $0.500v_P$ and $1.50v_P$. The temperature is then increased to 500 K. What are (d) the most probable speed v_P of the molecules and (e) the maximum value P_{max} of the probability distribution function $P(v)$? Did (f) v_P and (g) P_{max} increase, decrease, or remain the same because of the temperature increase?

65 The temperature of 3.00 mol of a gas with $C_V = 6.00$ cal/mol · K is to be raised 50.0 K. If the process is at *constant volume*, what are (a) the energy transferred as heat Q, (b) the work W done by the gas, (c) the change ΔE_{int} in internal energy of the gas, and (d) the change ΔK in the total translational kinetic energy? If the process is at *constant pressure*, what are (e) Q, (f) W, (g) ΔE_{int}, and (h) ΔK? If the process is *adiabatic*, what are (i) Q, (j) W, (k) ΔE_{int}, and (l) ΔK?

66 An ideal gas is suddenly allowed to expand freely so that the ratio of its new volume V_1 to its initial volume V_0 is $V_1/V_0 = 5.00$. The gas is then adiabatically compressed back to its initial volume V_0, leaving it with a pressure p_2 that is $(5.00)^{0.40}$ times its initial pressure p_0. (a) Is the gas monatomic, diatomic with no rotation of the molecules, diatomic with rotating molecules, or polyatomic? What is the ratio of the final value of the average kinetic energy per molecule to the initial value (b) after the free expansion and (c) after the adiabatic compression?

67 For adiabatic processes in an ideal gas, show that (a) the bulk modulus is given by

$$B = -V\frac{dp}{dV} = \gamma p,$$

and therefore (b) the speed of sound in the gas is

$$v_s = \sqrt{\frac{\gamma p}{\rho}} = \sqrt{\frac{\gamma RT}{M}}.$$

See Eqs. 17-2 and 17-3.

68 Air at 0.000°C and 1.00 atm pressure has a density of 1.29×10^{-3} g/cm³, and the speed of sound in air is 331 m/s at that temperature. Use those data and the results of Problem 67 to compute the ratio γ of the molar specific heats of air.

69 The molar mass of iodine is 127 g/mol. A standing wave in a tube filled with iodine gas at 400 K has nodes that are 6.77 cm apart when the frequency is 1400 Hz. (a) What is γ for iodine gas? (b) Is iodine gas monatomic or diatomic? (*Hint:* See Problem 67.)

70 Oxygen (O_2) gas at 273 K and 1.0 atm is confined to a cubical container 10 cm on a side. Calculate $\Delta U_g/K_{avg}$, where ΔU_g is the change in the gravitational potential energy of an oxygen molecule falling the height of the box and K_{avg} is the molecule's average translational kinetic energy.

71 The temperature of 2.00 mol of an ideal monatomic gas is raised 15.0 K in an adiabatic process. What are (a) the work W done by the gas, (b) the energy transferred as heat Q, (c) the change ΔE_{int} in internal energy of the gas, and (d) the change ΔK in the average kinetic energy per atom?

72 An ideal diatomic gas, with rotation but no oscillation, undergoes an adiabatic expansion. Its initial pressure and volume are 1.20 atm and 0.200 m³. Its final pressure is 2.40 atm. How much work is done by the gas?

73 During a compression at a constant pressure of 250 Pa, the volume of an ideal gas decreases from 0.80 m³ to 0.20 m³. The initial temperature is 360 K, and the gas loses 210 J as heat. What are (a) the change in the internal energy of the gas and (b) the final temperature of the gas?

74 An ideal gas consists of 1.50 mol of diatomic molecules that rotate but do not oscillate. The molecular diameter is 250 pm. The gas is expanded at a constant pressure of 1.50×10^5 Pa, with a transfer of 200 J as heat. What is the change in the mean free path of the molecules?

75 At what frequency do molecules collide in oxygen gas (O_2) at temperature 400 K and pressure 2.00 atm? Assume that the molecular diameter is 290 pm and that the gas is ideal.

76 An ideal monatomic gas initially has a temperature of 330 K and a pressure of 6.00 atm. It is to expand from volume 500 cm³ to volume 1500 cm³. If the expansion is isothermal, what are (a) the final pressure and (b) the work done by the gas? If, instead, the expansion is adiabatic, what are (c) the final pressure and (d) the work done by the gas?

77 An ideal gas with 3.00 mol is initially in state 1 with pressure $p_1 = 20.0$ atm and volume $V_1 = 1500$ cm³. First it is taken to state 2 with pressure $p_2 = 1.50p_1$ and volume $V_2 = 2.00V_1$. Then it is taken to state 3 with pressure $p_3 = 2.00p_1$ and volume $V_3 = 0.500V_1$. What is the temperature of the gas in (a) state 1 and (b) state 2? (c) What is the net change in internal energy from state 1 to state 3?

78 An ideal gas undergoes an adiabatic compression from $p = 1.0$ atm, $V = 1.0 \times 10^6$ L, $T = 0.0°C$ to $p = 1.0 \times 10^5$

atm, $V = 1.0 \times 10^3$ L. (a) Is the gas monatomic, diatomic, or polyatomic? (b) What is its final temperature? (c) How many moles of gas are present? What is the total translational kinetic energy per mole (d) before and (e) after the compression? (f) What is the ratio of the squares of the rms speeds before and after the compression?

79 A sample of ideal gas expands from an initial pressure and volume of 32 atm and 1.0 L to a final volume of 4.0 L. The initial temperature is 300 K. If the gas is monatomic and the expansion isothermal, what are the (a) final pressure p_f, (b) final temperature T_f, and (c) work W done by the gas? If the gas is monatomic and the expansion adiabatic, what are (d) p_f, (e) T_f, and (f) W? If the gas is diatomic and the expansion adiabatic, what are (g) p_f, (h) T_f, and (i) W?

80 An ideal gas, at initial temperature T_1 and initial volume 2.0 m^3, is expanded adiabatically to a volume of 4.0 m^3, then expanded isothermally to a volume of 10 m^3, and then compressed adiabatically back to T_1. What is its final volume?

81 Calculate the work done by an external agent during an isothermal compression of 1.00 mol of oxygen from a volume of 22.4 L at 0°C and 1.00 atm to a volume of 16.8 L.

82 A steel tank contains 300 g of ammonia gas (NH$_3$) at a pressure of 1.35×10^6 Pa and a temperature of 77°C. (a) What is the volume of the tank in liters? (b) Later the temperature is 22°C and the pressure is 8.7×10^5 Pa. How many grams of gas have leaked out of the tank?

83 The mass of a helium atom is 6.66×10^{-27} kg. Compute the specific heat at constant volume for (monatomic) helium gas (in kJ/kg · K) from the molar specific heat C_V.

84 (a) What is the molar volume (volume per mole) of an ideal gas at standard conditions (0.00°C, 1.00 atm)? (b) Calculate the ratio of the rms speed of helium atoms to that of neon atoms under these conditions. (c) What is the mean free path of helium atoms under these conditions? Assume the atomic diameter d of helium to be 1.00×10^{-8} cm. (d) What is the mean free path of neon atoms under these conditions? Assume they have the same atomic diameter as helium. (e) Comment on the results of (c) and (d) in view of the fact that the helium atoms are traveling faster than the neon atoms.

85 Figure 19-26 shows a cycle consisting of five paths: AB is isothermal at 300 K, BC is adiabatic with work = 5.0 J, CD is at a constant pressure of 5 atm, DE is isothermal, and EA is adiabatic with a change in internal energy of 8.0 J. What is the change in internal energy of the gas along path CD?

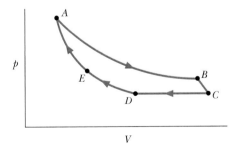

Fig. 19-26
Problem 85.

86 At what temperature do atoms of helium gas have the same rms speed as molecules of hydrogen gas at 20.0°C? (The molar masses are given in Table 19-1.)

87 A quantity of an ideal monatomic gas consists of n moles initially at temperature T_1. The pressure and volume are then slowly doubled in such a manner as to trace out a straight line on a p-V diagram. For this process, what are the ratios (a) W/nRT_1, (b) $\Delta E_{int}/nRT_1$, and (c) Q/nRT_1? (d) If C is defined as the molar specific heat for the process, what is C/R?

88 An ideal gas initially has a volume of 4.00 m^3, a pressure of 5.67 Pa, and a temperature of −56.0°C. The gas is then expanded to 7.00 m^3, leaving it with a temperature of 40.0°C. What then is the pressure?

89 Figure 19-27 shows a hypothetical speed distribution for particles of a certain gas: $P(v) = Cv^2$ for $0 < v \le v_0$ and $P(v) = 0$ for $v > v_0$. Find (a) an expression for C in terms of v_0, (b) the average speed of the particles, and (c) their rms speed.

Fig. 19-27   Problem 89.

90 The temperature of 3.00 mol of an ideal diatomic gas is increased by 40.0 C° without the pressure of the gas changing. The molecules in the gas rotate but do not oscillate. (a) How much energy is transferred to the gas as heat? (b) What is the change in the internal energy of the gas? (c) How much work is done by the gas? (d) By how much does the translational kinetic energy of the gas increase?

91 An ideal gas undergoes isothermal compression from an initial volume of 4.00 m^3 to a final volume of 3.00 m^3. There is 3.50 mol of the gas, and its temperature is 10.0°C. (a) How much work is done by the gas? (b) How much energy is transferred as heat between the gas and its environment?

92 (a) What is the number of molecules per cubic meter in air at 20°C and at a pressure of 1.0 atm ($=1.01 \times 10^5$ Pa)? (b) What is the mass of 1.0 m^3 of this air? Assume that 75% of the molecules are nitrogen (N$_2$) and 25% are oxygen (O$_2$).

93 An ideal gas initially at 300 K is compressed at a constant pressure of 25 N/m^2 from a volume of 3.0 m^3 to a volume of 1.8 m^3. In the process, 75 J is lost by the gas as heat. What are (a) the change in internal energy of the gas and (b) the final temperature of the gas?

94 Figure 19-28 represents an adiabatic compression of 2.0 mol of an ideal gas from 15 m^3 to 12 m^3, followed by an isothermal compression at 300 K to a final volume of 3.0 m^3. What is the total energy transferred as heat?

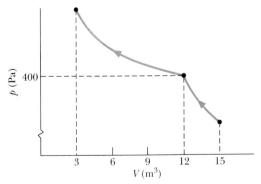

Fig. 19-28   Problem 94.

20

Entropy and the Second Law of Thermodynamics

When a bag of popcorn is heated in a microwave oven, the corn kernels explode into fluffy, edible structures that are just the right snack for a football game. If, however, you then decided to remove thermal energy from the popped kernels by sticking them in a refrigerator, they would never change back to their original unpopped state. Why can't the process that occurred in the oven be reversed, like a videotape run backwards?

What in the world gives direction to time?

The answer is in this chapter.

Time has direction, the direction in which we age. We are accustomed to one-way processes—that is, processes that can occur only in a certain sequence (the right way) and never in the reverse sequence (the wrong way). An egg is dropped onto a floor, a pizza is baked, a car is driven into a lamppost, large waves erode a sandy beach—these one-way processes are **irreversible,** meaning that they cannot be reversed by means of only small changes in their environment.

One goal of physics is to understand why time has direction and why one-way processes are irreversible. Although this physics might seem disconnected from the practical issues of everyday life, it is in fact at the heart of any engine, such as a car engine, because it determines how well an engine can run.

The key to understanding why one-way processes cannot be reversed involves a quantity known as *entropy.*

20-2 *Irreversible Processes and Entropy*

The one-way character of irreversible processes is so pervasive that we take it for granted. If these processes were to occur *spontaneously* (on their own) in the wrong way, we would be astonished. Yet *none* of these wrong-way events would violate the law of conservation of energy.

For example, if you were to wrap your hands around a cup of hot coffee, you would be astonished if your hands got cooler and the cup got warmer. That is obviously the wrong way for the energy transfer, but the total energy of the closed system (*hands + cup of coffee*) would be the same as the total energy if the process had run in the right way. For another example, if you popped a helium balloon, you would be astonished if, later, all the helium molecules were to gather together in the original shape of the balloon. That is obviously the wrong way for molecules to spread, but the total energy of the closed system (*molecules + room*) would be the same as for the right way.

Thus, changes in energy within a closed system do not set the direction of irreversible processes. Rather, that direction is set by another property that we shall discuss in this chapter—the *change in entropy* ΔS of the system. The change in entropy of a system is defined in the next section, but we can here state its central property, often called the *entropy postulate:*

> If an irreversible process occurs in a *closed* system, the entropy S of the system always increases; it never decreases.

Entropy differs from energy in that entropy does *not* obey a conservation law. The *energy* of a closed system is conserved; it always remains constant. For irreversible processes, the *entropy* of a closed system always increases. Because of this property, the change in entropy is sometimes called "the arrow of time." For example, we associate the explosion of a popcorn kernel with the forward direction of time and with an increase in entropy. The backward direction of time (a videotape run backwards) would correspond to the exploded popcorn re-forming the original kernel. Because this backward process would result in an entropy decrease, it never happens.

There are two equivalent ways to define the change in entropy of a system: (1) in terms of the system's temperature and the energy the system gains or loses as heat, and (2) by counting the ways in which the atoms or molecules that make up the system can be arranged. We use the first approach in the next section and the second in Section 20-8.

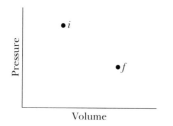

Fig. 20-1 The free expansion of an ideal gas. (*a*) The gas is confined to the left half of an insulated container by a closed stopcock. (*b*) When the stopcock is opened, the gas rushes to fill the entire container. This process is irreversible; that is, it does not occur in reverse, with the gas spontaneously collecting itself in the left half of the container.

20-3 Change in Entropy

Let's approach this definition of *change in entropy* by looking again at a process that we described in Sections 18-11 and 19-11: the free expansion of an ideal gas. Figure 20-1*a* shows the gas in its initial equilibrium state *i*, confined by a closed stopcock to the left half of a thermally insulated container. If we open the stopcock, the gas rushes to fill the entire container, eventually reaching the final equilibrium state *f* shown in Fig. 20-1*b*. This is an irreversible process; all the molecules of the gas will never return to the left half of the container.

The *p-V* plot of the process, in Fig. 20-2, shows the pressure and volume of the gas in its initial state *i* and final state *f*. Pressure and volume are *state properties,* properties that depend only on the state of the gas and not on how it reached that state. Other state properties are temperature and energy. We now assume that the gas has still another state property—its entropy. Furthermore, we define the **change in entropy** $S_f - S_i$ of a system during a process that takes the system from an initial state *i* to a final state *f* as

$$\Delta S = S_f - S_i = \int_i^f \frac{dQ}{T} \qquad \text{(change in entropy defined).} \qquad (20\text{-}1)$$

Here Q is the energy transferred as heat to or from the system during the process, and T is the temperature of the system in kelvins. Thus, an entropy change depends not only on the energy transferred as heat but also on the temperature at which the transfer takes place. Because T is always positive, the sign of ΔS is the same as that of Q. We see from Eq. 20-1 that the SI unit for entropy and entropy change is the joule per kelvin.

There is a problem, however, in applying Eq. 20-1 to the free expansion of Fig. 20-1. As the gas rushes to fill the entire container, the pressure, temperature, and volume of the gas fluctuate unpredictably. In other words, they do not have a sequence of well-defined equilibrium values during the intermediate stages of the change from initial equilibrium state *i* to final equilibrium state *f*. Thus, we cannot trace a pressure–volume path for the free expansion on the *p-V* plot of Fig. 20-2 and, more important, we cannot find a relation between Q and T that allows us to integrate as Eq. 20-1 requires.

However, if entropy is truly a state property, the difference in entropy between states *i* and *f* must depend *only on those states* and not at all on the way the system went from one to the other. Suppose, then, that we replace the irreversible free expansion of Fig. 20-1 with a *reversible* process that connects states *i* and *f*. With a reversible process we can trace a pressure–volume path on a *p-V* plot, and we can find a relation between Q and T that allows us to use Eq. 20-1 to obtain the entropy change.

We saw in Section 19-11 that the temperature of an ideal gas does not change during a free expansion: $T_i = T_f = T$. Thus, points *i* and *f* in Fig. 20-2 must be on the same isotherm. A convenient replacement process is then a reversible isothermal expansion from state *i* to state *f*, which actually proceeds *along* that isotherm. Furthermore, because T is constant throughout a reversible isothermal expansion, the integral of Eq. 20-1 is greatly simplified.

Figure 20-3 shows how to produce such a reversible isothermal expansion. We confine the gas to an insulated cylinder that rests on a thermal reservoir maintained at the temperature T. We begin by placing just enough lead shot on the movable piston so that the pressure and volume of the gas are those of the

Fig. 20-2 A *p-V* diagram showing the initial state *i* and the final state *f* of the free expansion of Fig. 20-1. The intermediate states of the gas cannot be shown because they are not equilibrium states.

initial state *i* of Fig. 20-1*a*. We then remove shot slowly (piece by piece) until the pressure and volume of the gas are those of the final state *f* of Fig. 20-1*b*. The temperature of the gas does not change because the gas remains in thermal contact with the reservoir throughout the process.

The reversible isothermal expansion of Fig. 20-3 is physically quite different from the irreversible free expansion of Fig. 20-1. However, *both processes have the same initial state and the same final state and thus must have the same change in entropy*. Because we removed the lead shot slowly, the intermediate states of the gas are equilibrium states, so we can plot them on a *p-V* diagram (Fig. 20-4).

To apply Eq. 20-1 to the isothermal expansion, we take the constant temperature *T* outside the integral, obtaining

$$\Delta S = S_f - S_i = \frac{1}{T} \int_i^f dQ.$$

Because $\int dQ = Q$, where Q is the total energy transferred as heat during the process, we have

$$\Delta S = S_f - S_i = \frac{Q}{T} \qquad \text{(change in entropy, isothermal process).} \qquad (20\text{-}2)$$

To keep the temperature *T* of the gas constant during the isothermal expansion of Fig. 20-3, heat *Q* must have been energy transferred *from* the reservoir *to* the gas. Thus, *Q* is positive and the entropy of the gas *increases* during the isothermal process and during the free expansion of Fig. 20-1.

To summarize:

> To find the entropy change for an irreversible process occurring in a *closed* system, replace that process with any reversible process that connects the same initial and final states. Calculate the entropy change for this reversible process with Eq. 20-1.

When the temperature change ΔT of a system is small relative to the temperature (in kelvins) before and after the process, the entropy change can be approximated as

$$\Delta S = S_f - S_i \approx \frac{Q}{T_{avg}}, \qquad (20\text{-}3)$$

where T_{avg} is the average temperature of the system in kelvins during the process.

✔ **CHECKPOINT 1** Water is heated on a stove. Rank the entropy changes of the water as its temperature rises (a) from 20°C to 30°C, (b) from 30°C to 35°C, and (c) from 80°C to 85°C, greatest first.

Fig. 20-3 The isothermal expansion of an ideal gas, done in a reversible way. The gas has the same initial state *i* and same final state *f* as in the irreversible process of Figs. 20-1 and 20-2.

Fig. 20-4 A *p-V* diagram for the reversible isothermal expansion of Fig. 20-3. The intermediate states, which are now equilibrium states, are shown.

Sample Problem 20-1

Suppose 1.0 mol of nitrogen gas is confined to the left side of the container of Fig. 20-1*a*. You open the stopcock, and the volume of the gas doubles. What is the entropy change of the gas for this irreversible process? Treat the gas as ideal.

Solution: We need two **Key Ideas** here. One is that we can determine the entropy change for the irreversible process by calculating it for a reversible process that provides the same change in volume. The other is that the temperature of the gas does not change in the free expansion. Thus, the reversible process should be an isothermal expansion—namely, the one of Figs. 20-3 and 20-4.

From Table 19-4, the energy Q added as heat to the gas as it expands isothermally at temperature T from an initial volume V_i to a final volume V_f is

$$Q = nRT \ln \frac{V_f}{V_i},$$

in which n is the number of moles of gas present. From Eq. 20-2 the entropy change for this reversible process is

$$\Delta S_{rev} = \frac{Q}{T} = \frac{nRT \ln(V_f/V_i)}{T} = nR \ln \frac{V_f}{V_i}.$$

Substituting $n = 1.00$ mol and $V_f/V_i = 2$, we find

$$\Delta S_{rev} = nR \ln \frac{V_f}{V_i} = (1.00 \text{ mol})(8.31 \text{ J/mol} \cdot \text{K})(\ln 2)$$

$$= +5.76 \text{ J/K}.$$

Thus, the entropy change for the free expansion (and for all other processes that connect the initial and final states shown in Fig. 20-2) is

$$\Delta S_{irrev} = \Delta S_{rev} = +5.76 \text{ J/K}. \qquad \text{(Answer)}$$

Because ΔS is positive, the entropy increases, in accordance with the entropy postulate of Section 20-2.

CHECKPOINT 2   An ideal gas has temperature T_1 at the initial state i shown in the p-V diagram here. The gas has a higher temperature T_2 at final states a and b, which it can reach along the paths shown. Is the entropy change along the path to state a larger than, smaller than, or the same as that along the path to state b?

Sample Problem 20-2

Figure 20-5a shows two identical copper blocks of mass $m = 1.5$ kg: block L at temperature $T_{iL} = 60°C$ and block R at temperature $T_{iR} = 20°C$. The blocks are in a thermally insulated box and are separated by an insulating shutter. When we lift the shutter, the blocks eventually come to the equilibrium temperature $T_f = 40°C$ (Fig. 20-5b). What is the net entropy change of the two-block system during this irreversible process? The specific heat of copper is 386 J/kg · K.

Solution: The Key Idea here is that to calculate the entropy change, we must find a reversible process that takes the system from the initial state of Fig. 20-5a to the final state of Fig. 20-5b. We can calculate the net entropy change ΔS_{rev} of the reversible process using Eq. 20-1, and then the entropy change for the irreversible process is equal to ΔS_{rev}. For such a reversible process we need a thermal reservoir whose temperature can be changed slowly (say, by turning a knob). We then take the blocks through the following two steps, illustrated in Fig. 20-6.

Step 1   With the reservoir's temperature set at 60°C, put block L on the reservoir. (Since block and reservoir are at the same temperature, they are already in thermal

(a) Step 1   (b) Step 2

Fig. 20-6   The blocks of Fig. 20-5 can proceed from their initial state to their final state in a reversible way if we use a reservoir with a controllable temperature (a) to extract heat reversibly from block L and (b) to add heat reversibly to block R.

equilibrium.) Then slowly lower the temperature of the reservoir and the block to 40°C. As the block's temperature changes by each increment dT during this process, energy dQ is transferred as heat *from* the block to the reservoir. Using Eq. 18-14, we can write this transferred energy as $dQ = mc \, dT$, where c is the specific heat of copper. According to Eq. 20-1, the entropy change ΔS_L of block L during the full temperature change from initial temperature T_{iL} ($= 60°C = 333$ K) to final temperature T_f ($= 40°C = 313$ K) is

$$\Delta S_L = \int_i^f \frac{dQ}{T} = \int_{T_{iL}}^{T_f} \frac{mc \, dT}{T} = mc \int_{T_{iL}}^{T_f} \frac{dT}{T}$$

$$= mc \ln \frac{T_f}{T_{iL}}.$$

Inserting the given data yields

$$\Delta S_L = (1.5 \text{ kg})(386 \text{ J/kg} \cdot \text{K}) \ln \frac{313 \text{ K}}{333 \text{ K}}$$

$$= -35.86 \text{ J/K}.$$

Step 2   With the reservoir's temperature now set at 20°C, put block R on the reservoir. Then slowly raise the temperature of the reservoir and the block to 40°C. With the

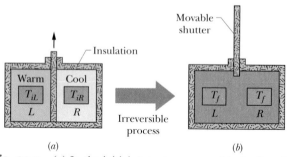

(a)   (b)

Fig. 20-5   (a) In the initial state, two copper blocks L and R, identical except for their temperatures, are in an insulating box and are separated by an insulating shutter. (b) When the shutter is removed, the blocks exchange energy as heat and come to a final state, both with the same temperature T_f.

same reasoning used to find ΔS_L, you can show that the entropy change ΔS_R of block R during this process is

$$\Delta S_R = (1.5 \text{ kg})(386 \text{ J/kg} \cdot \text{K}) \ln \frac{313 \text{ K}}{293 \text{ K}}$$

$$= +38.23 \text{ J/K}.$$

The net entropy change ΔS_{rev} of the two-block system undergoing this two-step reversible process is then

$$\Delta S_{\text{rev}} = \Delta S_L + \Delta S_R$$
$$= -35.86 \text{ J/K} + 38.23 \text{ J/K} = 2.4 \text{ J/K}.$$

Thus, the net entropy change ΔS_{irrev} for the two-block system undergoing the actual irreversible process is

$$\Delta S_{\text{irrev}} = \Delta S_{\text{rev}} = 2.4 \text{ J/K}. \qquad \text{(Answer)}$$

This result is positive, in accordance with the entropy postulate of Section 20-2.

Sample Problem 20-3

The *pericarp* part of a popcorn kernel is a small container of liquid water. When the popcorn is heated to about 180°C, the pericarp walls burst, and the liquid vaporizes so rapidly that the vapor explodes the rest of the kernel to many times its original volume. The explosion produces the audible *pop* of the popcorn. If the water in a pericarp has a mass of about 4 mg, what is the change in entropy of the water due to the vaporization and expansion?

Solution: The vaporization and expansion of the water in the pericarp constitute an irreversible process in which values for temperature and pressure vary in an unknown manner. A **Key Idea** here is that we can calculate the change ΔS in the water's entropy in the irreversible process by calculating ΔS for two reversible processes: (1) vaporization at 180°C and (2) adiabatic expansion of the vapor. (We assume the expansion is adiabatic because it takes place so rapidly that the vapor cannot exchange energy as heat with its environment.)

Process 1: Vaporization. Because we assume the vaporization occurs reversibly at a constant temperature, we can calculate the change in entropy from Eq. 20-2 ($\Delta S = Q/T$).

Substituting for Q with Eq. 18-16 ($Q = Lm$), using the latent heat of vaporization $L_V = 2256 \text{ kJ/kg}$ from Eq. 18-17, and converting 180°C to 453 K, we find

$$\Delta S_1 = \frac{Q}{T} = \frac{L_V m}{T} = \frac{(2256 \times 10^3 \text{ J/kg})(4 \times 10^{-6} \text{ kg})}{453 \text{ K}}$$

$$= 1.99 \times 10^{-2} \text{ J/K}.$$

Process 2: Adiabatic expansion. Because we assume the expansion does not allow any exchange of energy as heat Q between the water vapor and its environment, Q is zero. From Eq. 20-1, we see that change in the vapor's entropy is also zero. Thus, $\Delta S_2 = 0$.

The total change in entropy for the vaporization and expansion is

$$\Delta S = \Delta S_1 + \Delta S_2 = 1.99 \times 10^{-2} \text{ J/K} + 0$$

$$\approx 0.02 \text{ J/K}. \qquad \text{(Answer)}$$

Each audible pop in popping popcorn signals an increase in the water's entropy of about 0.02 J/K.

Entropy as a State Function

We have assumed that entropy, like pressure, energy, and temperature, is a property of the state of a system and is independent of how that state is reached. That entropy is indeed a *state function* (as state properties are usually called) can be deduced only by experiment. However, we can prove it is a state function for the special and important case in which an ideal gas is taken through a reversible process.

To make the process reversible, it is done slowly in a series of small steps, with the gas in an equilibrium state at the end of each step. For each small step, the energy transferred as heat to or from the gas is dQ, the work done by the gas is dW, and the change in internal energy is dE_{int}. These are related by the first law of thermodynamics in differential form (Eq. 18-27):

$$dE_{\text{int}} = dQ - dW.$$

Because the steps are reversible, with the gas in equilibrium states, we can use Eq. 18-24 to replace dW with $p\, dV$ and Eq. 19-45 to replace dE_{int} with $nC_V\, dT$. Solving for dQ then leads to

$$dQ = p\, dV + nC_V\, dT.$$

Using the ideal gas law, we replace p in this equation with nRT/V. Then we divide each term in the resulting equation by T, obtaining

$$\frac{dQ}{T} = nR\frac{dV}{V} + nC_V\frac{dT}{T}.$$

Now let us integrate each term of this equation between an arbitrary initial state i and an arbitrary final state f to get

$$\int_i^f \frac{dQ}{T} = \int_i^f nR \frac{dV}{V} + \int_i^f nC_V \frac{dT}{T}.$$

The quantity on the left is the entropy change ΔS ($= S_f - S_i$) defined by Eq. 20-1. Substituting this and integrating the quantities on the right yield

$$\Delta S = S_f - S_i = nR \ln \frac{V_f}{V_i} + nC_V \ln \frac{T_f}{T_i}. \qquad (20\text{-}4)$$

Note that we did not have to specify a particular reversible process when we integrated. Therefore, the integration must hold for all reversible processes that take the gas from state i to state f. Thus, the change in entropy ΔS between the initial and final states of an ideal gas depends only on properties of the initial state (V_i and T_i) and properties of the final state (V_f and T_f); ΔS does not depend on how the gas changes between the two states.

20-4 The Second Law of Thermodynamics

Here is a puzzle. We saw in Sample Problem 20-1 that if we cause the reversible process of Fig. 20-3 to proceed from (*a*) to (*b*) in that figure, the change in entropy of the gas—which we take as our system—is positive. However, because the process is reversible, we can just as easily make it proceed from (*b*) to (*a*), simply by slowly adding lead shot to the piston of Fig. 20-3*b* until the original volume of the gas is restored. In this reverse process, energy must be extracted as heat *from the gas* to keep its temperature from rising. Hence Q is negative and so, from Eq. 20-2, the entropy of the gas must decrease.

Doesn't this decrease in the entropy of the gas violate the entropy postulate of Section 20-2, which states that entropy always increases? No, because that postulate holds only for *irreversible* processes occurring in closed systems. The procedure suggested here does not meet these requirements. The process is *not* irreversible, and (because energy is transferred as heat from the gas to the reservoir) the system—which is the gas alone—is *not* closed.

However, if we include the reservoir, along with the gas, as part of the system, then we do have a closed system. Let's check the change in entropy of the enlarged system *gas + reservoir* for the process that takes it from (*b*) to (*a*) in Fig. 20-3. During this reversible process, energy is transferred as heat from the gas to the reservoir—that is, from one part of the enlarged system to another. Let $|Q|$ represent the absolute value (or magnitude) of this heat. With Eq. 20-2, we can then calculate separately the entropy changes for the gas (which loses $|Q|$) and the reservoir (which gains $|Q|$). We get

$$\Delta S_{\text{gas}} = -\frac{|Q|}{T}$$

and

$$\Delta S_{\text{res}} = +\frac{|Q|}{T}.$$

The entropy change of the closed system is the sum of these two quantities, *which is zero*.

With this result, we can modify the entropy postulate of Section 20-2 to include both reversible and irreversible processes:

> If a process occurs in a *closed* system, the entropy of the system increases for irreversible processes and remains constant for reversible processes. It never decreases.

Although entropy may decrease in part of a closed system, there will always be an equal or larger entropy increase in another part of the system, so that the entropy of the system as a whole never decreases. This fact is one form of the **second law of thermodynamics** and can be written as

$$\Delta S \geq 0 \qquad \text{(second law of thermodynamics)}, \qquad (20\text{-}5)$$

where the greater-than sign applies to irreversible processes and the equals sign to reversible processes. Equation 20-5 applies only to closed systems.

In the real world almost all processes are irreversible to some extent because of friction, turbulence, and other factors, so the entropy of real closed systems undergoing real processes always increases. Processes in which the system's entropy remains constant are always idealizations.

Fig. 20-7 The elements of a Carnot engine. The two black arrowheads on the central loop suggest the working substance operating in a cycle, as if on a p-V plot. Energy $|Q_H|$ is transferred as heat from the high-temperature reservoir at temperature T_H to the working substance. Energy $|Q_L|$ is transferred as heat from the working substance to the low-temperature reservoir at temperature T_L. Work W is done by the engine (actually by the working substance) on something in the environment.

20-5 Entropy in the Real World: Engines

A **heat engine,** or more simply, an **engine,** is a device that extracts energy from its environment in the form of heat and does useful work. At the heart of every engine is a *working substance.* In a steam engine, the working substance is water, in both its vapor and its liquid form. In an automobile engine the working substance is a gasoline–air mixture. If an engine is to do work on a sustained basis, the working substance must operate in a *cycle*; that is, the working substance must pass through a closed series of thermodynamic processes, called *strokes,* returning again and again to each state in its cycle. Let us see what the laws of thermodynamics can tell us about the operation of engines.

A Carnot Engine

We have seen that we can learn much about real gases by analyzing an ideal gas, which obeys the simple law $pV = nRT$. This is a useful plan because, although an ideal gas does not exist, any real gas approaches ideal behavior as closely as you wish if its density is low enough. In much the same spirit we choose to study real engines by analyzing the behavior of an **ideal engine.**

> In an ideal engine, all processes are reversible and no wasteful energy transfers occur due to, say, friction and turbulence.

We shall focus on a particular ideal engine called a **Carnot engine** after the French scientist and engineer N. L. Sadi Carnot (pronounced "car-no"), who first proposed the engine's concept in 1824. This ideal engine turns out to be the best (in principle) at using energy as heat to do useful work. Surprisingly, Carnot was able to analyze the performance of this engine before the first law of thermodynamics and the concept of entropy had been discovered.

Figure 20-7 shows schematically the operation of a Carnot engine. During each cycle of the engine, the working substance absorbs energy $|Q_H|$ as heat from a thermal reservoir at constant temperature T_H and discharges energy $|Q_L|$ as heat to a second thermal reservoir at a constant lower temperature T_L.

Figure 20-8 shows a p-V plot of the *Carnot cycle*—the cycle followed by the working substance. As indicated by the arrows, the cycle is traversed in the clockwise direction. Imagine the working substance to be a gas, confined to an insulating cylinder with a weighted, movable piston. The cylinder may be placed at will on either of the two thermal reservoirs, as in Fig. 20-6, or on an insulating slab. Figure 20-8 shows that, if we place the cylinder in contact with the high-temperature reservoir at temperature T_H, heat $|Q_H|$ is transferred *to* the working substance *from* this reservoir as the gas undergoes an isothermal *expansion* from volume V_a to volume V_b. Similarly, with the working substance in contact with the low-temperature reservoir at temperature T_L, heat $|Q_L|$ is transferred *from*

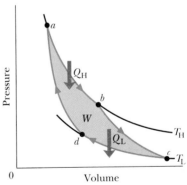

Fig. 20-8 A pressure–volume plot of the cycle followed by the working substance of the Carnot engine in Fig. 20-7. The cycle consists of two isothermal (ab and cd) and two adiabatic processes (bc and da). The shaded area enclosed by the cycle is equal to the work W per cycle done by the Carnot engine.

the working substance *to* the low-temperature reservoir as the gas undergoes an isothermal *compression* from volume V_c to volume V_d.

In the engine of Fig. 20-7, we assume that heat transfers to or from the working substance can take place *only* during the isothermal processes *ab* and *cd* of Fig. 20-8. Therefore, processes *bc* and *da* in that figure, which connect the two isotherms at temperatures T_H and T_L, must be (reversible) adiabatic processes; that is, they must be processes in which no energy is transferred as heat. To ensure this, during processes *bc* and *da* the cylinder is placed on an insulating slab as the volume of the working substance is changed.

During the consecutive processes *ab* and *bc* of Fig. 20-8, the working substance is expanding and thus doing positive work as it raises the weighted piston. This work is represented in Fig. 20-8 by the area under curve *abc*. During the consecutive processes *cd* and *da*, the working substance is being compressed, which means that it is doing negative work on its environment or, equivalently, that its environment is doing work on it as the loaded piston descends. This work is represented by the area under curve *cda*. The *net work per cycle*, which is represented by *W* in both Figs. 20-7 and 20-8, is the difference between these two areas and is a positive quantity equal to the area enclosed by cycle *abcda* in Fig. 20-8. This work *W* is performed on some outside object, such as a load to be lifted.

Equation 20-1 ($\Delta S = \int dQ/T$) tells us that any energy transfer as heat must involve a change in entropy. To illustrate the entropy changes for a Carnot engine, we can plot the Carnot cycle on a temperature-entropy (*T-S*) diagram as shown in Fig. 20-9. The lettered points *a*, *b*, *c*, and *d* in Fig. 20-9 correspond to the lettered points in the *p-V* diagram in Fig. 20-8. The two horizontal lines in Fig. 20-9 correspond to the two isothermal processes of the Carnot cycle (because the temperature is constant). Process *ab* is the isothermal expansion of the cycle. As the working substance (reversibly) absorbs energy $|Q_H|$ as heat at constant temperature T_H during the expansion, its entropy increases. Similarly, during the isothermal compression *cd*, the working substance (reversibly) loses energy $|Q_L|$ as heat at constant temperature T_L, and its entropy decreases.

The two vertical lines in Fig. 20-9 correspond to the two adiabatic processes of the Carnot cycle. Because no energy is transferred as heat during the two processes, the entropy of the working substance is constant during them.

The Work   To calculate the net work done by a Carnot engine during a cycle, let us apply Eq. 18-26, the first law of thermodynamics ($\Delta E_{int} = Q - W$), to the working substance. That substance must return again and again to any arbitrarily selected state in the cycle. Thus, if *X* represents any state property of the working substance, such as pressure, temperature, volume, internal energy, or entropy, we must have $\Delta X = 0$ for every cycle. It follows that $\Delta E_{int} = 0$ for a complete cycle of the working substance. Recalling that *Q* in Eq. 18-26 is the *net* heat transfer per cycle and *W* is the *net* work, we can write the first law of thermodynamics for the Carnot cycle as

$$W = |Q_H| - |Q_L|. \tag{20-6}$$

Entropy Changes   In a Carnot engine, there are *two* (and only two) reversible energy transfers as heat, and thus two changes in the entropy of the working substance—one at temperature T_H and one at T_L. The net entropy change per cycle is then

$$\Delta S = \Delta S_H + \Delta S_L = \frac{|Q_H|}{T_H} - \frac{|Q_L|}{T_L}. \tag{20-7}$$

Here ΔS_H is positive because energy $|Q_H|$ is *added to* the working substance as heat (an increase in entropy) and ΔS_L is negative because energy $|Q_L|$ is *removed from* the working substance as heat (a decrease in entropy). Because entropy is a state function, we must have $\Delta S = 0$ for a complete cycle. Putting $\Delta S = 0$ in

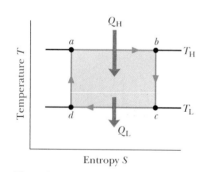

Fig. 20-9   The Carnot cycle of Fig. 20-8 plotted on a temperature–entropy diagram. During processes *ab* and *cd* the temperature remains constant. During processes *bc* and *da* the entropy remains constant.

Eq. 20-7 requires that

$$\frac{|Q_H|}{T_H} = \frac{|Q_L|}{T_L}. \qquad (20\text{-}8)$$

Note that, because $T_H > T_L$, we must have $|Q_H| > |Q_L|$; that is, more energy is extracted as heat from the high-temperature reservoir than is delivered to the low-temperature reservoir.

We shall now use Eqs. 20-6 and 20-8 to derive an expression for the efficiency of a Carnot engine.

Efficiency of a Carnot Engine

The purpose of any engine is to transform as much of the extracted energy Q_H into work as possible. We measure its success in doing so by its **thermal efficiency** ε, defined as the work the engine does per cycle ("energy we get") divided by the energy it absorbs as heat per cycle ("energy we pay for"):

$$\varepsilon = \frac{\text{energy we get}}{\text{energy we pay for}} = \frac{|W|}{|Q_H|} \qquad \text{(efficiency, any engine).} \qquad (20\text{-}9)$$

For a Carnot engine we can substitute for W from Eq. 20-6 to write Eq. 20-9 as

$$\varepsilon_C = \frac{|Q_H| - |Q_L|}{|Q_H|} = 1 - \frac{|Q_L|}{|Q_H|}. \qquad (20\text{-}10)$$

Using Eq. 20-8 we can write this as

$$\varepsilon_C = 1 - \frac{T_L}{T_H} \qquad \text{(efficiency, Carnot engine),} \qquad (20\text{-}11)$$

where the temperatures T_L and T_H are in kelvins. Because $T_L < T_H$, the Carnot engine necessarily has a thermal efficiency less than unity—that is, less than 100%. This is indicated in Fig. 20-7, which shows that only part of the energy extracted as heat from the high-temperature reservoir is available to do work, and the rest is delivered to the low-temperature reservoir. We shall show in Section 20-7 that no real engine can have a thermal efficiency greater than that calculated from Eq. 20-11.

Inventors continually try to improve engine efficiency by reducing the energy $|Q_L|$ that is "thrown away" during each cycle. The inventor's dream is to produce the *perfect engine,* diagrammed in Fig. 20-10, in which $|Q_L|$ is reduced to zero and $|Q_H|$ is converted completely into work. Such an engine on an ocean liner, for example, could extract energy as heat from the water and use it to drive the propellers, with no fuel cost. An automobile fitted with such an engine could extract energy as heat from the surrounding air and use it to drive the car, again with no fuel cost. Alas, a perfect engine is only a dream: Inspection of Eq. 20-11 shows that we can achieve 100% engine efficiency (that is, $\varepsilon = 1$) only if $T_L = 0$ or $T_H \to \infty$, requirements that are impossible to meet. Instead, decades of practical engineering experience have led to the following alternative version of the second law of thermodynamics:

No series of processes is possible whose sole result is the transfer of energy as heat from a thermal reservoir and the complete conversion of this energy to work.

In short, *there are no perfect engines.*

To summarize: The thermal efficiency given by Eq. 20-11 applies only to Carnot engines. Real engines, in which the processes that form the engine cycle are not reversible, have lower efficiencies. If your car were powered by a Carnot

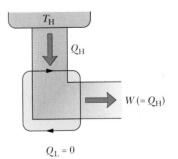

Fig. 20-10 The elements of a perfect engine—that is, one that converts heat Q_H from a high-temperature reservoir directly to work W with 100% efficiency.

Fig. 20-11 The North Anna nuclear power plant near Charlottesville, Virginia, which generates electric energy at the rate of 900 MW. At the same time, by design, it discards energy into the nearby river at the rate of 2100 MW. This plant and all others like it throw away more energy than they deliver in useful form. They are real counterparts of the ideal engine of Fig. 20-7.

engine, it would have an efficiency of about 55% according to Eq. 20-11; its actual efficiency is probably about 25%. A nuclear power plant (Fig. 20-11), taken in its entirety, is an engine. It extracts energy as heat from a reactor core, does work by means of a turbine, and discharges energy as heat to a nearby river. If the power plant operated as a Carnot engine, its efficiency would be about 40%; its actual efficiency is about 30%. In designing engines of any type, there is simply no way to beat the efficiency limitation imposed by Eq. 20-11.

Stirling Engine

Equation 20-11 applies not to all ideal engines but only to those that can be represented as in Fig. 20-8—that is, to Carnot engines. For example, Fig. 20-12 shows the operating cycle of an ideal **Stirling engine.** Comparison with the Carnot cycle of Fig. 20-8 shows that each engine has isothermal heat transfers at temperatures T_H and T_L. However, the two isotherms of the Stirling engine cycle are connected, not by adiabatic processes as for the Carnot engine but by constant-volume processes. To increase the temperature of a gas at constant volume reversibly from T_L to T_H (process *da* of Fig. 20-12) requires a transfer of energy as heat to the working substance from a thermal reservoir whose temperature can be varied smoothly between those limits. Also, a reverse transfer is required in process *bc*. Thus, reversible heat transfers (and corresponding entropy changes) occur in all four of the processes that form the cycle of a Stirling engine, not just two processes as in a Carnot engine. Thus, the derivation that led to Eq. 20-11 does not apply to an ideal Stirling engine. More important, the efficiency of an ideal Stirling engine is lower than that of a Carnot engine operating between the same two temperatures. Real Stirling engines have even lower efficiencies.

The Stirling engine was developed in 1816 by Robert Stirling. This engine, long neglected, is now being developed for use in automobiles and spacecraft. A Stirling engine delivering 5000 hp (3.7 MW) has been built.

> ✓**CHECKPOINT 3** Three Carnot engines operate between reservoir temperatures of (a) 400 and 500 K, (b) 600 and 800 K, and (c) 400 and 600 K. Rank the engines according to their thermal efficiencies, greatest first.

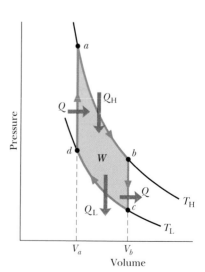

Fig. 20-12 A *p-V* plot for the working substance of an ideal Stirling engine, with the working substance assumed for convenience to be an ideal gas.

Sample Problem 20-4

Imagine a Carnot engine that operates between the temperatures $T_H = 850$ K and $T_L = 300$ K. The engine performs 1200 J of work each cycle, which takes 0.25 s.

(a) What is the efficiency of this engine?

Solution: The **Key Idea** here is that the efficiency ε of a Carnot engine depends only on the ratio T_L/T_H of the temperatures (in kelvins) of the thermal reservoirs to which it is connected. Thus, from Eq. 20-11, we have

$$\varepsilon = 1 - \frac{T_L}{T_H} = 1 - \frac{300 \text{ K}}{850 \text{ K}} = 0.647 \approx 65\%. \qquad \text{(Answer)}$$

(b) What is the average power of this engine?

Solution: Here the **Key Idea** is that the average power P of an engine is the ratio of the work W it does per cycle to the time t that each cycle takes. For this Carnot engine, we find

$$P = \frac{W}{t} = \frac{1200 \text{ J}}{0.25 \text{ s}} = 4800 \text{ W} = 4.8 \text{ kW}. \qquad \text{(Answer)}$$

(c) How much energy $|Q_H|$ is extracted as heat from the high-temperature reservoir every cycle?

Solution: Now the **Key Idea** is that for any engine, including a Carnot engine, the efficiency ε is the ratio of the work W that is done per cycle to the energy $|Q_H|$ that is extracted as heat from the high-temperature reservoir per cycle ($\varepsilon = W/|Q_H|$). Thus,

$$|Q_H| = \frac{W}{\varepsilon} = \frac{1200 \text{ J}}{0.647} = 1855 \text{ J}. \qquad \text{(Answer)}$$

(d) How much energy $|Q_L|$ is delivered as heat to the low-temperature reservoir every cycle?

Solution: The **Key Idea** here is that for a Carnot engine, the work W done per cycle is equal to the difference in the energy transfers as heat: $|Q_H| - |Q_L|$, as in Eq. 20-6. Thus, we have

$$|Q_L| = |Q_H| - W$$
$$= 1855 \text{ J} - 1200 \text{ J} = 655 \text{ J}. \qquad \text{(Answer)}$$

(e) By how much does the entropy of the working substance change as a result of the energy transferred to it from the high-temperature reservoir? From it to the low-temperature reservoir?

Solution: The **Key Idea** here is that the entropy change ΔS during a transfer of energy as heat Q at constant temperature T is given by Eq. 20-2 ($\Delta S = Q/T$). Thus, for the *positive* transfer of energy Q_H from the high-temperature reservoir at T_H, the change in the entropy of the working substance is

$$\Delta S_H = \frac{Q_H}{T_H} = \frac{1855 \text{ J}}{850 \text{ K}} = +2.18 \text{ J/K}. \qquad \text{(Answer)}$$

Similarly, for the *negative* transfer of energy Q_L to the low-temperature reservoir at T_L, we have

$$\Delta S_L = \frac{Q_L}{T_L} = \frac{-655 \text{ J}}{300 \text{ K}} = -2.18 \text{ J/K}. \qquad \text{(Answer)}$$

Note that the net entropy change of the working substance for one cycle is zero, as we discussed in deriving Eq. 20-8.

Sample Problem 20-5

An inventor claims to have constructed an engine that has an efficiency of 75% when operated between the boiling and freezing points of water. Is this possible?

Solution: The **Key Idea** here is that the efficiency of a real engine (with its irreversible processes and wasteful energy transfers) must be less than the efficiency of a Carnot engine operating between the same two temperatures. From Eq.

20-11, we find that the efficiency of a Carnot engine operating between the boiling and freezing points of water is

$$\varepsilon = 1 - \frac{T_L}{T_H} = 1 - \frac{(0 + 273) \text{ K}}{(100 + 273) \text{ K}} = 0.268 \approx 27\%.$$

Thus, the claimed efficiency of 75% for a real engine operating between the given temperatures is impossible.

PROBLEM-SOLVING TACTICS

TACTIC 1:
The Language of Thermodynamics

A rich but sometimes misleading language is used in scientific and engineering studies of thermodynamics. You may see statements that say heat is added, absorbed, subtracted, extracted, rejected, discharged, discarded, withdrawn, delivered, gained, lost, transferred, or expelled, or that it flows from one body to another (as if it were a liquid). You may also see statements that describe a body as *having* heat (as if heat can be held or possessed) or that its heat is increased or decreased. You should always keep in mind what is meant by the term *heat* in science and engineering:

Heat is energy that is transferred from one body to another body due to a difference in the temperatures of the bodies.

When we identify one of the bodies as being our system of interest, any such transfer of energy into the system is positive heat Q, and any such transfer out of the system is negative heat Q.

The term *work* also requires close attention. You may see statements that say work is produced or generated, or combined with heat or changed from heat. Here is what is meant by the term *work*:

> Work is energy that is transferred from one body to another body due to a force that acts between them.

When we identify one of the bodies as being our system of interest, any such transfer of energy out of the system is either positive work W done *by* the system or negative work W done *on* the system. Any such transfer of energy into the system is negative work done *by* the system or positive work done *on* the system. (The preposition that is used is important.) Obviously, this can be confusing—whenever you see the term *work*, read carefully.

20-6 Entropy in the Real World: Refrigerators

A **refrigerator** is a device that uses work to transfer energy from a low-temperature reservoir to a high-temperature reservoir as the device continuously repeats a set series of thermodynamic processes. In a household refrigerator, for example, work is done by an electrical compressor to transfer energy from the food storage compartment (a low-temperature reservoir) to the room (a high-temperature reservoir).

Air conditioners and heat pumps are also refrigerators. The differences are only in the nature of the high- and low-temperature reservoirs. For an air conditioner, the low-temperature reservoir is the room that is to be cooled and the high-temperature reservoir is the (presumably warmer) outdoors. A heat pump is an air conditioner that can be operated in reverse to heat a room; the room is the high-temperature reservoir, and heat is transferred to it from the (presumably cooler) outdoors.

Let us consider an *ideal refrigerator:*

> In an ideal refrigerator, all processes are reversible and no wasteful energy transfers occur as a result of, say, friction and turbulence.

Figure 20-13 shows the basic elements of an ideal refrigerator. Note that its operation is the reverse of how the Carnot engine of Fig. 20-7 operates. In other words, all the energy transfers, as either heat or work, are reversed from those of a Carnot engine. We can call such an ideal refrigerator a **Carnot refrigerator.**

The designer of a refrigerator would like to extract as much energy $|Q_L|$ as possible from the low-temperature reservoir (what we want) for the least amount of work $|W|$ (what we pay for). A measure of the efficiency of a refrigerator, then, is

$$K = \frac{\text{what we want}}{\text{what we pay for}} = \frac{|Q_L|}{|W|} \qquad \text{(coefficient of performance, any refrigerator),} \qquad (20\text{-}12)$$

where K is called the *coefficient of performance.* For a Carnot refrigerator, the first law of thermodynamics gives $|W| = |Q_H| - |Q_L|$, where $|Q_H|$ is the magnitude of the energy transferred as heat to the high-temperature reservoir. Equation 20-12 then becomes

$$K_C = \frac{|Q_L|}{|Q_H| - |Q_L|}. \qquad (20\text{-}13)$$

Because a Carnot refrigerator is a Carnot engine operating in reverse, we can combine Eq. 20-8 with Eq. 20-13; after some algebra we find

$$K_C = \frac{T_L}{T_H - T_L} \qquad \text{(coefficient of performance, Carnot refrigerator).} \qquad (20\text{-}14)$$

For typical room air conditioners, $K \approx 2.5$. For household refrigerators, $K \approx 5$. Perversely, the value of K is higher the closer the temperatures of the two reservoirs are to each other. That is why heat pumps are more effective in temperate climates than in climates where the outside temperature is much lower than the desired inside temperature.

Fig. 20-13 The elements of a refrigerator. The two black arrowheads on the central loop suggest the working substance operating in a cycle, as if on a p-V plot. Energy is transferred as heat Q_L to the working substance from the low-temperature reservoir. Energy is transferred as heat Q_H to the high-temperature reservoir from the working substance. Work W is done on the refrigerator (on the working substance) by something in the environment.

It would be nice to own a refrigerator that did not require some input of work—that is, one that would run without being plugged in. Figure 20-14 represents another "inventor's dream," a *perfect refrigerator* that transfers energy as heat Q from a cold reservoir to a warm reservoir without the need for work. Because the unit operates in cycles, the entropy of the working substance does not change during a complete cycle. The entropies of the two reservoirs, however, do change: The entropy change for the cold reservoir is $-|Q|/T_L$, and that for the warm reservoir is $+|Q|/T_H$. Thus, the net entropy change for the entire system is

$$\Delta S = -\frac{|Q|}{T_L} + \frac{|Q|}{T_H}.$$

Because $T_H > T_L$, the right side of this equation is negative and thus the net change in entropy per cycle for the closed system *refrigerator + reservoirs* is also negative. Because such a decrease in entropy violates the second law of thermodynamics (Eq. 20-5), a perfect refrigerator does not exist. (If you want your refrigerator to operate, you must plug it in.)

This result leads us to another (equivalent) formulation of the second law of thermodynamics:

> No series of processes is possible whose sole result is the transfer of energy as heat from a reservoir at a given temperature to a reservoir at a higher temperature.

In short, *there are no perfect refrigerators.*

Fig. 20-14 The elements of a perfect refrigerator—that is, one that transfers energy from a low-temperature reservoir to a high-temperature reservoir without any input of work.

> **CHECKPOINT 4** You wish to increase the coefficient of performance of an ideal refrigerator. You can do so by (a) running the cold chamber at a slightly higher temperature, (b) running the cold chamber at a slightly lower temperature, (c) moving the unit to a slightly warmer room, or (d) moving it to a slightly cooler room. The magnitudes of the temperature changes are to be the same in all four cases. List the changes according to the resulting coefficients of performance, greatest first.

20-7 *The Efficiencies of Real Engines*

Let ε_C be the efficiency of a Carnot engine operating between two given temperatures. In this section we prove that no real engine operating between those temperatures can have an efficiency greater than ε_C. If it could, the engine would violate the second law of thermodynamics.

Let us assume that an inventor, working in her garage, has constructed an engine X, which she claims has an efficiency ε_X that is greater than ε_C:

$$\varepsilon_X > \varepsilon_C \qquad \text{(a claim).} \tag{20-15}$$

Let us couple engine X to a Carnot refrigerator, as in Fig. 20-15a. We adjust the

Fig. 20-15 (a) Engine X drives a Carnot refrigerator. (b) If, as claimed, engine X is more efficient than a Carnot engine, then the combination shown in (a) is equivalent to the perfect refrigerator shown here. This violates the second law of thermodynamics, so we conclude that engine X *cannot* be more efficient than a Carnot engine.

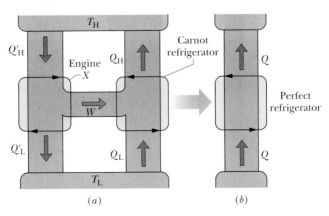

strokes of the Carnot refrigerator so that the work it requires per cycle is just equal to that provided by engine *X*. Thus, no (external) work is performed on or by the combination *engine + refrigerator* of Fig. 20-15*a*, which we take as our system.

If Eq. 20-15 is true, from the definition of efficiency (Eq. 20-9), we must have

$$\frac{|W|}{|Q'_H|} > \frac{|W|}{|Q_H|},$$

where the prime refers to engine *X* and the right side of the inequality is the efficiency of the Carnot refrigerator when it operates as an engine. This inequality requires that

$$|Q_H| > |Q'_H|. \tag{20-16}$$

Because the work done by engine *X* is equal to the work done on the Carnot refrigerator, we have, from the first law of thermodynamics as given by Eq. 20-6,

$$|Q_H| - |Q_L| = |Q'_H| - |Q'_L|,$$

which we can write as

$$|Q_H| - |Q'_H| = |Q_L| - |Q'_L| = Q. \tag{20-17}$$

Because of Eq. 20-16, the quantity *Q* in Eq. 20-17 must be positive.

Comparison of Eq. 20-17 with Fig. 20-15 shows that the net effect of engine *X* and the Carnot refrigerator working in combination is to transfer energy *Q* as heat from a low-temperature reservoir to a high-temperature reservoir without the requirement of work. Thus, the combination acts like the perfect refrigerator of Fig. 20-14, whose existence is a violation of the second law of thermodynamics.

Something must be wrong with one or more of our assumptions, and it can only be Eq. 20-15. We conclude that *no real engine can have an efficiency greater than that of a Carnot engine when both engines work between the same two temperatures.* At most, the real engine can have an efficiency equal to that of a Carnot engine. In that case, the real engine *is* a Carnot engine.

20-8   A Statistical View of Entropy

In Chapter 19 we saw that the macroscopic properties of gases can be explained in terms of their microscopic, or molecular, behavior. For one example, recall that we were able to account for the pressure exerted by a gas on the walls of its container in terms of the momentum transferred to those walls by rebounding gas molecules. Such explanations are part of a study called **statistical mechanics.**

Here we shall focus our attention on a single problem, one involving the distribution of gas molecules between the two halves of an insulated box. This problem is reasonably simple to analyze, and it allows us to use statistical mechanics to calculate the entropy change for the free expansion of an ideal gas. You will see in Sample Problem 20-7 that statistical mechanics leads to the same entropy change we obtained in Sample Problem 20-1 using thermodynamics.

Figure 20-16 shows a box that contains six identical (and thus indistinguishable) molecules of a gas. At any instant, a given molecule will be in either the left or the right half of the box; because the two halves have equal volumes, the molecule has the same likelihood, or probability, of being in either half.

Table 20-1 shows the seven possible *configurations* of the six molecules, each configuration labeled with a Roman numeral. For example, in configuration I, all six molecules are in the left half of the box ($n_1 = 6$) and none are in the right half ($n_2 = 0$). We see that, in general, a given configuration can be achieved in a number of different ways. We call these different arrangements of the molecules *microstates*. Let us see how to calculate the number of microstates that correspond to a given configuration.

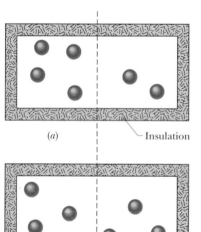

(a)   Insulation

(b)

Fig. 20-16   An insulated box contains six gas molecules. Each molecule has the same probability of being in the left half of the box as in the right half. The arrangement in (*a*) corresponds to configuration III in Table 20-1, and that in (*b*) corresponds to configuration IV.

TABLE 20-1
Six Molecules in a Box

Configuration			Multiplicity W (number of microstates)	Calculation of W (Eq. 20-18)	Entropy 10^{-23} J/K (Eq. 20-19)
Label	n_1	n_2			
I	6	0	1	$6!/(6!\,0!) = 1$	0
II	5	1	6	$6!/(5!\,1!) = 6$	2.47
III	4	2	15	$6!/(4!\,2!) = 15$	3.74
IV	3	3	20	$6!/(3!\,3!) = 20$	4.13
V	2	4	15	$6!/(2!\,4!) = 15$	3.74
VI	1	5	6	$6!/(1!\,5!) = 6$	2.47
VII	0	6	1	$6!/(0!\,6!) = 1$	0
			Total = 64		

Suppose we have N molecules, distributed with n_1 molecules in one half of the box and n_2 in the other. (Thus $n_1 + n_2 = N$.) Let us imagine that we distribute the molecules "by hand," one at a time. If $N = 6$, we can select the first molecule in six independent ways; that is, we can pick any one of the six molecules. We can pick the second molecule in five ways, by picking any one of the remaining five molecules; and so on. The total number of ways in which we can select all six molecules is the product of these independent ways, or $6 \times 5 \times 4 \times 3 \times 2 \times 1 = 720$. In mathematical shorthand we write this product as $6! = 720$, where 6! is pronounced "six factorial." Your hand calculator can probably calculate factorials. For later use you will need to know that $0! = 1$. (Check this on your calculator.)

However, because the molecules are indistinguishable, these 720 arrangements are not all different. In the case that $n_1 = 4$ and $n_2 = 2$ (which is configuration III in Table 20-1), for example, the order in which you put four molecules in one half of the box does not matter, because after you have put all four in, there is no way that you can tell the order in which you did so. The number of ways in which you can order the four molecules is $4! = 24$. Similarly, the number of ways in which you can order two molecules for the other half of the box is simply $2! = 2$. To get the number of *different* arrangements that lead to the (4, 2) split of configuration III, we must divide 720 by 24 and also by 2. We call the resulting quantity, which is the number of microstates that correspond to a given configuration, the *multiplicity W* of that configuration. Thus, for configuration III,

$$W_{\text{III}} = \frac{6!}{4!\,2!} = \frac{720}{24 \times 2} = 15.$$

Thus, Table 20-1 tells us there are 15 independent microstates that correspond to configuration III. Note that, as the table also tells us, the total number of microstates for six molecules distributed over the seven configurations is 64.

Extrapolating from six molecules to the general case of N molecules, we have

$$W = \frac{N!}{n_1!\,n_2!} \qquad \text{(multiplicity of configuration)}. \qquad (20\text{-}18)$$

You should verify that Eq. 20-18 gives the multiplicities for all the configurations listed in Table 20-1.

The basic assumption of statistical mechanics is

 All microstates are equally probable.

In other words, if we were to take a great many snapshots of the six molecules as they jostle around in the box of Fig. 20-16 and then count the number of times

Fig. 20-17 For a *large* number of molecules in a box, a plot of the number of microstates that require various percentages of the molecules to be in the left half of the box. Nearly all the microstates correspond to an approximately equal sharing of the molecules between the two halves of the box; those microstates form the *central configuration peak* on the plot. For $N \approx 10^{22}$, the central configuration peak is much too narrow to be drawn on this plot.

each microstate occurred, we would find that all 64 microstates would occur equally often. Thus the system will spend, on average, the same amount of time in each of the 64 microstates.

Because all microstates are equally probable but different configurations have different numbers of microstates, the configurations are *not* all equally probable. In Table 20-1 configuration IV, with 20 microstates, is the *most probable configuration*, with a probability of 20/64 = 0.313. This result means that the system is in configuration IV 31.3% of the time. Configurations I and VII, in which all the molecules are in one half of the box, are the least probable, each with a probability of 1/64 = 0.016 or 1.6%. It is not surprising that the most probable configuration is the one in which the molecules are evenly divided between the two halves of the box, because that is what we expect at thermal equilibrium. However, it *is* surprising that there is *any* probability, however small, of finding all six molecules clustered in half of the box, with the other half empty.

For large values of N there are extremely large numbers of microstates, but nearly all the microstates belong to the configuration in which the molecules are divided equally between the two halves of the box, as Fig. 20-17 indicates. Even though the measured temperature and pressure of the gas remain constant, the gas is churning away endlessly as its molecules "visit" all probable microstates with equal probability. However, because so few microstates lie outside the very narrow central configuration peak of Fig. 20-17, we might as well assume that the gas molecules are always divided equally between the two halves of the box. As we shall see, this is the configuration with the greatest entropy.

Sample Problem 20-6

Suppose that there are 100 indistinguishable molecules in the box of Fig. 20-16. How many microstates are associated with the configuration $n_1 = 50$ and $n_2 = 50$, and with the configuration $n_1 = 100$ and $n_2 = 0$? Interpret the results in terms of the relative probabilities of the two configurations.

Solution: The Key Idea here is that the multiplicity W of a configuration of indistinguishable molecules in a closed box is the number of independent microstates with that configuration, as given by Eq. 20-18. For the (n_1, n_2) configuration (50, 50), that equation yields

$$W = \frac{N!}{n_1! \, n_2!} = \frac{100!}{50! \, 50!}$$
$$= \frac{9.33 \times 10^{157}}{(3.04 \times 10^{64})(3.04 \times 10^{64})}$$
$$= 1.01 \times 10^{29}. \qquad \text{(Answer)}$$

Similarly, for the configuration (100, 0), we have

$$W = \frac{N!}{n_1! \, n_2!} = \frac{100!}{100! \, 0!} = \frac{1}{0!} = \frac{1}{1} = 1. \quad \text{(Answer)}$$

Thus, a 50–50 distribution is more likely than a 100–0 distribution by the enormous factor of about 1×10^{29}. If you could count, at one per nanosecond, the number of microstates that correspond to the 50–50 distribution, it would take you about 3×10^{12} years, which is about 200 times longer than the age of the universe. Even 100 molecules is *still* a very small number. Imagine what these calculated probabilities would be like for a mole of molecules, say about $N = 10^{24}$. Thus, you need never worry about suddenly finding all the air molecules clustering in one corner of your room, with you gasping for air in another corner.

Probability and Entropy

In 1877, Austrian physicist Ludwig Boltzmann (the Boltzmann of Boltzmann's constant k) derived a relationship between the entropy S of a configuration of a gas and the multiplicity W of that configuration. That relationship is

$$S = k \ln W \qquad \text{(Boltzmann's entropy equation)}. \tag{20-19}$$

This famous formula is engraved on Boltzmann's tombstone.

It is natural that S and W should be related by a logarithmic function. The total entropy of two systems is the *sum* of their separate entropies. The probability of occurrence of two independent systems is the *product* of their separate probabilities. Because $\ln ab = \ln a + \ln b$, the logarithm seems the logical way to connect these quantities.

Table 20-1 displays the entropies of the configurations of the six-molecule system of Fig. 20-16, computed using Eq. 20-19. Configuration IV, which has the greatest multiplicity, also has the greatest entropy.

When you use Eq. 20-18 to calculate W, your calculator may signal "OVER-FLOW" if you try to find the factorial of a number greater than a few hundred. Fortunately, there is a very good approximation, known as **Stirling's approximation,** not for $N!$ but for $\ln N!$, which is exactly what is needed in Eq. 20-19. Stirling's approximation is

$$\ln N! \approx N(\ln N) - N \quad \text{(Stirling's approximation)}. \quad (20\text{-}20)$$

The Stirling of this approximation was an English mathematician and not the Robert Stirling of engine fame.

CHECKPOINT 5   A box contains 1 mol of a gas. Consider two configurations: (a) each half of the box contains half the molecules and (b) each third of the box contains one-third of the molecules. Which configuration has more microstates?

Sample Problem 20-7

In Sample Problem 20-1 we showed that when n moles of an ideal gas doubles its volume in a free expansion, the entropy increase from the initial state i to the final state f is $S_f - S_i = nR \ln 2$. Derive this result with statistical mechanics.

Solution: One **Key Idea** here is that we can relate the entropy S of any given configuration of the molecules in the gas to the multiplicity W of microstates for that configuration, using Eq. 20-19 ($S = k \ln W$). We are interested in two configurations: the final configuration f (with the molecules occupying the full volume of their container in Fig. 20-1b) and the initial configuration i (with the molecules occupying the left half of the container).

A second **Key Idea** is that, because the molecules are in a closed container, we can calculate the multiplicity W of their microstates with Eq. 20-18. Here we have N molecules in the n moles of the gas. Initially, with the molecules all in the left half of the container, their (n_1, n_2) configuration is $(N, 0)$. Then, Eq. 20-18 gives their multiplicity as

$$W_i = \frac{N!}{N!\, 0!} = 1.$$

Finally, with the molecules spread through the full volume, their (n_1, n_2) configuration is $(N/2, N/2)$. Then, Eq. 20-18 gives their multiplicity as

$$W_f = \frac{N!}{(N/2)!\,(N/2)!}.$$

From Eq. 20-19, the initial and final entropies are

$$S_i = k \ln W_i = k \ln 1 = 0$$

and

$$S_f = k \ln W_f = k \ln(N!) - 2k \ln[(N/2)!]. \quad (20\text{-}21)$$

In writing Eq. 20-21, we have used the relation

$$\ln \frac{a}{b^2} = \ln a - 2 \ln b.$$

Now, applying Eq. 20-20 to evaluate Eq. 20-21, we find that

$$\begin{aligned}
S_f &= k \ln(N!) - 2k \ln[(N/2)!] \\
&= k[N(\ln N) - N] - 2k[(N/2)\ln(N/2) - (N/2)] \\
&= k[N(\ln N) - N - N \ln(N/2) + N] \\
&= k[N(\ln N) - N(\ln N - \ln 2)] = Nk \ln 2. \quad (20\text{-}22)
\end{aligned}$$

From Eq. 19-8 we can substitute nR for Nk, where R is the universal gas constant. Equation 20-22 then becomes

$$S_f = nR \ln 2.$$

The change in entropy from the initial state to the final is thus

$$\begin{aligned}
S_f - S_i &= nR \ln 2 - 0 \\
&= nR \ln 2, \quad \text{(Answer)}
\end{aligned}$$

which is what we set out to show. In Sample Problem 20-1 we calculated this entropy increase for a free expansion with thermodynamics by finding an equivalent reversible process and calculating the entropy change for *that* process in terms of temperature and heat transfer. In this sample problem, we calculate the same increase in entropy with statistical mechanics using the fact that the system consists of molecules.

Review & Summary

One-Way Processes   An **irreversible process** is one that cannot be reversed by means of small changes in the environment. The direction in which an irreversible process proceeds is set by the *change in entropy* ΔS of the system undergoing the process. Entropy S is a *state property* (or *state function*) of the system; that is, it depends only on the state of the system and not on the way in which the system reached that state. The *entropy postulate* states (in part): *If an irreversible process occurs in a closed system, the entropy of the system always increases.*

Calculating Entropy Change The **entropy change** ΔS for an irreversible process that takes a system from an initial state i to a final state f is exactly equal to the entropy change ΔS for *any reversible process* that takes the system between those same two states. We can compute the latter (but not the former) with

$$\Delta S = S_f - S_i = \int_i^f \frac{dQ}{T}. \quad (20\text{-}1)$$

Here Q is the energy transferred as heat to or from the system during the process, and T is the temperature of the system in kelvins during the process.

For a reversible isothermal process, Eq. 20-1 reduces to

$$\Delta S = S_f - S_i = \frac{Q}{T}. \quad (20\text{-}2)$$

When the temperature change ΔT of a system is small relative to the temperature (in kelvins) before and after the process, the entropy change can be approximated as

$$\Delta S = S_f - S_i \approx \frac{Q}{T_{avg}}, \quad (20\text{-}3)$$

where T_{avg} is the system's average temperature during the process.

When an ideal gas changes reversibly from an initial state with temperature T_i and volume V_i to a final state with temperature T_f and volume V_f, the change ΔS in the entropy of the gas is

$$\Delta S = S_f - S_i = nR \ln \frac{V_f}{V_i} + nC_V \ln \frac{T_f}{T_i}. \quad (20\text{-}4)$$

The Second Law of Thermodynamics This law, which is an extension of the entropy postulate, states: *If a process occurs in a closed system, the entropy of the system increases for irreversible processes and remains constant for reversible processes. It never decreases.* In equation form,

$$\Delta S \geq 0. \quad (20\text{-}5)$$

Engines An **engine** is a device that, operating in a cycle, extracts energy as heat $|Q_H|$ from a high-temperature reservoir and does a certain amount of work $|W|$. The *efficiency* ε of any engine is defined as

$$\varepsilon = \frac{\text{energy we get}}{\text{energy we pay for}} = \frac{|W|}{|Q_H|}. \quad (20\text{-}9)$$

In an **ideal engine,** all processes are reversible and no wasteful energy transfers occur due to, say, friction and turbulence. A **Carnot engine** is an ideal engine that follows the cycle of Fig. 20-8. Its efficiency is

$$\varepsilon_C = 1 - \frac{|Q_L|}{|Q_H|} = 1 - \frac{T_L}{T_H}, \quad (20\text{-}10, 20\text{-}11)$$

in which T_H and T_L are the temperatures of the high- and low-temperature reservoirs, respectively. Real engines always have an efficiency lower than that given by Eq. 20-11. Ideal engines that are not Carnot engines also have efficiencies lower than that given by Eq. 20-11.

A *perfect engine* is an imaginary engine in which energy extracted as heat from the high-temperature reservoir is converted completely to work. Such an engine would violate the second law of thermodynamics, which can be restated as follows: No series of processes is possible whose sole result is the absorption of energy as heat from a thermal reservoir and the complete conversion of this energy to work.

Refrigerators A refrigerator is a device that, operating in a cycle, has work W done on it as it extracts energy $|Q_L|$ as heat from a low-temperature reservoir. The coefficient of performance K of a refrigerator is defined as

$$K = \frac{\text{what we want}}{\text{what we pay for}} = \frac{|Q_L|}{|W|}. \quad (20\text{-}12)$$

A **Carnot refrigerator** is a Carnot engine operating in reverse. For a Carnot refrigerator, Eq. 20-12 becomes

$$K_C = \frac{|Q_L|}{|Q_H| - |Q_L|} = \frac{T_L}{T_H - T_L}. \quad (20\text{-}13, 20\text{-}14)$$

A *perfect refrigerator* is an imaginary refrigerator in which energy extracted as heat from the low-temperature reservoir is converted completely to heat discharged to the high-temperature reservoir, without any need for work. Such a refrigerator would violate the second law of thermodynamics, which can be restated as follows: No series of processes is possible whose sole result is the transfer of energy as heat from a reservoir at a given temperature to a reservoir at a higher temperature.

Entropy from a Statistical View The entropy of a system can be defined in terms of the possible distributions of its molecules. For identical molecules, each possible distribution of molecules is called a **microstate** of the system. All equivalent microstates are grouped into a **configuration** of the system. The number of microstates in a configuration is the **multiplicity** W of the configuration.

For a system of N molecules that may be distributed between the two halves of a box, the multiplicity is given by

$$W = \frac{N!}{n_1! \, n_2!}, \quad (20\text{-}18)$$

in which n_1 is the number of molecules in one half of the box and n_2 is the number in the other half. A basic assumption of **statistical mechanics** is that all the microstates are equally probable. Thus, configurations with a large multiplicity occur most often. When N is very large (say, $N = 10^{22}$ molecules or more), the molecules are nearly always in the configuration in which $n_1 = n_2$.

The multiplicity W of a configuration of a system and the entropy S of the system in that configuration are related by Boltzmann's entropy equation:

$$S = k \ln W, \quad (20\text{-}19)$$

where $k = 1.38 \times 10^{-23}$ J/K is the Boltzmann constant.

When N is very large (the usual case), we can approximate $\ln N!$ with *Stirling's approximation:*

$$\ln N! \approx N(\ln N) - N. \quad (20\text{-}20)$$

Questions

1 In four experiments, 2.5 mol of hydrogen gas undergoes reversible isothermal expansions, starting from the same volume but at different temperatures. The corresponding p-V plots are shown in Fig. 20-18. Rank the situations according to the change in the entropy of the gas, greatest first. (*Hint:* See Sample Problem 20-1.)

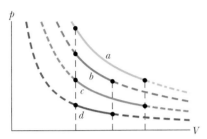

Fig. 20-18   Question 1.

2 An ideal gas, in contact with a controllable thermal reservoir, can be taken from initial state i to final state f along the four reversible paths in Fig. 20-19. Rank the paths according to the magnitudes of the resulting entropy changes of (a) the gas, (b) the reservoir, and (c) the gas–reservoir system, greatest first.

3 Point i in Fig. 20-20 represents the initial state of an ideal gas at temperature T. Taking algebraic signs into account, rank the entropy changes that the gas undergoes as it moves, successively and reversibly, from point i to points a, b, c, and d, greatest first.

Fig. 20-19   Question 2.

4 In four experiments, blocks A and B, starting at different initial temperatures, were brought together in an insulating box (as in Sample Problem 20-2) and allowed to reach a common final temperature. The entropy changes for the blocks in the four experiments had the following values (in joules per kelvin), but not necessarily in the order given. Determine which values for A go with which values for B.

Fig. 20-20   Question 3.

Block		Values		
A	8	5	3	9
B	−3	−8	−5	−2

5 A gas, confined to an insulated cylinder, is compressed ad-iabatically to half its volume. Does the entropy of the gas increase, decrease, or remain unchanged during this process?

6 An ideal monatomic gas at initial temperature T_0 (in kelvins) expands from initial volume V_0 to volume $2V_0$ by each of the five processes indicated in the T-V diagram of Fig. 20-21. In which process is the expansion (a) isothermal, (b) isobaric (constant pressure), and (c) adiabatic? Explain your answers. (d) In which processes does the entropy of the gas decrease?

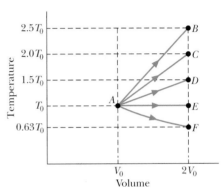

Fig. 20-21   Question 6.

7 An inventor claims to have invented four engines, each of which operates between constant-temperature reservoirs at 400 and 300 K. Data on each engine, per cycle of operation, are: engine A, $Q_H = 200$ J, $Q_L = -175$ J, and $W = 40$ J; engine B, $Q_H = 500$ J, $Q_L = -200$ J, and $W = 400$ J; engine C, $Q_H = 600$ J, $Q_L = -200$ J, and $W = 400$ J; engine D, $Q_H = 100$ J, $Q_L = -90$ J, and $W = 10$ J. Of the first and second laws of thermodynamics, which (if either) does each engine violate?

8 Three Carnot engines operate between temperature limits of (a) 400 and 500 K, (b) 500 and 600 K, and (c) 400 and 600 K. Each engine extracts the same amount of energy per cycle from the high-temperature reservoir. Rank the magnitudes of the work done by the engines per cycle, greatest first.

9 Does the entropy per cycle increase, decrease, or remain the same for (a) a Carnot engine, (b) a real engine, and (c) a perfect engine (which is, of course, impossible to build)?

10 Does the entropy per cycle increase, decrease, or remain the same for (a) a Carnot refrigerator, (b) a real refrigerator, and (c) a perfect refrigerator (which is, of course, impossible to build)?

11 A box contains 100 atoms in a configuration that has 50 atoms in each half of the box. Suppose that you could count the different microstates associated with this configuration at the rate of 100 billion states per second, using a supercomputer. Without written calculation, guess how much computing time you would need: a day, a year, or much more than a year.

Problems

sec. 20-3   Change in Entropy

•1 Suppose 4.00 mol of an ideal gas undergoes a reversible isothermal expansion from volume V_1 to volume $V_2 = 2.00V_1$ at temperature $T = 400$ K. Find (a) the work done by the gas and (b) the entropy change of the gas. (c) If the expansion is reversible and adiabatic instead of isothermal, what is the entropy change of the gas? **SSM**

•2 How much energy must be transferred as heat for a reversible isothermal expansion of an ideal gas at 132°C if the entropy of the gas increases by 46.0 J/K?

•3 A 2.50 mol sample of an ideal gas expands reversibly and isothermally at 360 K until its volume is doubled. What is the increase in entropy of the gas? **ILW**

•4 (a) What is the entropy change of a 12.0 g ice cube that melts completely in a bucket of water whose temperature is just above the freezing point of water? (b) What is the entropy change of a 5.00 g spoonful of water that evaporates completely on a hot plate whose temperature is slightly above the boiling point of water?

•5 Find (a) the energy absorbed as heat and (b) the change in entropy of a 2.00 kg block of copper whose temperature is increased reversibly from 25.0°C to 100°C. The specific heat of copper is 386 J/kg · K. **ILW**

•6 An ideal gas undergoes a reversible isothermal expansion at 77.0°C, increasing its volume from 1.30 L to 3.40 L. The entropy change of the gas is 22.0 J/K. How many moles of gas are present?

••7 In an experiment, 200 g of aluminum (with a specific heat of 900 J/kg · K) at 100°C is mixed with 50.0 g of water at 20.0°C, with the mixture thermally isolated. (a) What is the equilibrium temperature? What are the entropy changes of (b) the aluminum, (c) the water, and (d) the aluminum–water system? **SSM WWW**

••8 A gas sample undergoes a reversible isothermal expansion. Figure 20-22 gives the change ΔS in entropy of the gas versus the final volume V_f of the gas. How many moles are in the sample?

Fig. 20-22
Problem 8.

••9 In the irreversible process of Fig. 20-5, let the initial temperatures of identical blocks L and R be 305.5 and 294.5 K, respectively, and let 215 J be the energy that must be transferred between the blocks in order to reach equilibrium. For the reversible processes of Fig. 20-6, what is ΔS for (a) block L, (b) its reservoir, (c) block R, (d) its reservoir, (e) the two-block system, and (f) the system of the two blocks and the two reservoirs?

••10 A 364 g block is put in contact with a thermal reservoir. The block is initially at a lower temperature than the reservoir. Assume that the consequent transfer of energy as heat from the reservoir to the block is reversible. Figure 20-23 gives the change in entropy ΔS of the block until thermal equilibrium is reached. What is the specific heat of the block?

Fig. 20-23   Problem 10.

••11 For n moles of a diatomic ideal gas taken through the cycle in Fig. 20-24 with the molecules rotating but not oscillating, what are (a) p_2/p_1, (b) p_3/p_1, and (c) T_3/T_1? For path $1 \rightarrow 2$, what are (d) W/nRT_1, (e) Q/nRT_1, (f) $\Delta E_{int}/nRT_1$, and (g) $\Delta S/nR$? For path $2 \rightarrow 3$, what are (h) W/nRT_1, (i) Q/nRT_1, (j) $\Delta E_{int}/nRT_1$, (k) $\Delta S/nR$? For path $3 \rightarrow 1$, what are (l) W/nRT_1, (m) Q/nRT_1, (n) $\Delta E_{int}/nRT_1$, (o) $\Delta S/nR$?

Fig. 20-24   Problem 11.

••12 At very low temperatures, the molar specific heat C_V of many solids is approximately $C_V = AT^3$, where A depends on the particular substance. For aluminum, $A = 3.15 \times 10^{-5}$ J/mol · K⁴. Find the entropy change for 4.00 mol of aluminum when its temperature is raised from 5.00 K to 10.0 K.

••13 A 50.0 g block of copper whose temperature is 400 K is placed in an insulating box with a 100 g block of lead whose temperature is 200 K. (a) What is the equilibrium temperature of the two-block system? (b) What is the change in the internal energy of the system between the initial state and the equilibrium state? (c) What is the change in the entropy of the system? (See Table 18-3.) **ILW**

Fig. 20-25   Problem 14.

••14 A 2.0 mol sample of an ideal monatomic gas undergoes the reversible process shown in Fig. 20-25. (a) How much energy is absorbed as heat by the gas? (b) What is the change in the internal energy of the gas? (c) How much work is done by the gas?

••15 A 10 g ice cube at $-10°C$ is placed in a lake whose temperature is 15°C. Calculate the change in entropy of the cube–lake system as the ice cube comes to thermal equilibrium with the lake. The specific heat of ice is 2220 J/kg · K. (*Hint:* Will the ice cube affect the lake temperature?) **SSM**

••16 An 8.0 g ice cube at $-10°C$ is put into a Thermos flask containing 100 cm³ of water at 20°C. By how much has the entropy of the cube–water system changed when equilibrium is reached? The specific heat of ice is 2220 J/kg · K.

••17 A mixture of 1773 g of water and 227 g of ice is in an initial equilibrium state at 0.000°C. The mixture is then, in a reversible process, brought to a second equilibrium state where the water–ice ratio, by mass, is 1.00:1.00 at 0.000°C. (a) Calculate the entropy change of the system during this process. (The heat of fusion for water is 333 kJ/kg.) (b) The system is then returned to the initial equilibrium state in an irreversible process (say, by using a Bunsen burner). Calculate the entropy change of the system during this process. (c) Are your answers consistent with the second law of thermodynamics? **SSM**

••18 (a) For 1.0 mol of a monatomic ideal gas taken through the cycle in Fig. 20-26, what is W/p_0V_0 as the gas goes from state a to state c along path abc? What is $\Delta E_{int}/p_0V_0$ in going (b) from b to c and (c) through one full cycle? What is ΔS in going (d) from b to c and (e) through one full cycle?

Fig. 20-26   Problem 18.

•••19 Suppose 1.00 mol of a monatomic ideal gas is taken from initial pressure p_1 and volume V_1 through two steps: (1) an isothermal expansion to volume $2.00V_1$ and (2) a pressure increase to $2.00p_1$ at constant volume. What is Q/p_1V_1 for (a) step 1 and (b) step 2? What is W/p_1V_1 for (c) step 1 and (d) step 2? For the full process, what are (e) $\Delta E_{int}/p_1V_1$ and (f) ΔS? The gas is returned to its initial state and again taken to the same final state but now through these two steps: (1) an isothermal compression to pressure $2.00p_1$ and (2) a volume increase to $2.00V_1$ at constant pressure. What is Q/p_1V_1 for (g) step 1 and (h) step 2? What is W/p_1V_1 for (i) step 1 and (j) step 2? For the full process, what are (k) $\Delta E_{int}/p_1V_1$ and (l) ΔS?

•••20 Expand 1.00 mol of an monatomic gas initially at 5.00 kPa and 600 K from initial volume $V_i = 1.00$ m³ to final volume $V_f = 2.00$ m³. At any instant during the expansion, the pressure p and volume V of the gas are related by $p = 5.00 \exp[(V_i - V)/a]$, with p in kilopascals, V_i and V in cubic meters, and $a = 1.00$ m³. What are the final (a) pressure and (b) temperature of the gas? (c) How much work is done by the gas during the expansion? (d) What is ΔS for the expansion (*Hint:* Use two simple reversible processes to find ΔS.)

sec. 20-5   Entropy in the Real World: Engines

•21 A Carnot engine operates between 235°C and 115°C, absorbing 6.30×10^4 J per cycle at the higher temperature. (a) What is the efficiency of the engine? (b) How much work per cycle is this engine capable of performing? **SSM www**

•22 A Carnot engine absorbs 52 kJ as heat and exhausts 36 kJ as heat in each cycle. Calculate (a) the engine's efficiency and (b) the work done per cycle in kilojoules.

•23 A Carnot engine whose low-temperature reservoir is at 17°C has an efficiency of 40%. By how much should the temperature of the high-temperature reservoir be increased to increase the efficiency to 50%?

•24 In a hypothetical nuclear fusion reactor, the fuel is deuterium gas at a temperature of 7×10^8 K. If this gas could be used to operate a Carnot engine with $T_L = 100°C$, what would be the engine's efficiency? Take both temperatures to be exact and report your answer to seven significant figures.

•25 A Carnot engine has an efficiency of 22.0%. It operates between constant-temperature reservoirs differing in temperature by 75.0 C°. What is the temperature of the (a) lower-temperature and (b) higher-temperature reservoir? **SSM**

••26 A 500 W Carnot engine operates between constant-temperature reservoirs at 100°C and 60.0°C. What is the rate at which energy is (a) taken in by the engine as heat and (b) exhausted by the engine as heat?

••27 Figure 20-27 shows a reversible cycle through which 1.00 mol of a monatomic ideal gas is taken. Process bc is an adiabatic expansion, with $p_b = 10.0$ atm and $V_b = 1.00 \times 10^{-3}$ m³. For the cycle, find (a) the energy added to the gas as heat, (b) the energy leaving the gas as heat, (c) the net work done by the gas, and (d) the efficiency of the cycle. **SSM ILW**

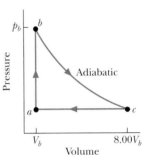

Fig. 20-27   Problem 27.

••28 A Carnot engine is set up to produce a certain work W per cycle. In each cycle, energy in the form of heat Q_H is transferred to the working substance of the engine from the higher-temperature thermal reservoir, which is at an adjustable temperature T_H. The lower-temperature thermal reservoir is maintained at temperature $T_L = 250$ K. Figure 20-28 gives Q_H for a range of T_H. If T_H is set at 550 K, what is Q_H?

Fig. 20-28   Problem 28.

••29 Figure 20-29 shows a reversible cycle through which 1.00 mol of a monatomic ideal gas is taken. Assume that $p = 2p_0$, $V = 2V_0$, $p_0 = 1.01 \times 10^5$ Pa, and $V_0 = 0.0225$ m³. Calculate (a) the work done during the cycle, (b) the energy added as heat during stroke abc, and (c) the efficiency of the cycle. (d) What is the efficiency of a Carnot engine operating between the highest

Fig. 20-29   Problem 29.

and lowest temperatures that occur in the cycle? (e) Is this greater than or less than the efficiency calculated in (c)?

••30 In the first stage of a two-stage Carnot engine, energy is absorbed as heat Q_1 at temperature T_1, work W_1 is done, and energy is expelled as heat Q_2 at a lower temperature T_2. The second stage absorbs that energy as heat Q_2, does work W_2, and expels energy as heat Q_3 at a still lower temperature T_3. Prove that the efficiency of the engine is $(T_1 - T_3)/T_1$.

••31 The efficiency of a particular car engine is 25% when the engine does 8.2 kJ of work per cycle. Assume the process is reversible. What are (a) the energy the engine gains per cycle as heat Q_{gain} from the fuel combustion and (b) the energy the engine loses per cycle as heat Q_{lost}. If a tune-up increases the efficiency to 31%, what are (c) Q_{gain} and (d) Q_{lost} at the same work value?

••32 An ideal gas (1.0 mol) is the working substance in an engine that operates on the cycle shown in Fig. 20-30. Processes BC and DA are reversible and adiabatic. (a) Is the gas monatomic, diatomic, or polyatomic? (b) What is the engine efficiency?

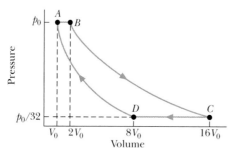

Fig. 20-30 Problem 32.

•••33 The cycle in Fig. 20-31 represents the operation of a gasoline internal combustion engine. Assume the gasoline–air intake mixture is an ideal gas with $\gamma = 1.30$. What are the ratios (a) T_2/T_1, (b) T_3/T_1, (c) T_4/T_1, (d) p_3/p_1, and (e) p_4/p_1? (f) What is the engine efficiency?

sec. 20-6 Entropy in the Real World: Refrigerators

•34 To make ice, a freezer that is a reverse Carnot engine extracts 42 kJ as heat at −15°C during each cycle, with coefficient of performance 5.7. The room temperature is 30.3°C. How much (a) energy per cycle is delivered as heat to the room and (b) work per cycle is required to run the freezer?

•35 A Carnot air conditioner takes energy from the thermal energy of a room at 70°F and transfers it as heat to the outdoors, which is at 96°F. For each joule of electric energy required to operate the air conditioner, how many joules are removed from the room? **SSM**

•36 The electric motor of a heat pump transfers energy as heat from the outdoors, which is at −5.0°C, to a room that is at 17°C. If the heat pump were a Carnot heat pump (a Carnot engine working in reverse), how much energy would be transferred as heat to the room for each joule of electric energy consumed?

•37 A heat pump is used to heat a building. The outside temperature is −5.0°C, and the temperature inside the building is to be maintained at 22°C. The pump's coefficient of performance is 3.8, and the heat pump delivers 7.54 MJ as heat to the building each hour. If the heat pump is a Carnot engine working in reverse, at what rate must work be done to run it? **SSM**

•38 How much work must be done by a Carnot refrigerator to transfer 1.0 J as heat (a) from a reservoir at 7.0°C to one at 27°C, (b) from a reservoir at −73°C to one at 27°C, (c) from a reservoir at −173°C to one at 27°C, and (d) from a reservoir at −223°C to one at 27°C?

••39 An air conditioner operating between 93°F and 70°F is rated at 4000 Btu/h cooling capacity. Its coefficient of performance is 27% of that of a Carnot refrigerator operating between the same two temperatures. What horsepower is required of the air conditioner motor? **ILW**

••40 (a) During each cycle, a Carnot engine absorbs 750 J as heat from a high-temperature reservoir at 360 K, with the low-temperature reservoir at 280 K. How much work is done per cycle? (b) The engine is then made to work in reverse to function as a Carnot refrigerator between those same two reservoirs. During each cycle, how much work is required to remove 1200 J as heat from the low-temperature reservoir?

••41 Figure 20-32 represents a Carnot engine that works between temperatures $T_1 = 400$ K and $T_2 = 150$ K and drives a Carnot refrigerator that works between temperatures $T_3 = 325$ K and $T_4 = 225$ K. What is the ratio Q_3/Q_1? **SSM**

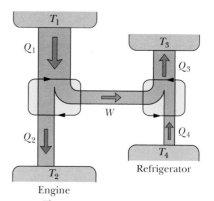

Fig. 20-32 Problem 41.

••42 The motor in a refrigerator has a power of 200 W. If the freezing compartment is at 270 K and the outside air is at 300 K, and assuming the efficiency of a Carnot refrigerator, what is the maximum amount of energy that can be extracted as heat from the freezing compartment in 10.0 min?

sec. 20-8 A Statistical View of Entropy

•43 Construct a table like Table 20-1 for eight molecules.

••44 A box contains N identical gas molecules equally divided between its two halves. For $N = 50$, what are (a) the multiplicity W of the central configuration, (b) the total number of microstates, and (c) the percentage of the time the system spends in the central configuration? For $N = 100$, what

Fig. 20-31 Problem 33.

are (d) W of the central configuration, (e) the total number of microstates, and (f) the percentage of the time the system spends in the central configuration? For $N = 200$, what are (g) W of the central configuration, (h) the total number of microstates, and (i) the percentage of the time the system spends in the central configuration? (j) Does the time spent in the central configuration increase or decrease with an increase in N?

•••45 A box contains N gas molecules. Consider the box to be divided into three equal parts. (a) By extension of Eq. 20-18, write a formula for the multiplicity of any given configuration. (b) Consider two configurations: configuration A with equal numbers of molecules in all three thirds of the box, and configuration B with equal numbers of molecules in each half of the box divided into two equal parts rather than three. What is the ratio W_A/W_B of the multiplicity of configuration A to that of configuration B? (c) Evaluate W_A/W_B for $N = 100$. (Because 100 is not evenly divisible by 3, put 34 molecules into one of the three box parts of configuration A and 33 in each of the other two parts.) SSM WWW

Additional Problems

46 Energy can be removed from water as heat at and even below the normal freezing point (0.0°C at atmospheric pressure) without causing the water to freeze; the water is then said to be *supercooled*. Suppose a 1.00 g water drop is supercooled until its temperature is that of the surrounding air, which is at −5.00°C. The drop then suddenly and irreversibly freezes, transferring energy to the air as heat. What is the entropy change for the drop? (*Hint:* Use a three-step reversible process as if the water were taken through the normal freezing point.) The specific heat of ice is 2220 J/kg · K.

47 Four particles are in the insulated box of Fig. 20-16. What are (a) the least multiplicity, (b) the greatest multiplicity, (c) the least entropy, and (d) the greatest entropy of the four-particle system?

48 A three-step cycle is undergone by 3.4 mol of an ideal diatomic gas: (1) the temperature of the gas is increased from 200 K to 500 K at constant volume; (2) the gas is then isothermally expanded to its original pressure; (3) the gas is then contracted at constant pressure back to its original volume. Throughout the cycle, the molecules rotate but do not oscillate. What is the efficiency of the cycle?

49 As a sample of nitrogen gas (N_2) undergoes a temperature increase at constant volume, the distribution of molecular speeds increases. That is, the probability distribution function $P(v)$ for the molecules spreads to higher speed values, as suggested in Fig. 19-7b. One way to report the spread in $P(v)$ is to measure the difference Δv between the most probable speed v_P and the rms speed v_{rms}. When $P(v)$ spreads to higher speeds, Δv increases. Assume that the gas is ideal and the N_2 molecules rotate but do not oscillate. For 1.5 mol, an initial temperature of 250 K, and a final temperature of 500 K, what are (a) the initial difference Δv_i, (b) the final difference Δv_f, and (c) the entropy change ΔS for the gas?

50 A brass rod is in thermal contact with a constant-temperature reservoir at 130°C at one end and a constant-temperature reservoir at 24.0°C at the other end. (a) Compute the total change in entropy of the rod–reservoirs system when 5030 J of energy is conducted through the rod, from one reservoir to the other. (b) Does the entropy of the rod change?

51 Suppose that a deep shaft were drilled in Earth's crust near one of the poles, where the surface temperature is −40°C, to a depth where the temperature is 800°C. (a) What is the theoretical limit to the efficiency of an engine operating between these temperatures? (b) If all the energy released as heat into the low-temperature reservoir were used to melt ice that was initially at −40°C, at what rate could liquid water at 0°C be produced by a 100 MW power plant (treat it as an engine)? The specific heat of ice is 2220 J/kg · K; water's heat of fusion is 333 kJ/kg. (Note that the engine can operate only between 0°C and 800°C in this case. Energy exhausted at −40°C cannot warm anything above −40°C.)

52 (a) A Carnot engine operates between a hot reservoir at 320 K and a cold one at 260 K. If the engine absorbs 500 J as heat per cycle at the hot reservoir, how much work per cycle does it deliver? (b) If the engine working in reverse functions as a refrigerator between the same two reservoirs, how much work per cycle must be supplied to remove 1000 J as heat from the cold reservoir?

53 An insulated Thermos contains 130 g of water at 80.0°C. You put in a 12.0 g ice cube at 0°C to form a system of *ice + original water*. (a) What is the equilibrium temperature of the system? What are the entropy changes of the water that was originally the ice cube (b) as it melts and (c) as it warms to the equilibrium temperature? (d) What is the entropy change of the original water as it cools to the equilibrium temperature? (e) What is the net entropy change of the *ice + original water* system as it reaches the equilibrium temperature?

54 A 600 g lump of copper at 80.0°C is placed in 70.0 g of water at 10.0°C in an insulated container. (See Table 18-3 for specific heats.) (a) What is the equilibrium temperature of the copper–water system? What entropy changes do (b) the copper, (c) the water, and (d) the copper–water system undergo in reaching the equilibrium temperature?

55 Suppose 0.550 mol of an ideal gas is isothermally and reversibly expanded in the four situations given below. What is the change in the entropy of the gas for each situation?

Situation	(a)	(b)	(c)	(d)
Temperature (K)	250	350	400	450
Initial volume (cm^3)	0.200	0.200	0.300	0.300
Final volume (cm^3)	0.800	0.800	1.20	1.20

56 What is the entropy change for 3.20 mol of an ideal monatomic gas undergoing a reversible increase in temperature from 380 K to 425 K at constant volume?

57 A 0.600 kg sample of water is initially ice at temperature −20°C. What is the sample's entropy change if its temperature is increased to 40°C?

58 A three-step cycle is undergone reversibly by 4.00 mol of an ideal gas: (1) an adiabatic expansion that gives the gas 2.00 times its initial volume, (2) a constant-volume process, (3) an isothermal compression back to the initial state of the gas. We do not know whether the gas is monatomic or diatomic; if it is diatomic, we do not know whether the molecules are rotating or oscillating. What are the entropy changes for (a) the cycle, (b) process 1, (c) process 3, and (d) process 2?

59 An apparatus that liquefies helium is in a room maintained at 300 K. If the helium in the apparatus is at 4.0 K, what is the minimum ratio Q_{to}/Q_{from}, where Q_{to} is the energy delivered as heat to the room and Q_{from} is the energy removed as heat from the helium?

60 Suppose 1.0 mol of a monatomic ideal gas initially at 10 L and 300 K is heated at constant volume to 600 K, allowed to expand isothermally to its initial pressure, and finally compressed at constant pressure to its original volume, pressure, and temperature. During the cycle, what are (a) the net energy entering the system (the gas) as heat and (b) the net work done by the gas? (c) What is the efficiency of the cycle?

61 System A of three particles and system B of five particles are in insulated boxes like that in Fig. 20-16. What is the least multiplicity W of (a) system A and (b) system B? What is the greatest multiplicity W of (c) A and (d) B? What is the greatest entropy of (e) A and (f) B?

62 Calculate the efficiency of a fossil-fuel power plant that consumes 380 metric tons of coal each hour to produce useful work at the rate of 750 MW. The heat of combustion of coal (the heat due to burning it) is 28 MJ/kg.

63 The temperature of 1.00 mol of a monatomic ideal gas is raised reversibly from 300 K to 400 K, with its volume kept constant. What is the entropy change of the gas?

64 Repeat Problem 63, with the pressure now kept constant.

65 Suppose that 260 J is conducted from a constant-temperature reservoir at 400 K to one at (a) 100 K, (b) 200 K, (c) 300 K, and (d) 360 K. What is the net change in entropy ΔS_{net} of the reservoirs in each case? (e) As the temperature difference of the two reservoirs decreases, does ΔS_{net} increase, decrease, or remain the same?

66 A 2.00 mol diatomic gas initially at 300 K undergoes this cycle: It is (1) heated at constant volume to 800 K, (2) then allowed to expand isothermally to its initial pressure, (3) then compressed at constant pressure to its initial state. Assuming the gas molecules neither rotate nor oscillate, find (a) the net energy transferred as heat to the gas, (b) the net work done by the gas, and (c) the efficiency of the cycle.

67 A Carnot engine whose high-temperature reservoir is at 400 K has an efficiency of 30.0%. By how much should the temperature of the low-temperature reservoir be changed to increase the efficiency to 40.0%?

68 A 45.0 g block of tungsten at 30.0°C and a 25.0 g block of silver at −120°C are placed together in an insulated container. (See Table 18-3 for specific heats.) (a) What is the equilibrium temperature? What entropy changes do (b) the tungsten, (c) the silver, and (d) the tungsten–silver system undergo in reaching the equilibrium temperature?

69 A box contains N molecules. Consider two configurations: configuration A with an equal division of the molecules between the two halves of the box, and configuration B with 60.0% of the molecules in the left half of the box and 40.0% in the right half. For N = 50, what are (a) the multiplicity W_A of configuration A, (b) the multiplicity W_B of configuration B, and (c) the ratio $f_{B/A}$ of the time the system spends in configuration B to the time its spends in configuration A? For N = 100, what are (d) W_A, (e) W_B, and (f) $f_{B/A}$? For N = 200, what are (g) W_A, (h) W_B, and (i) $f_{B/A}$? (j) With increasing N, does f increase, decrease, or remain the same?

70 A cylindrical copper rod of length 1.50 m and radius 2.00 cm is insulated to prevent heat loss through its curved surface. One end is attached to a thermal reservoir fixed at 300°C; the other is attached to a thermal reservoir fixed at 30.0°C. What is the rate at which entropy increases for the rod–reservoirs system?

71 An ideal refrigerator does 150 J of work to remove 560 J as heat from its cold compartment. (a) What is the refrigerator's coefficient of performance? (b) How much heat per cycle is exhausted to the kitchen?

72 Suppose 2.00 mol of a diatomic gas is taken reversibly around the cycle shown in the T-S diagram of Fig. 20-33. The molecules do not rotate or oscillate. What is the energy transferred as heat Q for (a) path $1 \rightarrow 2$, (b) path $2 \rightarrow 3$, and (c) the full cycle? (d) What is the work W for the isothermal process? The volume V_1 in state 1 is 0.200 m³. What is the volume in (e) state 2 and (f) state 3?

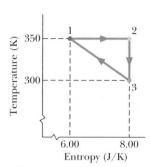

Fig. 20-33 Problem 72.

What is the change ΔE_{int} for (g) path $1 \rightarrow 2$, (h) path $2 \rightarrow 3$, and (i) the full cycle? (*Hint:* (h) can be done with one or two lines of calculation using Section 19-8 or with a page of calculation using Section 19-11.) (j) What is the work W for the adiabatic process?

73 A Carnot refrigerator extracts 35.0 kJ as heat during each cycle, operating with a coefficient of performance of 4.60. What are (a) the energy per cycle transferred as heat to the room and (b) the work done per cycle?

74 An inventor has built an engine X and claims that its efficiency ε_X is greater than the efficiency ε of an ideal engine operating between the same two temperatures. Suppose you couple engine X to an ideal refrigerator (Fig. 20-34a) and adjust the cycle of engine X so that the work per cycle it provides equals the work per cycle required by the ideal refrigerator. Treat this combination as a single unit and show that if the inventor's claim were true (if $\varepsilon_X > \varepsilon$), the combined unit would act as a perfect refrigerator (Fig. 20-34b), transferring energy as heat from the low-temperature reservoir to the high-temperature reservoir without the need for work.

Fig. 20-34 Problem 74.

The International System of Units (SI)*

TABLE 1
The SI Base Units

Quantity	Name	Symbol	Definition
length	meter	m	". . . the length of the path traveled by light in vacuum in 1/299,792,458 of a second." (1983)
mass	kilogram	kg	". . . this prototype [a certain platinum–iridium cylinder] shall henceforth be considered to be the unit of mass." (1889)
time	second	s	". . . the duration of 9,192,631,770 periods of the radiation corresponding to the transition between the two hyperfine levels of the ground state of the cesium-133 atom." (1967)
electric current	ampere	A	". . . that constant current which, if maintained in two straight parallel conductors of infinite length, of negligible circular cross section, and placed 1 meter apart in vacuum, would produce between these conductors a force equal to 2×10^{-7} newton per meter of length." (1946)
thermodynamic temperature	kelvin	K	". . . the fraction 1/273.16 of the thermodynamic temperature of the triple point of water." (1967)
amount of substance	mole	mol	". . . the amount of substance of a system which contains as many elementary entities as there are atoms in 0.012 kilogram of carbon-12." (1971)
luminous intensity	candela	cd	". . . the luminous intensity, in a given direction, of a source that emits monochromatic radiation of frequency 540×10^{12} hertz and that has a radiant intensity in that direction of 1/683 watt per steradian." (1979)

*Adapted from "The International System of Units (SI)," National Bureau of Standards Special Publication 330, 1972 edition. The definitions above were adopted by the General Conference of Weights and Measures, an international body, on the dates shown. In this book we do not use the candela.

TABLE 2
Some SI Derived Units

Quantity	Name of Unit	Symbol	
area	square meter	m^2	
volume	cubic meter	m^3	
frequency	hertz	Hz	s^{-1}
mass density (density)	kilogram per cubic meter	kg/m^3	
speed, velocity	meter per second	m/s	
angular velocity	radian per second	rad/s	
acceleration	meter per second per second	m/s^2	
angular acceleration	radian per second per second	rad/s^2	
force	newton	N	$kg \cdot m/s^2$
pressure	pascal	Pa	N/m^2
work, energy, quantity of heat	joule	J	$N \cdot m$
power	watt	W	J/s
quantity of electric charge	coulomb	C	$A \cdot s$
potential difference, electromotive force	volt	V	W/A
electric field strength	volt per meter (or newton per coulomb)	V/m	N/C
electric resistance	ohm	Ω	V/A
capacitance	farad	F	$A \cdot s/V$
magnetic flux	weber	Wb	$V \cdot s$
inductance	henry	H	$V \cdot s/A$
magnetic flux density	tesla	T	Wb/m^2
magnetic field strength	ampere per meter	A/m	
entropy	joule per kelvin	J/K	
specific heat	joule per kilogram kelvin	$J/(kg \cdot K)$	
thermal conductivity	watt per meter kelvin	$W/(m \cdot K)$	
radiant intensity	watt per steradian	W/sr	

TABLE 3
The SI Supplementary Units

Quantity	Name of Unit	Symbol
plane angle	radian	rad
solid angle	steradian	sr

Some Fundamental Constants of Physics* B

Constant	Symbol	Computational Value	Best (1998) Value	
			Value[a]	Uncertainty[b]
Speed of light in a vacuum	c	3.00×10^8 m/s	2.997 924 58	exact
Elementary charge	e	1.60×10^{-19} C	1.602 176 462	0.039
Gravitational constant	G	6.67×10^{-11} m^3/s$^2 \cdot$ kg	6.673	1500
Universal gas constant	R	8.31 J/mol $\cdot$ K	8.314 472	1.7
Avogadro constant	N_A	6.02×10^{23} mol^{-1}	6.022 141 99	0.079
Boltzmann constant	k	1.38×10^{-23} J/K	1.380 650 3	1.7
Stefan–Boltzmann constant	σ	5.67×10^{-8} W/m$^2 \cdot$ K^4	5.670 400	7.0
Molar volume of ideal gas at STP[d]	V_m	2.27×10^{-2} m^3/mol	2.271 098 1	1.7
Permittivity constant	ϵ_0	8.85×10^{-12} F/m	8.854 187 817 62	exact
Permeability constant	μ_0	1.26×10^{-6} H/m	1.256 637 061 43	exact
Planck constant	h	6.63×10^{-34} J $\cdot$ s	6.626 068 76	0.078
Electron mass[c]	m_e	9.11×10^{-31} kg	9.109 381 88	0.079
		5.49×10^{-4} u	5.485 799 110	0.0021
Proton mass[c]	m_p	1.67×10^{-27} kg	1.672 621 58	0.079
		1.0073 u	1.007 276 466 88	1.3×10^{-4}
Ratio of proton mass to electron mass	m_p/m_e	1840	1836.152 667 5	0.0021
Electron charge-to-mass ratio	e/m_e	1.76×10^{11} C/kg	1.758 820 174	0.040
Neutron mass[c]	m_n	1.68×10^{-27} kg	1.674 927 16	0.079
		1.0087 u	1.008 664 915 78	5.4×10^{-4}
Hydrogen atom mass[c]	m_{1H}	1.0078 u	1.007 825 031 6	0.0005
Deuterium atom mass[c]	m_{2H}	2.0141 u	2.014 101 777 9	0.0005
Helium atom mass[c]	m_{4He}	4.0026 u	4.002 603 2	0.067
Muon mass	m_μ	1.88×10^{-28} kg	1.883 531 09	0.084
Electron magnetic moment	μ_e	9.28×10^{-24} J/T	9.284 763 62	0.040
Proton magnetic moment	μ_p	1.41×10^{-26} J/T	1.410 606 663	0.041
Bohr magneton	μ_B	9.27×10^{-24} J/T	9.274 008 99	0.040
Nuclear magneton	μ_N	5.05×10^{-27} J/T	5.050 783 17	0.040
Bohr radius	a	5.29×10^{-11} m	5.291 772 083	0.0037
Rydberg constant	R	1.10×10^7 m^{-1}	1.097 373 156 854 8	7.6×10^{-6}
Electron Compton wavelength	λ_C	2.43×10^{-12} m	2.426 310 215	0.0073

[a]Values given in this column should be given the same unit and power of 10 as the computational value.

[b]Parts per million.

[c]Masses given in u are in unified atomic mass units, where 1 u = 1.660 538 73 $\times 10^{-27}$ kg.

[d]STP means standard temperature and pressure: 0°C and 1.0 atm (0.1 MPa).

*The values in this table were selected from the 1998 CODATA recommended values (www.physics.nist.gov).

C

Some Astronomical Data

Some Distances from Earth

To the Moon*	3.82×10^8 m	To the center of our galaxy	2.2×10^{20} m
To the Sun*	1.50×10^{11} m	To the Andromeda Galaxy	2.1×10^{22} m
To the nearest star (Proxima Centauri)	4.04×10^{16} m	To the edge of the observable universe	$\sim\!10^{26}$ m

*Mean distance.

The Sun, Earth, and the Moon

Property	Unit	Sun	Earth	Moon
Mass	kg	1.99×10^{30}	5.98×10^{24}	7.36×10^{22}
Mean radius	m	6.96×10^8	6.37×10^6	1.74×10^6
Mean density	kg/m^3	1410	5520	3340
Free-fall acceleration at the surface	m/s^2	274	9.81	1.67
Escape velocity	km/s	618	11.2	2.38
Period of rotation[a]	—	37 d at poles[b] 26 d at equator[b]	23 h 56 min	27.3 d
Radiation power[c]	W	3.90×10^{26}		

[a]Measured with respect to the distant stars.

[b]The Sun, a ball of gas, does not rotate as a rigid body.

[c]Just outside Earth's atmosphere solar energy is received, assuming normal incidence, at the rate of 1340 W/m^2.

Some Properties of the Planets

	Mercury	Venus	Earth	Mars	Jupiter	Saturn	Uranus	Neptune	Pluto
Mean distance from Sun, 10^6 km	57.9	108	150	228	778	1430	2870	4500	5900
Period of revolution, y	0.241	0.615	1.00	1.88	11.9	29.5	84.0	165	248
Period of rotation,[a] d	58.7	−243[b]	0.997	1.03	0.409	0.426	−0.451[b]	0.658	6.39
Orbital speed, km/s	47.9	35.0	29.8	24.1	13.1	9.64	6.81	5.43	4.74
Inclination of axis to orbit	<28°	≈3°	23.4°	25.0°	3.08°	26.7°	97.9°	29.6°	57.5°
Inclination of orbit to Earth's orbit	7.00°	3.39°		1.85°	1.30°	2.49°	0.77°	1.77°	17.2°
Eccentricity of orbit	0.206	0.0068	0.0167	0.0934	0.0485	0.0556	0.0472	0.0086	0.250
Equatorial diameter, km	4880	12 100	12 800	6790	143 000	120 000	51 800	49 500	2300
Mass (Earth = 1)	0.0558	0.815	1.000	0.107	318	95.1	14.5	17.2	0.002
Density (water = 1)	5.60	5.20	5.52	3.95	1.31	0.704	1.21	1.67	2.03
Surface value of g,[c] m/s^2	3.78	8.60	9.78	3.72	22.9	9.05	7.77	11.0	0.5
Escape velocity,[c] km/s	4.3	10.3	11.2	5.0	59.5	35.6	21.2	23.6	1.1
Known satellites	0	0	1	2	60 + ring	31 + rings	21 + rings	11 + rings	1

[a]Measured with respect to the distant stars.

[b]Venus and Uranus rotate opposite their orbital motion.

[c]Gravitational acceleration measured at the planet's equator.

Conversion Factors

D

Conversion factors may be read directly from these tables. For example, 1 degree = 2.778 × 10^{-3} revolutions, so 16.7° = 16.7 × 2.778 × 10^{-3} rev. The SI units are fully capitalized. Adapted in part from G. Shortley and D. Williams, *Elements of Physics,* 1971, Prentice-Hall, Englewood Cliffs, NJ.

Plane Angle

	°	′	″	RADIAN	rev
1 degree =	1	60	3600	1.745×10^{-2}	2.778×10^{-3}
1 minute =	1.667×10^{-2}	1	60	2.909×10^{-4}	4.630×10^{-5}
1 second =	2.778×10^{-4}	1.667×10^{-2}	1	4.848×10^{-6}	7.716×10^{-7}
1 RADIAN =	57.30	3438	2.063×10^{5}	1	0.1592
1 revolution =	360	2.16×10^{4}	1.296×10^{6}	6.283	1

Solid Angle

1 sphere = 4π steradians = 12.57 steradians

Length

	cm	METER	km	in.	ft	mi
1 centimeter =	1	10^{-2}	10^{-5}	0.3937	3.281×10^{-2}	6.214×10^{-6}
1 METER =	100	1	10^{-3}	39.37	3.281	6.214×10^{-4}
1 kilometer =	10^{5}	1000	1	3.937×10^{4}	3281	0.6214
1 inch =	2.540	2.540×10^{-2}	2.540×10^{-5}	1	8.333×10^{-2}	1.578×10^{-5}
1 foot =	30.48	0.3048	3.048×10^{-4}	12	1	1.894×10^{-4}
1 mile =	1.609×10^{5}	1609	1.609	6.336×10^{4}	5280	1

1 angström = 10^{-10} m

1 nautical mile = 1852 m
= 1.151 miles = 6076 ft

1 fermi = 10^{-15} m

1 light-year = 9.460×10^{12} km

1 parsec = 3.084×10^{13} km

1 fathom = 6 ft

1 Bohr radius = 5.292×10^{-11} m

1 yard = 3 ft

1 rod = 16.5 ft

1 mil = 10^{-3} in.

1 nm = 10^{-9} m

Area

	METER^2	cm^2	ft^2	in.^{2}
1 SQUARE METER =	1	10^{4}	10.76	1550
1 square centimeter =	10^{-4}	1	1.076×10^{-3}	0.1550
1 square foot =	9.290×10^{-2}	929.0	1	144
1 square inch =	6.452×10^{-4}	6.452	6.944×10^{-3}	1

1 square mile = 2.788×10^7 ft^2 = 640 acres

1 barn = 10^{-28} m^2

1 acre = 43 560 ft^2

1 hectare = 10^4 m^2 = 2.471 acres

Volume

	METER^3	cm^3	L	ft^3	in.^{3}
1 CUBIC METER = 1		10^6	1000	35.31	6.102×10^4
1 cubic centimeter = 10^{-6}		1	1.000×10^{-3}	3.531×10^{-5}	6.102×10^{-2}
1 liter = 1.000×10^{-3}		1000	1	3.531×10^{-2}	61.02
1 cubic foot = 2.832×10^{-2}		2.832×10^4	28.32	1	1728
1 cubic inch = 1.639×10^{-5}		16.39	1.639×10^{-2}	5.787×10^{-4}	1

1 U.S. fluid gallon = 4 U.S. fluid quarts = 8 U.S. pints = 128 U.S. fluid ounces = 231 in.^{3}

1 British imperial gallon = 277.4 in.^{3} = 1.201 U.S. fluid gallons

Mass

Quantities in the colored areas are not mass units but are often used as such. For example, when we write 1 kg "=" 2.205 lb, this means that a kilogram is a *mass* that *weighs* 2.205 pounds at a location where g has the standard value of 9.80665 m/s^2.

	g	KILOGRAM	slug	u	oz	lb	ton
1 gram = 1	0.001	6.852×10^{-5}	6.022×10^{23}	3.527×10^{-2}	2.205×10^{-3}	1.102×10^{-6}
1 KILOGRAM = 1000	1	6.852×10^{-2}	6.022×10^{26}	35.27	2.205	1.102×10^{-3}
1 slug = 1.459×10^4	14.59	1	8.786×10^{27}	514.8	32.17	1.609×10^{-2}
1 atomic mass unit = 1.661×10^{-24}	1.661×10^{-27}	1.138×10^{-28}	1	5.857×10^{-26}	3.662×10^{-27}	1.830×10^{-30}
1 ounce = 28.35	2.835×10^{-2}	1.943×10^{-3}	1.718×10^{25}	1	6.250×10^{-2}	3.125×10^{-5}
1 pound = 453.6	0.4536	3.108×10^{-2}	2.732×10^{26}	16	1	0.0005
1 ton = 9.072×10^5	907.2	62.16	5.463×10^{29}	3.2×10^4	2000	1

1 metric ton = 1000 kg

Density

Quantities in the colored areas are weight densities and, as such, are dimensionally different from mass densities. See the note for the mass table.

	slug/ft^3	KILOGRAM/ METER^3	g/cm^3	lb/ft^3	lb/in.^{3}
1 slug per foot^3 = 1		515.4	0.5154	32.17	1.862×10^{-2}
1 KILOGRAM per METER^3 = 1.940×10^{-3}		1	0.001	6.243×10^{-2}	3.613×10^{-5}
1 gram per centimeter^3 = 1.940		1000	1	62.43	3.613×10^{-2}
1 pound per foot^3 = 3.108×10^{-2}		16.02	16.02×10^{-2}	1	5.787×10^{-4}
1 pound per inch^3 = 53.71		2.768×10^4	27.68	1728	1

Time

	y	d	h	min	SECOND
1 year = 1	365.25	8.766×10^3	5.259×10^5	3.156×10^7
1 day = 2.738×10^{-3}	1	24	1440	8.640×10^4
1 hour = 1.141×10^{-4}	4.167×10^{-2}	1	60	3600
1 minute = 1.901×10^{-6}	6.944×10^{-4}	1.667×10^{-2}	1	60
1 SECOND = 3.169×10^{-8}	1.157×10^{-5}	2.778×10^{-4}	1.667×10^{-2}	1

Speed

	ft/s	km/h	METER/SECOND	mi/h	cm/s
1 foot per second =	1	1.097	0.3048	0.6818	30.48
1 kilometer per hour = 0.9113		1	0.2778	0.6214	27.78
1 METER per SECOND = 3.281		3.6	1	2.237	100
1 mile per hour = 1.467		1.609	0.4470	1	44.70
1 centimeter per second = 3.281×10^{-2}		3.6×10^{-2}	0.01	2.237×10^{-2}	1

1 knot = 1 nautical mi/h = 1.688 ft/s   1 mi/min = 88.00 ft/s = 60.00 mi/h

Force

Force units in the colored areas are now little used. To clarify: 1 gram-force (= 1 gf) is the force of gravity that would act on an object whose mass is 1 gram at a location where g has the standard value of 9.80665 m/s².

	dyne	NEWTON	lb	pdl	gf	kgf
1 dyne = 1		10^{-5}	2.248×10^{-6}	7.233×10^{-5}	1.020×10^{-3}	1.020×10^{-6}
1 NEWTON = 10^5		1	0.2248	7.233	102.0	0.1020
1 pound = 4.448×10^5		4.448	1	32.17	453.6	0.4536
1 poundal = 1.383×10^4		0.1383	3.108×10^{-2}	1	14.10	1.410×10^2
1 gram-force = 980.7		9.807×10^{-3}	2.205×10^{-3}	7.093×10^{-2}	1	0.001
1 kilogram-force = 9.807×10^5		9.807	2.205	70.93	1000	1

1 ton = 2000 lb

Pressure

	atm	dyne/cm²	inch of water	cm Hg	PASCAL	lb/in.²	lb/ft²
1 atmosphere = 1		1.013×10^6	406.8	76	1.013×10^5	14.70	2116
1 dyne per centimeter² = 9.869×10^{-7}		1	4.015×10^{-4}	7.501×10^{-5}	0.1	1.405×10^{-5}	2.089×10^{-3}
1 inch of water^a at 4°C = 2.458×10^{-3}		2491	1	0.1868	249.1	3.613×10^{-2}	5.202
1 centimeter of mercury^a at 0°C = 1.316×10^{-2}		1.333×10^4	5.353	1	1333	0.1934	27.85
1 PASCAL = 9.869×10^{-6}		10	4.015×10^{-3}	7.501×10^{-4}	1	1.450×10^{-4}	2.089×10^{-2}
1 pound per inch² = 6.805×10^{-2}		6.895×10^4	27.68	5.171	6.895×10^3	1	144
1 pound per foot² = 4.725×10^{-4}		478.8	0.1922	3.591×10^{-2}	47.88	6.944×10^{-3}	1

^{a}Where the acceleration of gravity has the standard value of 9.80665 m/s².

1 bar = 10^6 dyne/cm² = 0.1 MPa   1 millibar = 10^3 dyne/cm² = 10^2 Pa   1 torr = 1 mm Hg

Energy, Work, Heat

Quantities in the colored areas are not energy units but are included for convenience. They arise from the relativistic mass–energy equivalence formula $E = mc^2$ and represent the energy released if a kilogram or unified atomic mass unit (u) is completely converted to energy (bottom two rows) or the mass that would be completely converted to one unit of energy (rightmost two columns).

	Btu	erg	ft·lb	hp·h	JOULE	cal	kW·h	eV	MeV	kg	u
1 British thermal unit =	1	1.055×10^{10}	777.9	3.929×10^{-4}	1055	252.0	2.930×10^{-4}	6.585×10^{21}	6.585×10^{15}	1.174×10^{-14}	7.070×10^{12}
1 erg =	9.481×10^{-11}	1	7.376×10^{-8}	3.725×10^{-14}	10^{-7}	2.389×10^{-8}	2.778×10^{-14}	6.242×10^{11}	6.242×10^{5}	1.113×10^{-24}	670.2
1 foot-pound =	1.285×10^{-3}	1.356×10^{7}	1	5.051×10^{-7}	1.356	0.3238	3.766×10^{-7}	8.464×10^{18}	8.464×10^{12}	1.509×10^{-17}	9.037×10^{9}
1 horsepower-hour =	2545	2.685×10^{13}	1.980×10^{6}	1	2.685×10^{6}	6.413×10^{5}	0.7457	1.676×10^{25}	1.676×10^{19}	2.988×10^{-11}	1.799×10^{16}
1 JOULE =	9.481×10^{-4}	10^{7}	0.7376	3.725×10^{-7}	1	0.2389	2.778×10^{-7}	6.242×10^{18}	6.242×10^{12}	1.113×10^{-17}	6.702×10^{9}
1 calorie =	3.968×10^{-3}	4.1868×10^{7}	3.088	1.560×10^{-6}	4.1868	1	1.163×10^{-6}	2.613×10^{19}	2.613×10^{13}	4.660×10^{-17}	2.806×10^{10}
1 kilowatt-hour =	3413	3.600×10^{13}	2.655×10^{6}	1.341	3.600×10^{6}	8.600×10^{5}	1	2.247×10^{25}	2.247×10^{19}	4.007×10^{-11}	2.413×10^{16}
1 electron-volt =	1.519×10^{-22}	1.602×10^{-12}	1.182×10^{-19}	5.967×10^{-26}	1.602×10^{-19}	3.827×10^{-20}	4.450×10^{-26}	1	10^{-6}	1.783×10^{-36}	1.074×10^{-9}
1 million electron-volts =	1.519×10^{-16}	1.602×10^{-6}	1.182×10^{-13}	5.967×10^{-20}	1.602×10^{-13}	3.827×10^{-14}	4.450×10^{-20}	10^{-6}	1	1.783×10^{-30}	1.074×10^{-3}
1 kilogram =	8.521×10^{13}	8.987×10^{23}	6.629×10^{16}	3.348×10^{10}	8.987×10^{16}	2.146×10^{16}	2.497×10^{10}	5.610×10^{35}	5.610×10^{29}	1	6.022×10^{26}
1 unified atomic mass unit =	1.415×10^{-13}	1.492×10^{-3}	1.101×10^{-10}	5.559×10^{-17}	1.492×10^{-10}	3.564×10^{-11}	4.146×10^{-17}	9.320×10^{8}	932.0	1.661×10^{-27}	1

Power

	Btu/h	ft·lb/s	hp	cal/s	kW	WATT
1 British thermal unit per hour =	1	0.2161	3.929×10^{-4}	6.998×10^{-2}	2.930×10^{-4}	0.2930
1 foot-pound per second =	4.628	1	1.818×10^{-3}	0.3239	1.356×10^{-3}	1.356
1 horsepower =	2545	550	1	178.1	0.7457	745.7
1 calorie per second =	14.29	3.088	5.615×10^{-3}	1	4.186×10^{-3}	4.186
1 kilowatt =	3413	737.6	1.341	238.9	1	1000
1 WATT =	3.413	0.7376	1.341×10^{-3}	0.2389	0.001	1

Magnetic Field

	gauss	TESLA	milligauss
1 gauss =	1	10^{-4}	1000
1 TESLA =	10^{4}	1	10^{7}
1 milligauss =	0.001	10^{-7}	1

1 tesla = 1 weber/meter^2

Magnetic Flux

	maxwell	WEBER
1 maxwell =	1	10^{-8}
1 WEBER =	10^{8}	1

Mathematical Formulas

E

Geometry

Circle of radius r: circumference $= 2\pi r$; area $= \pi r^2$.
Sphere of radius r: area $= 4\pi r^2$; volume $= \frac{4}{3}\pi r^3$.
Right circular cylinder of radius r and height h:
  area $= 2\pi r^2 + 2\pi rh$; volume $= \pi r^2 h$.
Triangle of base a and altitude h: area $= \frac{1}{2}ah$.

Quadratic Formula

If $ax^2 + bx + c = 0$, then $x = \dfrac{-b \pm \sqrt{b^2 - 4ac}}{2a}$.

Trigonometric Functions of Angle θ

$$\sin \theta = \frac{y}{r} \quad \cos \theta = \frac{x}{r}$$

$$\tan \theta = \frac{y}{x} \quad \cot \theta = \frac{x}{y}$$

$$\sec \theta = \frac{r}{x} \quad \csc \theta = \frac{r}{y}$$

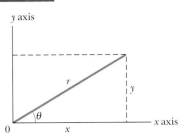

Pythagorean Theorem

In this right triangle,
  $a^2 + b^2 = c^2$

Triangles

Angles are A, B, C
Opposite sides are a, b, c
Angles $A + B + C = 180°$

$$\frac{\sin A}{a} = \frac{\sin B}{b} = \frac{\sin C}{c}$$

$c^2 = a^2 + b^2 - 2ab \cos C$

Exterior angle $D = A + C$

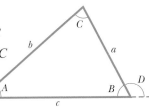

Mathematical Signs and Symbols

$=$ equals
$\approx$ equals approximately
$\sim$ is the order of magnitude of
$\neq$ is not equal to
$\equiv$ is identical to, is defined as
$>$ is greater than ($\gg$ is much greater than)
$<$ is less than ($\ll$ is much less than)
$\geq$ is greater than or equal to (or, is no less than)
$\leq$ is less than or equal to (or, is no more than)
$\pm$ plus or minus
$\propto$ is proportional to
Σ the sum of
x_{avg} the average value of x

Trigonometric Identities

$\sin(90° - \theta) = \cos \theta$

$\cos(90° - \theta) = \sin \theta$

$\sin \theta / \cos \theta = \tan \theta$

$\sin^2 \theta + \cos^2 \theta = 1$

$\sec^2 \theta - \tan^2 \theta = 1$

$\csc^2 \theta - \cot^2 \theta = 1$

$\sin 2\theta = 2 \sin \theta \cos \theta$

$\cos 2\theta = \cos^2 \theta - \sin^2 \theta = 2 \cos^2 \theta - 1 = 1 - 2 \sin^2 \theta$

$\sin(\alpha \pm \beta) = \sin \alpha \cos \beta \pm \cos \alpha \sin \beta$

$\cos(\alpha \pm \beta) = \cos \alpha \cos \beta \mp \sin \alpha \sin \beta$

$\tan(\alpha \pm \beta) = \dfrac{\tan \alpha \pm \tan \beta}{1 \mp \tan \alpha \tan \beta}$

$\sin \alpha \pm \sin \beta = 2 \sin \frac{1}{2}(\alpha \pm \beta) \cos \frac{1}{2}(\alpha \mp \beta)$

$\cos \alpha + \cos \beta = 2 \cos \frac{1}{2}(\alpha + \beta) \cos \frac{1}{2}(\alpha - \beta)$

$\cos \alpha - \cos \beta = -2 \sin \frac{1}{2}(\alpha + \beta) \sin \frac{1}{2}(\alpha - \beta)$

Binomial Theorem

$$(1 + x)^n = 1 + \frac{nx}{1!} + \frac{n(n-1)x^2}{2!} + \cdots \qquad (x^2 < 1)$$

Exponential Expansion

$$e^x = 1 + x + \frac{x^2}{2!} + \frac{x^3}{3!} + \cdots$$

Logarithmic Expansion

$$\ln(1 + x) = x - \tfrac{1}{2}x^2 + \tfrac{1}{3}x^3 - \cdots \qquad (|x| < 1)$$

Trigonometric Expansions (θ in radians)

$$\sin \theta = \theta - \frac{\theta^3}{3!} + \frac{\theta^5}{5!} - \cdots$$

$$\cos \theta = 1 - \frac{\theta^2}{2!} + \frac{\theta^4}{4!} - \cdots$$

$$\tan \theta = \theta + \frac{\theta^3}{3} + \frac{2\theta^5}{15} + \cdots$$

Cramer's Rule

Two simultaneous equations in unknowns x and y,

$$a_1 x + b_1 y = c_1 \quad \text{and} \quad a_2 x + b_2 y = c_2,$$

have the solutions

$$x = \frac{\begin{vmatrix} c_1 & b_1 \\ c_2 & b_2 \end{vmatrix}}{\begin{vmatrix} a_1 & b_1 \\ a_2 & b_2 \end{vmatrix}} = \frac{c_1 b_2 - c_2 b_1}{a_1 b_2 - a_2 b_1}$$

and

$$y = \frac{\begin{vmatrix} a_1 & c_1 \\ a_2 & c_2 \end{vmatrix}}{\begin{vmatrix} a_1 & b_1 \\ a_2 & b_2 \end{vmatrix}} = \frac{a_1 c_2 - a_2 c_1}{a_1 b_2 - a_2 b_1}.$$

Products of Vectors

Let $\hat{i}$, $\hat{j}$, and $\hat{k}$ be unit vectors in the x, y, and z directions. Then

$$\hat{i} \cdot \hat{i} = \hat{j} \cdot \hat{j} = \hat{k} \cdot \hat{k} = 1, \qquad \hat{i} \cdot \hat{j} = \hat{j} \cdot \hat{k} = \hat{k} \cdot \hat{i} = 0,$$

$$\hat{i} \times \hat{i} = \hat{j} \times \hat{j} = \hat{k} \times \hat{k} = 0,$$

$$\hat{i} \times \hat{j} = \hat{k}, \quad \hat{j} \times \hat{k} = \hat{i}, \quad \hat{k} \times \hat{i} = \hat{j}.$$

Any vector $\vec{a}$ with components a_x, a_y, and a_z along the x, y, and z axes can be written as

$$\vec{a} = a_x \hat{i} + a_y \hat{j} + a_z \hat{k}.$$

Let $\vec{a}$, $\vec{b}$, and $\vec{c}$ be arbitrary vectors with magnitudes a, b, and c. Then

$$\vec{a} \times (\vec{b} + \vec{c}) = (\vec{a} \times \vec{b}) + (\vec{a} \times \vec{c})$$

$$(s\vec{a}) \times \vec{b} = \vec{a} \times (s\vec{b}) = s(\vec{a} \times \vec{b}) \qquad (s = \text{a scalar}).$$

Let θ be the smaller of the two angles between $\vec{a}$ and $\vec{b}$. Then

$$\vec{a} \cdot \vec{b} = \vec{b} \cdot \vec{a} = a_x b_x + a_y b_y + a_z b_z = ab \cos \theta$$

$$\vec{a} \times \vec{b} = -\vec{b} \times \vec{a} = \begin{vmatrix} \hat{i} & \hat{j} & \hat{k} \\ a_x & a_y & a_z \\ b_x & b_y & b_z \end{vmatrix}$$

$$= \hat{i} \begin{vmatrix} a_y & a_z \\ b_y & b_z \end{vmatrix} - \hat{j} \begin{vmatrix} a_x & a_z \\ b_x & b_z \end{vmatrix} + \hat{k} \begin{vmatrix} a_x & a_y \\ b_x & b_y \end{vmatrix}$$

$$= (a_y b_z - b_y a_z)\hat{i} + (a_z b_x - b_z a_x)\hat{j} + (a_x b_y - b_x a_y)\hat{k}$$

$$|\vec{a} \times \vec{b}| = ab \sin \theta$$

$$\vec{a} \cdot (\vec{b} \times \vec{c}) = \vec{b} \cdot (\vec{c} \times \vec{a}) = \vec{c} \cdot (\vec{a} \times \vec{b})$$

$$\vec{a} \times (\vec{b} \times \vec{c}) = (\vec{a} \cdot \vec{c})\vec{b} - (\vec{a} \cdot \vec{b})\vec{c}$$

Derivatives and Integrals

In what follows, the letters u and v stand for any functions of x, and a and m are constants. To each of the indefinite integrals should be added an arbitrary constant of integration. The *Handbook of Chemistry and Physics* (CRC Press Inc.) gives a more extensive tabulation.

1. $\dfrac{dx}{dx} = 1$

2. $\dfrac{d}{dx}(au) = a\dfrac{du}{dx}$

3. $\dfrac{d}{dx}(u + v) = \dfrac{du}{dx} + \dfrac{dv}{dx}$

4. $\dfrac{d}{dx}x^m = mx^{m-1}$

5. $\dfrac{d}{dx}\ln x = \dfrac{1}{x}$

6. $\dfrac{d}{dx}(uv) = u\dfrac{dv}{dx} + v\dfrac{du}{dx}$

7. $\dfrac{d}{dx}e^x = e^x$

8. $\dfrac{d}{dx}\sin x = \cos x$

9. $\dfrac{d}{dx}\cos x = -\sin x$

10. $\dfrac{d}{dx}\tan x = \sec^2 x$

11. $\dfrac{d}{dx}\cot x = -\csc^2 x$

12. $\dfrac{d}{dx}\sec x = \tan x \sec x$

13. $\dfrac{d}{dx}\csc x = -\cot x \csc x$

14. $\dfrac{d}{dx}e^u = e^u\dfrac{du}{dx}$

15. $\dfrac{d}{dx}\sin u = \cos u\dfrac{du}{dx}$

16. $\dfrac{d}{dx}\cos u = -\sin u\dfrac{du}{dx}$

1. $\displaystyle\int dx = x$

2. $\displaystyle\int au\, dx = a\int u\, dx$

3. $\displaystyle\int (u + v)\, dx = \int u\, dx + \int v\, dx$

4. $\displaystyle\int x^m\, dx = \dfrac{x^{m+1}}{m + 1} \quad (m \neq -1)$

5. $\displaystyle\int \dfrac{dx}{x} = \ln |x|$

6. $\displaystyle\int u\dfrac{dv}{dx}\, dx = uv - \int v\dfrac{du}{dx}\, dx$

7. $\displaystyle\int e^x\, dx = e^x$

8. $\displaystyle\int \sin x\, dx = -\cos x$

9. $\displaystyle\int \cos x\, dx = \sin x$

10. $\displaystyle\int \tan x\, dx = \ln |\sec x|$

11. $\displaystyle\int \sin^2 x\, dx = \tfrac{1}{2}x - \tfrac{1}{4}\sin 2x$

12. $\displaystyle\int e^{-ax}\, dx = -\dfrac{1}{a}e^{-ax}$

13. $\displaystyle\int xe^{-ax}\, dx = -\dfrac{1}{a^2}(ax + 1)e^{-ax}$

14. $\displaystyle\int x^2 e^{-ax}\, dx = -\dfrac{1}{a^3}(a^2x^2 + 2ax + 2)e^{-ax}$

15. $\displaystyle\int_0^\infty x^n e^{-ax}\, dx = \dfrac{n!}{a^{n+1}}$

16. $\displaystyle\int_0^\infty x^{2n}e^{-ax^2}\, dx = \dfrac{1 \cdot 3 \cdot 5 \, \cdots \, (2n - 1)}{2^{n+1}a^n}\sqrt{\dfrac{\pi}{a}}$

17. $\displaystyle\int \dfrac{dx}{\sqrt{x^2 + a^2}} = \ln(x + \sqrt{x^2 + a^2})$

18. $\displaystyle\int \dfrac{x\, dx}{(x^2 + a^2)^{3/2}} = -\dfrac{1}{(x^2 + a^2)^{1/2}}$

19. $\displaystyle\int \dfrac{dx}{(x^2 + a^2)^{3/2}} = \dfrac{x}{a^2(x^2 + a^2)^{1/2}}$

20. $\displaystyle\int_0^\infty x^{2n+1}e^{-ax^2}\, dx = \dfrac{n!}{2a^{n+1}} \quad (a > 0)$

21. $\displaystyle\int \dfrac{x\, dx}{x + d} = x - d\ln(x + d)$

F _____

Properties of the Elements

All physical properties are for a pressure of 1 atm unless otherwise specified.

Element	Symbol	Atomic Number Z	Molar Mass, g/mol	Density, g/cm³ at 20°C	Melting Point, °C	Boiling Point, °C	Specific Heat, J/(g · °C) at 25°C
Actinium	Ac	89	(227)	10.06	1323	(3473)	0.092
Aluminum	Al	13	26.9815	2.699	660	2450	0.900
Americium	Am	95	(243)	13.67	1541	—	—
Antimony	Sb	51	121.75	6.691	630.5	1380	0.205
Argon	Ar	18	39.948	1.6626×10^{-3}	−189.4	−185.8	0.523
Arsenic	As	33	74.9216	5.78	817 (28 atm)	613	0.331
Astatine	At	85	(210)	—	(302)	—	—
Barium	Ba	56	137.34	3.594	729	1640	0.205
Berkelium	Bk	97	(247)	14.79	—	—	—
Beryllium	Be	4	9.0122	1.848	1287	2770	1.83
Bismuth	Bi	83	208.980	9.747	271.37	1560	0.122
Bohrium	Bh	107	262.12	—	—	—	—
Boron	B	5	10.811	2.34	2030	—	1.11
Bromine	Br	35	79.909	3.12 (liquid)	−7.2	58	0.293
Cadmium	Cd	48	112.40	8.65	321.03	765	0.226
Calcium	Ca	20	40.08	1.55	838	1440	0.624
Californium	Cf	98	(251)	—	—	—	—
Carbon	C	6	12.01115	2.26	3727	4830	0.691
Cerium	Ce	58	140.12	6.768	804	3470	0.188
Cesium	Cs	55	132.905	1.873	28.40	690	0.243
Chlorine	Cl	17	35.453	3.214×10^{-3} (0°C)	−101	−34.7	0.486
Chromium	Cr	24	51.996	7.19	1857	2665	0.448
Cobalt	Co	27	58.9332	8.85	1495	2900	0.423
Copper	Cu	29	63.54	8.96	1083.40	2595	0.385
Curium	Cm	96	(247)	13.3	—	—	—
Darmstadtium	Ds	110	(271)	—	—	—	—
Dubnium	Db	105	262.114	—	—	—	—
Dysprosium	Dy	66	162.50	8.55	1409	2330	0.172
Einsteinium	Es	99	(254)	—	—	—	—
Erbium	Er	68	167.26	9.15	1522	2630	0.167
Europium	Eu	63	151.96	5.243	817	1490	0.163
Fermium	Fm	100	(237)	—	—	—	—
Fluorine	F	9	18.9984	1.696×10^{-3} (0°C)	−219.6	−188.2	0.753
Francium	Fr	87	(223)	—	(27)	—	—
Gadolinium	Gd	64	157.25	7.90	1312	2730	0.234
Gallium	Ga	31	69.72	5.907	29.75	2237	0.377
Germanium	Ge	32	72.59	5.323	937.25	2830	0.322
Gold	Au	79	196.967	19.32	1064.43	2970	0.131

Element	Symbol	Atomic Number Z	Molar Mass, g/mol	Density, g/cm³ at 20°C	Melting Point, °C	Boiling Point, °C	Specific Heat, J/(g·°C) at 25°C
Hafnium	Hf	72	178.49	13.31	2227	5400	0.144
Hassium	Hs	108	(265)	—	—	—	—
Helium	He	2	4.0026	0.1664×10^{-3}	−269.7	−268.9	5.23
Holmium	Ho	67	164.930	8.79	1470	2330	0.165
Hydrogen	H	1	1.00797	0.08375×10^{-3}	−259.19	−252.7	14.4
Indium	In	49	114.82	7.31	156.634	2000	0.233
Iodine	I	53	126.9044	4.93	113.7	183	0.218
Iridium	Ir	77	192.2	22.5	2447	(5300)	0.130
Iron	Fe	26	55.847	7.874	1536.5	3000	0.447
Krypton	Kr	36	83.80	3.488×10^{-3}	−157.37	−152	0.247
Lanthanum	La	57	138.91	6.189	920	3470	0.195
Lawrencium	Lr	103	(257)	—	—	—	—
Lead	Pb	82	207.19	11.35	327.45	1725	0.129
Lithium	Li	3	6.939	0.534	180.55	1300	3.58
Lutetium	Lu	71	174.97	9.849	1663	1930	0.155
Magnesium	Mg	12	24.312	1.738	650	1107	1.03
Manganese	Mn	25	54.9380	7.44	1244	2150	0.481
Meitnerium	Mt	109	(266)	—	—	—	—
Mendelevium	Md	101	(256)	—	—	—	—
Mercury	Hg	80	200.59	13.55	−38.87	357	0.138
Molybdenum	Mo	42	95.94	10.22	2617	5560	0.251
Neodymium	Nd	60	144.24	7.007	1016	3180	0.188
Neon	Ne	10	20.183	0.8387×10^{-3}	−248.597	−246.0	1.03
Neptunium	Np	93	(237)	20.25	637	—	1.26
Nickel	Ni	28	58.71	8.902	1453	2730	0.444
Niobium	Nb	41	92.906	8.57	2468	4927	0.264
Nitrogen	N	7	14.0067	1.1649×10^{-3}	−210	−195.8	1.03
Nobelium	No	102	(255)	—	—	—	—
Osmium	Os	76	190.2	22.59	3027	5500	0.130
Oxygen	O	8	15.9994	1.3318×10^{-3}	−218.80	−183.0	0.913
Palladium	Pd	46	106.4	12.02	1552	3980	0.243
Phosphorus	P	15	30.9738	1.83	44.25	280	0.741
Platinum	Pt	78	195.09	21.45	1769	4530	0.134
Plutonium	Pu	94	(244)	19.8	640	3235	0.130
Polonium	Po	84	(210)	9.32	254	—	—
Potassium	K	19	39.102	0.862	63.20	760	0.758
Praseodymium	Pr	59	140.907	6.773	931	3020	0.197
Promethium	Pm	61	(145)	7.22	(1027)	—	—
Protactinium	Pa	91	(231)	15.37 (estimated)	(1230)	—	—
Radium	Ra	88	(226)	5.0	700	—	—
Radon	Rn	86	(222)	9.96×10^{-3} (0°C)	(−71)	−61.8	0.092
Rhenium	Re	75	186.2	21.02	3180	5900	0.134
Rhodium	Rh	45	102.905	12.41	1963	4500	0.243
Rubidium	Rb	37	85.47	1.532	39.49	688	0.364
Ruthenium	Ru	44	101.107	12.37	2250	4900	0.239
Rutherfordium	Rf	104	261.11	—	—	—	—

Element	Symbol	Atomic Number Z	Molar Mass, g/mol	Density, g/cm³ at 20°C	Melting Point, °C	Boiling Point, °C	Specific Heat, J/(g · °C) at 25°C
Samarium	Sm	62	150.35	7.52	1072	1630	0.197
Scandium	Sc	21	44.956	2.99	1539	2730	0.569
Seaborgium	Sg	106	263.118	—	—	—	—
Selenium	Se	34	78.96	4.79	221	685	0.318
Silicon	Si	14	28.086	2.33	1412	2680	0.712
Silver	Ag	47	107.870	10.49	960.8	2210	0.234
Sodium	Na	11	22.9898	0.9712	97.85	892	1.23
Strontium	Sr	38	87.62	2.54	768	1380	0.737
Sulfur	S	16	32.064	2.07	119.0	444.6	0.707
Tantalum	Ta	73	180.948	16.6	3014	5425	0.138
Technetium	Tc	43	(99)	11.46	2200	—	0.209
Tellurium	Te	52	127.60	6.24	449.5	990	0.201
Terbium	Tb	65	158.924	8.229	1357	2530	0.180
Thallium	Tl	81	204.37	11.85	304	1457	0.130
Thorium	Th	90	(232)	11.72	1755	(3850)	0.117
Thulium	Tm	69	168.934	9.32	1545	1720	0.159
Tin	Sn	50	118.69	7.2984	231.868	2270	0.226
Titanium	Ti	22	47.90	4.54	1670	3260	0.523
Tungsten	W	74	183.85	19.3	3380	5930	0.134
Unnamed	Uuu	111	(272)	—	—	—	—
Unnamed	Uub	112	(285)	—	—	—	—
Unnamed	Uut	113	—	—	—	—	—
Unnamed	Unq	114	(289)	—	—	—	—
Unnamed	Uup	115	—	—	—	—	—
Unnamed	Uuh	116	—	—	—	—	—
Unnamed	Uus	117	—	—	—	—	—
Unnamed	Uuo	118	(293)	—	—	—	—
Uranium	U	92	(238)	18.95	1132	3818	0.117
Vanadium	V	23	50.942	6.11	1902	3400	0.490
Xenon	Xe	54	131.30	5.495×10^{-3}	−111.79	−108	0.159
Ytterbium	Yb	70	173.04	6.965	824	1530	0.155
Yttrium	Y	39	88.905	4.469	1526	3030	0.297
Zinc	Zn	30	65.37	7.133	419.58	906	0.389
Zirconium	Zr	40	91.22	6.506	1852	3580	0.276

The values in parentheses in the column of molar masses are the mass numbers of the longest-lived isotopes of those elements that are radioactive. Melting points and boiling points in parentheses are uncertain.

The data for gases are valid only when these are in their usual molecular state, such as H_2, He, O_2, Ne, etc. The specific heats of the gases are the values at constant pressure.

Source: Adapted from J. Emsley, *The Elements,* 3rd ed., 1998, Clarendon Press, Oxford. See also www.webelements.com for the latest values and newest elements.

Periodic Table of the Elements

G

Metals
Metalloids
Nonmetals

Noble gases 0

Transition metals

Alkali metals IA

THE HORIZONTAL PERIODS

	IA	IIA	IIIB	IVB	VB	VIB	VIIB	VIIIB			IB	IIB	IIIA	IVA	VA	VIA	VIIA	0
1	1 H																	2 He
2	3 Li	4 Be											5 B	6 C	7 N	8 O	9 F	10 Ne
3	11 Na	12 Mg											13 Al	14 Si	15 P	16 S	17 Cl	18 Ar
4	19 K	20 Ca	21 Sc	22 Ti	23 V	24 Cr	25 Mn	26 Fe	27 Co	28 Ni	29 Cu	30 Zn	31 Ga	32 Ge	33 As	34 Se	35 Br	36 Kr
5	37 Rb	38 Sr	39 Y	40 Zr	41 Nb	42 Mo	43 Tc	44 Ru	45 Rh	46 Pd	47 Ag	48 Cd	49 In	50 Sn	51 Sb	52 Te	53 I	54 Xe
6	55 Cs	56 Ba	57-71 *	72 Hf	73 Ta	74 W	75 Re	76 Os	77 Ir	78 Pt	79 Au	80 Hg	81 Tl	82 Pb	83 Bi	84 Po	85 At	86 Rn
7	87 Fr	88 Ra	89-103 †	104 Rf	105 Db	106 Sg	107 Bh	108 Hs	109 Mt	110 Ds	111	112	113	114	115	116	117	118

Inner transition metals

Lanthanide series *	57 La	58 Ce	59 Pr	60 Nd	61 Pm	62 Sm	63 Eu	64 Gd	65 Tb	66 Dy	67 Ho	68 Er	69 Tm	70 Yb	71 Lu
Actinide series †	89 Ac	90 Th	91 Pa	92 U	93 Np	94 Pu	95 Am	96 Cm	97 Bk	98 Cf	99 Es	100 Fm	101 Md	102 No	103 Lr

Elements 111, 112, 114, and 116 have been discovered but, as of 2003, have not yet been named. See www.webelements.com for the latest information and newest elements.

Answers

chapter 12

CP **1.** c, e, f **2.** directly below the rod (torque on the apple due to $\vec{F}_g$, about the suspension, is zero) **3.** (a) no; (b) at site of $\vec{F}_1$, perpendicular to plane of figure; (c) 45 N **4.** (a) at C (to eliminate forces there from a torque equation); (b) plus; (c) minus; (d) equal **5.** d **6.** (a) equal; (b) B; (c) B **Q** **1.** (a) yes; (b) yes; (c) yes; (d) no **3.** (a) 1 and 3 tie, then 2; (b) all tie; (c) 1 and 3 tie, then 2 (zero) **5.** (a) 12 kg; (b) 3 kg; (c) 1 kg **7.** increase **9.** 4, 1, then 2 and 3 tie   **P** **1.** (a) 1.00 m; (b) 2.00 m; (c) 0.987 m; (d) 1.97 m **3.** 7.92 kN **5.** (a) 9.4 N; (b) 4.4 N **7.** (a) 1.2 kN; (b) down; (c) 1.7 kN; (d) up; (e) left; (f) right **9.** 74.4 g **11.** (a) 2.8×10^2 N; (b) 8.8×10^2 N; (c) 71° **13.** (a) 5.0 N; (b) 30 N; (c) 1.3 m **15.** (a) 2.7 kN; (b) up; (c) 3.6 kN; (d) down **17.** (a) 0.64 m; (b) increased **19.** 13.6 N **21.** (a) 192 N; (b) 96.1 N; (c) 55.5 N **23.** (a) 6.63 kN; (b) 5.74 kN; (c) 5.96 kN **25.** 2.20 m **27.** (a) 1.50 m; (b) 433 N; (c) 250 N **29.** (a) 445 N; (b) 0.50; (c) 315 N **31.** 0.34 **33.** (a) slides; (b) 31°; (c) tips; (d) 34° **35.** (a) 211 N; (b) 534 N; (c) 320 N **37.** (a) 6.5×10^6 N/m²; (b) 1.1×10^{-5} m **39.** (a) 866 N; (b) 143 N; (c) 0.165 **41.** (a) 1.4×10^9 N; (b) 75 **43.** 76 N **45.** (a) 8.01 kN; (b) 3.65 kN; (c) 5.66 kN **47.** 71.7 N **49.** (a) 1.38 kN; (b) 180 N **51.** (a) $\mu < 0.57$; (b) $\mu > 0.57$ **53.** (a) $L/2$; (b) $L/4$; (c) $L/6$; (d) $L/8$; (e) $25L/24$ **55.** 0.29 **57.** 8.7 N **59.** (a) 15 N; (b) 29 N **61.** 60° **63.** 2.4×10^9 N/m² **65.** $L/4$ **67.** (a) 88 N; (b) $(30\hat{i} + 97\hat{j})$ N **69.** (a) 1.8×10^7 N; (b) 1.4×10^7 N; (c) 16 **71.** (a) 1.5 kN; (b) 1.9 kN **73.** (a) $1.16\hat{j}$ kN; (b) $1.74\hat{j}$ kN **75.** (a) $(35\hat{i} + 200\hat{j})$ N; (b) $(-45\hat{i} + 200\hat{j})$ N; (c) 1.9×10^2 N **77.** 0.19 m **79.** (a) $a_1 = L/2$, $a_2 = 5L/8$, $h = 9L/8$; (b) $b_1 = 2L/3$, $b_2 = L/2$, $h = 7L/6$ **81.** (a) 2.5 m; (b) 7.3°

chapter 13

CP **1.** all tie. **2.** (a) 1, tie of 2 and 4, then 3; (b) line d **3.** negative y direction **4.** (a) increase; (b) negative **5.** (a) 2; (b) 1 **6.** (a) path 1 (decreased E (more negative) gives decreased a); (b) less (decreased a gives decreased T)   **Q** **1.** (a) between, closer to less massive particle; (b) no; (c) no **3.** Gm^2/r^2, upward **5.** (a) c, b, a; (b) a, b, c **7.** yes, in second quadrant, closer to y axis, at a distance that depends on its mass **9.** $b, d,$ and f all tie, then e, c, a **11.** (a) all tie; (b) all tie   **P** **1.** 19 m **3.** 1/2 **5.** 2.60×10^5 km **7.** $-5.00d$ **9.** (a) $M = m$; (b) 0 **11.** 8.31×10^{-9} N **13.** (a) $-1.88d$; (b) $-3.90d$; (c) $0.489d$ **15.** 2.6×10^6 m **17.** (a) 7.6 m/s²; (b) 4.2 m/s² **19.** 5×10^{24} kg **21.** (a) $(3.0 \times 10^{-7}$ N/kg)m; (b) $(3.3 \times 10^{-7}$ N/kg)m; (c) $(6.7 \times 10^{-7}$ N/kg·m)mr **23.** (a) 9.83 m/s²; (b) 9.84 m/s²; (c) 9.79 m/s² **25.** (a) 0.74; (b) 3.8 m/s²; (c) 5.0 km/s **27.** (a) 0.0451; (b) 28.5 **29.** 5.0×10^9 J **31.** (a) 0.50 pJ; (b) -0.50 pJ **33.** (a) 1.7 km/s; (b) 2.5×10^5 m; (c) 1.4 km/s **35.** (a) 82 km/s; (b) 1.8×10^4 km/s **37.** -4.82×10^{-13} J **39.** 6.5×10^{23} kg **41.** 5×10^{10} stars **43.** (a) 7.82 km/s; (b) 87.5 min

45. (a) 6.64×10^3 km; (b) 0.0136 **47.** (a) 1.9×10^{13} m; (b) $3.6R_P$ **51.** 0.71 y **53.** $(GM/L)^{0.5}$ **55.** (a) 2.8 y; (b) 1.0×10^{-4} **57.** (a) 3.19×10^3 km; (b) lifting **59.** (a) $r^{1.5}$; (b) r^{-1}; (c) $r^{0.5}$; (d) $r^{-0.5}$ **61.** (a) 7.5 km/s; (b) 97 min; (c) 4.1×10^2 km; (d) 7.7 km/s; (e) 93 min; (f) 3.2×10^{-3} N; (g) no; (h) yes **63.** (a) 1.0×10^3 kg; (b) 1.5 km/s **65.** 7.2×10^{-9} N **67.** $-0.044\hat{j}$ μN **69.** (a) 2.15×10^4 s; (b) 12.3 km/s; (c) 12.0 km/s; (d) 2.17×10^{11} J; (e) -4.53×10^{11} J; (f) -2.35×10^{11} J; (g) 4.04×10^7 m; (h) 1.22×10^3 s; (i) elliptical **71.** (a) 1.3×10^{12} m/s²; (b) 1.6×10^6 m/s **73.** $0.37\hat{j}$ μN **75.** 29 pN **77.** 2.5×10^4 km **79.** (a) 2.2×10^{-7} rad/s; (b) 89 km/s **81.** 3.2×10^{-7} N **83.** (a) 38.3 MJ; (b) 1.03×10^3 km **85.** (a) 0; (b) 1.8×10^{32} J; (c) 1.8×10^{32} J; (d) 0.99 km/s **87.** (a) 1.9×10^{11} m; (b) 4.6×10^4 m/s **89.** 7.9 km/s **91.** $-Gm(M_E/R + M_m/r)$ **93.** 1.1% **95.** $GM_Em/12R_E$ **103.** $2\pi r^{1.5}G^{-0.5}(M + m/4)^{-0.5}$

chapter 14

CP **1.** all tie **2.** (a) all tie (the gravitational force on the penguin is the same); (b) $0.95\rho_0$, ρ_0, $1.1\rho_0$ **3.** 13 cm³/s, outward **4.** (a) all tie; (b) 1, then 2 and 3 tie, 4 (wider means slower); (c) 4, 3, 2, 1 (wider and lower mean more pressure)   **Q** **1.** e, then b and d tie, then a and c tie **3.** (a) 2; (b) 1, less; 3, equal; 4, greater **5.** all tie **7.** a, b, c **9.** B, C, A **P** **1.** 2.9×10^4 N **3.** 1.1×10^5 Pa **5.** 0.074 **7.** (b) 26 kN **9.** 1.90×10^4 Pa **11.** 7.2×10^5 N **13.** 2.80 m **15.** 0.635 J **17.** 4.69×10^5 N **19.** (a) 1.88×10^9 N; (b) 2.20×10^{10} N·m; (c) 11.7 m **21.** (a) 7.9 km; (b) 16 km **23.** 8.50 kg **25.** (a) 2.04×10^{-2} m³; (b) 1.57 kN **27.** five **29.** (a) 6.7×10^2 kg/m³; (b) 7.4×10^2 kg/m³ **31.** (a) 1.2 kg; (b) 1.3×10^3 kg/m³ **33.** 57.3 cm **35.** 0.126 m³ **37.** (a) 1.80 m³; (b) 4.75 m³ **39.** (a) 637.8 cm³; (b) 5.102 m³; (c) 5.102×10^3 kg **41.** 8.1 m/s **43.** 66 W **45.** (a) 2.5 m/s; (b) 2.6×10^5 Pa **47.** (a) 3.9 m/s; (b) 88 kPa **49.** (a) 1.6×10^{-3} m³/s; (b) 0.90 m **51.** 1.4×10^5 J **53.** (a) 74 N; (b) 1.5×10^2 m³ **55.** (a) 35 cm; (b) 30 cm; (c) 20 cm **57.** (a) 0.0776 m³/s; (b) 69.8 kg/s **59.** (b) 2.0×10^{-2} m³/s **61.** (b) 63.3 m/s **63.** 7.8 cm/h **65.** 4.00 cm **67.** 45.3 cm³ **69.** (a) 3.2 m/s; (b) 9.2×10^4 Pa; (c) 10.3 m **71.** (a) 1.8×10^2 kN; (b) 81 kN; (c) 20 kN; (d) 0; (e) 78 kPa; (f) no **73.** (a) 3.1 m/s; (b) 9.5 m/s **75.** 4.4 mm **77.** 1.00×10^{-2} m/s **79.** (a) 0.38 m; (b) C, D, B, A **81.** 1.07×10^3 g **83.** 6.0×10^2 kg/m³ **85.** 43 cm/s **87.** 60 MPa **89.** (a) 41 h; (b) yes **91.** 1.5 g/cm³ **93.** (a) 6.06×10^9 N; (b) 20 atm; (c) no **95.** 0.50 m **97.** (a) 2.2; (b) 3.6

chapter 15

CP **1.** (sketch x versus t) (a) $-x_m$; (b) $+x_m$; (c) 0 **2.** a (F must have form of Eq. 15-10) **3.** (a) 5 J; (b) 2 J; (c) 5 J **4.** all tie (in Eq. 15-29, m is included in I) **5.** 1, 2, 3 (the ratio m/b matters; k does not)   **Q** **1.** c **3.** a and b **5.** (a) $-\pi$, $-180°$; (b) $-\pi/2$, $-90°$; (c) $+\pi/2$, $+90°$ **7.** (a) between B and C; (b) between $\pi/2$ rad and π rad

9. (a) *A, B, C;* (b) *C, B, A* **11.** *b* (infinite period, does not oscillate), *c, a* **P 1.** (a) 1.0 mm; (b) 0.75 m/s; (c) 5.7 × 10^2 m/s² **3.** (a) 6.28 × 10^5 rad/s; (b) 1.59 mm **5.** (a) 0.500 s; (b) 2.00 Hz; (c) 12.6 rad/s; (d) 79.0 N/m; (e) 4.40 m/s; (f) 27.6 N **7.** (a) 0.50 s; (b) 2.0 Hz; (c) 18 cm **9.** (a) 3.0 m; (b) −49 m/s; (c) −2.7 × 10^2 m/s²; (d) 20 rad; (e) 1.5 Hz; (f) 0.67 s **11.** 39.6 Hz **13.** (a) 498 Hz; (b) greater **15.** 3.1 cm **17.** (a) 5.58 Hz; (b) 0.325 kg; (c) 0.400 m **19.** (a) 0.500 m; (b) −0.251 m; (c) 3.06 m/s **21.** (a) 0.18*A*; (b) same direction **23.** (a) 2.2 Hz; (b) 56 cm/s; (c) 0.10 kg; (d) 20.0 cm **25.** 22 cm **27.** (a) 0.525 m; (b) 0.686 s **29.** 37 mJ **31.** (a) 0.75; (b) 0.25; (c) $2^{-0.5}x_m$ **33.** (a) 2.25 Hz; (b) 125 J; (c) 250 J; (d) 86.6 cm **35.** (a) 3.1 ms; (b) 4.0 m/s; (c) 0.080 J; (d) 80 N; (e) 40 N **37.** (a) 1.1 m/s; (b) 3.3 cm **39.** (a) 39.5 rad/s; (b) 34.2 rad/s; (c) 124 rad/s² **43.** (a) 0.205 kg·m²; (b) 47.7 cm; (c) 1.50 s **45.** 5.6 cm **47.** 1.83 s **49.** 0.0653 s **51.** (a) 0.53 m; (b) 2.1 s **53.** (a) 0.845 rad; (b) 0.0602 rad **55.** (a) 2.26 s; (b) increases; (c) same **57.** 0.39 **59.** (a) 14.3 s; (b) 5.27 **61.** d and e **63.** 5.0 cm **65.** (a) 1.2 J; (b) 50 **67.** 1.53 m **69.** (a) 1.72 ms; (b) 11.2 ms **71.** (a) 16.6 cm; (b) 1.23% **73.** (a) 8.11 × 10^{-5} kg·m²; (b) 3.14 rad/s **75.** (a) 1.23 kN/m; (b) 76.0 N **77.** 7.2 m/s **79.** (a) 1.1 Hz; (b) 5.0 cm **81.** (a) 1.3 × 10^2 N/m; (b) 0.62 s; (c) 1.6 Hz; (d) 5.0 cm; (e) 0.51 m/s **83.** (a) 0.873 s; (b) 6.3 cm **85.** (a) 0.35 Hz; (b) 0.39 Hz; (c) 0 (no oscillation) **87.** (a) 4.0 s; (b) $\pi/2$ rad/s; (c) 0.37 cm; (d) (0.37 cm) cos($\pi t/2$); (e) (−0.58 cm/s) sin($\pi t/2$); (f) 0.58 cm/s; (g) 0.91 cm/s²; (h) 0; (i) 0.58 cm/s **89.** (a) 147 N/m; (b) 0.733 s **91.** (a) 1.6 Hz; (b) 1.0 m/s; (c) 0; (d) 10 m/s²; (e) ±10 cm; (f) (−10 N/m)*x* **93.** (a) 0.20 s; (b) 0.20 kg; (c) −0.20 m; (d) −2.0 × 10^2 m/s²; (e) 4.0 J **95.** (a) 3.5 m; (b) 0.75 s **97.** (a) 3.2 Hz; (b) 0.26 m; (c) *x* = (0.26 m) cos(20*t* − $\pi/2$), with *t* in seconds **99.** 0.079 kg·m² **101.** (a) 0.45 s; (b) 0.10 m above and 0.20 m below; (c) 0.15 m; (d) 2.3 J **103.** (a) 0.30 m; (b) 30 m/s²; (c) 0; (d) 4.4 s **105.** (a) 245 N/m; (b) 0.284 s **107.** (a) 1.0 × 10^2 N/m; (b) 0.45 s **109.** (a) 8.3 s; (b) no **111.** 50 cm

chapter 16

CP 1. a, 2; b, 3; c, 1 (compare with phase in Eq. 16-2, then see Eq. 16-5) **2.** (a) 2, 3, 1 (see Eq. 16-12); (b) 3, then 1 and 2 tie (find amplitude of *dy/dt*) **3.** (a) same (independent of *f*); (b) decrease ($\lambda = v/f$); (c) increase; (d) increase **4.** (a) increase; (b) increase; (c) increase **5.** 0.20 and 0.80 tie, then 0.60, 0.45 **6.** (a) 1; (b) 3; (c) 2 **7.** (a) 75 Hz; (b) 525 Hz **Q 1.** (a) 3, then 1 and 2 tie; (b) all tie; (c) 1 and 2 tie, then 3 **3.** (a) 1, 4, 2, 3; (b) 1, 4, 2, 3 **5.** intermediate (closer to fully destructive) **7.** a and d tie, then b and c tie **9.** (a) 8; (b) antinode; (c) longer; (d) lower **11.** (a) node; (b) antinode **P 1.** (a) 0.680 s; (b) 1.47 Hz; (c) 2.06 m/s **3.** 1.1 ms **5.** (a) 11.7 cm; (b) π rad **7.** (a) 64 Hz; (b) 1.3 m; (c) 4.0 cm; (d) 5.0 m⁻¹; (e) 4.0 × 10^2 s⁻¹; (f) $\pi/2$ rad; (g) minus **9.** (a) 3.0 mm; (b) 16 m⁻¹; (c) 2.4 × 10^2 s⁻¹; (d) minus **11.** (a) negative; (b) 4.0 cm; (c) 0.31 cm⁻¹; (d) 0.63 s⁻¹; (e) π rad; (f) minus; (g) 2.0 cm/s; (h) −2.5 cm/s **13.** 129 m/s **15.** (a) 15 m/s; (b) 0.036 N **17.** (a) 0.12 mm; (b) 141 m⁻¹; (c) 628 s⁻¹; (d) plus **19.** (a) 5.0 cm; (b) 40 cm; (c) 12 m/s; (d) 0.033 s; (e) 9.4 m/s; (f) 16 m⁻¹; (g) 1.9 × 10^2 s⁻¹; (h) 0.93 rad; (i) plus **21.** 2.63 m **25.** 3.2 mm **27.** 0.20 m/s **29.** $1.41y_m$ **31.** (a) 9.0 mm; (b) 16 m⁻¹; (c) 1.1 × 10^3 s⁻¹; (d) 2.7 rad; (e) plus **33.** 5.0 cm **35.** (a) π rad;

(b) 3.0 mm; (c) 0 rad; (d) 13 mm; (e) 9.4 mm **37.** (a) 3.29 mm; (b) 1.55 rad; (c) 1.55 rad **39.** (a) 7.91 Hz; (b) 15.8 Hz; (c) 23.7 Hz **41.** (a) 82.0 m/s; (b) 16.8 m; (c) 4.88 Hz **43.** (a) 144 m/s; (b) 60.0 cm; (c) 241 Hz **45.** (a) 105 Hz; (b) 158 m/s **47.** (a) 0.25 cm; (b) 1.2 × 10^2 cm/s; (c) 3.0 cm; (d) 0 **49.** (a) 0.50 cm; (b) 3.1 m⁻¹; (c) 3.1 × 10^2 s⁻¹; (d) minus **51.** (a) +4.0 cm; (b) 0; (c) 0; (d) −0.13 m/s **53.** (a) 2.00 Hz; (b) 2.00 m; (c) 4.00 m/s; (d) 50.0 cm; (e) 150 cm; (f) 250 cm; (g) 0; (h) 100 cm; (i) 200 cm **55.** (a) 324 Hz; (b) eight **57.** 2.8 rad or −3.5 rad **59.** (a) 5.0 cm/s; (b) +*x* **61.** (a) $0.83y_1$; (b) 37° **63.** 1.2 rad **65.** (a) 0.16 m; (b) 2.4 × 10^2 N; (c) *y*(*x, t*) = (0.16 m) sin[(1.57 m⁻¹)*x*] sin[(31.4 s⁻¹)*t*] **67.** (a) $2\pi y_m/\lambda$; (b) no **69.** (a) 1.00 cm; (b) 3.46 × 10^3 s⁻¹; (c) 10.5 m⁻¹; (d) plus **71.** (a) 6.7 mm; (b) 45° **73.** (a) 75 Hz; (b) 13 ms **75.** (a) 240 cm; (b) 120 cm; (c) 80 cm **77.** (a) 144 m/s; (b) 3.00 m; (c) 1.50 m; (d) 48.0 Hz; (e) 96.0 Hz **79.** (a) 2.0 mm; (b) 95 Hz; (c) +30 m/s; (d) 31 cm; (e) 1.2 m/s **81.** (a) 0.52 m; (b) 40 m/s; (c) 0.40 m **83.** 36 N **85.** (a) 8.0 cm; (b) 1.0 cm **87.** (a) $[k\,\Delta\ell\,(\ell + \Delta\ell)/m]^{0.5}$ **89.** (a) 4.3 × 10^{14} Hz to 7.5 × 10^{14} Hz; (b) 1.0 m to 2.0 × 10^2 m; (c) 6.0 × 10^{16} Hz to 3.0 × 10^{19} Hz **91.** (a) 2.0 m/s; (d) −*x* **93.** (b) kinetic energy of transversely moving flat sections of string

chapter 17

CP 1. beginning to decrease (example: mentally move the curves of Fig. 17-7 rightward past the point at *x* = 42 m) **2.** (a) 0, fully constructive; (b) 4λ, fully constructive **3.** (a) 1 and 2 tie, then 3 (see Eq. 17-28); (b) 3, then 1 and 2 tie (see Eq. 17-26). **4.** second (see Eqs. 17-39 and 17-41) **5.** *a*, greater; *b*, less; *c*, can't tell; *d*, can't tell; *e*, greater; *f*, less **6.** (measure speeds relative to the air) (a) 222 m/s; (b) 222 m/s **Q 1.** (a) 2.0 wavelengths; (b) 1.5 wavelengths; (c) fully constructive in (a), fully destructive in (b) **3.** (a) 0, 0.2 wavelength, 0.5 wavelength (zero); (b) $4P_{avg,1}$ **5.** 150 Hz and 450 Hz **7.** *d*, fundamental **9.** 1, 4, 3, 2 **P 1.** 1.7 × 10^2 m **3.** (a) 79 m; (b) 41 m; (c) 89 m **5.** 40.7 m **7.** 1.9 × 10^3 km **9.** (a) 76.2 μm; (b) 0.333 mm **11.** (a) 2.3 × 10^2 Hz; (b) higher **13.** (a) 6.1 nm; (b) 9.2 m⁻¹; (c) 3.1 × 10^3 s⁻¹; (d) 5.9 nm; (e) 9.8 m⁻¹; (f) 3.1 × 10^3 s⁻¹ **15.** (a) 14; (b) 14 **17.** (a) 343 Hz; (b) 3; (c) 5; (d) 686 Hz; (e) 2; (f) 3 **19.** (a) 143 Hz; (b) 3; (c) 5; (d) 286 Hz; (e) 2; (f) 3 **21.** 15.0 mW **23.** 36.8 nm **25.** (a) 1.0 × 10^3; (b) 32 **27.** (a) 10 μW/m²; (b) 0.10 μW/m²; (c) 70 nm; (d) 7.0 nm **29.** (a) 5 dB; (b) 3.2 **31.** (a) 0.34 nW; (b) 0.68 nW; (c) 1.4 nW; (d) 0.88 nW; (e) 0 **33.** (a) 2; (b) 1 **35.** (a) 833 Hz; (b) 0.418 m **37.** (a) 405 m/s; (b) 596 N; (c) 44.0 cm; (d) 37.3 cm **39.** (a) 3; (b) 1129 Hz; (c) 1506 Hz **41.** 12.4 m **43.** 45.3 N **45.** 387 Hz **47.** 0.020 **49.** (a) 526 Hz; (b) 555 Hz **51.** 4.61 m/s **53.** 155 Hz **55.** (a) 1.58 kHz; (b) 0.208 m; (c) 2.16 kHz; (d) 0.152 m **57.** 41 kHz **59.** (a) 485.8 Hz; (b) 500.0 Hz; (c) 486.2 Hz; (d) 500.0 Hz **61.** (a) 598 Hz; (b) 608 Hz; (c) 589 Hz **63.** (a) 42°; (b) 11 s **65.** 0.250 **67.** (a) 0; (b) 0.572 m; (c) 1.14 m **69.** 0 **71.** (a) 2; (b) 6; (c) 10 **73.** 0.25 **75.** (a) $L(v_m − v)/v_m v$; (b) 364 m **77.** (a) 9.7 × 10^2 Hz; (b) 1.0 kHz; (c) 60 Hz, no **79.** (a) 5.5 × 10^2 m/s; (b) 1.1 × 10^3 m/s; (c) 1 **81.** (a) 2.00; (b) 1.41; (c) 1.73; (d) 1.85 **83.** (a) 10 W; (b) 0.032 W/m²; (c) 99 dB **85.** (a) 7.70 Hz; (b) 7.70 Hz **87.** 0.76 μm **89.** 400 Hz **91.** 3 **93.** (a) 59.7; (b) 2.81 × 10^{-4} **95.** (a) 5.2 kHz; (b) 2 **97.** 30° **99.** (a) 0.50 m; (b) 0.34 m; (c) 0.66 m **101.** (a) 88

mW/m²; (b) 0.75   **103.** 2.1 m   **105.** 171 m
107. (a) rightward; (b) 0.90 m/s; (c) less   **109.** 1 cm
111. (a) 482 Hz; (b) 660 Hz

chapter 18

CP   1. (a) all tie; (b) 50°X, 50°Y, 50°W   **2.** (a) 2 and 3 tie,
then 1, then 4; (b) 3, 2, then 1 and 4 tie (from Eqs. 18-9 and
18-10, assume that change in area is proportional to initial
area)   **3.** *A* (see Eq. 18-14)   **4.** *c* and *e* (maximize area enclosed by a clockwise cycle)   **5.** (a) all tie (ΔE_{int} depends on
i and *f*, not on path); (b) 4, 3, 2, 1 (compare areas under
curves); (c) 4, 3, 2, 1 (see Eq. 18-26)   **6.** (a) zero (closed cycle); (b) negative (W_{net} is negative; see Eq. 18-26)   **7.** *b* and
d tie, then *a*, *c* (P_{cond} identical; see Eq. 18-32)   **Q   1.** *Z*,
X, *Y*   **3.** *B*, then *A* and *C* tie   **5.** (a) both clockwise;
(b) both clockwise   **7.** (a) all tie; (b) all tie   **9.** (a) greater;
(b) 1, 2, 3; (c) 1, 3, 2; (d) 1, 2, 3; (e) 2, 3, 1   **11.** (a) *f*, because ice temperature will not rise to freezing point and
then drop; (b) *b* and *c* at freezing point, *d* above, *e* below;
(c) in *b* liquid partly freezes and no ice melts; in *c* no liquid
freezes and no ice melts; in *d* no liquid freezes and ice
fully melts; in *e* liquid fully freezes and no ice melts
P   1. (a) 0.06 kPa; (b) nitrogen   **3.** 348 K   **5.** (a) −96°F;
(b) 56.7°C   **7.** −92.1°X   **9.** 2.731 cm   **11.** 29 cm³   **13.** 11
cm²   **15.** 0.26 cm³   **17.** 360°C   **19.** 0.13 mm   **21.** 7.5 cm
23. (a) 523 J/kg · K; (b) 26.2 J/mol · K; (c) 0.600 mol   **25.** 42.7
kJ   **27.** 160 s   **29.** 3.0 min   **31.** 33 g   **33.** 33 m²
35. 742 kJ   **37.** (a) 5.3°C; (b) 0; (c) 0°C; (d) 60 g   **39.** (a) 0°C;
(b) 2.5°C   **41.** 8.71 g   **43.** (a) 1.2 × 10² J; (b) 75 J; (c) 30 J
45. −30 J   **47.** 60 J   **49.** (a) 6.0 cal; (b) −43 cal; (c) 40 cal; (d)
18 cal; (e) 18 cal   **51.** 1.66 kJ/s   **53.** (a) 16 J/s; (b) 0.048 g/s
55. 0.50 min   **57.** (a) 1.7 × 10⁴ W/m²; (b) 18 W/m²
59. −4.2°C   **61.** 0.40 cm/h   **63.** 1.5   **65.** (a) 6.61 mm;
(b) 3.006 606 m; (c) 6.62 mm; (d) 2.999 985 m; (e) 1.45 ×
10⁻⁵ m   **67.** (a) 90 W; (b) 2.3 × 10² W; (c) 3.3 × 10² W
69. (a) 10 000°F; (b) 37.0°C; (c) −57°C; (d) −297°F   **71.** 0.41
kJ/kg · K   **73.** (a) 1.87 × 10⁴; (b) 10.4 h   **75.** 1.7 × 10² km
77. (a) $11p_1V_1$; (b) $6p_1V_1$ **79.** 4.83 × 10⁻² cm³   **81.** 23 J
83. (a) 80 J; (b) 80 J   **85.** 1.17 C°   **87.** −6.1 nW   **89.** 10.5°C
91. 20 MJ   **93.** 79.5°C   **95.** (a) 13 × 10⁻⁶/F°; (b) 4.2 mm
97. 2.16 × 10⁻⁵ m²   **99.** (a) 1.8 W; (b) 0.024 C°   **101.** 333 J

chapter 19

CP   1. all but *c*   **2.** (a) all tie; (b) 3, 2, 1   **3.** gas *A*   **4.** 5
(greatest change in *T*), then tie of 1, 2, 3, and 4   **5.** 1, 2, 3
($Q_3 = 0$, Q_2 goes into work W_2, but Q_1 goes into greater
work W_1 and increases gas temperature)   **Q   1.** *d*, then *a*
and *b* tie, then *c*   **3.** 20 J   **5.** constant-volume process
7. (a) 3; (b) 1; (c) 4; (d) 2; (e) yes   **9.** (a) 1, 2, 3, 4; (b) 1, 2, 3
P   1. 0.933 kg   **3.** (a) 5.47 × 10⁻⁸ mol; (b) 3.29 × 10¹⁶ molecules   **5.** (a) 0.0388 mol; (b) 220°C   **7.** (a) 106 mol;
(b) 0.892 m³   **9.** 0.2   **11.** 207 J   **13.** 5.60 kJ   **15.** 2.0 ×
10⁵ Pa   **17.** 1.8 × 10² m/s   **19.** 511 m/s; (b) −200°C;
(c) 899°C   **21.** 1.9 kPa   **23.** 3.3 × 10⁻²⁰ J   **25.** (a) 6.76 ×
10⁻²⁰ J; (b) 10.7   **27.** (a) 6 × 10⁹ km   **29.** (a) 3.27 × 10¹⁰
molecules/cm³; (b) 172 m   **31.** (a) 6.5 km/s; (b) 7.1 km/s
33. (a) 420 m/s; (b) 458 m/s; (c) yes   **35.** (a) 1.0 × 10⁴ K;
(b) 1.6 × 10⁵ K; (c) 4.4 × 10² K; (d) 7.0 × 10³ K; (e) no;

(f) yes   **37.** (a) 7.0 km/s; (b) 2.0 × 10⁻⁸ cm; (c) 3.5 × 10¹⁰
collisions/s   **39.** (a) 0.67; (b) 1.2; (c) 1.3; (d) 0.33   **41.** (a) 0;
(b) +374 J; (c) +374 J; (d) + 3.11 × 10⁻²² J   **43.** 15.8
J/mol · K   **45.** (a) 6.6 × 10⁻²⁶ kg; (b) 40 g/mol   **47.** (a)
7.72 × 10⁴ J; (b) 5.46 × 10⁴ J; (c) 5.17 J/mol · K; (d) 4.32 ×
10⁴ J; (e) 8.86 × 10⁴ J; (f) 8.38 J/mol · K   **49.** 8.0 kJ
51. (a) 6.98 kJ; (b) 4.99 kJ; (c) 1.99 kJ; (d) 2.99 kJ   **53.** (a)
14 atm; (b) 6.2 × 10² K   **55.** − 15 J   **57.** − 20 J   **59.** (a)
3.74 kJ; (b) 3.74 kJ; (c) 0; (d) 0; (e) −1.81 kJ; (f) 1.81 kJ;
(g) − 3.22 kJ; (h) − 1.93 kJ; (i) − 1.29 kJ; (j) 520 J; (k) 0;
(l) 520 J; (m) 0.0246 m³; (n) 2.00 atm; (o) 0.0373 m³; (p) 1.00
atm   **61.** (a) 22.4 L   **63.** 349 K   **65.** (a) 900 cal; (b) 0;
(c) 900 cal; (d) 450 cal; (e) 1200 cal; (f) 300 cal; (g) 900 cal;
(h) 450 cal; (i) 0; (j) − 900 cal; (k) 900 cal; (l) 450 cal
69. (a) 1.37; (b) diatomic   **71.** (a) − 374 J; (b) 0; (c) + 374 J;
(d) + 3.11 × 10⁻²² J   **73.** (a) − 60 J; (b) 90 K   **75.** 7.03 ×
10⁹ s⁻¹   **77.** (a) 122 K; (b) 365 K; (c) 0   **79.** (a) 8.0 atm;
(b) 300 K; (c) 4.4 kJ; (d) 3.2 atm; (e) 120 K; (f) 2.9 kJ; (g) 4.6
atm; (h) 170 K; (i) 3.4 kJ   **81.** 653 J   **83.** 3.11 kJ/kg · K
85. − 3.0 J   **87.** (a) 1.5; (b) 4.5; (c) 6; (d) 2   **89.** (a) $3/v_0^3$;
(b) $0.750v_0$; (c) $0.775v_0$ **91.** (a) − 2.37 kJ; (b) 2.37 kJ
93. (a) − 45 J; (b) 1.8 × 10² K

chapter 20

CP   1. a, b, c   **2.** smaller (*Q* is smaller)   **3.** c, b, a   **4.** a, d,
c, b   **5.** b   **Q   1.** *a* and *c* tie, then *b* and *d* tie   **3.** *b*, *a*,
c, *d*   **5.** unchanged   **7.** A, first; B, first and second; C,
second; D, neither   **9.** (a) same; (b) increase; (c) decrease
11. more than the age of the universe   **P   1.** (a) 9.22 kJ;
(b) 23.1 J/K; (c) 0   **3.** 14.4 J/K   **5.** (a) 5.79 × 10⁴ J; (b) 173
J/K   **7.** (a) 57.0°C; (b) −22.1 J/K; (c) +24.9 J/K; (d) +2.8
J/K   **9.** (a) −710 mJ/K; (b) +710 mJ/K; (c) +723 mJ/K;
(d) −723 mJ/K; (e) +13 mJ/K; (f) 0   **11.** (a) 0.333; (b) 0.215;
(c) 0.644; (d) 1.10; (e) 1.10; (f) 0; (g) 1.10; (h) 0; (i) −0.889;
(j) −0.889; (k) −1.10; (l) −0.889; (m) 0; (n) 0.889; (o) 0
13. (a) 320 K; (b) 0; (c) +1.72 J/K   **15.** +0.76 J/K
17. (a) −943 J/K; (b) +943 J/K; (c) yes   **19.** (a) 0.693;
(b) 4.50; (c) 0.693; (d) 0; (e) 4.50; (f) 23.0 J/K; (g) −0.693;
(h) 7.50; (i) −0.693; (j) 3.00; (k) 4.50; (l) 23.0 J/K
21. (a) 23.6%; (b) 1.49 × 10⁴ J   **23.** 97 K   **25.** (a) 266 K;
(b) 341 K   **27.** (a) 1.47 kJ; (b) 554 J; (c) 918 J; (d) 62.4%
29. (a) 2.27 kJ; (b) 14.8 kJ; (c) 15.4%; (d) 75.0%; (e) greater
31. (a) 33 kJ; (b) 25 kJ; (c) 26 kJ; (d) 18 kJ   **33.** (a) 3.00;
(b) 1.98; (c) 0.660; (d) 0.495; (e) 0.165; (f) 34.0%   **35.** 20 J
37. 440 W   **39.** 0.25 hp   **41.** 2.03   **45.** (a) *W* =
$N!/(n_1! \, n_2! \, n_3!)$; (b) [(N/2)! (N/2!)]/[(N/3)! (N/3)! (N/3)!];
(c) 4.2 × 10¹⁶   **47.** (a) 1; (b) 6; (c) 0; (d) 2.47 × 10⁻²³ J/K
49. (a) 87 m/s; (b) 1.2 × 10² m/s; (c) 22 J/K   **51.** (a) 78%;
(b) 82 kg/s   **53.** (a) 66.5°C; (b) 14.6 J/K; (c) 11.0 J/K;
(d) −21.2 J/K; (e) 4.39 J/K   **55.** (a) 6.34 J/K; (b) 6.34
J/K; (c) 6.34 J/K; (d) 6.34 J/K   **57.** 1.18 × 10³ J/K
59. 75   **61.** (a) 1; (b) 1; (c) 3; (d) 10; (e) 1.5 × 10⁻²³ J/K;
(f) 3.2 × 10⁻²³ J/K   **63.** +3.59 J/K   **65.** (a) 1.95 J/K;
(b) 0.650 J/K; (c) 0.217 J/K; (d) 0.072 J/K; (e) decrease
67. −40 K   **69.** (a) 1.26 × 10¹⁴; (b) 4.71 × 10¹³; (c) 0.37;
(d) 1.01 × 10²⁹; (e) 1.37 × 10²⁸; (f) 0.14; (g) 9.05 × 10⁵⁸;
(h) 1.64 × 10⁵⁷; (i) 0.018; (j) decrease
71. (a) 3.73; (b) 710 J   **73.** (a) 42.6 kJ; (b) 7.61 kJ

Photo Credits

Index

Figures are noted by page numbers in *italics*, tables are indicated by t following the page number.